Basic Mathematics for Economists is a classic of its genre, and this fourth edition continues to build on the success of previous editions. Suitable for students who may only have a basic mathematics background, as well as students who may have followed more advanced mathematics courses but who still want a clear explanation of fundamental concepts, this book covers all the basic tenets required for an understanding of mathematics and how it is applied in economics, finance and business.

Starting with revisions of the essentials of arithmetic and algebra, students are then taken through to more advanced topics in calculus, comparative statics, dynamic analysis and matrix algebra, with all topics explained in the context of relevant applications. This fourth edition includes updated/additional real-world applications and examples of concepts and techniques throughout, with fuller analysis of topics such as savings and pension schemes, and asset valuation techniques.

Including digital supplements for both students and lecturers, this book is the most logical, user-friendly book on the market and is suitable for mathematics of economics, finance and business courses globally.

Piotr Lis is Deputy Head of School, Associate Professor in Economics at Coventry University, UK.

Mike Rosser is a former Principal Lecturer in Economics at Coventry University, UK.

"This book is my go-to guide for my intermediate microeconomics class. The author is able to explain well mathematical concepts which result in an extraordinary support to improve the mathematical ability of my students."

— Gloria L. Bernal, *Assistant Professor, Economics Department at Pontificia Universidad Javeriana, Columbia*

BASIC MATHEMATICS FOR ECONOMISTS

Fourth Edition

Piotr Lis and Mike Rosser

LONDON AND NEW YORK

Designed cover image: kertlis / E+ / Getty Images®

Fourth edition published 2025
by Routledge
4 Park Square, Milton Park, Abingdon, Oxon, OX14 4RN

and by Routledge
605 Third Avenue, New York, NY 10158

Routledge is an imprint of the Taylor & Francis Group, an informa business

First edition published by HarperCollins 1993
Third edition published by Routledge 2016

British Library Cataloguing-in-Publication Data
A catalogue record for this book is available from the British Library

Library of Congress Cataloging-in-Publication Data
Names: Lis, Piotr, author. | Rosser, M. J., 1949– author.
Title: Basic mathematics for economists / Piotr Lis and Mike Rosser.
Description: Fourth edition. | Abingdon, Oxon; New York, NY: Routledge, 2025. | Originally entered under: Rosser, M. J. | Includes bibliographical references and index.
Identifiers: LCCN 2024046315 | ISBN 9781032420196 (hbk) | ISBN 9781032420165 (pbk) | ISBN 9781003360827 (ebk)
Subjects: LCSH: Economics, Mathematical. | Business mathematics.
Classification: LCC HB135 .R665 2025 | DDC 510—dc23/eng/20250115
LC record available at https://lccn.loc.gov/2024046315

ISBN: 978-1-032-42019-6 (hbk)
ISBN: 978-1-032-42016-5 (pbk)
ISBN: 978-1-003-36082-7 (ebk)

DOI: 10.4324/9781003360827

Typeset in Times New Roman
by codeMantra

Access the Support Material: www.routledge.com/9781032420165

Contents

List of symbols and terminology *xiii*
Preface *xv*
Acknowledgements *xvii*

1 Introduction **1**

2 Arithmetic **8**

3 Introduction to algebra **33**

4 Graphs and functions **63**

5 Simultaneous linear equations **107**

6 Linear programming **147**

7 Quadratic equations **169**

8 Financial mathematics – series, time and investment **189**

9 Introduction to calculus **275**

10 Unconstrained optimization **299**

11 Partial differentiation **318**

12 Constrained optimization **363**

13 Further topics in differentiation and integration **395**

14 Dynamics and difference equations **433**

15 Exponential functions, continuous growth and differential equations **471**

16 Matrix algebra **507**

Answers 569
Index 581

Detailed contents

List of symbols and terminology *xiii*
Preface *xv*
Acknowledgements *xvii*

1 Introduction **1**
1.1 Why study mathematics? 1
1.2 Calculators and computers 3
1.3 Using this book 5

2 Arithmetic **8**
2.1 Revision of basic concepts 8
2.2 Multiple operations 9
2.3 Brackets 11
2.4 Fractions 12
2.5 Elasticity of demand 15
2.6 Decimals 19
2.7 Negative numbers 22
2.8 Powers 23
2.9 Roots and fractional powers 26
2.10 Logarithms 29

3 Introduction to algebra **33**
3.1 Representation 33
3.2 Evaluation 36
3.3 Simplification: addition and subtraction 38
3.4 Simplification: multiplication 40
3.5 Simplification: factorizing 44

3.6 Simplification: division 48
3.7 Solving simple equations 50
3.8 The summation sign $\sum$ and price indexes 55
3.9 Inequality signs 59

4 Graphs and functions **63**
4.1 Functions 63
4.2 Inverse functions 66
4.3 Graphs of linear functions 69
4.4 Fitting linear functions 73
4.5 Slope 76
4.6 Budget constraints 81
4.7 Non-linear functions 85
4.8 Composite functions 88
4.9 Using a spreadsheet to plot functions 92
4.10 Functions with two independent variables 97
4.11 Summing functions horizontally 102

5 Simultaneous linear equations **107**
5.1 Systems of simultaneous linear equations 107
5.2 Solving simultaneous linear equations 108
5.3 Graphical solution 108
5.4 The equating to same variable method 110
5.5 The substitution method 112
5.6 The row operations method 114
5.7 More than two unknowns 116
5.8 Which method? 119
5.9 Comparative statics and the reduced form of an economic model 123
5.10 Price discrimination 133
5.11 Multiplant monopoly 140

6 Linear programming **147**
6.1 Constrained maximization 147
6.2 Constrained minimization 159
6.3 Mixed constraints 166
6.4 More than two variables 168

7 Quadratic equations **169**
7.1 Solving quadratic equations 169
7.2 Graphical solution 171
7.3 Factorization 175
7.4 The quadratic formula 177
7.5 Quadratic simultaneous equations 178
7.6 Polynomials 182

8 Financial mathematics – series, time and investment **189**
8.1 Discrete and continuous growth 189
8.2 Interest 191
8.3 Part year investment and the annual equivalent rate 196
8.4 Time periods, initial amounts and interest rates 201
8.5 Investment appraisal: net present value 207
8.6 The internal rate of return 216
8.7 Geometric series and annuities 221
8.8 Perpetual annuities 227
8.9 Pension pots, annuity income and drawdown pensions 231
8.10 Drawdown pension income 239
8.11 Loan repayments and mortgages 241
8.12 Savings schemes 249
8.13 Sinking fund savings schemes 253
8.14 Other applications of growth and decline 257
8A Appendix: asset valuation 263
8A.1 Valuation of bonds 263
8A.2 Valuation of shares 269

9 Introduction to calculus **275**
9.1 Differential calculus 275
9.2 Rules for differentiation 277
9.3 Marginal revenue and total revenue 281
9.4 Marginal cost and total cost 285
9.5 Profit maximization 287
9.6 Re-specifying functions 290
9.7 Point elasticity of demand 291
9.8 Tax yield 294
9.9 The Keynesian multiplier 296

10 Unconstrained optimization **299**
10.1 First-order conditions for a maximum 299
10.2 Second-order conditions for a maximum 300
10.3 Second-order conditions for a minimum 303
10.4 Summary of second-order conditions 304
10.5 Profit maximization 307
10.6 Inventory control 309
10.7 Comparative static effects of taxes 312

11 Partial differentiation **318**
11.1 Partial differentiation and the marginal product 318
11.2 Further applications of partial differentiation 323
11.3 Second-order partial derivatives 334
11.4 Unconstrained optimization: functions with two variables 340
11.5 Total differentials and total derivatives 353

12 Constrained optimization **363**
12.1 Constrained optimization and resource allocation 363
12.2 Constrained optimization by substitution 364
12.3 The Lagrange multiplier: constrained maximization with two variables 371
12.4 The Lagrange multiplier: second-order conditions 377
12.5 Constrained minimization using the Lagrange multiplier 380
12.6 Constrained optimization with more than two variables 385

13 Further topics in differentiation and integration **395**
13.1 Overview 395
13.2 The chain rule 395
13.3 The product rule 403
13.4 The quotient rule 408
13.5 Integration 414
13.6 Definite integrals 420
13.7 Integration by substitution and integration by parts 427

14 Dynamics and difference equations **433**
14.1 Dynamic economic analysis 433
14.2 The cobweb: iterative solutions 434
14.3 The cobweb: difference equation solutions 444
14.4 The lagged Keynesian macroeconomic model 453
14.5 Duopoly price adjustment 465

15 Exponential functions, continuous growth and differential equations **471**
15.1 Continuous growth and the exponential function 471
15.2 Accumulated final values after continuous growth 474
15.3 Continuous growth rates and initial amounts 477
15.4 Natural logarithms 481
15.5 Differentiation of logarithmic functions 487
15.6 Continuous time and differential equations 488
15.7 Solution of homogeneous differential equations 489
15.8 Solution of non-homogeneous differential equations 493
15.9 Continuous adjustment of market price 497
15.10 Continuous adjustment in a Keynesian macroeconomic model 502

16 Matrix algebra **507**
16.1 Introduction to matrices and vectors 507
16.2 Basic principles of matrix multiplication 512
16.3 Matrix multiplication – the general case 515
16.4 The matrix inverse and the solution of simultaneous equations 521
16.5 Determinants 525
16.6 Minors, cofactors and the Laplace expansion 258

16.7 The transpose matrix, the cofactor matrix, the adjoint and the matrix inverse formula 531
16.8 Application of the matrix inverse to the solution of linear simultaneous equations 536
16.9 Cramer's rule 542
16.10 Second-order conditions and the Hessian matrix 544
16.11 Constrained optimization and the bordered Hessian 551
16.12 Input-output analysis 555
16.13 Multiple industry input-output models 561

Answers 569
Index 581

Symbols and terminology

$\lvert x \rvert$	absolute value
$[\,]_0^m$	definite integral
$\frac{dy}{dx}$	derivative
e^x	exponential function
f_1	first-order derivative
$>$	greater than inequality
$\geq$	greater than or equal to weak inequality
$\equiv$	identity
∞	infinity
$\int$	integral
λ	Lagrange multiplier
$<$	less than inequality
$\leq$	less than or equal to weak inequality
log	logarithm (base 10)
ln	natural logarithm
$\frac{\partial y}{\partial x}$	partial derivative
f_{11}	second-order partial derivative
Δx	small change in x
$\surd$	square root
$\sum$	summation
$\sqrt[x]{y}$	x^{th} root of y

Preface

Many students who enrol on economics and finance degree courses have not studied mathematics beyond GCSE or an equivalent level. It is mainly for these students that this book is intended. It aims to develop their mathematical ability up to the level required for a general economics degree course (i.e. one not specializing in mathematical economics) or for a modular degree course in economics and related subjects, such as business studies. To achieve this aim it has several objectives.

First, it provides a revision of arithmetical and algebraic methods that students studied at school but may have largely forgotten. It is a misconception to assume that, just because a GCSE mathematics syllabus includes certain topics, students who passed examinations on that syllabus two or more years ago are all still familiar with the material. They usually require some revision exercises to jog their memories and to get into the habit of using the different mathematical techniques again. The first few chapters are mainly devoted to this revision, set out where possible in the context of applications in economics.

Second, this book introduces mathematical techniques that will be new to most students through examples of their application to economic concepts. It also tries to get students tackling problems in economics using these techniques as soon as possible so that they can see how useful they are. Students are not required to work through unnecessary proofs, or wrestle with complicated special cases that they are unlikely ever to encounter again. For example, when covering the topic of calculus, some other textbooks require students to plough through abstract theoretical applications of the technique of differentiation to every conceivable type of function and special case before any mention of its uses in economics is made. In this book, however, we introduce the basic concept of differentiation followed by examples of economic applications in Chapter 9. Further developments of the topic, such as the second-order conditions for optimization, partial differentiation and the rules for differentiation of composite functions, are then gradually brought in over the next few chapters, again in the context of economics application.

Third, this book tries to cover those mathematical techniques that will be relevant to students' economics degree programmes. Most applications are in the field of microeconomics, rather than macroeconomics, given the increased emphasis on business economics within many degree courses. Furthermore, Chapter 8 reflects the increased emphasis on finance in many economics and related courses and concentrates on mathematical techniques that are relevant to finance and investment decision-making, with fuller analysis of topics such as savings, pension schemes and asset valuation.

Given that students now have access to computing facilities, ways of using a spreadsheet package to solve certain problems that are extremely difficult or time consuming to solve manually are also explained.

Although it starts at a gentle pace through fairly elementary material, so that the students who gave up mathematics some years ago because they thought that they could not cope with A-level maths are able to build up their confidence, this is *not* a watered-down 'mathematics without tears or effort' type of textbook. As the book progresses, the pace is increased and students are expected to put in a serious amount of time and effort to master the material. However, given the way in which this material is developed, it is hoped that students will be motivated to do so. Not everyone finds mathematics easy, but at least it helps if you can see the reason for having to study it.

Acknowledgements

Microsoft® Windows and Microsoft® Excel® are registered trademarks of the Microsoft Corporation.

1 Introduction

Learning objectives

After completing this chapter students should be able to:

- understand why mathematics is needed for the study of economics and related areas, such as finance
- adopt a successful approach to their further study of mathematics.

1.1 WHY STUDY MATHEMATICS?

Economics is a social science that attempts to explain both the operation of the economy as a whole and the behaviour of individuals, firms and institutions who participate in it. Economists often make predictions about what may happen to specified economic variables if certain changes take place, e.g. what effect an increase in sales tax will have on the price of finished goods, or what will happen to unemployment if government expenditure is changed by a specified amount. Economics also provides guidelines that firms, governments or other economic agents may follow if they wish to allocate resources efficiently. Mathematics is often considered as a language of economics and is fundamental to any serious application of economics to these areas. The need for mathematics is even more obvious in financial economics, where numeric data is needed to make decisions on investments and the pricing of financial products.

Quantification

Introductory economic analysis is often performed with the aid of sketch diagrams. For example, supply and demand analysis predicts that if the supply of a good falls in a competitive market, then its price will rise. This is simply common sense, as any market trader will tell you. However, an economist needs to be able to say by how much price is expected to rise if supply contracts by a specified amount. This quantification of economic predictions requires the use of mathematics.

DOI: 10.4324/9781003360827-1

Although non-mathematical economic analysis may sometimes be useful for making qualitative predictions (i.e. predicting the direction of any expected changes), it cannot by itself provide the quantification that users of economic predictions require. A firm needs to know how much the quantity sold is expected to change in response to a price increase. The government wants to know by how much consumer demand will change if it increases a sales tax.

Simplification

Sometimes students believe that mathematics makes economics more complicated. However, algebraic notations, which are essentially a form of shorthand, can make certain concepts much clearer to understand than if they were just set out in words. It can also save a great deal of time and effort in writing out tedious verbal explanations.

For example, the relationship between the quantity of apples consumers wish to buy and the price of apples might be expressed as: 'the quantity of apples demanded per day is 1,200 kg when price is zero and then decreases by 10 kg for every 1p rise in the price of a kilo of apples'. It is much easier, however, to express this mathematically as:

$$q = 1{,}200 - 10p$$

where q is the quantity of apples demanded per day in kilograms and p is the price in pence per kilogram of apples.

This is a very simple example. The relationships between economic variables can be much more complex and mathematical formulation then becomes the only feasible method for dealing with the analysis.

Scarcity and choice

Problems dealt with in economics are concerned with the most efficient way of allocating limited resources. These are known as 'optimization' problems. For example, a firm may wish to maximize the output it can produce within a fixed budget for expenditure on inputs. Mathematics must be used to obtain answers to these problems.

Many economics graduates will enter employment in industry, commerce or the public sector, where very real resource allocation decisions have to be made. Mathematical methods are used as a basis for many of these decisions. Even if students do not go on to specialize in subjects such as managerial economics or operational research where the applications of these decision-making techniques are studied in more depth, it is essential that they gain an understanding of the sort of resource allocation problems that can be tackled and the information that is needed to enable them to be solved.

Economic statistics and estimating relationships

As well as using mathematics to work out predictions from economic models where the relationships are already quantified, one also needs mathematics to estimate the parameters of the models in the first place. For example, if the demand relationship in an actual market is described by the economic model $q = 1{,}200 - 10p$, then this would mean that the parameters (i.e. the numbers 1,200 and 10) had been estimated from statistical data.

The study of how the parameters of economic models can be estimated from statistical data is known as econometrics. Although this is not one of the topics covered in this book, you will find that several of the mathematical techniques that are covered are necessary to understand the methods used in econometrics. Students using this book will probably also study an introductory statistics course, which may be a prerequisite for econometrics, and here again certain basic mathematical tools will be essential.

Mathematics and business

Some students using this book may be on courses that have more emphasis on business or financial studies than pure economics. Two criticisms that these students sometimes make are that:

(a) Simple economic models do not bear much resemblance to the real-world financial or business decisions that are made in practice.
(b) Even if the models are relevant to financial or business decisions, there is not always enough actual data available on the relevant variables to make use of these economic models.

Criticism (a) should be answered in the first few lectures of your economics course when the methodology of economic theory is explained. In summary, one needs to start with a simplified model that can explain how firms (and other economic agents) behave in general before looking at more complex situations only relevant to specific firms.

Criticism (b) may be partially true, but a lack of complete data does not mean that one should not try to make the best decision using the information that is available. Just because some mathematical methods can be difficult to understand, this does not mean that efficient decision-making should be abandoned in favour of guesswork, rule of thumb and intuition.

1.2 CALCULATORS AND COMPUTERS

Some students may ask, 'what's the point in spending a great deal of time and effort studying mathematics when nowadays everyone uses calculators and computers for calculations?' There are several answers to this question.

Rubbish in, rubbish out

Perhaps the most important point is that calculators and computers can only calculate what they are told to. They are machines that can perform arithmetic computations much faster than you can do by hand, and this speed does indeed make them very useful tools. However, if you feed in useless information you will get useless information back – hence the well-known phrase 'rubbish in, rubbish out'.

For example, consider the previously shown demand relationship

$$q = 1{,}200 - 10p$$

where q is the quantity of apples demanded. What would the quantity demanded be if price p was 150? A computer would give the answer – 300, but this is nonsense as you cannot have a negative quantity of apples. It only makes sense for the mathematical relationship to apply to positive values of p and q. Therefore, if price is 120, quantity sold will be zero, and if any price higher than 120 is charged, such as 150, quantity sold will still be zero. This case illustrates why you must take care to interpret mathematical answers sensibly and not blindly assume that any numbers produced by a computer will always be correct even if the 'correct' numbers have been fed into it.

Algebra

Much economic analysis involves algebraic notation, with letters representing concepts that are capable of taking on different values (see Chapter 3). The manipulation of these algebraic expressions cannot usually be carried out by calculators and computers.

When should you use calculators and computers?

Calculators are useful for basic arithmetic operations that take a long time to do manually, such as long division or finding square roots. Nonetheless, care needs to be taken to ensure that individual calculations are done in the correct order so that the fundamental rules of mathematics are satisfied and needless inaccuracies through rounding are avoided.

However, the level of mathematics in this book requires more than basic arithmetic functions. You should use a mathematical calculator that can at least calculate powers, roots, logarithms and exponential function values. These may be shown as the function keys:

$$\left[y^x\right] \quad [{}^x\sqrt{y}] \quad [\text{LOG}] \quad \left[10^x\right] \quad [\text{LN}] \quad \left[e^x\right]$$

although other display formats are sometimes used. The meaning and use of these functions will be explained in the following chapters, but it will help if you first learn to use all the function keys on your calculator.

Most students will have access to computing facilities and your lecturer will advise whether or not you have access to specialist computer software packages that can be used to tackle specific types of mathematical problems. For example, you may have access to a graphics package that tells you when certain lines intersect or solves linear programming problems (see Chapter 6). However, all students will normally have access to standard spreadsheet programs, such as Excel, which can be particularly useful, especially for the sort of financial problems covered in Chapter 8 and for performing the mathematical operations on matrices explained in Chapter 16.

However, you will still need to understand the method of solution so that you will understand the answer that the computer gives you. Also, many economic problems have to be set up in the form of a mathematical problem before they can be fed into a computer. For example, although Excel has some built-in formulae, you may also have to create your own formulae when creating spreadsheets to tackle certain problems, particularly in financial economics. An understanding of relevant mathematical methods and algebraic formulation is required to create such spreadsheets.

Most problems and exercises in this book can be tackled without using computers although solving some problems only using a calculator would be very time consuming. Many of the problems requiring a large number of calculations are in Chapter 8 where methods of solution using Excel spreadsheets are suggested. However, specialist financial calculators can also help with this type of calculation.

As Excel is probably the spreadsheet program most commonly used by students, the spreadsheet suggested solutions to certain problems are given in Excel format. It is assumed that students will be familiar with the basic operational functions of this program (e.g. entering data, using the copy command, etc.), and the solutions in this book only suggest a set of commands necessary to solve specific problems.

1.3 USING THIS BOOK

Students using this book will probably be on the first year of a degree course in economics, finance or a related subject, and some may not have studied A-level mathematics. Some of you will be following a mathematics course specifically designed for people without A-level mathematics, whilst others will be mixed in with more mathematically experienced students. Addressing the needs of such a mixed audience poses a great challenge and, to ensure that no one is left behind, this book starts from some very basic mathematical principles. You should already have covered most of these for GCSE mathematics, or for an equivalent qualification, but only you can judge whether you are sufficiently competent in a technique to be able to skip some of the sections.

It would be advisable, however, to start at the beginning of the book and work through all the set problems. Many of you will have had at least a two-year break since last studying mathematics and will benefit from some revision. If you cannot easily answer all the questions in a section, then you obviously need to spend more time working through the topic. You should find that a lot of the basic mathematical material is familiar to you, although new applications of mathematics to economics and financial topics are introduced as the book progresses.

It is assumed that students using this book will also be studying economics, either as their main degree subject or as part of a business or finance degree. Because some students may not have had any prior study of economics at school, the examples in the first few chapters only use some basic economic theory, such as supply and demand analysis. By the time you get to the later chapters, it will be assumed that you have covered additional topics in economic analysis, such as production and cost theory. If you come across problems that assume knowledge of economics topics that you have not yet covered, then you should leave them until you understand these topics or consult your lecturer. In some instances, the basic analysis of certain economic concepts is explained before the mathematical application of these concepts, but this should not be considered a complete coverage of the topic.

Practise, practise

You will not learn mathematics by reading this book, or any other book for that matter. The only way you will learn mathematics is by practising working through problems. It may be more hard work than just reading through the pages of a book, but your effort will be rewarded when you master the different techniques. As with many other skills that people acquire, such as riding a bike or driving a car, a book can help you to understand how something is supposed to be done, but you will only be able to do it yourself if you spend time and effort practising.

You cannot acquire a skill just by sitting down in front of a book and hoping that you can 'absorb' what you read. Rather than memorizing a series of facts, you need to become competent in using methods of analysis yourself, and practice is the only way to be competent at doing that.

Group working

Your lecturer will make it clear to you which problems you must do by yourself as part of your course assessment, and which problems you may confer over with others. Asking others for help makes sense if you are stuck and cannot understand a topic. However, you should make every effort to work through all the problems that you are set by yourself before asking your lecturer or fellow students for help. When you do ask for help it should be to find out *how* to tackle a problem, not just to get the answer.

Some students who have difficulty with mathematics tend to copy answers from other students without really understanding what they are doing, or when a lecturer runs through an answer in class they just write down a verbatim copy of the answer without asking for clarification of points they do not follow. They are only fooling themselves. The point of studying mathematics is to learn how to be able to apply it to various topics in economics, finance or other areas. Students who pretend that they have no difficulty with something they do not properly understand will not get very far.

What is important is that you understand the *method* of solving different types of problems. There is no point in having a set of answers to problems if you do not understand how these answers were obtained.

Don't give up!

Do not get disheartened if you do not understand a topic the first time it is explained to you. Mathematics can be a difficult subject and you will need to work through some sections several times before they become clear to you. If you make the effort to try all the set problems and consult your lecturer when you really get stuck then you will eventually master the subject.

Because the topics follow on from each other, each chapter assumes that students are familiar with material covered in previous chapters. It is therefore very important that you keep up to date with your work. You cannot 'skip' a topic that you find difficult and hope to get through without answering examination questions on it, as it is sometimes possible to do in other subjects.

About half of all students on economics and finance-based degree courses gave up mathematics at school at the age of 16, many of them because they thought that they were not good enough at mathematics to take it for A-level. However, most of them usually manage to complete their first-year mathematics for economics course successfully and go on to achieve an honours degree. There is no reason why you should not do likewise if you are prepared to put in the effort.

2 Arithmetic

Learning objectives

After completing this chapter students should be able to:

- use basic arithmetic operations learned at school, including the use of brackets, fractions, decimals, percentages, negative numbers, powers, roots and logarithms
- apply some of these arithmetic operations to simple economic problems
- calculate arc elasticity of demand values by dividing a fraction by another fraction.

2.1 REVISION OF BASIC CONCEPTS

Most students will have previously covered all, or nearly all, of the topics in this chapter. They are included here for revision purposes and to ensure that everyone is familiar with basic arithmetical operations before going on to further mathematical topics. Therefore, only a fairly brief explanation is given for most of the arithmetic methods set out in this chapter.

As a starting point, it is assumed that everyone is familiar with the basic operations of addition, subtraction, multiplication and division, as applied to whole numbers (or integers) at least. The usual ways of expressing the notation for these operations are:

Example 2.1

Addition(+):	$24+204=228$
Subtraction(–):	$9,089-393=8,696$
Multiplication(× or•):	$12\times24=288$
Division(÷ or/):	$4,448\div16=278$

DOI: 10.4324/9781003360827-2

The sign '•' is sometimes used for multiplication when using algebraic notation but, as you will see from Chapter 2 onwards, there is usually no need to use any multiplication sign to signify that two algebraic variables are being multiplied together, e.g. a times b is simply written as ab.

Most students will have learned at school how to perform these operations, but if you cannot answer the following questions, either by hand or using a calculator, then you should refer to an elementary arithmetic text or see your lecturer for advice.

QUESTIONS 2.1

1. 323 + 3,232 =
2. 1,012 − 147 =
3. 460 × 202 =
4. 1,288 / 56 =

2.2 MULTIPLE OPERATIONS

Consider the following problem involving only addition and subtraction.

Example 2.2

A bus sets off with 22 passengers aboard. At the first stop, 7 passengers get off and 12 get on. At the second stop, 18 get off and 4 get on. How many passengers remain on the bus?

You would probably answer this by saying 22 − 7 = 15, 15 + 12 = 27, 27 − 18 = 9, 9 + 4 = 13 passengers remaining, which is the correct answer.

If you were faced with an abstract mathematical problem

$$22-7+12-18+4=?$$

you should answer it in the same way, i.e. working from left to right.

If you performed the addition operations first, then you would get 22 − 19 − 22 = −19 which is clearly not the correct answer to the bus passenger problem!

If we now consider an example involving only multiplication and division we can see that the same rule applies.

Example 2.3

A restaurant catering for a large party sits 6 people to a table. Each table requires 2 dishes of vegetables. How many dishes of vegetables are required for a party of 60?

Most people would answer this by saying 60 ÷ 6 = 10 tables, 10 × 2 = 20 dishes, which is correct.

If this is set out as the calculation 60 ÷ 6 × 2 = ?

then the left-to-right rule must be used.

If you did *not* use this rule, then you might get $60 \div 6 \times 2 = 60 \div 12 = 5$ which is incorrect.

Thus, the general rule to use when a calculation involves several arithmetical operations and

(i) only addition and subtraction are involved, or
(ii) only multiplication and division are involved

is that the operations should be performed by working from left to right.

Example 2.4

(i) $48 - 18 + 6 = 30 + 6 = 36$
(ii) $6 + 16 - 7 = 22 - 7 = 15$
(iii) $68 + 5 - 32 - 6 + 14 = 73 - 32 - 6 + 14 = 41 - 6 + 14 = 35 + 14 = 49$
(iv) $22 \times 8 \div 4 = 176 \div 4 = 44$
(v) $460 \div 5 \times 4 = 92 \times 4 = 368$
(vi) $200 \div 25 \times 8 \times 3 \div 4 = 8 \times 8 \times 3 \div 4 = 64 \times 3 \div 4 = 192 \div 4 = 48$

When a calculation involves both addition/subtraction and multiplication/division, then the rule is: **multiplication and division calculations must be done before addition and subtraction calculations** (except when brackets are involved – see Section 2.3). To illustrate the rationale for this rule, consider the following simple example.

Example 2.5

How much change do you get from £5 if you buy 6 oranges at 40p each?

Solution

All calculations must be done using the same units and so, converting the £5 to 500 pence,

$$\text{change} = 500 - 6 \times 40 = 500 - 240 = 260\text{p} = £2.60$$

Note that the multiplication must be done before the subtraction in order to arrive at the correct answer.

QUESTIONS 2.2

1. $962 - 88 + 312 - 267 =$
2. $240 - 20 \times 3 \div 4 =$
3. $300 \times 82 \div 6 \div 25 =$
4. $360 \div 4 \times 7 - 3 =$
5. $6 \times 12 \times 4 + 48 \times 3 + 8 =$
6. $420 \div 6 \times 2 - 64 + 25 =$

2.3 BRACKETS

If a calculation involves brackets, then the operations inside the brackets must be done first. Thus, brackets take precedence over the rule for multiple operations set out previously in Section 2.2.

Example 2.6

A firm produces 220 units of a good which cost an average of £8.25 per unit to produce and sells them at a price of £9.95 per unit. What is the firm's total profit?

Solution

$$\text{profit per unit} = £9.95 - £8.25$$
$$\text{total profit} = 220 \times (£9.95 - £8.25) = 220 \times £1.70 = £374$$

The brackets can be removed in a calculation that only involves addition or subtraction. However, you must remember that if there is a minus sign before a set of brackets then all the terms within the brackets must be multiplied by −1 if the brackets are removed, i.e. all + and − signs are reversed. (See Section 2.7 if you are not familiar with the concept of negative numbers.)

Example 2.7

$$(92 - 24) - (20 - 2) = ?$$

Solution

$68 - 18 = 50$ doing calculations within brackets first

Or

$92 - 24 - 20 + 2 = 50$ by removing the brackets

QUESTIONS 2.3

1. (12 × 3 − 8) × (44 − 14) =
2. (68 − 32) − (100 − 84 + 3) =
3. 60 + (36 − 8) × 4 =
4. 4 × (62 ÷ 2) − 8 ÷ (12 ÷ 3) =

5. If a firm produces 600 units of a good at an average cost of £76 and sells them all at £99 per unit, what is its total profit?
6. (124 + 6 × 81) − (42 − 2 × 15) =
7. How much net (i.e. after tax) profit does a firm make if it produces 440 units of a good at an average cost of £3.40 each, and pays 15p tax to the government on each unit sold at the market price of £3.95, assuming it sells everything it produces?

2.4 FRACTIONS

If computers and calculators use decimals when dealing with portions of whole numbers, why bother with fractions? There are several reasons:

1. Certain operations, particularly multiplication and division, can sometimes be done more quickly by fractions if some numbers can be cancelled out.
2. When using algebraic notation instead of actual numbers, the operations on formulae have to be performed using fractions.
3. In some cases fractions can give a more accurate answer than a calculator owing to rounding error (see Example 2.15).

A fraction is written as

$$\frac{\text{numerator}}{\text{denominator}}$$

and is just another way of saying that the numerator is divided by the denominator. Thus,

$$\frac{120}{960} = 120 \div 960$$

Before carrying out any arithmetical operations, it is best to simplify individual fractions. Both numerator and denominator can be divided by any whole number that they are both a multiple of. It therefore usually helps if any large numbers are 'factorized', i.e. broken down into the smaller numbers that they are a multiple of.

Example 2.8

$$\frac{168}{104} = \frac{21 \times 8}{13 \times 8} = \frac{21}{13}$$

In this example it is obvious that the 8s cancel out top and bottom, i.e. the numerator and denominator can both be divided by 8.

Example 2.9

$$\frac{120}{960}=\frac{12\times 10}{12\times 8\times 10}=\frac{1}{8}$$

Addition and subtraction of fractions are carried out by converting all fractions so that they have a common denominator (usually the largest one) and then adding or subtracting the different quantities with this common denominator. To convert fractions to the common (largest) denominator, one multiplies both the top and bottom of the fraction by whatever number is necessary to get the required denominator. For example, to convert 1/6 to a fraction with 12 as its denominator, one simply multiplies the top and bottom by 2. Thus,

$$\frac{1}{6}=\frac{2\times 1}{2\times 6}=\frac{2}{12}$$

Example 2.10

$$\frac{1}{6}+\frac{5}{12}=\frac{2}{12}+\frac{5}{12}=\frac{2+5}{12}=\frac{7}{12}$$

Any numbers that have an integer (i.e. a whole number) in them should be converted into fractions with the same denominator before carrying out addition or subtraction operations. This is done by multiplying the integer by the denominator of the fraction and then adding.

Example 2.11

$$1\frac{3}{5}=\frac{1\times 5}{5}+\frac{3}{5}=\frac{5}{5}+\frac{3}{5}=\frac{8}{5}$$

Example 2.12

$$2\frac{3}{7}-\frac{24}{63}=\frac{17}{7}-\frac{8}{21}=\frac{51-8}{21}=\frac{43}{21}=2\frac{1}{21}$$

Multiplication of fractions is carried out by multiplying the numerators of the different fractions (3×5 in the following example) and then multiplying the denominators (8×7 in the following example).

Example 2.13

$$\frac{3}{8}\times\frac{5}{7}=\frac{15}{56}$$

The exercise can be simplified if one first cancels out any whole numbers that can be divided into both the numerator and the denominator.

Example 2.14

$$\frac{20}{3}\times\frac{12}{35}\times\frac{4}{5}=\frac{(4\times5)\times(4\times3)\times4}{3\times35\times5}=\frac{4\times4\times4}{35}=\frac{64}{35}$$

The usual way of performing this operation is simply to cross through numbers that cancel:

$$\frac{\overset{4}{\cancel{20}}}{\underset{1}{\cancel{3}}}\times\frac{\overset{4}{\cancel{12}}}{\underset{7}{\cancel{35}}}\times\frac{4}{5}=\frac{64}{35}$$

Multiplying out fractions may provide a more accurate answer than by working out the decimal value of a fraction with a calculator before multiplying, especially if you write down calculated values and then re-enter them before arriving at the final answer. However, if you use a calculator and store the answer to each part, you should avoid rounding errors.

Example 2.15

$$\frac{4}{7}\times\frac{7}{2}=?$$

Solution

$$\frac{4}{7}\times\frac{7}{2}=\frac{4}{2}=2 \text{ using fractions}$$

Using a mathematical calculator, if you enter the numbers and commands

4[÷]7[×]7[÷]2[=]

you should also get the correct answer of 2.

However, if you were to perform the operation 4 [÷] 7, note the answer of 0.5714286 and then re-enter this number and multiply by 3.5 (that is 7 [÷] 2), you would get the slightly inaccurate answer of 2.0000001.

To divide by a fraction, one simply multiplies by its inverse.

Example 2.16

$$3 \div \frac{1}{6} = 3 \times \frac{6}{1} = 18$$

Example 2.17

$$\frac{44}{7} \div \frac{8}{49} = \frac{44}{7} \times \frac{49}{8} = \frac{11}{1} \times \frac{7}{2} = \frac{77}{2} = 38\frac{1}{2}$$

QUESTIONS 2.4

1. $\frac{1}{6} + \frac{1}{7} + \frac{1}{8} =$
2. $\frac{3}{7} + \frac{2}{9} - \frac{1}{4} =$
3. $\frac{2}{5} \times \frac{60}{7} \times \frac{21}{15} =$
4. $\frac{4}{5} \div \frac{24}{19} =$
5. $4\frac{2}{7} - 1\frac{2}{3} =$
6. $2\frac{1}{6} + 3\frac{1}{4} - \frac{4}{5} =$
7. $3\frac{1}{4} \times 4\frac{1}{3} =$
8. $8\frac{1}{2} \div 2\frac{1}{6} =$
9. $20\frac{1}{4} - \frac{3}{5} \times 2\frac{1}{8} =$
10. $6 - \frac{2}{3} \div \frac{1}{12} + 3\frac{1}{2} =$

2.5 ELASTICITY OF DEMAND

The arithmetic operation of dividing a fraction by a fraction is usually the first technique that students on an economics course need to brush up on if their mathematics is a bit rusty. It is needed to calculate elasticity of demand, which is a concept you should encounter fairly early in your microeconomics course, where its uses should be explained. Price elasticity of demand is a measure of the responsiveness of demand to changes in price. It is usually defined as

$$e = \frac{\text{\% change in quantity demanded}}{\text{\% change in price}}$$

When there are relatively large changes in price and quantity it is best to use the concept of arc elasticity to measure elasticity along a section of a demand schedule, otherwise the percentage changes on the same section of a demand schedule can vary depending on whether price rises or falls. For example, a price rise from £15 to £20 is an increase of 33.33%, but a price fall from £20 to £15 is a decrease of 25%.

The arc elasticity measure takes the changes in quantity and price as percentages of the averages of their values before and after the change. Thus, arc elasticity is usually defined as

$$\text{arc } e = \frac{\dfrac{\text{change in quantity}}{0.5(\text{1st quantity} + \text{2nd quantity})} \times 100}{\dfrac{\text{change in price}}{0.5(\text{1st price} + \text{2nd price})} \times 100}$$

The 0.5 and the 100 cancel out. Thus, we are left with

$$\text{arc } e = \frac{\dfrac{\text{change in quantity}}{(\text{1st quantity} + \text{2nd quantity})}}{\dfrac{\text{change in price}}{(\text{1st price} + \text{2nd price})}}$$

as the formula actually used for calculating price arc elasticity of demand.

Note that arc elasticity will normally have a negative value because a positive price change will correspond to a negative quantity change, and vice versa.

Example 2.18

Calculate the arc elasticity of demand between points A and B on the demand schedule shown in Figure 2.1.

Solution

Between points A and B price falls by 5 from 20 to 15 and quantity rises by 20 from 40 to 60. Thus, using the arc elasticity formula defined earlier

$$\text{arc } e = \frac{\dfrac{20}{40+60}}{\dfrac{-5}{20+15}} = \frac{\dfrac{20}{100}}{\dfrac{-5}{35}} = \frac{20}{100} \times \frac{35}{-5} = \frac{1}{5} \times \frac{7}{-1} = -\frac{7}{5} = -1\frac{2}{5} = -1.4$$

Figure 2.1 Linear demand schedule

The elasticity value is less than −1 (with an absolute value greater than 1), therefore the demand schedule is price elastic over this range. In other words, quantity demanded is sensitive to changes in price and an increase in price will lead to a relatively larger proportionate drop in quantity demanded.

Example 2.19

When the price of a good is lowered from £350 to £200 the quantity demanded increases from 600 to 750 units. Calculate the elasticity of demand over this section of its demand schedule.

Solution

Price falls and the change is −£150 and quantity rises by 150 units. Therefore, the arc elasticity is

$$e = \frac{\dfrac{150}{600+750}}{\dfrac{-150}{350+200}} = \frac{\dfrac{150}{1,350}}{\dfrac{-150}{550}} = \frac{150}{1,350} \times \frac{550}{-150} = \frac{1}{27} \times \frac{11}{-1} = -\frac{11}{27}$$

This value of elasticity is greater than −1 (with an absolute value less than 1), therefore the demand schedule is price inelastic over this range. Thus, quantity demanded is relatively insensitive to changes in price and an increase in price will lead to a relatively smaller proportionate drop in quantity demanded.

QUESTIONS 2.5

1. With reference to the demand schedule in Figure 2.2, calculate the arc elasticity of demand between the prices of (a) £3 and £6, (b) £6 and £9, (c) £9 and £12, (d) £12 and £15, and (e) £15 and £18.
2. A city bus service charges a uniform fare for every journey made. When this fare is increased from 50p to £1, the number of journeys made drops from 80,000 a day to 40,000. Calculate the arc elasticity of demand over this section of the demand schedule for bus journeys.
3. Calculate the arc elasticity of demand between (a) £5 and £10, and (b) between £10 and £15, for the demand schedule shown in Figure 2.3.
4. The data show the quantity demanded of a good at various prices. Calculate the arc elasticity of demand for each £5 increment along the demand schedule.

Price	£40	£35	£30	£25	£20	£15	£10	£5	£0
Quantity	0	50	100	150	200	250	300	350	400

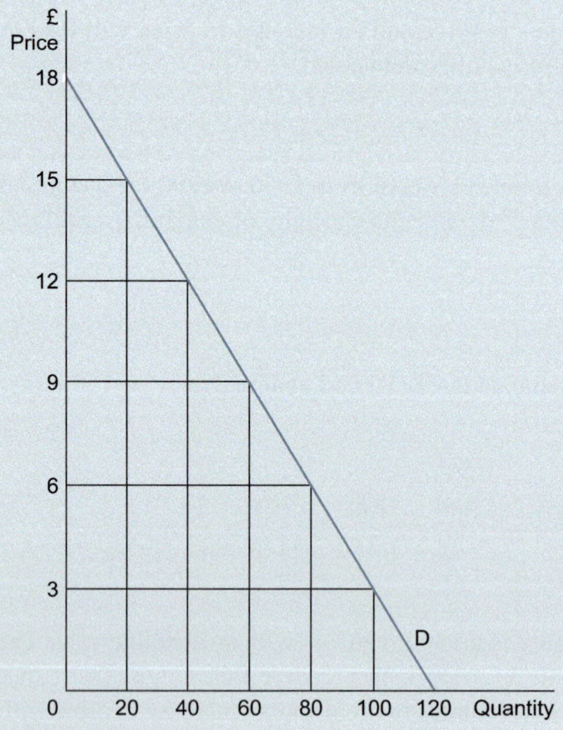

Figure 2.2 Linear demand schedule

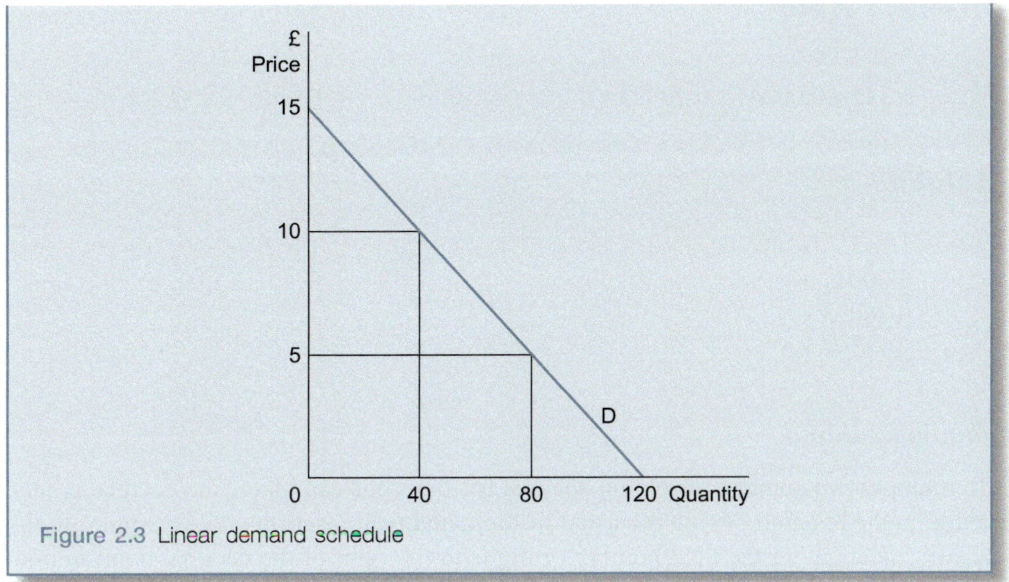

Figure 2.3 Linear demand schedule

2.6 DECIMALS

Decimals are just another way of expressing fractions.

$$0.1 = 1/10$$
$$0.01 = 1/100$$
$$0.001 = 1/1{,}000 \text{ etc.}$$

Thus, 0.234 is equivalent to 234 / 1,000.

Most of the time you will be able to perform operations involving decimals by using a calculator and so only a very brief summary of the manual methods of performing arithmetic operations using decimals is given here.

Addition and subtraction

When adding or subtracting decimals, only 'like terms' must be added or subtracted. The easiest way to do this is to write any list of decimal numbers to be added so that the decimal points are all in a vertical column, in a similar fashion to the way that you may have been taught in primary school to add whole numbers by putting them in columns for hundreds, tens and units. You then add all the numbers that are the same number of digits away from the decimal point, carrying units over to the next column when the total is more than 9.

Example 2.20

$$1.345 + 0.00041 + 0.20023 = ?$$

Solution

$$\begin{array}{ll} 1.345 & + \\ 0.00041 & + \\ \underline{0.20023} & + \\ 1.54564 & = \end{array}$$

Multiplication

To multiply two numbers involving decimal fractions one can ignore the decimal points, multiply the two numbers in the usual fashion, and then insert the decimal point in the answer by counting the total number of digits to the right of the decimal point in both numbers that were multiplied.

Example 2.21

$$2.463 \times 0.38 = ?$$

Solution

Removing the decimal places and multiplying the whole numbers remaining gives

$$\begin{array}{rl} 2,463 & \times \\ 38 & \\ \hline 19,704 & + \\ 73,890 & \\ \hline 93,954 & = \end{array}$$

There were a total of 5 digits to the right of the decimal place in the two numbers to be multiplied and so the answer is 0.93594.

Division

When dividing by a decimal fraction, one first multiplies the fraction by the multiple of 10 that will convert it into a whole number. Then the number that is being divided is multiplied by the same multiple of 10 and the normal division operation is applied.

Example 2.22

$$360.54 \div 0.04 = ?$$

Solution

Multiplying both terms by 100 the problem becomes

$$36{,}054 \div 4 = 9{,}013.5$$

Given that actual arithmetic operations involving decimals can usually be performed with a calculator, perhaps one of the most common problems you are likely to face is how to express quantities as decimals before setting up a calculation.

Example 2.23

Express 0.01p as a decimal fraction of £1.

Solution

$$\begin{aligned} & 1\text{p} = £0.01 \\ \text{Therefore}\quad & 0.01\text{p} = £0.0001 \end{aligned}$$

In mathematics, a decimal format is often required for a value that is usually specified as a percentage in everyday usage. For example, interest rates are usually specified as percentages. A percentage format is just another way of specifying a decimal fraction, e.g.

$$62\% = \frac{62}{100} = 0.62$$

Thus, percentages can easily be converted into decimal fractions by dividing by 100.

Example 2.24

$$\begin{array}{lll} 22\% = 0.22 & 0.24\% = 0.0024 & \\ 24.56\% = 0.2456 & 0.02\% = 0.0002 & 2.4\% = 0.024 \end{array}$$

You will need to convert interest rate percentages to their decimal equivalent when you learn about investment appraisal methods and other aspects of financial mathematics in Chapter 8.

Because some fractions cannot be expressed exactly in decimals, one may need to 'round off' an answer for convenience. In many of the problems in this book there is not much point in taking answers beyond two decimal places. Where this is done, then the note '(to 2 dp)' can be put after the answer. For example, 1/7 as a percentage is 14.29% (to 2 dp). However, students should be careful not to round answers until the problem they are working on is completely finished. If answers are rounded at intermediate stages, this may lead to inaccurate final answers.

QUESTIONS 2.6

(Try to answer these without using a calculator.)

1. 53.024 − 16.11 =
2. 44.2 × 17 =
3. 602.025 + 34.1006 − 201.016 =
4. 432.984 ÷ 0.012 =
5. 64.5 × 0.0015 =
6. 18.3 ÷ 0.03 =
7. How many pencils costing 30p each can be bought for £42.00?
8. What is 1 millimetre as a decimal fraction of
   (a) 1 centimetre (b) 1 metre (c) 1 kilometre?
9. Specify the following percentages as decimal fractions:
   (a) 45.2% (c) 7.5%
   (b) 243.15 (d) 0.2%
10. Specify the fraction 5/8 as both a decimal and a percentage.

2.7 NEGATIVE NUMBERS

There are numerous instances where one comes across negative quantities, such as temperatures below zero or bank overdrafts. For example, if you have £35 in your bank account and withdraw £60 then, assuming you are allowed an overdraft, your bank balance becomes −£25. There are instances, however, where it is not usually possible to have negative quantities. For example, a firm's production level cannot be negative.

To add negative numbers, one simply subtracts the number after the negative sign. In the examples that follow the negative numbers are written with brackets around them to help you distinguish between the addition of negative numbers and the subtraction of positive numbers.

Example 2.25

$$45+(-32)+(-6)=45-32-6=7$$

If it is required to subtract a negative number then the two negatives will cancel out and one adds the absolute value of the number.

Example 2.26

$$0.5-(-0.45)-(-0.1)=0.5+0.45+0.1=1.05$$

The rules for multiplication and division of negative numbers are:

- A negative multiplied (or divided) by a positive gives a negative.
- A negative multiplied (or divided) by a negative gives a positive.

Example 2.27

Eight students have an overdraft of £210 each. What is their total bank balance?

Solution

$$\text{total balance} = 8\times(-210) = -£1,680$$

Example 2.28

$$\frac{24}{-5}\div\frac{-32}{-10}=\frac{24}{-5}\times\frac{-10}{-32}=\frac{3}{1}\times\frac{2}{-4}=\frac{6}{-4}=-\frac{3}{2}$$

QUESTIONS 2.7

1. Subtract −4 from −6.
2. Multiply −4 by 6.
3. −48 + 6 − 21 + 30 =
4. −0.55 + 1.0 =
5. 1.2 + (−0.65) − 0.2 =
6. −26 × 4.5 −
7. 30 × (4 − 15) =
8. (−60) × (−60) =
9. $-\frac{1}{4}\times\frac{9}{7}-\frac{4}{5}=$
10. $(-1)\dfrac{\frac{4}{30+34}}{\frac{-2}{16+18}}=$

2.8 POWERS

We have all come across terms such as 'square metres' or 'cubic capacity'. A square metre is a rectangular area with each side equal to 1 metre. If a square room had all walls 5 metres long, then its area would be 5 × 5 = 25 square metres.

When we multiply a number by itself in this fashion, then we say we are 'squaring' it. The mathematical notation for this operation is the superscript 2. Thus, '12 squared' is written as 12^2.

Example 2.29

$$2.5^2 = 2.5 \times 2.5 = 6.25$$

We find the cubic capacity of a room, in cubic metres, by multiplying length × width × height. If all these distances are equal, say at 3 metres (i.e. the room is a perfect cube), then cubic capacity is 3 × 3 × 3 = 27 cubic metres. When a number is cubed in this fashion, the notation used is the superscript 3, e.g. 12^3.

These superscripts are known as 'powers' or 'exponents' and denote the number of times a number is multiplied by itself. Although there are no physical analogies for powers other than 2 and 3, in mathematics one can encounter powers of any value.

Example 2.30

$$12^4 = 12 \times 12 \times 12 \times 12 = 20,736$$
$$12^5 = 12 \times 12 \times 12 \times 12 \times 12 = 248,832$$

To multiply numbers which are expressed as powers of the same number one adds all the powers together.

Example 2.31

$$3^3 \times 3^5 = (3 \times 3 \times 3) \times (3 \times 3 \times 3 \times 3 \times 3) = 3^8 = 6,561$$

To divide numbers in terms of powers of the same base number, one subtracts the superscript of the denominator from the superscript of the numerator.

Example 2.32

$$\frac{6^6}{6^3} = \frac{6 \times 6 \times 6 \times 6 \times 6 \times 6}{6 \times 6 \times 6} = 6 \times 6 \times 6 = 6^{6-3} = 6^3 = 216$$

In the previous two examples the multiplication and division processes are set out in full to illustrate how these processes work with exponents. In practice, of course, one need not do this and it is just necessary to add or subtract the powers.

Any number to the power of 1 is simply the number itself. Although we do not normally write in the power 1 for single numbers, we must not forget to include it in calculations involving powers.

Example 2.33

$$4.6 \times 4.6^3 \times 4.6^2 = 4.6^{1+3+2} = 4.6^6 = 9{,}474.3 \text{ (to 1 dp)}$$

In this example, the first term 4.6 is counted as 4.6^1 when the powers are added up in the multiplication process.

Any number to the power of 0 is equal to 1. For example, $8^2 \times 8^0 = 8^{(2+0)} = 8^2$ so 8^0 must be 1.

Powers can also take negative values or can be fractions (see Section 2.9). A negative superscript indicates the number of times that one is dividing by the given number.

Example 2.34

$$3^6 \times 3^{-4} = \frac{3^6}{3^4} = \frac{3 \times 3 \times 3 \times 3 \times 3 \times 3}{3 \times 3 \times 3 \times 3} = 3^2$$

Multiplying by a number with a negative power (when both quantities are expressed as powers of the same number) simply involves adding the (negative) power to the power of the number being multiplied.

Example 2.35

$$8^4 \times 8^{-2} = 8^2 = 64$$

Example 2.36

$$14^7 \times 14^{-9} \times 14^6 = 14^4 = 38{,}416$$

To evaluate a number using the [y^x] or [^] function on most calculators, the usual procedure is to enter y, the number to be multiplied, hit the [y^x] or [^] function key, then enter x, the exponent, and finally hit the [=] key. For example, to find 14^4 enter 14 [y^x] 4 [=] and you should get 38,416 as your answer.

Care has to be taken if using the [y^x] function to evaluate powers of negative numbers, which can be entered in brackets. Remembering that a negative multiplied by a positive gives a negative number, and a negative multiplied by a negative gives a positive, we can work out that if a negative number has an even whole number exponent then the whole term will be positive.

Example 2.37

$$(-3)^4 = (-3)^2 \times (-3)^2 = 9 \times 9 = 81$$

However, if the exponent is an odd number the term will be negative.

Example 2.38

$$(-3)^5 = 3^5 \times (-1)^5 = 243 \times (-1) = -243$$

Therefore, when using a calculator to find the values of negative numbers taken to powers, one works with the absolute value and then puts in the negative sign if the power value is an odd number.

Example 2.39

$$(-2)^{-2} \times (-2)^{-1} = (-2)^{-3} = \frac{1}{(-2)^3} = \frac{1}{-8} = -0.125$$

QUESTIONS 2.8

1. $4^2 \div 4^3 =$
2. $123^7 \times 123^{-6} =$
3. $6^4 \div (6^2 \times 6) =$
4. $(-2)^3 \times (-2)^3 =$
5. $1.42^4 \times 1.42^3 =$
6. $9^5 \times 9^{-3} \times 9^4 =$
7. $8.673^3 \div 8.673^6 =$
8. $(-6)^5 \times (-6)^{-3} =$
9. $(-8.52)^4 \times (-8.52)^{-1} =$
10. $(-2.5)^{-8} + (0.2)^6 \times (0.2)^{-8} =$

2.9 ROOTS AND FRACTIONAL POWERS

The **square root** of a number is the quantity which when squared gives the original number. There are different forms of notation. The square root of 16 can be written

$$\sqrt{16} = 4 \quad \text{or} \quad 16^{0.5} = 4$$

Writing the root in the exponential format shows us that roots are just fractional powers. We can check this exponential format of $16^{0.5}$ using the rule for multiplying powers.

$$\left(16^{0.5}\right)^2 = 16^{0.5} \times 16^{0.5} = 16^{0.5+0.5} = 16^1 = 16$$

Even basic calculators have a root function and so it is not normally worth bothering with the rather tedious manual method of calculating roots when the root is not obvious, as it is in the previous example.

Example 2.40

$$2246^{0.5} = \sqrt{2246} = 47.391982 \quad \text{(using a calculator)}$$

Although the positive square root of a number is perhaps the most obvious one, there will also be a negative square root. For example,

$$(-4)\times(-4)=16$$

and so (−4) is a square root of 16, as well as 4. The negative square root is often important in the mathematical analysis of economic problems and it should not be neglected. The usual convention is to use the ± sign which means 'plus or minus'. Therefore, we really ought to say

$$\sqrt{16} = \pm 4$$

There are other roots. For example, $\sqrt[3]{27}$ or $27^{1/3}$ is the cube root, which is the number which when multiplied by itself three times equals 27. This is easily checked as

$$\left(27^{1/3}\right)^3 = 27^{1/3} \times 27^{1/3} \times 27^{1/3} = 27^1 = 27$$

When multiplying roots they need to be expressed in the exponential format, e.g. $6^{0.5}$, so that the rules for multiplying powers can be applied.

Example 2.41

$$47^{0.5} \times 47^{0.5} = 47$$

Example 2.42

$$\begin{aligned} 15\times 9^{0.75}\times 9^{0.75} &= 15\times 9^{1.5} \\ &= 15\times 9^{1.0}\times 9^{0.5} \\ &= 15\times 9\times 3 = 405 \end{aligned}$$

These basic rules for multiplying numbers with powers as fractions will prove very useful when we get to algebra in Chapter 3.

Roots other than square roots can be evaluated using the $\sqrt[x]{y}$ function key on a calculator.

Example 2.43

To evaluate $\sqrt[5]{261}$ the usual procedure is to enter

$$261\left[\sqrt[x]{y}\right]5[=]$$

which should give 3.0431832.

Not all fractional powers correspond to an exact root in the same sense as square or cube roots, e.g. $6^{0.625}$ is not any particular root. To evaluate these other fractional powers in this exponent format you can use the $[y^x]$ or [^] function key on a calculator.

Example 2.44

To evaluate $452^{0.85}$ most calculators require you to enter

$$452\left[y^x\right]0.85[=] \quad \text{which should give the answer } 180.66236$$

Some roots and other powers less than one cannot be evaluated for negative numbers. For example, a negative number cannot be the product of two positive or two negative numbers, so the square root of a negative number does not exist. Some other roots for negative numbers do exist, e.g. $\sqrt[3]{-1} = -1$. Some applications of these rules to financial problems are explained in Chapter 8.

Example 2.45

$$24^{0.45} \times 24^{-1} = 24^{-0.55} = 0.1741341$$

Note that you may need to use the [+/−] key on your calculator after entering 0.55 when evaluating this power. Alternatively, you could have calculated

$$\frac{1}{24^{0.55}} = \frac{1}{5.7427007} = 0.1741341$$

Example 2.46

$$20 \times 8^{0.3} \times 8^{0.25} = 20 \times 8^{0.55} = 20 \times 3.1383364 = 62.766728$$

Sometimes it may help to multiply together two numbers with a common power. Both numbers can be put inside brackets with the common power outside the brackets.

Example 2.47

$$18^{0.5} \times 2^{0.5} = (18 \times 2)^{0.5} = 36^{0.5} = 6$$

QUESTIONS 2.9

Rewrite the following problems as powers, and then evaluate.

1. $\sqrt{625} =$
2. $\sqrt[3]{8} =$
3. $5^{0.5} \times 5^{-1.5} =$
4. $(7)^{0.5} \times (7)^{0.5} =$
5. $6^{0.3} \times 6^{-0.2} \times 6^{0.4} =$
6. $12 \times 4^{0.8} \times 4^{0.7} =$
7. $20^{0.5} \times 5^{0.5} =$
8. $16^{0.4} \times 16^{0.2} =$
9. $462^{-0.83} \times 462^{0.48} \div 46^{2-0.2} =$
10. $76^{0.62} \times 18^{0.62} =$

2.10 LOGARITHMS

The logarithm of a number is simply the power to which the 'logarithm base number' must be raised to equal that number. Logarithms are frequently used in economics and finance to calculate growth rates and describe social processes, as will be shown in Chapters 8 and 15.

In general, we write

$$\log_a x = b$$

which can be read as 'the logarithm of x to base a' equals b. We then say that $\log_a x$ is defined for every positive number x such that $a^b = x$.

Example 2.48

$\log_2 16 = 4$ because you need to take 2 to the power of 4 to get 16 ($2^4 = 16$),

$\log_3 \frac{1}{9} = -2$ because $3^{-2} = \frac{1}{9}$.

The most commonly encountered logarithm is the base 10 logarithm. What this means is that the logarithm of any number is the power to which 10 must be raised to equal that number. The usual notation for logarithms to base 10 is simply 'log', i.e. instead of writing $\log_{10} x$, we skip the base 10 and write $\log x$. Thus,

if $\log x = b$ then $x = a^b$

Example 2.49

$$\begin{aligned} \log 100 &= 2 \textit{ since } 100 = 10^2 \\ \log 10 &= 1 \\ \log 1{,}000 &= 3 \\ \log 3.1622777 &= 0.5 \text{ (since the square root of 10 is 3.1622777)} \end{aligned}$$

Another commonly encountered logarithm is the natural logarithm, which has the base of e (e is a constant equal to approximately 2.718, discussed in more detail in Chapter 15). This is denoted using the symbol 'ln':

$$\text{if } \ln x = b \text{ then } x = e^b$$

Example 2.50

$$\begin{aligned} \ln e &= 1 \text{ since } e^1 = e \\ \ln 100 &= 4.605 \text{ since } 100 = 3^{4.605} \\ \ln 1{,}000 &= 6.908 \\ \ln 1.6487 &= 0.5 \end{aligned}$$

As you may expect, calculators come in handy when evaluating logarithms of any base. To evaluate logarithms to the base of 10, use the [log] button; for the natural logarithm, the [ln] button should be used. Not all calculators have a function that allows to evaluate logarithms to any arbitrary base. In that case, it might be necessary to change the base of the logarithm from an arbitrary value a to any desired value c. This can be done using the following rule:

$$\log a = \frac{\log_c x}{\log_c a}$$

Example 2.51

We want to evaluate log $_4$12, but first we need to change the logarithm's base to base 10. Thus,

$$\log_4 12 = \frac{\log 12}{\log 4}$$

Now we have a ratio or division of two logs to base 10 which can be easily evaluated using the [log] function on a calculator.

$$\begin{aligned} \log 12 &= 1.07918 \\ \log 4 &= 0.60206 \end{aligned}$$

Thus,

$$\frac{1.07918}{0.60206} = 1.792481 \text{ which is the same as } \log_4 12$$

This and some other useful rules when working with logarithms are summarized in the following box.

RULES FOR THE LOGARITHMS

Rule	Example
$\log_a(xy) = \log_a x + \log_a y$	$\log_5(3 \times 4) = \log_5 3 + \log_5 4$
$\log_a \dfrac{x}{y} = \log_a x - \log_a y$	$\log_5 \dfrac{3}{4} = \log_5 3 - \log_5 4$
$\log_a x^p = p \log_a x$	$\log_2 8^4 = 4 \times \log_2 8$
$\log_a x = \dfrac{\log_c x}{\log_c a}$	$\log_3 81 = \dfrac{\log_9 81}{\log_9 3}$

Let us now briefly outline some of the economic applications of logs. It can help in the estimation of the parameters of non-linear functions or equations. This application is explained further in Section 4.9. Logarithms can also be used to help solve equations involving unknown exponent values.

Example 2.52

If $460(1.08)^n = 925$, what is n?

Solution

$$460(1.08)^n = 925$$

$$(1.08)^n = 2.0108696$$

Putting in log form $n\log 1.08 = \log 2.0108696$

$$n = \frac{\log 2.0108696}{\log 1.08} = \frac{0.3033839}{0.0334238} = 9.0768945$$

We shall return to this type of problem in Chapter 8 when we consider for how long a sum of money needs to be invested at any given rate of interest to accumulate to a specified sum.

QUESTIONS 2.10

Use logs to answer the following.

1. $\log_6 216 =$
2. $\ln e^{12} =$
3. $\log 10{,}000 =$
4. $\log 100^8 =$
5. $\ln(7 \times 8) =$
6. $\log_4 \frac{64}{256} =$
7. If $(1.06)^n = 235$ what is n?
8. If $825(1.22)^n = 1{,}972$ what is n?
9. If $4{,}350(1.14)^n = 8{,}523$ what is n?

3 Introduction to algebra

Learning objectives

After completing this chapter students should be able to:

- construct algebraic expressions for economic concepts involving unknown values
- simplify and reformulate basic algebraic expressions
- solve single linear equations with one unknown variable
- use the summation sign $\sum$ and construct simple price index measures of inflation
- perform basic mathematical operations on algebraic expressions that involve inequality signs.

3.1 REPRESENTATION

Algebra is basically a system of shorthand notations. Symbols are used to represent concepts and variables that can take different values. For example, suppose that a biscuit manufacturer uses 0.2 kg of flour, 0.05 kg of sugar and 0.1 kg of butter for each packet of biscuits produced. Then, we could specify the total amount of flour used as '0.2 kg times the number of packets of biscuits produced'. However, it is much simpler if we let the letter q represent the number of packets of biscuits produced and then the amount of flour required in kilograms will be 0.2 times q, which we write as $0.2q$.

Thus, we can also say

$$\text{amount of sugar required} = 0.05q \text{ kilograms}$$
$$\text{amount of butter required} = 0.1q \text{ kilograms}$$

Sometimes an algebraic expression will contain several terms with different algebraic symbols representing the unknown quantities of different variables.

DOI: 10.4324/9781003360827-3

Consider the total expenditure on ingredients (inputs) by the firm in the previous above. Let the price (in £) per kilogram of flour be denoted by the letter a. The total cost of the amount of flour the firm uses will therefore be $0.2q$ times a, written as $0.2qa$.

If the price per kilogram of sugar is denoted by the letter b and the price per kilogram of butter is c then the total expenditure (in £) on inputs for biscuit production will be

$$0.2qa + 0.05qb + 0.1qc$$

When two algebraic symbols are multiplied together, it does not matter in which order they are written, e.g. $xy = yx$. This, of course, is the same rule that applies when multiplying numbers. For example:

$$5 \times 7 = 7 \times 5$$

Any operation that can be carried out with numbers (such as division or deriving the square root) can be carried out with algebraic symbols. The difference is that in most cases the answer will also be in terms of algebraic symbols rather than numbers.

An algebraic expression cannot be evaluated until values have been given to the variables that the symbols represent (see Section 3.2). For example, an expression for the length of fencing (in metres) needed to enclose a square plot of land of unknown size can be constructed as follows:

If the length of each side is A.
All squares have four sides.
Therefore, the length of fencing = 4 × (length of one side) = $4A$

Without information on the value of A we cannot say any more. Once the value of A is specified we can work out the value of the expression using basic arithmetic. For example, if the length of one side is 10 metres, then we just substitute 10 for A and so

$$\text{length of fencing} = 4A = 4 \times 10 = 40 \text{ metres}$$

An advantage of writing an expression in an algebraic form is that it is not necessary to work out a solution for every possible value of the unknown variables, and different values can just be substituted in as needed. In this section we start with some fairly simple expressions but later, as more complex relationships are dealt with, the usefulness of algebraic representation will become more obvious.

Example 3.1

You are tiling a bathroom with 10 cm square tiles. The number of square metres to be tiled is as yet unknown and is represented by q. Because you may break some tiles and will have to cut some to fit around corners, you work to the rule of thumb that you should buy enough tiles to cover the specified area plus 10%. Derive an expression for the number of tiles to be bought in terms of q.

Solution

One hundred square tiles with side length of 10 cm will cover 1 square metre and 110% written as a decimal is 1.1. Therefore, the number of tiles required is

$$100q \times 1.1 = 110q$$

QUESTIONS 3.1

1. An engineering firm makes metal components. Each component requires 0.01 tonnes of steel, 0.5 hours of labour plus 0.5 hours of machine time. Let the number of components produced be denoted by x. Using your own notation for relevant variables, derive algebraic expressions for:
   (a) the amount of steel required;
   (b) the amount of labour required;
   (c) the amount of machine time required.
2. If the price per tonne of steel is given by r, the price per hour of labour is given by w and the price per hour of machine time is given by m, then derive an expression for the total production costs of the firm in question 1.
3. The petrol consumption of your car is 12 miles per litre. Let x be the distance you travel in miles and p the price per litre of petrol in pence. Write expressions for (a) the amount of petrol you use, and (b) your expenditure on petrol.
4. Suppose that you are cooking a dinner for a number of people. You only know how to cook one dish, and this requires you to buy 0.1 kg of meat plus 0.3 kg of potatoes for each person. (Assume you already have a plentiful supply of any other ingredients.) Define your own algebraic symbols for relevant unknown quantities, and then write expressions for:
   (a) the amount of meat you need to buy;
   (b) the amount of potatoes you need to buy;
   (c) your total shopping bill.
5. You are cooking again! This time it's a turkey. The cookery book recommends a cooking time of 30 minutes for every kilogram weight of turkey, plus another quarter of an hour. Write an expression for the total cooking time (in hours) for your turkey in terms of its weight (w).
6. Someone is booking a meal in a restaurant for a group of people. They are told that there is a set menu that costs £19.50 per adult and £9.50 per child, and there is also a fixed service charge of £1 per head for each meal served.

Derive an expression for the total cost of the meal, in pounds, if there are x adults and y children.

7. A firm produces a good which it can sell any amount of at £12 per unit. Its costs are a fixed outlay of £6,000 plus £9 in variable costs for each unit produced. Write an expression for the firm's profit in terms of the number of units produced.

3.2 EVALUATION

An expression can be evaluated when the variables represented by algebraic symbols are given specific numerical values.

Example 3.2

Evaluate the expression $6.5x$ when $x = 8$.

Solution

$$6.5x = 6.5(8) = 52$$

Example 3.3

A firm's total costs are given by the expression

$$0.2qa + 0.05qb + 0.1qc$$

where q is output and a, b and c are the per-unit costs (in £) of the three different inputs used. Evaluate these costs if $q = 1{,}000$, $a = 0.6$, $b = 1.3$ and $c = 2.1$.

Solution

$$\begin{aligned} 0.2qa + 0.05qb + 0.1qc &= 0.2(1000 \times 0.6) + 0.05(1000 \times 1.3) + 0.1(1000 \times 2.1) \\ &= 0.2(600) + 0.05(1300) + 0.1(2100) \\ &= 120 + 65 + 210 = 395 \end{aligned}$$

Therefore, the total cost is £395.

Example 3.4

Evaluate the expression $(3^x + 4)y$ when $x = 2$ and $y = 6$.

Solution

$$\left(3^x + 4\right)y = \left(3^2 + 4\right)6 = (9+4)6 = 13 \times 6 = 78$$

Example 3.5

A bureau de change will sell euros at an exchange rate of €1.17 to the pound and charges a flat rate commission of £2 on all transactions.

(i) Write an expression for the number of euros that can be bought for x pounds (any given quantity of sterling), and
(ii) evaluate it for x = £250.

Solution

(i) Number of euros bought for x is given by $1.17(x - 2)$.
(ii) £250 will therefore buy

$$1.17(250-2) = 1.17(248) = €290.16$$

QUESTIONS 3.2

1. Evaluate the expression $2x^3 + 4x$ when $x = 6$.
2. Evaluate the expression $(6x + 2y))y^2$ when $x = 4.5$ and $y = 1.6$.
3. After the UK government privatized the water authorities it decided that annual percentage price increases for water would be limited to the rate of inflation plus z, where z was a figure to be determined by the government. Write an algebraic expression for the maximum annual percentage price increase and evaluate it for an inflation rate of 2.6% and $z = 3\%$.
4. Evaluate the expression $1.02^x + x^{-3.2}$ when $x = 2.8$.
5. A firm's average production costs (AC) are given by the expression

   $$AC = 450q^{-1} + 0.2q^{1.5}$$

   where q is the quantity of output. What will AC be when output is 175?
6. A firm's profit (in £) is given by the expression $7.5q - 1650$ where q is output sold. What profit will it make when q is 500?
7 If income tax is levied at a rate of 20% on annual income over £12,000 then:
   (a) write an expression for net monthly salary in terms of gross monthly salary (assumed to be greater than £1,000), and
   (b) evaluate it if gross monthly salary is £3,000.

3.3 SIMPLIFICATION: ADDITION AND SUBTRACTION

Simplifying an expression means rearranging the terms in it so that the expression becomes easier to work with. Before setting out the different rules for simplification, let us work through an example.

Example 3.6

A businesswoman driving her own car on her employer's business gets paid a set fee per mile travelled. During one week she records one journey of 234 miles, one of 166 miles and one of 90 miles. Derive an expression for total travelling expenses.

Solution

If the rate per mile is denoted by the letter M then her expenses will be $234M$ for the first journey and $166M$ and $90M$ for the second and third journeys, respectively. Total travelling expenses for the week will thus be

$$234M + 166M + 90M$$

We could, instead, simply add up the total number of miles travelled during the week and then multiply by the rate per mile. This would give

$$(234 + 166 + 90) \times M = 490M$$

It is therefore obvious, as both methods should give the same answer, that

$$234M + 166M + 90M = 490M$$

In other words, in an expression with different terms all in the same format of

$$(\text{a number}) \times M$$

all the terms can be added together.

The general rule is that like terms can be added or subtracted to simplify an expression. 'Like terms' have the same algebraic symbol or symbols, usually multiplied by a number.

Example 3.7

$$3x + 14x + 7x = 24x$$

Example 3.8

$$45A - 32A = 13A$$

Only terms that have exactly the same algebraic notation can be added or subtracted in this way. For example, the terms x, y^2 and xy are all different and cannot be added together or subtracted from each other.

Example 3.9

Simplify the expression $5x^2 + 6xy - 32x + 3yx - x^2 + 4x$.

Solution

Adding/subtracting all the terms in x^2 gives $4x^2$
Adding/subtracting all the terms in x gives $-28x$
Adding/subtracting all the terms in xy gives $9xy$
(The terms in xy and yx can be added together since $xy = yx$)
Putting all these terms together gives the simplified expression

$$4x^2 - 28x + 9xy$$

All the basic rules of arithmetic apply when algebraic symbols are used instead of actual numbers. The difference is that the simplified expression will still be in a format of algebraic terms.

The following example applies the rule that if there is a negative sign in front of a set of brackets, then the positive and negative signs of the terms within the brackets are reversed if the brackets are removed.

Example 3.10

Simplify the expression

$$16q + 33q - 2q - (15q - 6q)$$

Solution

Removing brackets, the previous expression becomes

$$16q + 33q - 2q - 15q + 6q = 38q$$

QUESTIONS 3.3

1. Simplify the expression $6x - (6 - 24x) + 10$.
2. Simplify the expression $4xy + (24x - 13y) - 12 + 3yx - 5y$.
3. A firm produces two goods, X and Y, which it sells at prices per unit of £26 and £22, respectively. Good X requires an initial outlay of £400 and then an expenditure of £16 on labour and £4 on raw materials for each unit produced. Good Y requires a fixed outlay of £250 plus £14 of labour and £3 of raw material for each unit. If the quantities produced of X and Y are x and y, respectively, write an expression in terms of x and y for the firm's total profit and then simplify it.
4. A worker earns £18 per hour for the first 40 hours a week and £27 per hour for any extra hours above 40. Assuming that she works at least 40 hours, write an expression for her gross weekly wage in terms of H, the total hours worked per week, and then simplify it.

3.4 SIMPLIFICATION: MULTIPLICATION

When a set of brackets containing different terms is multiplied by a symbol or a number it may be possible to simplify an expression by multiplying out, i.e. multiplying each term within the brackets by the term outside. In some circumstances, though, it may be preferable to leave brackets in the expression if it makes it clearer to work with.

Example 3.11

$$x(4+x) = 4x + x^2$$

Example 3.12

$$\begin{aligned} 5(7x^2 - x) - 3(3x^2 + 6x) &= 35x^2 - 5x - 9x^2 - 18x \\ &= 26x^2 - 23x \end{aligned}$$

Example 3.13

$$\begin{aligned} 6y(8+3x) - 2xy + 12y &= 48y + 18xy - 2xy + 12y \\ &= 60y + 16xy \end{aligned}$$

Example 3.14

The basic hourly rate for a weekly paid worker is £16 and any hours above 40 are paid at £24. Tax is paid at a rate of 25% on any earnings above £160 a week. Assuming hours worked per week (H) exceed 40, write an expression for net weekly wage in terms of H and then simplify it.

Solution

$$\begin{aligned}
\text{gross weekly pay} &= 40\times 16+(H-40)24\\
&= 640+24H-960\\
&= 24H-320\\
\text{net weekly pay} &= 0.75(\text{gross weekly pay}-160)+160\\
&= 0.75(24H-320-160)+160\\
&= 18H-240-120+160\\
&= 18H-200
\end{aligned}$$

If you are not sure whether the expression you have derived is correct, you can try to check it by substituting numerical values for unknown variables. In the previous example, if 50 hours per week were worked, then:

$$\text{Gross pay} = (40 \text{ hours @ £16}) + (10 \text{ hours @ £24}) = £640 + £240 = £880$$
$$\text{Tax payable} = 0.25(£880 - £160) = 0.25(£720) = £180$$
$$\text{Therefore, net pay} = £880 - £180 = £700$$

Using the expression derived in Example 3.14, if $H = 50$, then:

$$\text{net pay} = 18H - 200 = 18(50) - 200 = 900 - 200 = £700$$

This checks out with the previous answer and so we know our expression works.

It is rather more complicated to multiply pairs of brackets together. One method that can be used is the long multiplication that you probably learned at school, but instead of keeping all units, tens, hundreds, etc., in the same column, it is the same algebraic terms that are kept in the same column during the multiplying process so that they can be added together.

Example 3.15

Simplify $(6 + 2x)(4 - 2x)$.

Solution

Writing this as a long multiplication problem:

	$6+2x \quad \times$
	$\underline{4-2x}$
Multiplying $(6+2x)$ by $-2x$	$\quad -12x-4x^2$
Multiplying $(6+2x)$ by 4	$\underline{24+8x}$
Adding together gives the answer	$\underline{24-4x-4x^2}$

You do not have to use long multiplication. It is usually quicker to follow the basic principle that each term in one set of brackets must be multiplied by each term in the other set. Like terms can then be collected together to simplify the resulting expression.

Example 3.16

Multiply and simplify $(3x + 4y)(5x - 2y)$.

Solution

Multiplying the terms in the second set of brackets by $3x$ gives:

$$15x^2 - 6xy \qquad (1)$$

Multiplying the terms in the second set of brackets by $4y$ gives:

$$20xy - 8y^2 \qquad (2)$$

Therefore, adding (1) and (2) the whole expression is

$$15x^2 - 6xy + 20xy - 8y^2 = 15x^2 + 14xy - 8y^2$$

Example 3.17

Simplify $(x + y)^2$.

Solution

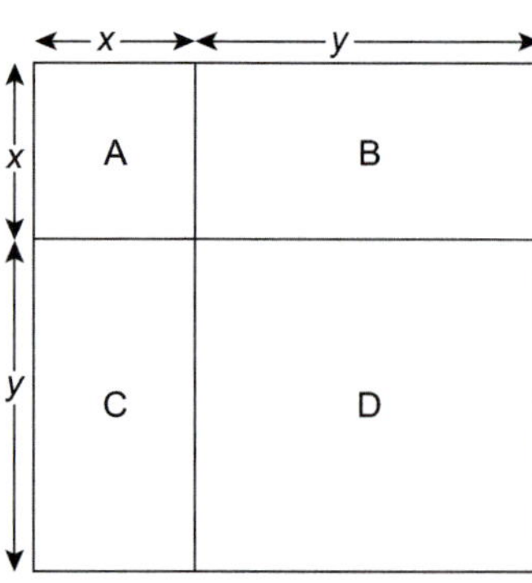

Figure 3.1 Graphical illustration of simplifying $(x + y)^2$

$$\begin{aligned}(x+y)^2 &= (x+y)(x+y)\\ &= x^2 + xy + yx + y^2\\ &= x^2 + 2xy + y^2\end{aligned}$$

The previous answer can be checked by referring to Figure 3.1. The area enclosed in the square with sides of length $x + y$ can be calculated by squaring the lengths of the sides, i.e. finding $(x + y)^2$. One can also see that this square is made up of the four rectangles A, B, C and D whose areas are x^2, xy, xy and y^2, respectively, in other words, $x^2 + 2xy + y^2$, which is the answer obtained previously.

Example 3.18

Multiply and simplify $(6 - 5x)(10 - 2x + 3y)$.

Solution

Multiplying out gives

$$60 - 12x + 18y - 50x + 10x^2 - 15xy = 60 - 62x + 18y + 10x^2 - 15xy$$

Some expressions may be best left with the brackets still in.

Example 3.19

If a sum of x is invested at an interest rate of r % write an expression for the value of the investment at the end of 2 years.

Solution

After 1 year the investment's value is $x\left(1+\frac{r}{100}\right)$

After 2 years the investment's value is $x\left(1+\frac{r}{100}\right)^2$

One could multiply out, but in this case the expression is probably clearer, and easier to evaluate if the brackets are left in. The next section explains how some expressions may be 'simplified' by reformatting into two expressions in brackets multiplied together. This is called 'factorization'.

QUESTIONS 3.4

Simplify the following expressions.

1. $6x(x - 4)$
2. $(x + 3)^2 - 2x$
3. $(2x + y)(x + 3)$
4. $(6x + 2y)(7x - 8y) + 4y + 2y$
5. $(4x - y + 7)(2y - 3) + (9x - 3y)(5 + 6y)$
6. $(12 - x + 3y + 4z)(10 + x + 2y)$

7. A good costs £180 a unit but if an order is made for more than 10 units this price is reduced by a discount of £2 for every additional unit, up to a maximum of 60 units purchased, e.g. price per unit is £178 if the order size is 11 and £176 if the order size is 12.

   Assuming the order size is between 10 and 60 units, write an expression for the total cost of an order in terms of order size and simplify it.
8. A holiday excursion costs £38 per person for transport plus £15 per adult and £8 per child for meals. Write an expression for the total cost of an excursion for x adults and y children and simplify it.
9. A firm is building a car park for its employees. Assume that a car park to accommodate x cars must have a length (in metres) of $4x + 10$ and a width of $2x + 10$. If 24 square metres will be specifically allocated for visitors' cars, write an expression for the amount of space available for the cars of the workforce in terms of x, the planned capacity of the car park.
10. A firm buys a raw material that costs £220 a tonne for the first 40 tonnes, £180 a tonne for the next 40 tonnes and £150 for any further quantities. Write an expression for the firm's total expenditure on this input in terms of the total amount used (which can be assumed to be greater than 80 tonnes) and simplify.

3.5 SIMPLIFICATION: FACTORIZING

For some purposes (see, for example, Section 3.6) it may be helpful if an algebraic expression can be simplified into a format of two sets of brackets multiplied together. For example

$$x^2 + 4x + 4 = (x + 2)(x + 2)$$
which can also be written as $(x + 2)^2$

This is rather like the arithmetical process of factorizing a number, which means finding all the prime numbers which when multiplied together equal that number, e.g.

$$126 = 2 \times 3 \times 3 \times 7$$

As we will see later in Chapter 7, it can also be helpful when solving quadratic equations if expressions in the format

$$ax^2 + bx + c$$

can be factorized. However, not all expressions in this form can be factorized into sets of brackets that only involve integers, i.e. whole numbers. There are no set rules for working out if and how an expression may be factorized, although if the term in x does not have a number in front of it (i.e. $a = 1$ in the previous format) then the expression can be factorized if there are two numbers which

(i) give c when multiplied together, and
(ii) give b when added together.

Example 3.20

Attempt to factorize the expression $x^2 + 6x + 9$.

Solution

In this example $a = 1$, $b = 6$ and $c = 9$.

Since $3 \times 3 = 9$ and $3 + 3 = 6$, this expression can be factorized as

$$x^2 + 6x + 9 = (x+3)(x+3)$$

This can be checked as

$$\begin{array}{r} x+3 \ \times \\ x+3 \\ \hline 3x+9 \\ x^2+3x \quad \\ \hline x^2+6x+9 \end{array}$$

Example 3.21

Attempt to factorize the expression $x^2 - 2x - 80$.

Solution

Since $(-10) \times 8 = -80$ and $(-10) + 8 = -2$ then the expression can be factorized as

$$x^2 - 2x - 80 = (x-10)(x+8)$$

Check this answer yourself by multiplying out the two brackets.

Example 3.22

Attempt to factorize the expression $x^2 + 3x + 11$.

Solution

There are no two numbers which when multiplied together give 11 and when added together give 3. Therefore, this expression cannot be factorized.

It is sometimes possible to simplify an expression before factorizing if all the terms are divisible by the same number.

Example 3.23

Attempt to factorize the expression $2x^2 - 10x + 12$.

Solution

$$2x^2 - 10x + 12 = 2\left(x^2 - 5x + 6\right) = 2(x-3)(x-2)$$

In expressions in the format $ax^2 + bx + c$ where a is not equal to 1, then one still has to find two numbers which multiply together to give c. However, one also has to find two numbers for the coefficients of the two terms in x within the two sets of brackets that when multiplied together equal a, and allow the coefficient b to be derived when multiplying out.

Example 3.24

Attempt to factorize the expression $30x^2 + 52x + 14$.

Solution

If we use the results that $6 \times 5 = 30$ and $2 \times 7 = 14$ we can try multiplying

$$\begin{array}{r} 5x + 7 \times \\ 6x + 2 \\ \hline 10x + 14 \\ 30x^2 + 42x \quad \\ \hline 30x^2 + 52x + 14 \end{array}$$

Thus, $\quad 30x^2 + 52x + 14 = (5x+7)(6x+2)$

Similar rules apply when one attempts to factorize an expression with two unknown variables, x and y. This may be in the format

$$ax^2 + bxy + cy^2$$

where a, b and c are given parameters.

Example 3.25

Attempt to factorize the expression $x^2 - y^2$.

Solution

In this example $a = 1$, $b = 0$ and $c = -1$. The two numbers -1 and 1 give -1 when multiplied together and 0 when added. Thus,

$$x^2 - y^2 = (x - y)(x + y)$$

To check this, multiply out:

$$(x - y)(x + y) = x^2 - xy + y^2 - yx = x^2 - y^2$$

Example 3.26

Attempt to factorize the expression $3x^2 + 8x + 23$.

Solution

As 23 is a positive prime number, the only pairs of positive integers that could possibly be multiplied together to give 23 are 1 and 23. Thus, whatever permutations of combinations with terms in x that we try, the term in x when brackets are multiplied out will be at least $24x$, e.g. $(3x + 23)(x + 1) = 3x^2 + 26x + 23$, whereas the given expression contains the term $8x$. It is therefore **not** possible to factorize this expression.

Unfortunately, it is not always obvious whether an expression can be factorized.

Example 3.27

Attempt to factorize the expression $3x^2 + 24 + 16$.

Solution

Although the numbers look promising, if you try various permutations, you will find that this expression does not factorize.

Factorizing expressions is often just a matter of trial and error. Do not despair though! As you will see later on, factorizing may help you to use shortcut methods of solving certain problems, but if you spend ages trying to factorize an expression, then this will defeat the object of using the shortcut method. If it is not obvious how an expression can be factorized after a few minutes of thought and experimentation with some potential possible solutions, then it is usually more efficient to forget factorization and use some other method of solving the problem. We shall return to this topic in Chapter 7.

QUESTIONS 3.5

Attempt to factorize the following expressions.

1. $x^2 + 8x + 16$
2. $x^2 - 6xy + 9y^2$
3. $x^2 + 7x + 22$
4. $8x^2 - 10x + 33$

Check your answer by multiplying out.

3.6 SIMPLIFICATION: DIVISION

To divide an algebraic expression by a number, one divides every term in the expression by the number, cancelling where appropriate.

Example 3.28

$$\frac{15x^2 + 2xy + 90}{3} = 5x^2 + \frac{2}{3}xy + 30$$

To divide by an unknown variable the same rule is used although where the numerator of a fraction does not contain that variable it cannot be simplified any further.

Example 3.29

$$\frac{2x^2}{x} = 2x$$

Example 3.30

$$\frac{4x^3 - 2x^2 + 10x}{x} = \frac{x\left(4x^2 - 2x + 10\right)}{x}$$
$$= 4x^2 - 2x + 10$$

Example 3.31

$$\frac{16x + 120}{x} = 16 + \frac{120}{x}$$

Example 3.32

A firm's total costs are $25x + 2x^2$, where x is output. Write an expression for average cost.

Solution

Average cost is total cost divided by output. Therefore,

$$\text{AC} = \frac{25x + 2x^2}{x} = 25 + 2x$$

If one expression is divided by another expression with more than one term in it, then terms can only be cancelled top and bottom if the numerator and denominator are both multiples of the same factor.

Example 3.33

$$\frac{x^2 + 2x}{x + 2} = \frac{x(x + 2)}{x + 2} = x$$

Example 3.34

$$\frac{x^2 + 5x + 6}{x + 3} = \frac{(x + 3)(x + 2)}{(x + 3)} = x + 2$$

Example 3.35

$$\frac{x^2 + 5x + 6}{x^2 + x - 2} = \frac{(x + 2)(x + 3)}{(x + 2)(x - 1)} = \frac{x + 3}{x - 1}$$

QUESTIONS 3.6

1. Simplify

   $$\frac{6x^2 + 14x - 40}{2x}$$

2. Simplify

   $$\frac{x^2 + 12x + 27}{x + 3}$$

3. Simplify

   $$\frac{8xy + 2x^2 + 24x}{2x}$$

4. A firm has to pay fixed costs of £200 and then £16 labour plus £5 raw materials for each unit produced of good X. Write an expression for average cost and simplify.
5. A firm sells 40% of its output at £200 a unit, 30% at £180 and 30% at £150. Write an expression for average revenue received on each unit sold and then simplify it.
6. You have probably come across this sort of party trick. Think of a number. Add 3. Double it. Add 4. Take away the number you first thought of. Take away 3. Take away the number you thought of again. Add 2. Your answer is 9. Show how this answer can be derived by algebraic simplification by letting x equal the number first thought of.

3.7 SOLVING SIMPLE EQUATIONS

We have seen that evaluating an expression means calculating its value for specific values of the unknown variables. This section explains how it is possible to work backwards to discover the value of an unknown variable when the total value of the expression is given. When an algebraic expression is known to equal a number, or another algebraic expression, we can write an equation, i.e. the two concepts are written on either side of an equality sign. For example,

$$45 = 24 + 3x$$

We have already written some equations when simplifying algebraic expressions in this chapter. However, the ones we have come across so far have usually not been in a format where the value of the unknown variables can be worked out. Take, for example, the simplification exercise

$$3x + 14x - 5x = 12x$$

The expressions on either side of the equality sign are equal, but x cannot be calculated from the information given.

Some equations are what are known as '**identities**', which means that they must always be true. For example, a firm's total costs (TC) can be split into the two components total fixed costs (TFC) and total variable costs (TVC). It must therefore always be the case that

$$\text{TC} = \text{TFC} + \text{TVC}$$

Identities are sometimes written with the three-bar equality sign '≡' instead of '=', but usually only when it is necessary to distinguish them from other forms of equations, such as functions.

A **function** is a relationship between two or more variables such that a unique value of one variable is determined by the values taken by the other variables in the function. (Functions are explained more fully in Chapter 4.) For example, statistical analysis may show that a demand function takes the form

$$q = 450 - 3p$$

where p is price and q is quantity demanded. Thus, the expected quantity demanded can be predicted for any given value of p. For example, if $p = 60$ then

$$q = 450 - 3(60) = 450 - 180 = 270$$

In this section, we shall not distinguish between equations that are identities and those that relate to specific values of functions, since the method of solution is the same for both. We shall also mainly confine the analysis to linear equations with one unknown variable whose value can be deduced from the information given. A **linear equation** is one where the unknown variable does not take any powers other than 1, i.e. there may be terms in x but not x^2, x^{-2}, $\frac{1}{x}$, etc.

Before setting out the formal rules for solving single linear equations, let us work through some simple examples.

Example 3.36

You go into a foreign exchange bureau to buy US dollars for your holiday. You exchange £200 and receive \$245. When you get home, you discover that you have lost your receipt. How can you find out the exchange rate used for your money if you know that the bureau charges a fixed £4 fee on all transactions?

Solution

After allowing for the fixed fee, the amount actually exchanged into dollars will be

$$£200 - £4 = £196$$

Let x be the exchange rate of pounds into dollars. Therefore,

$$\begin{aligned} 196x &= 245 \\ x &= \frac{245}{196} = 1.25 \end{aligned}$$

Thus, the exchange rate is \$1.25 to the pound.

This example illustrates the fundamental principle that one can divide both sides of an equation by the same number.

Example 3.37

If $62 = 34 + 4x$, what is x?

Solution

Subtracting 34 from both sides gives

$$28 = 4x$$

then dividing both sides by 4 gives the solution

$$7 = x$$

This example illustrates the principle that one can subtract the same amount from both sides of an equation and also divide by the same amount.

The **basic principles for solving equations** are that all the terms in the unknown variable have to be brought together on one side of the equation. To do this, one can add, subtract, multiply or divide both sides of an equation by the same number or algebraic term. One can also perform other arithmetical operations, such as finding the square root of both sides of an equality sign. Once the equation is in the form

$$ax = b$$

where a and b are numbers, then x can be found by dividing b by a.

Example 3.38

Solve for x if $16x - 4 = 68 + 7x$.

Solution

Subtracting $7x$ from both sides

$$9x - 4 = 68$$

Adding 4 to both sides

$$9x = 72$$

Dividing both sides by 9 gives the solution

$$x = 8$$

Example 3.39

Solve for x if $4 = 96x^{-1}$

Solution

Multiplying both sides by x

$$4x = 96$$

Dividing both sides by 4 gives the solution

$$x = 24$$

Example 3.40

Solve for x if $6x^2 + 12 = 162$

Solution

Subtracting 12 from both sides

$$6x^2 = 150$$

Dividing both sides by 6

$$x^2 = 25$$

Taking square roots of both sides gives the solution

$$x = 5 \text{ or } x = -5$$

(Note that both 5 and −5 give 25 when taken to the power of 2.)

Example 3.41

A firm has to pay fixed costs of £1,500 plus variable costs of £60 for each unit produced. How many units can it produce for a budget of £4,800?

Solution

$$\text{budget} = \text{total expenditure on production}$$

Therefore, if q is output level

$$4{,}800 = 1{,}500 + 60q$$

Subtracting 1,500 from both sides

$$3{,}300 = 60q$$

Dividing by 60 gives the solution

$$55 = q$$

Thus, the firm can produce 55 units for a budget of £4,800.

Example 3.42

You sell 500 shares in a company via a stockbroker who charges a flat £20 commission rate on all transactions under £1,000. Your bank account is credited with £692 from the sale of the shares. What price were your shares sold at?

Solution

Let price per share be x. Therefore,

$$692 = 500x - 20$$

Adding 20 to both sides

$$712 = 500x$$

Dividing both sides by 500 gives the solution

$$1.424 = x$$

Thus, the share price is £1.424.

QUESTIONS 3.7

1. Solve for x when $16x = 2x + 56$.
2. Solve for x when

$$14 = \frac{6 + 4x}{5x}$$

3. Solve for x when $45 = 24 + 3x$.
4. Solve for x if $5x^2 + 20 = 1{,}000$.
5. If $q = 560 - 3p$, solve for p when $q = 314$.
6. You get paid travelling expenses according to the distance you drive in your car plus a weekly sum of £21. You put in a claim for 420 miles travelled and receive an expenses payment of £273. What is the payment rate per mile?
7. In one module that you are studying, the overall module mark is calculated on the basis of a 30:70 weighting between coursework and examination marks. If you have scored 57% for coursework, what examination mark do you need to get to achieve an overall mark of 40%?
8. You sell 900 shares via your broker who charges a flat rate of commission of £20 on all transactions of less than £1,000. Your bank account is credited with £340 from the share sale. What price were your shares sold at?

9. Your net monthly salary is £2,280. You know that National Insurance and pension contributions take 15% of your gross salary and that income tax is levied at a rate of 20% on gross annual earnings above a £12,000 exemption limit. What is your gross monthly salary?
10. You have 64 square paving stones, each having side length 0.5 metres, and you wish to lay them to form a square patio in your garden. What will the length of a side of your patio be?
11. A firm faces the marginal revenue schedule MR = $80 - 2q$ and the marginal cost schedule MC = $15 + 0.5q$ where q is quantity produced. You know that a firm maximizes profit when MC = MR. What will the profit-maximizing output be?

3.8 THE SUMMATION SIGN Σ AND PRICE INDEXES

The summation sign Σ is used as a shorthand means of expressing the sum of a number of different terms added together. (Σ is the Greek capital letter sigma.) There are two ways in which it can be used. The first is when one variable increases its value by 1 in each successive term, as the following example illustrates.

Example 3.43

A new firm sells 30 units in the first week of business. Sales then increase at the rate of 30 units per week. If it continues in business for 5 weeks, its total cumulative sales will therefore be

$$(30\times 1)+(30\times 2)+(30\times 3)+(30\times 4)+(30\times 5)$$

You can see that the number representing the week is increased by 1 in each successive term. This is rather a cumbersome expression to work with. We can instead write

$$\text{total sales revenue} = \Sigma_{i=1}^{5} 30i$$

This means that one is summing all the terms $30i$ for values of i from 1 to 5.

If the number of weeks of business was not determined and denoted by n, we could instead write

$$\text{total sales revenue} = \Sigma_{i=1}^{n} 30i$$

To evaluate an expression containing a summation sign, it may still be necessary to calculate the value of each term separately and then add up. However, spreadsheets can be used to do repetitive calculations like these, and in some cases short-cut formulae may be used. (See Chapter 8.)

Example 3.44

Evaluate

$$\Sigma_{i=3}^{n}(20+3i) \quad \text{for } n=6$$

Solution

Note that in this example i starts at 3. Thus,

$$\begin{aligned}\Sigma_{i=3}^{6}(20+3i) &= (20+3\times3)+(20+3\times4)+(20+3\times5)+(20+3\times6)\\ &= 29+32+35+38=134\end{aligned}$$

The second way in which the summation sign can be used requires a set of data where observations are specified in numerical order.

Example 3.45

Assume that a researcher finds a random group of 12 students and observes their weight and height as shown in Table 3.1. If we let H_i represent the height and W_i represent the weight of student i, then the total weight of the first six students can be specified as

$$\Sigma_{i=1}^{6} W_t$$

In this method, i identifies the number of the observation and so the value of i is *not* incorporated into the actual calculations.

Staying with the same example, the average weight of the first n students could be specified as

$$\frac{1}{n}\Sigma_{i=1}^{n} W_i$$

When no superscript or subscripts are shown with the Σ sign, it usually means that all possible values are summed. For example, a price index to measure inflation is constructed by working out how much a weighted average of prices of all goods in a basket rises over time.

Table 3.1 Data on students' weight and height

Student no.	***1***	***2***	***3***	***4***	***5***	***6***	***7***	***8***	***9***	***10***	***11***	***12***
Height (cm)	178	175	170	166	168	185	169	189	175	181	177	180
Weight (kg)	72	68	58	52	55	82	55	86	70	71	65	68

LASPEYRE PRICE INDEX

One method of measuring how much, on average, prices rise between year 0 and year 1 is the **Laspeyre price index**

$$\frac{\Sigma p_i^1 x_i^0}{\Sigma p_i^0 x_i^0}$$

where p_i^1 is the price of good i in year 1, p_i^0 is the price of good i in year 0 and x_i^0 is the percentage of consumer expenditure on good i in year 0. If all goods are in the index, then by definition $\Sigma x_i^0 = 100$. The Laspeyre price index therefore tells us how much expenditure on the same basket of goods will change from year 0 to year 1.

Example 3.46

Given the figures in Table 3.2 for prices and expenditure proportions, calculate the rate of inflation between year 0 and year 1 and compare the price rise of food with the weighted average price rise.

Solution

Note that in this example we are just assuming one price for each category of expenditure. In reality, of course, the prices of numerous individual goods are included in a price index. It must be stressed that these are *prices* and not measures of expenditure on these goods and services.

The weighted average price increase will be

$$\frac{\Sigma p_i^1 x_i^0}{\Sigma p_i^0 x_i^0} = \frac{1{,}944+1{,}666+1{,}012+910+2{,}160+781+1{,}176+1{,}500}{1{,}800+1{,}360+770+840+2{,}120+682+1{,}128+1{,}300}$$

$$= \frac{11{,}149}{10{,}000} = 1.115$$

Table 3.2 Expenditure shares and prices over two years

	Percentage of expenditure (x_i)	*Prices, year 0* (p_i^0)	*Prices, year I* (p_i^1)
Durable goods	9	200	216
Food	17	80	98
Alcohol and tobacco	11	70	92
Footwear and clothing	7	120	130
Energy	8	265	270
Other goods	11	62	71
Rent, rates, water	12	94	98
Other services	25	52	60
	100		

Note: All prices are in £.

This means that, on average, prices in year 1 are 111.5% of prices in year 0 so, subtracting 100%, this means that the inflation rate is 11.5%. The price of food went up from 80 to 98, i.e. by 22.5% $\left(\frac{98}{80} = 1.225 \text{ or } 122.5\%\right)$, which is greater than the inflation rate. Adjusted for inflation, the real price increase for food is thus,

$$100\left(\frac{1.225}{1.115} - 1\right) = 100(1.099 - 1) = 9.9\%$$

There are other methods that can be used to construct price indexes. For example, the **Paasche price index** calculates how much the expenditure proportions from year 1 would have cost in year 0 and then calculates the overall weighted price rise of this expenditure mix by year 1. The Paasche price index formula is thus,

$$\frac{\Sigma p_i^1 x_i^1}{\Sigma p_i^0 x_i^1}$$

Where p_i^1 is the price of good i in year 1, p_i^0 is the price of good i in year 0 and x_i^1 is the percentage of consumer expenditure on good i in year 1.

Example 3.47

Assume that for the five major categories of consumer expenditure the proportions of current (year 1) expenditure are:

Housing 36%, Food and drink 28%, Travel 8%, Energy 9%, Other 19%.

The following price changes occur from year 0 to year 1 for each of these categories:

Housing from 410 to 485; Food and drink from 184 to 191; Travel from 152 to 135; Energy from 70 to 93; Other from 210 to 218.

Calculate the rate of inflation from year 0 to year 1 using the Paasche price index, to 1 dp.

Solution

$$\frac{\Sigma p_i^1 x_i^1}{\Sigma p_i^0 x_i^1} = \frac{147.6 + 51.52 + 12.16 + 6.3 + 39.9}{174.6 + 53.48 + 10.8 + 8.37 + 41.42} = \frac{257.48}{288.67} = 1.121136$$

Therefore, the rate of inflation is 12.1%.

QUESTIONS 3.8

1. Refer to Table 3.1 and write an expression for the average height of the first n students observed and evaluate for $n = 6$.
2. Evaluate

$$\Sigma_{i=1}^{5}(4+i)$$

3. Noting that i is an exponent, evaluate

$$\Sigma_{i=2}^{5} 2^i$$

4. A firm sells 6,000 tonnes of its output in its first year of operation. Sales then decrease each year by 10% of the previous year's sales figure. Write an expression for the firm's total sales over n years and evaluate for $n = 3$.
5. Observations of a firm's sales revenue per month (in £'000) are as follows:

Month	1	2	3	4	5	6	7	8	9	10	11	12
Revenue	4.5	4.2	4.6	4.4	5.0	5.3	5.2	4.9	4.7	5.4	5.3	5.8

   (a) Write an expression for average monthly sales revenue for the first n months and evaluate for $n = 4$.
   (b) Write an expression for average monthly sales revenue over the preceding 3 months for any given month n, assuming that n is not less than 4. Evaluate for $n = 10$.
6. Assume that the expenditure and price data given in Example 3.46 all still hold except that the price of alcohol and tobacco rises to £108 in year 1. Work out the new inflation rate according to the Laspeyre price index method and find the new real price increase in the price of food.
7. If the proportions of main consumer expenditure are currently (in year 1): Housing 36%, Food and drink 28%, Travel 8%, Energy 9%, Other 19%, and the following price changes occur from year 0 to year 1 for each of these categories:

   Housing from 730 to 759; Food and drink from 565 to 570; Travel from 242 to 231; Energy from 110 to 138; Other from 285 to 297.

   Calculate the rate of inflation from year 0 to year 1 using the Paasche price index.

3.9 INEQUALITY SIGNS

As well as the equality sign (=), the following four inequality signs are used in algebra:

- $>$ which means 'is always greater than'
- $<$ which means 'is always less than'
- $\geq$ which means 'is greater than or equal to'
- $\leq$ which means 'is less than or equal to'

The last two are sometimes called 'weak inequality' signs.

Example 3.48

If we let the number of days in any given month be represented by N, then whatever month is chosen it must be true that

$$N > 27 \qquad N < 32$$
$$N \geq 28 \qquad N \leq 31$$

Special care must be taken when using inequality signs if unknown variables can take negative values. For example, the inequality $2x < 3x$ only holds if $x > 0$.

If x took a negative value, then the inequality would be reversed. For example, if $x = -5$, then $2x = -10$ and $3x = -15$ and so $2x > 3x$.

When considering inequality relationships, it can be useful to work in terms of the absolute value of a variable x. This is written $|x|$ and is defined as

$$|x| = x \text{ when } x \geq 0 \quad \text{and} \quad |x| = -x \text{ when } x < 0$$

i.e. the absolute value of a positive number is the number itself and the absolute value of a negative number is the same number but without the negative sign.

If an inequality is specified in terms of the absolute value of an unknown variable, then the inequality will not be reversed if the variable takes on a negative value. For example,

$$|2x| < |3x| \qquad \text{for all positive and negative non-zero values of } x.$$

It is possible to simplify an inequality relationship by performing the same arithmetical operation on both sides of the inequality sign. However, sometimes the rules for doing this differ from those that apply when manipulating both sides of an ordinary equality sign.

One can add any number to or subtract any number from both sides of an inequality sign.

Example 3.49

If $\quad x + 6 > y + 2$

then $\quad x + 4 > y$

One can multiply or divide both sides of an inequality sign by a positive number.

Example 3.50

If $\quad x < y \quad$ then $\quad 8x < 8y \quad$ (multiplying both sides by 8)

However, if one multiplies or divides by a negative number then the direction of the inequality sign is reversed.

Example 3.51

If $3x < 18y$ then $-x > -6y$ (dividing both sides by -3)

If both sides of an inequality sign are squared, the same inequality sign only holds if both sides are initially positive values. This is because a negative number squared becomes a positive number.

Example 3.52

If $x+3 < y$

Then $(x+3)^2 < y^2$ if $(x+3) \geq 0$ and $y > 0$

Example 3.53

$$-6 < -4$$

but $(-6)^2 > (-4)^2$

since $36 > 16$

If both sides of an inequality sign are positive and are raised to the same negative power, then the direction of the inequality will be reversed.

Example 3.54

If $x > y$

then $x^{-1} < y^{-1}$ if x and y are positive.

Example 3.55

Two leisure park owners A and B have the same weekly running costs of £8,000. The numbers of customers visiting the two parks are x and y, respectively. If $x > y$, what can be said about comparative average costs per customer?

Solution

Since $x > y$

then $x^{-1} < y^{-1}$

thus $\dfrac{£8,000}{x} < \dfrac{£8,000}{x}$

and so average cost for A < average cost for B

QUESTIONS 3.9

1. You are studying a subject which is assessed by coursework and examination with the total mark for the course being calculated on a 30:70 weighting between these two components. Assuming you score 60% in coursework, insert the appropriate inequality sign between your possible overall mark for the course and the following percentage figures.
   (a) 18% … overall mark
   (b) 16% … overall mark
   (c) 88% … overall mark
   (d) 90% … overall mark
2. If $x \geq 1$, insert the appropriate inequality sign between:
   (a) $(x + 2)^2$ and 3
   (b) $(x + 2)^2$ and 9
   (c) $(x + 2)^2$ and $3x$
   (d) $(x + 2)^2$ and $6x$
3. If Q_1 and Q_2 represent positive production levels of a good and the equality $Q_2 = Z^n Q_1$ always holds where $Z > 1$, what can be said about the relationship between Q_1 and Q_2 if
   (a) $n > 0$, (b) $n = 0$, (c) $n < 0$?
4. If a monopolist can operate price discrimination and charge separate prices P_1 and P_2 in two different markets, it can be proved that for profit maximization the monopolist should choose values for P_1 and P_2 that satisfy the equation

$$P_1\left(1 - \frac{1}{e_1}\right) = P_2\left(1 - \frac{1}{e_2}\right)$$

   where e_1 and e_2 are price elasticities of demand in the two markets. In which market should price be higher if market 1 has the more elastic demand, i.e. $|e_1| > |e_2|$?

4 Graphs and functions

Learning objectives

After completing this chapter students should be able to:

- Interpret the meaning of functions and inverse functions.
- Draw graphs that correspond to linear, non-linear and composite functions.
- Find the slopes of linear functions.
- Use the slope of a linear demand function to calculate point elasticity.
- Show what happens to budget constraints when parameters change.
- Interpret the meaning of functions with two independent variables.
- Deduce the degree of returns to scale from the parameters of a Cobb-Douglas production function.
- Construct a spreadsheet to plot the values of different functional formats.
- Sum marginal revenue and marginal cost functions to help find solutions to price discrimination and multiplant monopoly problems.

4.1 FUNCTIONS

Suppose that average weekly household expenditure on food (C), measured in £, depends on average net household weekly income (Y) according to the relationship

$$C = 42 + 0.2Y$$

For any given value of Y, one can evaluate what C will be. For example,

if $\quad Y = 500$
then expenditure on food is $\quad C = 42 + 0.2(500) = 42 + 100 = £142$

Whatever value Y takes, there will be one unique corresponding value of C. This is an example of a function.

DOI: 10.4324/9781003360827-4

> **FUNCTION**
>
> A function is an expression of a relationship between the values of two or more variables. For a function to be defined a **unique** value of one of the variables must be determined by the value of the other variable or variables.

If the precise mathematical form of the relationship is not known, then a function may be written in a **general form**. For example, a general form demand function is

$$Q_d = f(P)$$

This just tells us that quantity demanded of a good (Q_d) depends on its price (P). The 'f' is not an algebraic symbol in the usual sense and so f(P) means 'is a function of P' and not 'f multiplied by P'. In this case, P is what is known as the **independent** or **exogenous** variable because its value is given and is not dependent on the value of Q_d, i.e. it is exogenously determined. On the other hand Q_d is the **dependent variable** because its value depends on the value of P. It may also be referred to as an **endogenous** variable as it is determined within the function.

Functions may have more than one independent variable. For example, the general form production function

$$Q = f(K, L)$$

tells us that the quantity of output (Q) depends on the values of the two independent variables capital (K) and labour (L).

The **specific form of a function** tells us exactly how the value of the dependent variable is determined from the values of the independent variable or variables. A specific form for a demand function might be

$$Q_d = 120 - 2P$$

For any given value of P this specific function allows us to calculate the value of Q_d. For example,

$$\text{when } P = 10, \text{ then } Q_d = 120 - 2(10) = 120 - 20 = 100$$
$$\text{when } P = 45, \text{ then } Q_d = 120 - 2(45) = 120 - 90 = 30$$

In economic applications it may make sense to restrict the 'domain' or 'range' of a function, i.e. the range of possible values that variables can take. For example, variables that represent price or output may be restricted to positive values only. Strictly speaking, the **domain limits the values of the independent variables** and the **range governs the possible values of the dependent variable**.

For more complex functions with more than one independent variable it may be helpful to draw up a table to show how different values of the independent variables affect the value of the dependent variable. Assuming Q, K, and L only take positive values, some possible different values for the specific form production function $Q = 4K^{0.5}L^{0.5}$ are shown in Table 4.1.

Table 4.1 Different values of independent variables and corresponding values of the dependent variable

K	L	$K^{0.5}$	$L^{0.5}$	Q
1	1	1	1	4
4	1	2	1	8
9	25	3	5	60
7	11	2.64575	3.31662	35.0998

When defining the specific form of a function, it is important to make sure that only **one unique value of the dependent variable is determined** from each given value of the independent variable(s). Consider the equation

$$y = 80 + x^{0.5}$$

This does **not** define a function because any given value of x corresponds to two possible values for y. For example, if $x = 25$, then $25^{0.5} = 5$ or -5 and so $y = 75$ or 85. However, if we define

$$y = 80 + x^{0.5} \quad \text{for} \quad x^{0.5} \geq 0$$

then this does constitute a function.

When domains are not specified then one should assume a sensible range for functions representing economic variables. For example, it is usually assumed $K \geq 0$ and $L \geq 0$ in a production function, as in Table 4.1.

QUESTIONS 4.1

1. An economist researching the market for tea assumes that

$$Q = \mathrm{f}(P, Y, A, N, C)$$

where Q is the quantity of tea demanded, P is the price of tea, Y is average household income, A is advertising expenditure on tea, N is population and C is the price of coffee.

(a) What does $Q = \mathrm{f}(P, Y, A, N, C)$ mean in words?

(b) Identify the dependent and independent variables and then, using your knowledge of economics, deduce whether the coefficients of the different independent variables should be positive or negative.

2. If a firm faces the total cost function

$$\mathrm{TC} = 6 + x^2$$

where x is output, what is TC when x is (a) 14? (b) 1? (c) 0?
What restrictions on the domain of this function would it be reasonable to make?

3. A firm's total expenditure E on inputs is determined by the formula

$$E = P_K K + P_L L$$

where K is the amount of input K used, L is the amount of input L used, P_K is the price per unit of K and P_L is the price per unit of L. Is one unique value for E determined by any given set of values for K, L, P_K and P_L? Does this mean that any one particular value for E must always correspond to the same set of values for K, L, P_K and P_L?

4.2 INVERSE FUNCTIONS

An inverse function reverses the relationship in a function. If we confine the analysis to functions with only one independent variable, x, this means that if y is a function of x, i.e.

$$y = \mathrm{f}(x)$$

then in the inverse function x will be a function of y, i.e.

$$x = \mathrm{g}(y)$$

The letter g is used instead of f to show that we are talking about a different function. Another way of denoting the inverse of a function is by using the superscript −1, e.g. f^{-1}.

Example 4.1

If the original function is	y	=	$4 + 5x$
then	$y - 4$	=	$5x$
dividing both sides by 5 gives	$0.2y - 0.8$	=	x
and so the inverse function is	x	=	$0.2y - 0.8$

Not all functions have an inverse. The mathematical condition necessary for a function to have a corresponding inverse function is that the original function must be

monotonic. This means that, as the value of the independent variable x is increased, the value of the dependent variable y must either always increase or always decrease. It cannot first increase and then decrease, or vice versa. This will ensure that as well as there being one unique value of y for any given value of x, there will also be one unique value of x for any given value of y. This point will probably become clearer to you in the following sections on graphs of functions, but it can be illustrated here with a simple example.

Example 4.2

Consider the function $y = 9x - x^2$ restricted to the domain $0 \leq x \leq 9$.

Each value of x will determine a unique value of y. However, some values of y will correspond to two values of x, e.g.

$$\text{when } x = 3 \text{ then } y = 27 - 9 = 18$$
$$\text{when } x = 6 \text{ then } y = 54 - 36 = 18$$

This is because the function $y = 9x - x^2$ is not monotonic. This can be established by calculating y for a few selected values of x:

x	1	2	3	4	5	6	7
y	8	14	18	20	20	18	14

These values show that y first increases and then decreases as x gets larger, and so there is no inverse for this non-monotonic function.

Although mathematically it may be possible to derive an inverse function, it may not always make sense to derive the inverse of an economic function. For example, if we take the geometric function that the area A of a square is related to the length L of its sides by the function $A = L^2$, then we can also write the inverse function that relates the length of a square's side to its area: $L = A^{0.5}$ (assuming that L can only take non-negative values). Once one value is known, then the other is determined by it. However, suppose that someone investigating expenditure on holidays abroad (H) finds that the level of average annual household income (M) is the main influence and the relationship can be explained by the function

$$H = 0.01M + 100 \text{ (for } M \geq £15{,}000)$$

This mathematical equation could be rearranged to give

$$M = 100H - 10{,}000$$

but to say that H determines M does not necessarily make sense. The amount spent on holidays abroad is unlikely to determine the level of average household income in most cases.

A cause-and-effect relationship within economic models may not always run in one direction. Consider the relationship between price and quantity in a demand function. A monopoly may set a product's price and then see how much consumers are willing to buy, i.e. $Q = \mathrm{f}(P)$. On the other hand, in a competitive industry firms may first decide how much they are going to produce and then see what price they can get for this output, i.e. $P = \mathrm{f}(Q)$.

Example 4.3

Given the demand function $Q = 200 - 4P$, derive the inverse demand function.

Solution

$$\begin{aligned} Q &= 200-4P \\ 4P+Q &= 200 \\ 4P &= 200-Q \\ P &= 50-0.25Q \end{aligned}$$

QUESTIONS 4.2

1. If temperature in degrees Fahrenheit (°F) can be converted to degrees Celsius (°C) using function below, find the inverse of this function.

$$°\mathrm{C} = \frac{5}{9}(°\mathrm{F} - 32)$$

2. Derive the inverse of the demand function Q = 1,200 − 0.5P.
3. The total revenue (TR) that a monopoly receives from selling different levels of output (q) is given by the function $\mathrm{TR} = 60q - 4q^2$ for $0 \le q \le 15$. Explain why one cannot derive the inverse function $q = \mathrm{f}(\mathrm{TR})$.
4. An empirical study suggests that a brewery's weekly sales of beer are determined by the average air temperature given that the price of beer, income, adult population and most other variables are constant in the short run. This functional relationship is estimated as

$$X = 400 + 16T^{0.5} \text{ (for } T > 0)$$

where X is the number of barrels sold per week and T is the mean average air temperature, in °F. What is the mathematical inverse of this function? Does it make sense to specify such an inverse function in economics?

4.3 GRAPHS OF LINEAR FUNCTIONS

We are all familiar with graphs of the sort illustrated in Figure 4.1. This shows a firm's annual sales figures. To find what its sales were in 2025 you first find 2025 on the horizontal axis, move vertically up to the line marked 'sales' and read off the corresponding figure on the vertical axis, which in this case is £120,000. Graphs are often used as an alternative to tables of data as they make trends in the numbers easier to identify visually. However, in this case this is *not* a graph of a function. Sales are not determined by time.

Mathematical functions are mapped out on what is known as a set of Cartesian axes, as shown in Figure 4.2. Variable x is measured by equal increments on the horizontal axis and variable y by equal increments on the vertical axis. Both x and y can be measured in positive or negative directions. Although obviously only a limited range of values can be shown on the page of a book, the Cartesian axes theoretically range from $+\infty$ to $-\infty$ (i.e. from plus to minus infinity).

Any point on the graph will have two coordinates, i.e. corresponding values on the x and y axes. For example, to find the coordinates of point A one needs to draw a vertical line down to the x axis and read off the value of 20 and draw a horizontal line across to the y axis and read off the value 17. The coordinates (20, 17) determine point A.

As only two variables can be measured on the two axes in Figure 4.2, this means that only functions with one independent variable can be illustrated by a graph on a two-dimensional sheet of paper. One axis measures the dependent variable and the other measures the independent variable. (A method of illustrating functions with two independent variables is explained in Section 4.9.)

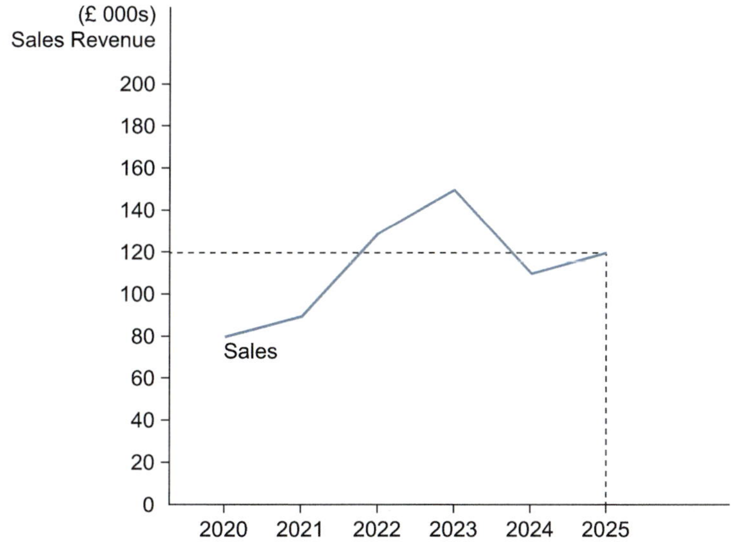

Figure 4.1 Annual sales revenue

Having set up the Cartesian axes in Figure 4.2, let us use it to determine the shape of the function

$$y = 5 + 0.6x$$

Calculating a few values of y for different values of x we get:

when $x = 0$	then	$y = 5 + 0.6(0) = 5$
when $x = 10$	then	$y = 5 + 0.6(10) = 5 + 6 = 11$
when $x = 20$	then	$y = 5 + 0.6(20) = 5 + 12 = 17$

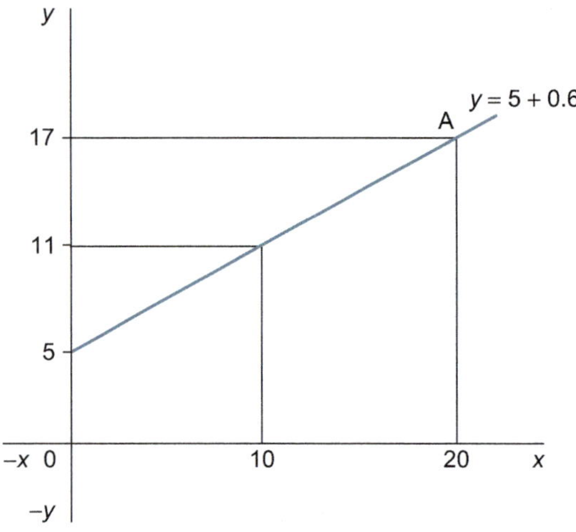

Figure 4.2 Cartesian axes and the function of two variables

These points are plotted in Figure 4.2, and it is obvious that they lie along a straight line. The rest of the function can be shown by drawing a straight line through the points that have been plotted.

Any function that takes the format $y = a + bx$ will correspond to a straight line when represented by a graph (where a and b can be any positive or negative numbers). This is because the value of y will change by the same amount, b, for every one unit increment in x. For example, in the function $y = 5 + 0.6x$ the value of y increases by 0.6 every time x increases by one unit.

Usually, the easiest way to **plot a linear function** is to find the points where it cuts the two axes and draw a straight line through them.

Example 4.4

Plot the graph of the function $y = 6 + 2x$.

Solution

The y axis is a vertical line through the point where x is zero.
When $x = 0$ then $y = 6$ and so this function must cut the y axis at $y = 6$.
The x axis is a horizontal line through the point where y is zero.

When $y = 0$ then

$$0 = 6 + 2x$$
$$-6 = 2x$$
$$-3 = x$$

and so this function must cut the x axis at $x = -3$.

The function $y = 6 + 2x$ is linear. Therefore, if we join up the points where it cuts the x and y axes by a straight line, we get the graph as shown in Figure 4.3.

If no restrictions are placed on the domain of the independent variable in a function, then the range of values of the dependent variable could possibly take any positive or negative value, depending on the nature of the function. However, in economics some variables may only take on positive values. A linear function that applies only to positive values of all the variables may sometimes only intercept with one axis. In such cases, all one has to do is simply plot another point and draw a line through the two points obtained.

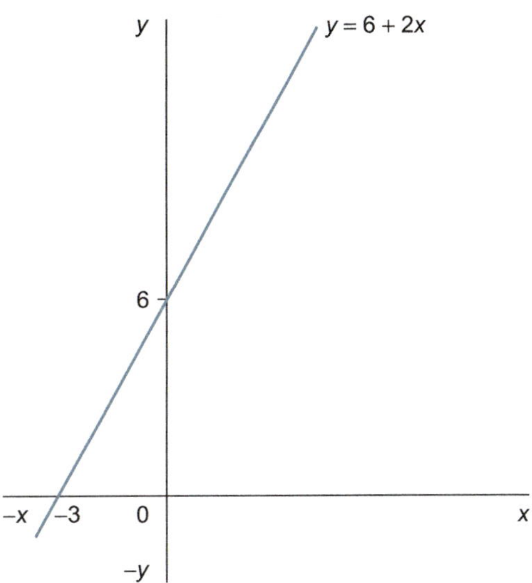

Figure 4.3 Graph of the function $y = 6 + 2x$

Example 4.5

Draw the graph of the function

$$C = 200 + 0.6Y$$

where C is consumer spending and Y is income, which cannot be negative ($Y \geq 0$).

Solution

Before plotting the shape of this function, you need to note that C is the dependent variable, measured on the vertical axis, and Y is the independent variable, measured on the horizontal axis.

When $Y = 0$, then $C = 200$, and so the line cuts the vertical axis at 200.

However, when $C = 0$, then

$$\begin{aligned} 0 &= 200 + 0.6Y \\ -0.6Y &= 200 \\ Y &= -\frac{200}{0.6} \end{aligned}$$

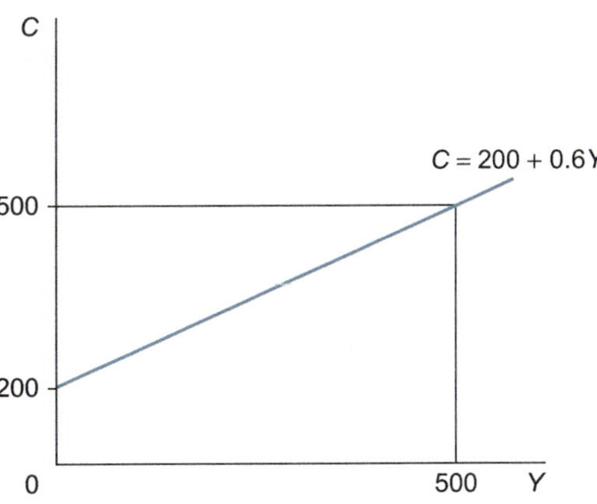

Figure 4.4 Graph of a consumption function

As negative values of income Y are unacceptable, find another point to help you plot the graph by choosing another pair of values.

For example, when $Y = 500$ then $C = 200 + 0.6(500) = 200 + 300 = 500$.

This graph is shown in Figure 4.4.

In mathematics the usual convention when drawing graphs is to measure the independent variable x along the horizontal axis and the dependent variable y along the vertical axis. However, in the economic analysis of supply and demand the usual convention is to measure price P on the vertical axis and quantity Q along the horizontal axis.

This may confuse some students when a function has Q as the dependent variable, such as the demand function

$$Q = 800 - 4P$$

but is then illustrated by a graph such as that in Figure 4.5. (Before you proceed, check that you understand why the intercepts on the two axes are as shown.)

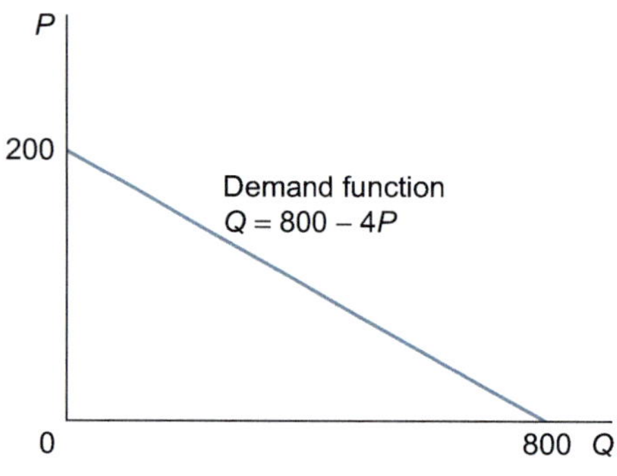

Figure 4.5 Linear demand function and its intercepts

Theoretically, it does not matter which axis is used to measure which variable. However, one of the main reasons for using graphs is to make analysis clearer to understand. Therefore, if one always has to keep checking which axis measures which variable this defeats the objective of the exercise. Thus, even though it may upset some mathematical purists, in this text we shall stick to the economists' convention of measuring quantity on the horizontal axis and price on the vertical axis, even if price is the independent variable in a function.

This means that care has to be taken when performing certain operations on functions. If it helps the analysis, one can transform monotonic functions to obtain the inverse function (as already explained). For example, for the demand function

$$Q = 800 - 4P$$

the inverse function is

$$P = \frac{800 - Q}{4} = 200 - 0.25Q$$

Check again in Figure 4.5 for the intercepts of the graph based on this inverse function.

In the analysis of demand and supply we often refer to functions with Q on the left-hand side of the equation sign as demand and supply functions, and when P is on the left-hand side then we talk about inverse demand and inverse supply functions. However, as graphs normally measure P on the vertical axis the term 'demand schedule' is sometimes used to refer to an inverse demand function.

QUESTIONS 4.3

Sketch the graphs of the linear functions 1 to 9 as follows, identifying the relevant intercepts on the axes. Assume that variables represented by letters that suggest they are economic variables (i.e. all variables except x and y) are restricted to non-negative values.

1. $y = 6 + 0.5x$
2. $y = 12x - 40$
3. $P = 60 - 0.2Q$
4. $Q = 750 - 5P$
5. $1200 = 50K + 30L$ (note that this equation represents a budget constraint for a firm and is an accounting identity rather than a function, although a given value of K will still determine a unique value of L, and vice versa.)
6. $\text{TR} = 8Q$
7. $\text{TC} = 200 + 5Q$
8. $\text{TFC} = 75$
9. Which of the following functions do you think realistically represents the supply schedule of a competitive industry? Why? (Assume $P \geq 0$ and $Q \geq 0$ in all cases.)
   (a) $P = 0.6Q + 2$
   (b) $P = 0.5Q - 10$
   (c) $P = 4Q$
   (d) $Q = -24 + 0.2P$

4.4 FITTING LINEAR FUNCTIONS

If you know the positions of two points on a straight line, then you can draw the rest of the line. You simply put your ruler on the page, join the two points and extend the line in either direction as far as you need to go. For example, suppose that a firm faces a linear demand schedule and that 400 units of output Q are sold when price is £40 and 500 units are sold when price is £20. Once these two price and quantity combinations have been marked as points A and B in Figure 4.6, then the rest of the demand schedule can be drawn in.

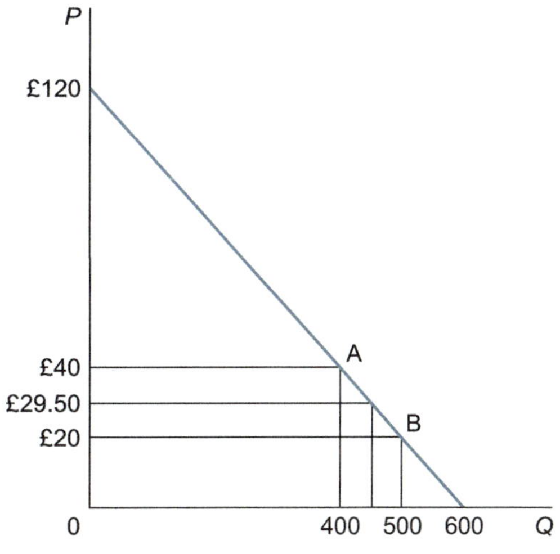

Figure 4.6 Fitting a linear demand function

This graph can then be used to predict the amounts sold at other prices. For example, when price is £29.50, the corresponding quantity can be read off as approximately 450. However, more accurate predictions of quantities demanded at different prices can be made if the information that is initially given is used to determine the algebraic format of the function.

As P is normally measured on the vertical axis it is clearer if we work with the inverse demand function which, if linear, is always in the format

$$P = a - bQ$$

where a and b are parameters that we wish to determine the value of.

From Figure 4.6 we can see that:

$$\text{when } P = 40 \text{ then } Q = 400 \text{ and so } 40 = a - 400b \qquad (1)$$

$$\text{when } P = 20 \text{ then } Q = 500 \text{ and so } 20 = a - 500b \qquad (2)$$

Equations (1) and (2) are what is known as simultaneous linear equations. Various methods of solving such sets of simultaneous equations (i.e. finding the values of a and b) are explained later in Chapter 5. Here we shall just use an intuitively obvious method of deducing the values of a and b from the graph in Figure 4.6.

Between points B and A, we can see that a £20 rise in price causes a 100 unit decrease in quantity demanded. As this is a linear function, we know that further price rises of £20 will also cause quantity demanded to fall by 100 units. At A, quantity is 400 units. Therefore, a rise in price of £80 is required to reduce quantity demanded from 400 to zero, i.e. a rise in price of 4 × £20 = £80 will reduce quantity demanded by 4 × 100 = 400 units. This means that the intercept of this function on the price axis is £80 plus £40 (the price at A), which is £120. This is the value of the parameter a.

To find the value of the parameter b, we need to ask 'what will be the fall in price necessary to cause quantity demanded to increase by one unit?' Given that a £20 price fall causes quantity to rise by 100 units, it must be the case that a price fall of £20/100 = £0.2 will cause quantity to rise by one unit. Conversely, a price rise of £0.2 will cause quantity demanded to fall by one unit. Therefore, $b = 0.2$. As we have already worked out that a is 120, our function can now be written as

$$P = 120 - 0.2Q$$

We can check that this is correct by substituting the original values of Q back into the function.

$$\text{If } Q = 400 \text{ then } P = 120 - 0.2(400) = 120 - 80 = 40$$
$$\text{If } Q = 500 \text{ then } P = 120 - 0.2(500) = 120 - 100 = 20$$

These are the values of P originally specified and so we are satisfied that the line that passes through points A and B in Figure 4.6 is the linear function $P = 120 - 0.2Q$. The inverse of this function will be the demand function $Q = 600 - 5P$.

Precise values of Q can now be derived for any given values of P. For example,

$$\text{when} \quad P = £29.50 \quad \text{then} \quad Q = 600 - 5(29.50) = 452.5$$

This is a more accurate value than the one read off the graph as approximately 450.

Having learned how to deduce the parameters of a linear downward-sloping demand function, let us now try to fit an upward-sloping linear function.

Example 4.6

It is assumed that consumption C depends on income Y and that this relationship takes the form of the linear function $C = a + bY$. When Y is £600, C is observed to be £660. When Y is £1,000, C is observed to be £900. What are the values of a and b in this function?

Solution

We expect b to be positive, i.e. consumption increases with income, and so our function will slope upwards, as shown in Figure 4.7. As this is a linear function, equal changes in Y will cause equal changes in C.

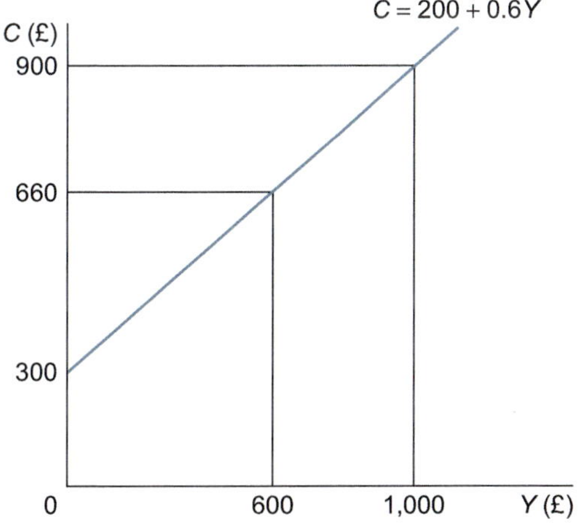

Figure 4.7 Linear consumption function

A decrease in Y of £400, from £1,000 to £600, causes C to fall by £240, from £900 to £660.

If Y is decreased by a further £600 (i.e. to zero) then the corresponding fall in C will be 1.5 times the fall caused by an income decrease of £400, since £600 = 1.5 × £400.

Therefore, the fall in C is 1.5 × £240 = £360.

This means that the value of C when Y is zero is £660 − £360 = £300.

Thus, $a = 300$.

A rise in Y of £400 causes C to rise by £240. Therefore, a rise in Y of £1 will cause C to rise by £240 / 400 = £0.6. Thus, $b = 0.6$.

The function can therefore be specified as

$$C = 300 + 0.6Y$$

We can now check this function gives the specified original values:

$$\text{when } Y = 600 \text{ then } C = 300 + 0.6(600) = 300 + 360 = 660.$$
$$\text{when } Y = 1{,}000 \text{ then } C = 300 + 0.6(1{,}000) = 300 + 600 = 900.$$

QUESTIONS 4.4

1. A monopoly sells 30 units of output when price is £12 and 40 units when price is £10. If its demand schedule is linear, what is the specific form of the actual demand function? Use this function to predict quantity demanded when price is £8. What domain restrictions would you put on this demand function?
2. Assume that consumption C depends on income Y according to the function $C = a + bY$, where a and b are parameters. If C is £60 when Y is £40 and C is £90 when Y is £80, what are the values of the parameters a and b?
3. On a linear demand function, quantity sold falls from 90 to 30 when price rises from £40 to £80. How much further will price have to rise for quantity sold to fall to zero?
4. A firm knows that its demand schedule takes the form $P = a - bQ$. If 200 units are sold when price is £9 and 400 units are sold when price is £6, what are the values of the parameters a and b?
5. A firm notices that its total production costs are £3,200 when output is 85 and £4,820 when output is 130. If total cost is assumed to be a linear function of output what expenditure on production will be necessary to manufacture 175 units?

4.5 SLOPE

British road signs used to give warning of steep hills by specifying their slope in a format such as '1 in 10', meaning that the road rose vertically by 1 foot for every 10 feet travelled in a horizontal direction. Now the European format is used and so instead of '1 in 10' a road sign will say 10%. In mathematics the same concept of slope is used but it is expressed as a decimal fraction rather than in percentage terms.

The graph in Figure 4.8 shows the function $y = 2 + 0.1x$. The slope, or gradient, is obviously the same along the whole length of this straight line and so it does not matter where the slope is measured. To measure the slope along the stretch AB, draw a horizontal line across from A and drop a vertical line down from B. These intersect at C, forming the triangle ABC with a right angle at C. The horizontal distance AC is 20 and the vertical distance BC is 2, and so if this was a cross-section of a hill you would clearly say that the slope is 2 in 20, which is the same as 1 in 10, or 10%.

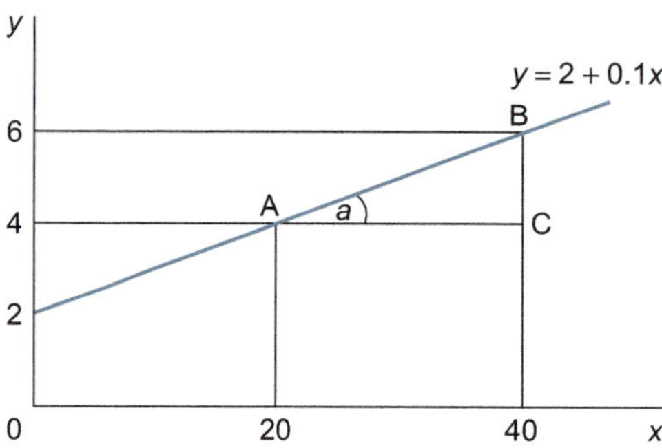

Figure 4.8 Finding a slope of a linear function

In mathematics the **slope of a line** that slopes upwards from left to right is defined as

$$\text{slope} = \frac{\text{height}}{\text{base}}$$

where height and base measure the sides of a right-angled triangle drawn in Figure 4.8. This is also known as the tangent of the angle *a*. Thus, in this example

$$\text{slope} = \frac{2}{20} = 0.1$$

The slope of this function (0.1) is the same as the coefficient of *x*. This is a general rule in linear functions. The slope is always represented by *b* in any linear function in the format $y = a + bx$.

Example 4.7

Find the slope of the function $y = -2 + 3x$.

Solution

The value of *y* increases by 3 for every 1 unit increase in *x* and so the slope of this linear function is 3.

When a line slopes downwards from left to right it has a negative slope. Thus, the slope coefficient *b* in the function $y = a + bx$ will take a negative value.

Consider the inverse demand function $P = 60 - 0.2Q$ where *P* is price and *Q* is quantity demanded, illustrated in Figure 4.9. As *P* and *Q* can be assumed not to take negative values, the whole function can be drawn by joining the intercepts on the two axes which are found as follows.

$$\begin{aligned}
&\text{When } Q = 0 \quad \text{then} \quad P = 60 \\
&\text{When } P = 0 \quad \text{then} \quad 0 = 60 - 0.2Q \\
&\qquad\qquad 0.2Q = 60 \\
&\qquad\qquad Q = \frac{60}{0.2} = 300
\end{aligned}$$

The slope of a function which slopes down from left to right is found by applying, to the relevant right-angled triangle, the formula

$$\text{slope} = (-1)\frac{\text{height}}{\text{base}}$$

Thus, using the triangle 0AB, the slope of the function in Figure 4.9 is

$$(-1)\frac{60}{300} = (-1)0.2 = -0.2$$

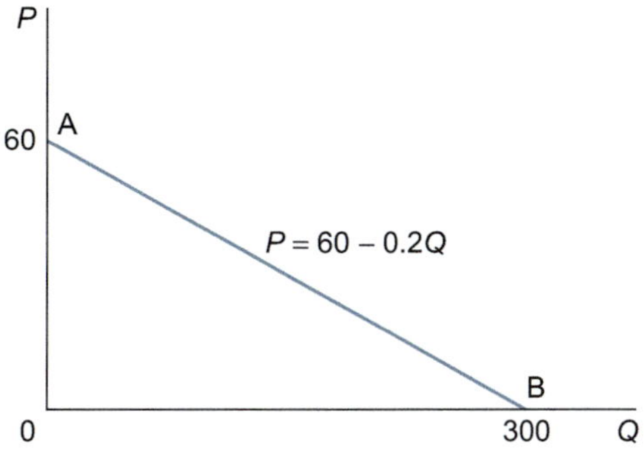

Figure 4.9 Linear demand function and its intercepts

This, of course, is the same as the slope coefficient of Q in the function $P = 60 - 0.2Q$.

Remember that in economics the usual convention is to measure P on the vertical axis of a graph. Therefore, if you are given a demand function in the format $Q = f(P)$ then first you would need to derive the inverse function $P = f(Q)$ to read off the slope.

Example 4.8

What is the slope of the demand function $Q = 830 - 2.5P$ when P is measured on the vertical axis of a graph?

Solution

$$\begin{aligned} \text{If} \quad & Q = 830 - 2.5P \\ \text{then} \quad & 2.5P = 830 - Q \\ & P = 332 - 0.4Q \end{aligned}$$

Therefore, the slope is the coefficient of Q, which is -0.4.

If the coefficient of x in a linear function is zero then the slope is also zero, i.e. the line is horizontal. For example, the function $y = 20$ means that y takes a value of 20 for every value of x.

Conversely, a vertical line will have an infinitely large slope. (Note, though, that a vertical line would not represent y as a function of x as no unique value of y is determined by a given value of x.)

POINT PRICE ELASTICITY OF DEMAND

In Chapter 2, the calculation of arc elasticity was explained. Because elasticity of demand can alter along the length of a demand schedule the arc elasticity measure is used as a sort of 'average'. However, now that you understand how the slope of a line is derived, we can examine how elasticity can be calculated at a specific point on a demand schedule. This is called the '**point price elasticity of demand**' and is defined as

$$e = \frac{P}{Q}\left(\frac{1}{\text{slope}}\right)$$

where P and Q are the price and quantity at the point in question. The slope refers to the slope of the demand schedule at this point although, of course, for a linear demand schedule the slope will be the same at all points. The derivation of this formula and its application to non-linear demand schedules is explained later in Chapter 9. Here, we shall just consider its application to linear demand schedules.

Example 4.9

Calculate the point elasticity of demand for the inverse demand function

$$P = 60 - 0.2Q$$

where price is (i) zero, (ii) £20, (iii) £40, (iv) £60.

Solution

This is the demand schedule referred to earlier and illustrated in Figure 4.9. Its slope is −0.2 at all points as it is a linear function and this is the coefficient of Q.

To find the values of Q corresponding to the given prices we need to derive the original underlying demand function, which is the inverse of this inverse demand function.

$$\begin{aligned} \text{Thus, given that} \quad P &= 60 - 0.2Q \\ \text{then} \quad 0.2Q &= 60 - P \\ Q &= 300 - 5P \end{aligned}$$

We can now use this result to find the values of Q for each of the prices in this question.

(i) When P is zero, at point B, then $Q = 300 - 5(0) = 300$.
The point elasticity of demand will therefore be

$$e = \frac{P}{Q}\left(\frac{1}{\text{slope}}\right) = \frac{0}{300}\left(\frac{1}{-0.2}\right) = 0$$

(ii) When $P = 20$ then $Q = 300 - 5(20) = 200$.

$$e = \frac{20}{200}\left(\frac{1}{-0.2}\right) = \frac{1}{10} \times \frac{1}{-0.2} = \frac{1}{-2} = -0.5$$

(iii) When $P = 40$ then $Q = 300 - 5(40) = 100$.

$$e = \frac{40}{100}\left(\frac{1}{-0.2}\right) = \frac{2}{5} \times \frac{1}{-0.2} = \frac{2}{-1} = -2$$

(iv) When $P = 60$ then $Q = 300 - 5(60) = 0$.
If $Q = 0$, then $P/Q \to \infty$. (i.e. it becomes infinitely large)

$$e = \frac{P}{Q}\left(\frac{1}{\text{slope}}\right) = \frac{60}{0}\left(\frac{1}{-0.2}\right) \to -\infty$$

QUESTIONS 4.5

1. In Figure 4.10, what are the slopes of the lines 0A, 0B, 0C and EF?

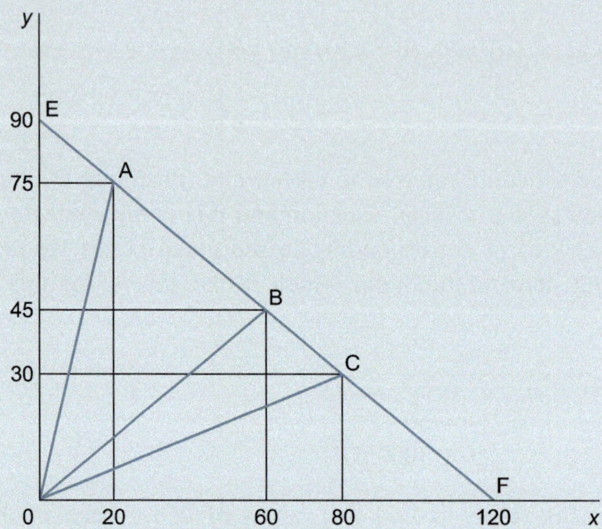

Figure 4.10 Finding slopes of straight lines

2. A market has a linear demand schedule with a slope of −0.3. When the price is £3, the quantity sold is 30 units. Where does this demand schedule hit the price and quantity axes? What is price if the quantity sold is 25 units? How much would be sold at the price of £9?
3. For the demand schedule $P = 60 - 0.2Q$ illustrated earlier in Figure 4.9, calculate point elasticity of demand when the price is (a) £24 and (b) £45.
4. Consider the three functions

   (a) $P = 8 - 0.75Q$ (b) $P = 8 - 1.25Q$ (c) $Q = 12 - 2P$

   Which has the flattest demand schedule, assuming that P is measured on the vertical axis? In which case is quantity sold the greatest when the price is

   (i) £1 and (ii) £5?
5. For positive values of x which, if any, of the following functions will intersect with the function $y = 1 + 0.5x$?

   (a) $y = 2 + 0.4x$ (c) $y = 2 + 1.5x$

   (b) $y = 4 + 0.5x$ (d) $y = 4$

6. In macroeconomics, the average propensity to consume (APC) and the marginal propensity to consume (MPC) can be defined as follows:

   ACP = C / Y where C = consumption, Y = income

   MPC = increase in C from a 1 unit increase in Y

   Explain why APC will always be greater than MPC if $C = 400 + 0.5\ Y$.
7. For the demand function $Q = 192 - 8P$, calculate point elasticity of demand when price is

   (a) £5 (b) £10 (c) £15

4.6 BUDGET CONSTRAINTS

A frequently used application of the concept of slope in economics is the relationship between prices and the slope of a budget constraint. A budget constraint shows the combinations of two goods (or inputs) that it is possible to buy with a given amount of money and a given set of prices.

Assume that a firm has a budget of £3,000 to spend on the two inputs capital, K, and labour, L, and that input K costs £50 and input L costs £30 a unit. If the firm spends the whole £3,000 on K, then it can buy

$$\frac{3{,}000}{50} = 60 \text{ units of K,}$$

and if it spends all its budget on L, then it can buy

$$\frac{3{,}000}{30} = 100 \text{ units of L}$$

These two quantities are marked on the axes of the graph in Figure 4.11. The firm could also split the budget between K and L. Many other combinations are possible, e.g.

30 of K and 50 of L

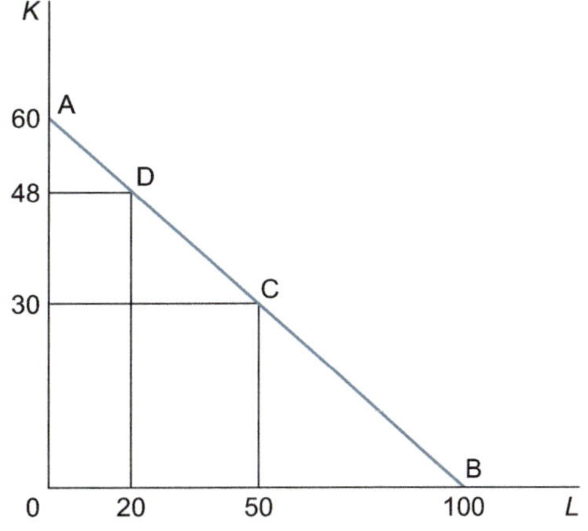

Figure 4.11 Budget constraint facing a firm

or

48 of K and 20 of L

If K and L are divisible into fractions of a unit, then all the combinations of K and L that can be bought with the given budget of £3,000 can be shown by the line AB, which is known as the **budget constraint** or **budget line**. The firm could, in fact, also purchase any of the combinations of K and L within the triangle 0AB but only combinations along the budget constraint AB would make it spend its entire budget.

Along the budget constraint any pairs of values of K and L must satisfy the equation

$$50K + 30L = 3,000$$

where K is the number of units of K bought and L is the number of units of L bought. This states that the total expenditure on K (price of K × amount bought) plus the total expenditure on L (price of L × amount bought) must sum to the total budget available. We can check that this holds for the combinations of K and L shown in Figure 4.11.

At A: $£50 \times 60 + £30 \times 0 = 3,000 + 0 = £3,000$
At B: $£50 \times 0 + £30 \times 100 = 0 + 3,000 = £3,000$
At C: $£50 \times 30 + £30 \times 50 = 1,500 + 1,500 = £3,000$
At D: $£50 \times 48 + £30 \times 20 = 2,400 + 600 = £3,000$

As budget lines usually slope down from left to right, they have a negative slope. From the graph it can be seen that this budget constraint has a slope of

$$\frac{-60}{100} = -0.6$$

The slope of a budget constraint can be deduced from the values of the prices of the two goods or inputs concerned. Consider the general case where the budget is M and the prices of the two goods X and Y are P_X and P_Y, respectively. The maximum amount of X that can be bought will be $M \div P_X$. This will be the intercept on the horizontal axis. Similarly, the maximum amount of Y that can be purchased will be $M \div P_Y$, which will be the intercept on the vertical axis. Therefore,

$$\text{slope of budget constraint} = (-)\frac{\left(\frac{M}{P_Y}\right)}{\left(\frac{M}{P_X}\right)} = (-)\frac{M}{P_Y}\frac{P_X}{M} = (-)\frac{P_X}{P_Y}$$

Thus, the budget constraint's slope will be the negative of the price ratio. However, you should note that it is the price of the good measured on the *horizontal* axis that is at the top in this formula.

From this result we can also see that

- if the price ratio changes the slope of the budget line changes
- if the budget alters but the prices remain unchanged, the slope of the budget line does not alter.

Example 4.10

A consumer has an income of £160 to spend on the two goods X and Y whose prices are £20 and £5 each, respectively.

(i) What is the slope of this consumer's budget constraint?

(ii) What happens to this slope if P_Y rises to £10?

(iii) What happens if income then falls to £100?

Y, 32 A, 16 C, 10 E, F, B, 0, 5, 8, X

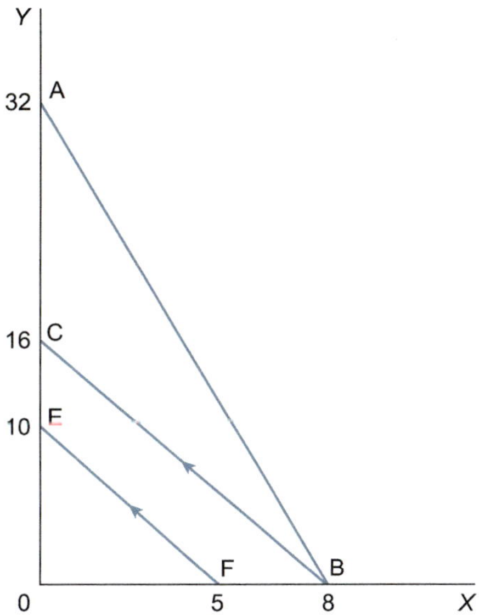

Figure 4.12 Effects of changes to prices and income on the budget constraint

Solution

(i) $\text{slope} = -\frac{P_X}{P_Y} = -\frac{20}{5} = -4$

This can be checked by considering the intercepts on the X and Y axes shown in Figure 4.12 by points B and A.

If the total budget of £160 is spent on X, then 160 ÷ 20 = 8 units are bought.

If the total budget is spent on Y, then 160 ÷ 5 = 32 units are bought.

Therefore,

$$\text{slope} = (-)\frac{\text{intercept on } Y \text{ axis}}{\text{intercept on } X \text{ axis}} = (-)\frac{32}{8} = -4$$

(ii) When P_Y rises to £10, the new slope of the budget constraint (shown by BC in Figure 4.12) becomes

$$-\frac{P_X}{P_Y} = -\frac{20}{10} = -2$$

(iii) The price ratio remains unchanged if income falls to £100. There is a parallel shift inwards of the budget constraint to EF. The new intercepts are

$$\frac{M}{P_Y} = \frac{100}{10} = 10 \text{ on the } Y \text{ axis} \quad \text{and} \quad \frac{M}{P_X} = \frac{100}{10} = 5 \text{ on the } X \text{ axis}$$

The slope is thus −10 ÷ 5 = −2, as before.

Example 4.11

A consumer can buy the two goods A and B at prices per unit of £6 and £4, respectively, and initially has a budget of £120.

(i) Show that a 25% rise in all prices will have a lesser effect on the consumer's purchasing possibilities than would a 25% reduction in budget with prices unchanged.
(ii) What is the opportunity cost of buying an extra unit of A? (Assume units of A and B are divisible.)

Solution

(i) The original intercept on the A is $\frac{120}{6} = 20$.

The original intercept on the B is $\frac{120}{4} = 30$.

If price of A rises by 25% to £7.50, the new intercept on the A axis becomes $\frac{120}{7.50} = 16$.

If price of B rises by 25% to £5, the new intercept on the B axis becomes $\frac{120}{5} = 24$.

Reducing income by 25% to £90 changes intercept on the A axis to $\frac{90}{6} = 15$ and changes the intercept on the B axis to $\frac{90}{4} = 22.5$.

Thus, the 25% fall in income shifts the budget constraint towards the origin slightly more than the 25% rise in prices, i.e. it reduces the consumer's purchasing possibilities by a greater amount.

(Note that the slope of the budget constraint always remains the same at $-6 \div 4 = -1.5$.)

(ii) The opportunity cost of something is the next best alternative that one has to forgo in order to obtain it. In this context, the opportunity cost of an extra unit of A will be the amount of B the consumer has to forgo.

One unit of A costs £6 and one unit of B costs £4. Therefore, the opportunity cost of A in terms of B is 1.5, which is the negative of the slope of the budget line.

QUESTIONS 4.6

1. A consumer can buy good A at £3 a unit and good B at £2 a unit and has a budget of £60. What is the slope of the budget constraint if the quantity of A is measured on the horizontal axis? What happens to this slope if:
(a) the price of A falls to £2?
(b) with A at its original price, the price of B rises to £3?
(c) both prices double?
(d) the budget is cut by 25%?

2. A firm has a budget of £800 per week to spend on the two inputs K and L. One week it is observed to buy 120 units of L and 25 of K. Another week it is observed to buy 80 units of L and 50 of K. Find out what the intercepts of its budget line on the *K* and *L* axes are and use this information to deduce the prices of K and L, which are assumed to be unchanged from one week to the next.
3. A firm can buy the two inputs K and L at £60 and £40 per unit, respectively, and has a budget of £480. Explain why it would not be able to purchase 6 units of K plus 4 units of L and then calculate what price reduction in L would make this input combination a feasible purchase.
4. If a firm buys the two inputs X and Y, what would the slope of its budget constraint be if the price of Y was £10 and
   (a) the price of X was £100?
   (b) the price of X was £10?
   (c) the price of X was £1?
   (d) the price of X was 25p?
   (e) X was free?
5. If a consumer's income doubles and the prices of the two goods that she spends her entire income on also double, what happens to her budget constraint?
6. An hourly paid worker can choose the number of hours per day worked, up to a maximum of 12, and gets paid £10 an hour. Leisure hours are assumed to be any hours not worked out of this 12. On a graph with leisure hours on the horizontal axis and total pay on the vertical axis, draw in the budget constraint showing the feasible combinations of leisure and pay that this worker might choose from. Show that the slope of this budget constraint equals −1 multiplied by the hourly wage rate.

4.7 NON-LINEAR FUNCTIONS

If the function $y = f(x)$ has a term with x to the power of anything other than 1, then it will be non-linear. For example,

$$y = x^2$$
$$y = 6 + x^{0.5}$$
$$y = 2 + x^{-1}$$

are all non-linear functions. However,

$$y = 5 + 0.2x$$

is a linear function because it contains x to the power of 1.

Non-linear functions can take a variety of shapes. We shall only consider a few possibilities that will be useful at a later stage when looking at functions of economic variables.

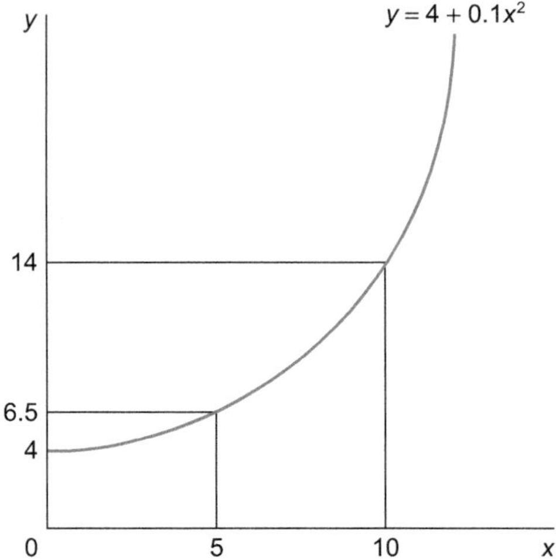

Figure 4.13 Graph of a quadratic function

Table 4.2 Values of x and quadratic and cubic functions of x

x	0	1	2	3	4	5	6
$y = x^2$	0	1	4	9	16	25	36
$y = x^3$	0	1	8	27	64	125	216

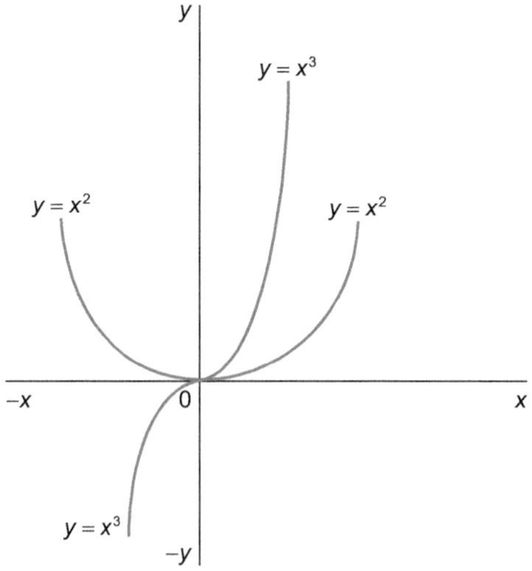

Figure 4.14 Comparing graphs of quadratic and cubic functions

If the function $y = f(x)$ has one term in x with x to a power greater than 1, then as long as x takes positive values, the function will rise at an increasing rate as x gets larger. This is obvious from Table 4.2. The graphs of the functions $y = x^2$ and $y = x^3$ will curve upwards since y increases at a faster rate than x. These functions all go through the origin, as y is zero when x is zero. The table shows that the greater the power of x then the more quickly y rises. Although the intercept may vary if there is a constant term in a function, and the rate of change of y may be modified if the term in x^2 (or x to any other power greater than 1) has a coefficient other than 1, the general shape of an upward-sloping curve will still be retained. For example, Figure 4.13 illustrates the function

$$y = 4 + 0.1x^2$$

In economics, the quantities one is working with are frequently constrained to positive values, e.g. price and quantity. However, if variables are allowed to take negative values, then the functions $y = x^2$ and $y = x^3$ will take the shapes shown in Figure 4.14. Note that when $x < 0$, $x^2 > 0$ but $x^3 < 0$.

If the power of x in a function lies between 0 and 1 then, as long as x is positive, the value of the function increases as x gets larger, but its rate of increase gets smaller and smaller. The values in Table 4.3 and Figure 4.15 illustrate this for the function $y = x^{0.5}$ (where only the positive square root is considered).

Table 4.3 Values of x and square-root function of x

x	0	1	2	3	4	5	6	7	8	9
$y = x^{0.5}$	0	1	1.414	1.732	2	2.236	2.449	2.646	2.828	3

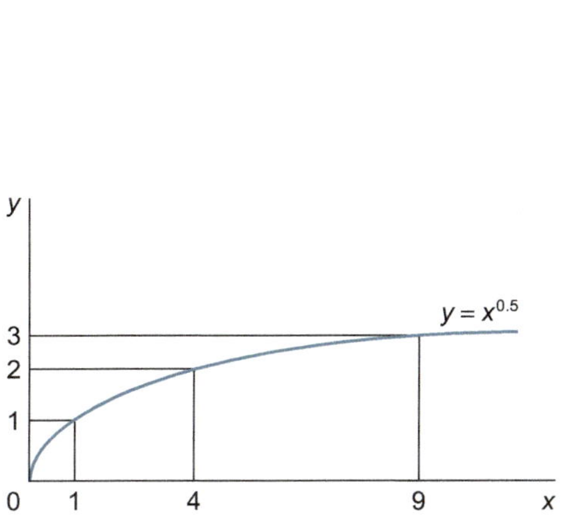

Figure 4.15 Function of x with a fractional exponent

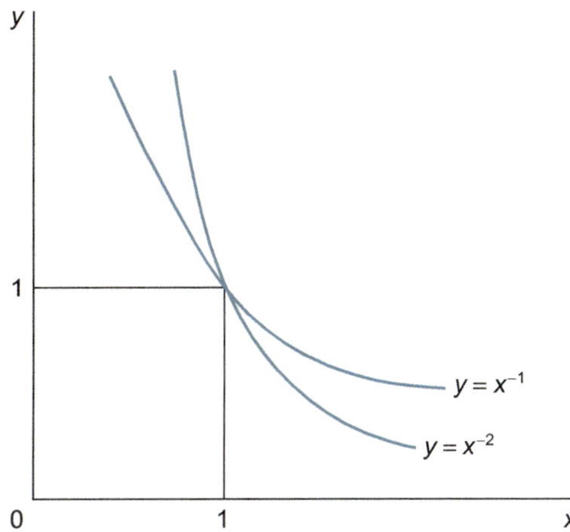

Figure 4.16 Function of x with a negative exponent

Table 4.4 Values of x and functions of x with negative exponents

x	0	0.1	1	2	3	4	5
$y = x^{-1}$	∞	10	1	0.5	0.33	0.25	0.2
$y = x^{-2}$	∞	100	1	0.25	0.11	0.0625	0.04

If the power of x in a function is negative then, as long as x is positive, the graph of the function will slope downwards and take the shape of a curve convex to the origin. The examples in Table 4.4 are illustrated in Figure 4.16 for positive values of x. Note that the value of y in these functions gets larger as x approaches zero.

A firm's average fixed cost (AFC) schedule typically takes a shape similar to the functions illustrated in Figure 4.16.

Example 4.12

A firm has to pay a fixed annual cost of £90,000 for leasing its premises. Derive its average fixed cost function (AFC).

Solution

$$\text{AFC} = \frac{\text{total fixed cost}}{Q} = \frac{90{,}000}{Q} = 90{,}000Q^{-1}$$

Although all values of Q^{-1} will be multiplied by 90,000, this will not alter the general shape of the function, which will be similar to the graph of $y = x^{-1}$ illustrated in Figure 4.16.

QUESTIONS 4.7

1. Sketch the approximate shapes of the following functions for positive values of x and y.
   (a) $y = -8 + 0.2x^3$
   (b) $y = 250 - 0.01x^2$
   (c) $y = x^{-1.5}$
   (d) $y = x^{-0.5}$
   (e) $y = 20 - 0.2x^{-1}$
2. Sketch the approximate shapes of the following functions when x and y are allowed to take both positive and negative values.
   (a) $y = 4 + 0.1x^2$ (b) $y = 0.01x^3$ (c) $y = 10 - x^{-1}$
3. Will the non-linear demand schedule $p = 570 - 0.4q^2$ get flatter or steeper as q rises?
4. A firm has to pay fixed costs of £65,000 before it starts production. What will its average fixed cost function look like? What will AFC be when output is 250?

4.8 COMPOSITE FUNCTIONS

When a function has more than one term then one can build up the shape of the overall function from its different components. We have already done this when showing how a constant term determines the starting point of a function on the vertical axis of a graph. Now some more complex functions are considered.

Example 4.13

A firm faces the average fixed cost function

$$AFC = 200x^{-1}$$

where x is output, and the average variable cost (AVC) function

$$AVC = 0.2x^2$$

What shape will its average total cost function (AC) take?

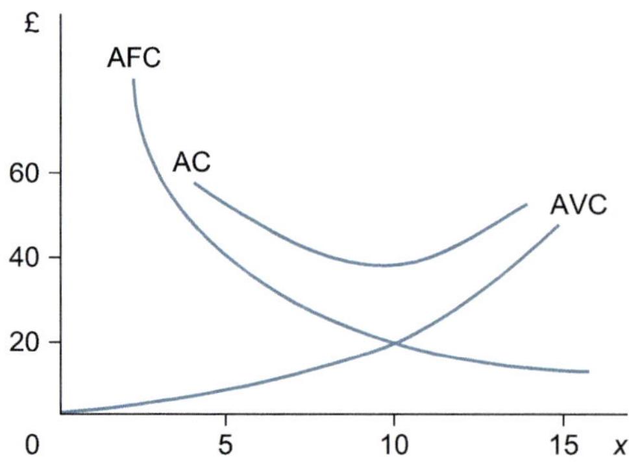

Figure 4.17 Average fixed cost, average variable cost and average cost

Solution

The graphs of AFC and AVC are illustrated in Figure 4.17. By definition,

$$AC = AFC + AVC$$

Therefore, substituting the given AFC and AVC functions, average total cost is

$$AC = 200x^{-1} + 0.2x^2$$

For any given value of x, this means that the position of the AC function can be found by vertically summing the corresponding values on the AFC and AVC schedules.

As x gets larger, the value of AFC gets closer and closer to zero and so the value of AC gets closer and closer to AVC. Therefore, the AC function will take the U-shape shown.

A composite function that takes the form

$$y = a_0 + a_1x + a_2x^2 + \dots + a_nx^n$$

where a_0, a_1, …, a_n are constants and n is a non-negative integer is what is called a 'polynomial'. The 'degree' of the polynomial is the highest power value of x. For example, the total cost (TC) function

$$TC = 4 + 6x - 0.2x^2 + 0.1x^3$$

is a polynomial of the third degree. (See if you can sketch the shape of this function. Don't worry if you can't. In Section 6.6 we will return to polynomials and Section 4.9 explains how computer spreadsheets can help to plot functions.)

To see how the graph of a composite function is constructed when one term has a negative value, an example of a total revenue function is worked through next.

Example 4.14

For the inverse demand function $P = 80 - 0.2Q$, what shape will the corresponding total revenue function take when TR is a function of Q?

Solution

Total revenue (TR) is simply the total amount of money raised by selling a good and so

$$TR = PQ$$

If we substitute the demand function $P = 80 - 0.2Q$ for P in this TR function, then

$$TR = (80 - 0.2Q)Q = 80Q - 0.2Q^2$$

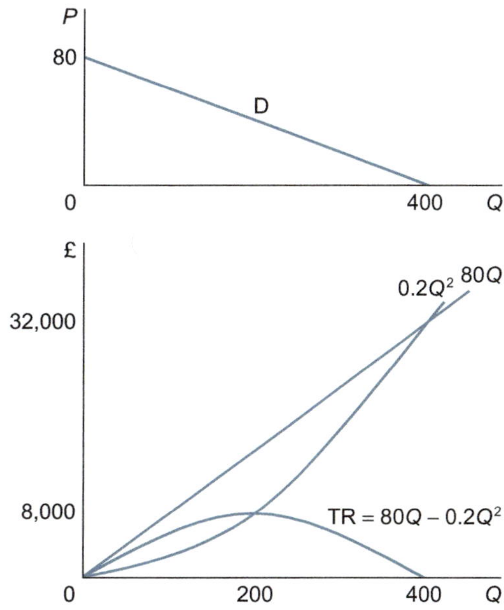

Figure 4.18 Composite functions: constructing a graph of a total revenue function

Now that we have derived the function for TR in terms of the single variable Q, its shape can be built up as shown in Figure 4.18. The component $80Q$ is clearly a straight line from the origin. The component $0.2Q^2$ is a curve rising at an increasing rate. One can easily see that, for low values of Q, $80Q > 0.2Q^2$. However, as Q becomes larger, the value of Q^2, and hence $0.2Q^2$, rapidly increases and eventually exceeds $80Q$.

Given that TR is the difference between $80Q$ and $0.2Q^2$, its value is the vertical distance between these two functions. This gets larger as Q increases to 200 and then decreases. It is zero when Q is 400 (when $80Q = 0.2Q^2$) and then becomes negative. Thus, we get the inverted U-shape shown.

We can check that this shape makes sense by referring to the demand schedule illustrated at the top of Figure 4.18 for $P = 80 - 0.2Q$. When Q is zero, nothing is sold and so TR must be zero. To sell 400, price must fall to zero and so again TR will be zero. Between these two output levels, TR will rise and then fall.

Slope of non-linear functions

We have seen how the slopes of non-linear composite functions can change along their length, but how can the slope of non-linear functions be measured from a graph? A mathematical method for finding the precise value of the slope of a function at any point using calculus is explained in Chapter 9. For now, we shall just consider an approximate geometrical method assuming that the graph of the function has already been drawn.

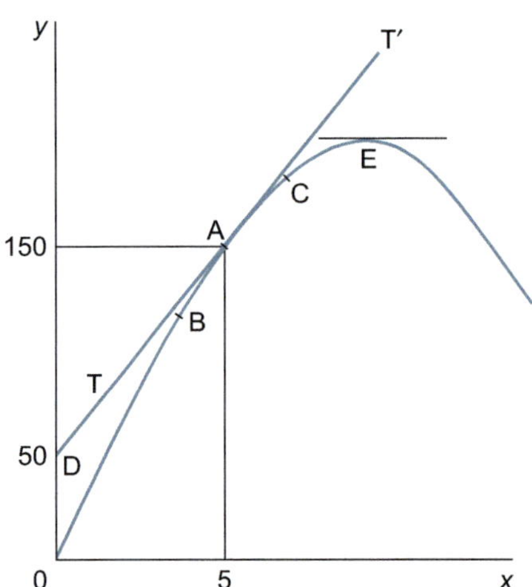

Figure 4.19 Quadratic function and tangent at point A

Example 4.15

The graph in Figure 4.19 illustrates the composite function $y = 40x - 2x^2$.

Find its slope at point A where $x = 5$.

Solution

First find the y coordinate of point A, where x is 5, which will be

$$y = 40x - 2x^2 = 40(5) - 2(5)^2 = 200 - 50 = 150$$

Draw a straight line that just touches the curve at point A. This line is known as a 'tangent' and is shown by TT′ in Figure 4.19. The slope of the line is the same as the slope of the function at A.

To understand why this is so, first consider point B which is slightly to the left of A. The function is steeper at B than at A and also has a greater slope than the tangent at

A. On the other hand, at point C (slightly to the right of A) the function is flatter than at A and has a slope less than that of the tangent TT′. If the slope of the tangent TT′ is less than the slope of the function for points slightly to the left of A and greater than the slope of the function for points slightly to the right of A, then it will be equal to the slope of the function at point A itself.

To determine the actual value of the slope of the tangent TT′ and hence the value of the slope of the function at A, it is necessary to find two sets of coordinates for the line TT′. We already know that it goes through A where $x = 5$ and $y = 150$. If TT′ is extended leftwards, it cuts the y axis at D where y is 50.

Using the method explained in Section 4.5, we can now fit an equation to this straight line in the format $y = a + bx$.

We have already found that the y intercept a is 50 when $x = 0$.

Between points D and A, the value of x increases from 0 to 5 and the value of y increases by 100 from 50 to 150. Thus, for every one unit increment in x the increase in y must be

$$\frac{100}{5} = 20 = b = \text{the slope of the tangent.}$$

Therefore, the tangent TT′ graphically represents the equation $y = 50 + 20x$.

The slope of this tangent is 20 and so at point A the slope of the function $y = 40 - 2x^2$ is also 20.

The slope of the function at other points can be determined in the same way by drawing other tangents. For example, the slope at the highest point of the function E will be the same as the slope of the tangent at E, which is a horizontal line. A horizontal line always has a slope of zero and so at its maximum point the slope of this function will also be zero.

Sometimes you may encounter composite functions with similar terms. The summed function may then be simplified so that it does not remain a composite function.

Example 4.16

A firm's manufacturing system requires two processes for each unit produced. Process A involves a fixed cost of £650 plus £15 for each unit produced, and process B involves a fixed cost of £220 plus £45 for each unit. What is the composite total cost function?

Solution

For process A total cost is $\quad TC_A = 650 + 15Q$

For process B total cost is $\quad TC_B = 220 + 45Q$

The overall summed total cost function is $\quad TC = TC_A + TC_B$

Therefore, $\quad TC = (650 + 15Q) + (220 + 45Q) = 870 + 60Q$

QUESTIONS 4.8

1. Sketch the approximate shape of the following composite functions for positive values of all independent variables.
   (a) $TR = 40q - 4q^2$
   (b) $TC = 12 + 4q + 0.2q^2$
   (c) $\pi = -12 + 36q - 3.8q^2$
   (d) $y = 15 - 2x - 1$
   (e) $AVC = 8 - 3q + 0.5q^2$
2. A firm is able to sell all its output at a fixed price of £50 per unit. Derive a function for profit (π) in terms of output q if its average cost of production is given by the function

$$AC = 100q^{-1} + 0.4q^2$$

   What approximate shape will this profit function take?
3. A small group of companies operate in an industry where all firms face the average cost function $AC = 40 + 1{,}250q^{-1}$, where q is output per week. This function refers only to production costs. They then decide to launch an advertising campaign, not just to try to increase sales but also to try to raise the total average cost of low output levels in order to deter potential smaller scale rival firms from competing in the same market. The cost of the advertising campaign is £2,000 per week per firm and any competitor would have to spend the same sum on advertising if it wished to compete in this market.
   (a) Derive a function for the new total average cost function including advertising and sketch its approximate shape.
   (b) Explain why this advertising campaign will deter competition if the original companies each sell 100 units a week at a price of £100 each and new competitors cannot produce more than 25 units a week.

4.9 USING A SPREADSHEET TO PLOT FUNCTIONS

It may not be immediately obvious what shape some composite functions take. If this is the case, then it may help to set up the function as a formula on a spreadsheet and then see how the value of the function changes over a range of values for the independent variable. Learning how to set up your own formulae on a spreadsheet can help you in a number of ways. In particular, spreadsheets can be very useful and save you a lot of time and effort when tackling problems that entail very complex and time-consuming numerical calculations. They can also be used to plot graphs to get a picture of how functions behave and to check that answers to mathematical problems derived from manual calculations are correct.

This book will not teach you how to use Excel, or any other spreadsheet package, from scratch. It is assumed that most students will already know the basics of entering data to spreadsheets. What we will do here is run through some methods of using spreadsheets to help solve, or illustrate and make clearer, aspects of some mathematical techniques that are developed. Spreadsheet applications will be explained when manual calculation would be very time consuming. The detailed instructions for constructing spreadsheets are given in Excel format, as this is probably the most commonly used spreadsheet package, but the basic principles for constructing the formulae can also be applied to other spreadsheet programs.

Although Excel offers a range of in-built formulae for commonly used functions, such as square root, you will still need to create your own formulae for many of the functions you will use.

HOW TO ENTER A FORMULA IN AN EXCEL SPREADSHEET

- Start with the sign =
- Use the usual arithmetic + and − signs on your keyboard, with * for multiplication and / for division.
- Do not leave any spaces between characters and make sure you use brackets properly.
- For powers and roots use the sign ^. Roots must be specified as fractional powers, e.g. use ^0.5 to denote square root.
- Arithmetic operations can be performed on numbers typed into a formula or on cell references that contain a number.
- When you copy a formula to another cell all the references to other cells change unless you anchor their row or column by typing the $ sign in front of it in the formula.
- The quickest way to copy cell contents in Excel is to
  (a) highlight the cells to be copied
  (b) hold the cursor over the bottom right corner of the cell (or block of cells) to be copied until the + sign appears
  (c) drag the highlighted cell or block over the cells where the copy is to go.

Example 4.17

Use an Excel spreadsheet to calculate values for TR for the function $\text{TR} = 80Q - 0.2Q^2$ from Example 4.14 for the range where both Q and TR take positive values and then plot these values on a graph.

Solution

To answer this question, the essential features of the required spreadsheet are:

- A column of values for Q.
- Another column that calculates the value of TR corresponding to the value in the Q column.

Table 4.5 Setting up a spreadsheet for Example 4.17

CELL	Enter	Explanation
A1	Ex. 4.17	Label to remind you what example this is
B1	TR = 80Q – 0.2Q^2	Label to remind you what the TR function is. NB This is NOT an actual Excel formula because it does not start with the sign =
A3	Q	Column heading label
B3	TR	Column heading label
A4	0	Initial value for Q
B4	=80*A4– 0.2*A4^2 (*The value 0 should appear*)	This formula calculates the value for TR that corresponds to the value of Q in cell A4.
A5	=A4+20	Calculates a 20 unit increase in Q.
A6 to A25	*Copy cell A5 formula down column* A	Calculates a series of values of Q in 20 unit increments (so we will only need 25 rows in the spreadsheet rather than 400 plus).
B5 to B25	*Copy cell B4 formula down column* B	Calculates values for TR in each row corresponding to the values of Q in column A.

Table 4.5 shows what to enter in the different cells of a spreadsheet to generate the relevant ranges of values, and also gives a brief explanation of what each entry means. Once a formula has been entered, only the calculated value appears in the cell where the formula is. However, when you put the cursor on a cell containing a formula, the full formula should always appear in the formula bar just above the spreadsheet.

When a formula is copied down a column, any cell references that the formula contains should also change. As the main formulae in this example are entered initially in row 4 and contain reference to cell A4, when they are copied to row 5 the reference should change to cell A5.

If you follow these instructions, you should end up with a spreadsheet that looks like Table 4.6. This clearly shows that TR increases as Q increases from 0 to 200 and then starts to decrease, equalling zero when Q reaches 400. We can also use this spreadsheet to calculate the value of TR for any given quantity, although we have only used increments of 20 units for Q to keep down the number of rows.

Table 4.6 Excel output based on Table 4.5

	A	B	C	D	E
1	Ex 4.17	TR = 80 – 0.2Q^2			
2					
3	Q	TR			
4	0	0			
5	20	1520			
6	40	2880			
7	60	4080			
8	80	5120			
9	100	6000			
10	120	6720			
11	140	7280			
12	160	7680			

(Continued)

Table 4.6 (*Continued*)

	A	B	C	D	E
13	180	7920			
14	200	8000			
15	220	7920			
16	240	7680			
17	260	7280			
18	280	6720			
19	300	6000			
20	320	5120			
21	340	4080			
22	360	2880			
23	380	1520			
24	400	0			
25	420	−1680			

Plotting a graph using Excel

Although it is obvious just by looking at the values of TR that this function rises and then falls, it is not quite so easy to get an idea of the exact shape of the function. It is easy, though, to use the Chart Tools facility in Excel to plot a graph for the columns of data for Q and TR generated in the spreadsheet. To do this:

1. Highlight the data that you want to draw a graph for, including the column heading. This will be the cell range B3 to B24 to plot the calculated TR values.
2. Go to **Insert** on the toolbar then click on Line in the Charts options (the one with coloured lines next to Column) so that you see various 2-D line graph options.
3. Click on the first of the Chart Sub-type examples. This will give a plain line graph showing a 'hill' shape for TR.
4. To put Q values along the horizontal axis, from the **Chart Tools** options click on **Select Data**. In the Chart data range box, click on Edit in the box on the right for **Horizontal (Category) Axis Labels** then, when prompted to insert a data range, enter $A$4:$A$24 by highlighting this range of values in the Q column, but not including the heading label Q. Then click on OK in this window and OK in the next one. The horizontal axis should now show the values of Q from 0 to 400. (If the Chart Tools tab disappears, just click once anywhere on the graph and it should appear, normally with green shading above the toolbar.)
5. The graph may still need a bit of tidying up. If the values on the horizontal axis do not lie exactly below the tick marks it may appear that the graph does not go exactly through the axis origin at zero. To rectify this, just go to the **Axis Options** window either by clicking on the Q axis on the actual graph or via the sequence of commands Layout > Axes > Primary Horizontal Axis > More options. Then in the box at the bottom ensure for **Position Axis** that the option **On tick marks** is chosen.

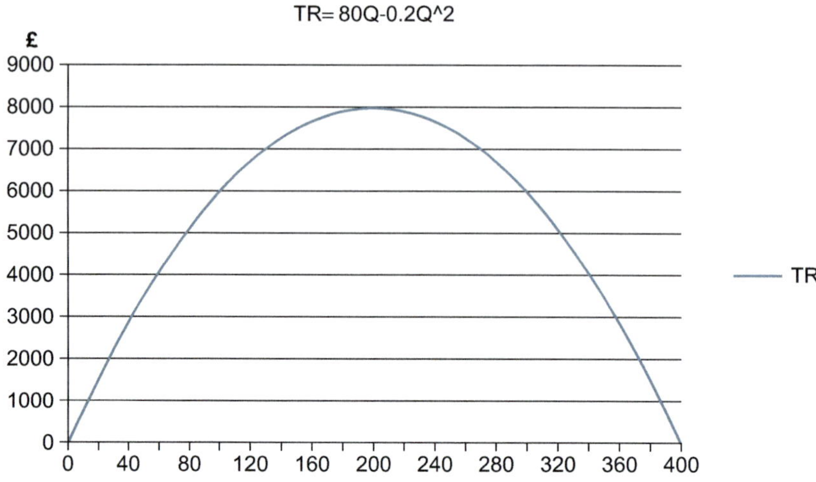

Figure 4.20 Total revenue

6. While you have this Primary Horizontal Axis window open you can also 'uncrowd' the labelling on the Q axis by changing the **Interval between labels** option from Automatic to **Specify interval** and then entering 2. This will mean that only every other value of Q will be shown on the axis, i.e. intervals will be 40 not 20.
7. To insert a label showing that Q is measured on the horizontal axis you can use the sequence of commands Layout > Axis Titles > Primary Horizontal Axis > **Title Below Axis**. This will produce a box under the Q axis that says 'Axis Title' which you can change to Q.
8. The Chart Title above the graph will initially just say 'TR' as this is the column of data used, but you can change this to 'TR = 80Q − 0.2Q^2' to make it clearer what function is being illustrated. (There should already be a Legend box at the side indicating that TR is illustrated by the blue line in the graph, which is the default colour.)
9. Your finished graph should look similar to Figure 4.20. This confirms that this function takes a smooth inverted U-shape. It has zero value when Q is 0 and 400 and has its maximum value of 8,000 when Q is 200. We will use this graph drawing tool again in Section 6.6 to help find solutions to polynomial equations.

If you want to enlarge the chart, just click on a corner or edge and drag to required size, or click on the chart itself and drag if you want to reposition it. Clicking on the line drawn on the graph itself will allow you to change colours. You can also click on Chart Tools in the toolbar at the top of the screen to go back and alter any of the formatting details, e.g. print font size. Try experimenting to learn how to get the chart format that suits you best. For example, try and insert the £ heading to the vertical axis shown in Figure 4.20.

QUESTIONS 4.9

Use a spreadsheet to plot values and draw graphs of the following functions:

1. $TR = 40q - 4q^2$
2. $TC = 12 + 4q + 0.2q^2$
3. $\pi = -12 + 36q - 3.8q^2$
4. $AC = 24q^{-1} + 8 - 3q + 0.5q^2$

4.10 FUNCTIONS WITH TWO INDEPENDENT VARIABLES

On a two-dimensional sheet of paper you cannot sketch a function with more than one independent variable as this would require more than two axes (one for the dependent variable and one each for the independent variables). However, in economics we often need to analyze functions that have two or more independent variables, e.g. production functions. When there are only two independent variables a 'contour line' graphing method can be used.

Consider the production function

$$Q = f(K, L)$$

Assume that the way in which Q depends on K and L is represented by the height above a two-dimensional surface on which K and L are measured. To show this production 'height', economics borrows the idea of contour lines from geography. On a map, contour lines join points of equal height and so, for example, a steep hill will be represented by closely spaced contour lines. In production theory a line that joins combinations of inputs K and L that will give the same production level (when used efficiently) is known as an 'isoquant'. An 'isoquant map' is shown in Figure 4.21. Isoquants normally show equal increments in output level, which enables one to get an idea of how quickly output responds to changes in the inputs. If isoquants are spaced far apart then output increases relatively slowly, and if they are spaced closely together then output increases relatively quickly.

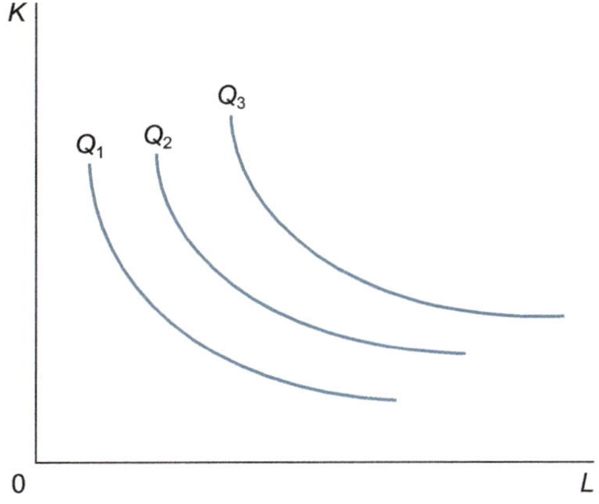

Figure 4.21 Isoquants

One can plot the position of an isoquant map from a production function although this is a rather tedious long-winded process. As we shall see later, it is not usually necessary to draw all the isoquants in order to tackle some of the resource allocation problems that this concept can be used for. Examples of some of the different combinations of K and L that would produce an output of 320 with the production function

$$Q = 20K^{0.5}L^{0.5}$$

are shown in Table 4.7. If plotted on a graph these would give a symmetrical curve known as a 'rectangular hyperbola' for the isoquant $Q = 320$.

A quicker way of finding out the shape of an isoquant is to transform it into a function with only two variables.

Table 4.7 Different values of inputs and corresponding output

K	L	$K^{0.5}$	$L^{0.5}$	Q
64	4	8	2	320
16	16	4	4	320
4	64	2	8	320
256	1	16	1	320
1	256	1	16	320

Example 4.18

For the production function $Q = 20K^{0.5} L^{0.5}$ derive a two-variable function for the isoquant $Q = 100$ in the format $K = f(L)$.

Solution

$$20K^{0.5}L^{0.5} = Q = 100$$

$$\text{Thus,} \quad K^{0.5}L^{0.5} = 5$$

$$K^{0.5} = \frac{5}{L^{0.5}}$$

Squaring both sides gives the required function for K in terms of L

$$K = \frac{25}{L} = 25L^{-1}$$

From Section 4.7 we know that this form of function will give a curve convex to the origin since the value of K gets closer to zero as L increases in value.

Example 4.19

For the production function $Q = 4.5K^{0.4} L^{0.7}$ derive a function in the form $K = f(L)$ for the isoquant representing an output of 54.

Solution

$$Q = 54 = 4.5K^{0.4}L^{0.7}$$

$$12 = K^{0.4}L^{0.7}$$

$$12L^{-0.7} = K^{0.4}$$

Taking both sides to the power of 2.5 (since 1 ÷ 0.4 = 2.5)

$$12^{2.5}L^{-1.75} = K$$
$$K = 498.83L^{-1.75}$$

This function will also give a curve convex to the origin since the value of $L^{-1.75}$ (and hence K) gets closer to zero as L increases.

The Cobb-Douglas production function

The production functions given in this section are examples of what are known as 'Cobb-Douglas' production functions. The general format of a Cobb-Douglas production function with two inputs K and L is

$$Q = AK^{\alpha}L^{\beta}$$

where A, α and β are positive parameters. (The Greek letter α is called 'alpha' and β 'beta'.) Many years ago, the two economists Cobb and Douglas found this form of function to be a good match to the statistical evidence on firms' input and output levels that they studied. Although economists have since developed more sophisticated forms of production functions, this basic Cobb-Douglas production function is still a good starting point for students to examine the relationship between a firm's output level and the inputs required, and hence costs.

Cobb-Douglas production functions fall into the mathematical category of **homogeneous** functions. In general terms, a function is said to be homogeneous of degree m if, when all inputs are multiplied by any given positive constant λ, the value of y increases by the proportion λ^m. (Note: λ is the Greek letter 'lambda'.)

Thus, if $y = f(x_1, x_2, \ldots, x_n)$

then $y\lambda^m = f(\lambda x_1, \lambda x_2, \ldots, \lambda x_n)$

An example of a function that is homogeneous of degree 1 is the production function

$$Q = 20K^{0.5}L^{0.5}$$

Assume that initially the input amounts are K_1 and L_1, giving production level

$$Q_1 = 20K_1^{0.5}L_1^{0.5}$$

If input amounts are doubled (i.e. $\lambda = 2$) then the new input amounts are

$$K_2 = 2K_1 \quad \text{and} \quad L_2 = 2L_1$$

giving the new output level

$$Q_2 = 20K_2^{0.5}L_2^{0.5}$$

This can be compared with the original output level by substituting $2K_1$ for K_2 and $2L_1$ for L_2. Thus,

$$Q_2 = 20(2K_1)^{0.5}(2L_1)^{0.5} = 20\left(2^{0.5}K_1^{0.5}2^{0.5}L_1^{0.5}\right) = 2\left(20K_1^{0.5}L_1^{0.5}\right) = 2Q_1$$

Therefore, when inputs are doubled, output doubles, and so this production function exhibits **constant returns to scale**.

The degree of homogeneity of a Cobb-Douglas production function can easily be determined by adding up the exponents on the input variables. This can be demonstrated for the two-input function

$$Q = AK^{\alpha}L^{\beta}$$

If we let initial input amounts be K_1 and L_1, then

$$Q_1 = AK_1^{\alpha}L_1^{\beta}$$

If all inputs are multiplied by the constant λ then new input amounts will be

$$K_2 = \lambda K_1 \quad \text{and} \quad L_2 = \lambda L_1$$

The new output level will then be

$$Q_2 = AK_2^{\alpha}L_2^{\beta} = A(\lambda K_1)^{\alpha}(\lambda L_1)^{\beta} = \lambda^{\alpha+\beta}AK_1^{\alpha}L_1^{\beta} = \lambda^{\alpha+\beta}Q_1$$

RETURNS TO SCALE

Since λ, α and β are all positive numbers, this result tells us the relationship between α and β and the three possible categories of returns to scale.

1. If $\alpha + \beta = 1$ then $\lambda^{\alpha+\beta} = \lambda$ and so $Q_2 = \lambda Q_1$, i.e. constant returns to scale.
2. If $\alpha + \beta > 1$ then $\lambda^{\alpha+\beta} > \lambda$ and so $Q_2 > \lambda Q_1$, i.e. increasing returns to scale.
3. If $\alpha + \beta < 1$ then $\lambda^{\alpha+\beta} < \lambda$ and so $Q_2 < \lambda Q_1$, i.e. decreasing returns to scale.

Example 4.20

What type of returns to scale does the production function $Q = 45K^{0.4}L^{0.4}$ exhibit?

Solution

Indices sum to 0.4 + 0.4 = 0.8. Thus, the degree of homogeneity is less than 1 and so there are decreasing returns to scale.

To estimate the parameters of functions with the format of Cobb-Douglas production functions using a basic linear regression model requires the use of logarithms. The standard linear regression analysis method (that you should learn about if you take a statistics module) allows you to use data on p and q to estimate the parameters a and b in linear functions such as the supply schedule

$$p = a + bq$$

If you have a non-linear function, logarithms can be used to transform it into a linear form so that linear regression analysis can be used to estimate parameters. For example, the parameters a and b can be estimated by linear regression analysis for the Cobb-Douglas production function

$$Q = AK^aL^b$$

by putting this function into log form as

$$\log Q = \log A + a\log K + b\log L$$

(See Section 2.10 if you are not sure how this transformation was obtained.)

In your economics course you should learn how the optimum input combination for a firm can be discovered using budget constraints, production functions and isoquant maps. We shall return to these concepts in Chapters 9 and 12, when mathematical solutions to optimization problems using calculus are explained.

QUESTIONS 4.10

1. For the following production functions, assume fractions of units of K and L can be used, and
   (a) derive a function for the isoquant representing the specified output level in the format $K = f(L)$
   (b) find the level of K required to achieve the given output level if $L = 100$, and
   (c) say what type of returns to scale are present.
   (1) $Q = 9K^{0.5}L^{0.5}$, $Q = 36$
   (2) $Q = 0.3K^{0.4}L^{0.6}$, $Q = 24$
   (3) $Q = 25K^{0.6}L^{0.6}$, $Q = 800$
   (4) $Q = 42K^{0.6}L^{0.75}$, $Q = 5{,}250$
   (5) $Q = 0.4K^{0.3}L^{0.5}$, $Q = 65$
   (6) $Q = 2.83K^{0.35}L^{0.62}$, $Q = 52$
2. Use logs to put the production function $Q = AK^{\alpha} L^{\beta} R^{\gamma}$ into a linear format.

4.11 SUMMING FUNCTIONS HORIZONTALLY

In economics, there are several occasions when theory requires certain functions to be summed 'horizontally'. Students are most likely to encounter this concept when studying the theories of price discrimination, multiplant monopoly (a monopoly firm with more than one production plant) and cartels. By 'horizontally' summing a function we mean summing it along the horizontal axis. This idea is best explained with an example.

Example 4.21

A price-discriminating monopolist sells in two separate markets at prices P_1 and P_2 (in £). The relevant demand and marginal revenue schedules are (for positive values of Q)

$$\begin{aligned} P_1 &= 12-0.15Q_1 \quad & P_2 &= 9-0.075Q_2 \\ \text{MR}_1 &= 12-0.3Q_1 \quad & \text{MR}_2 &= 9-0.15Q_2 \end{aligned}$$

It is assumed that output is allocated between the two markets according to the price discrimination revenue-maximizing criterion that $\text{MR}_1 = \text{MR}_2$.

Derive a formula for the aggregate marginal revenue schedule which is the horizontal sum of MR_1 and MR_2.

(In Chapter 5, we shall return to this example to find out how this summed MR schedule can help determine the profit-maximizing prices P_1 and P_2 when marginal cost is known.)

Solution

The two schedules MR_1 and MR_2 are illustrated in Figure 4.22. What we are required to do is find a formula for the summed schedule MR. This tells us what aggregate output will correspond to a given level of marginal revenue and vice versa, assuming that output is adjusted so that the marginal revenue from the last unit sold in each market is the same.

As you can see in Figure 4.22, the summed MR schedule is in fact kinked at point K. This is because the MR schedule sums the horizontal distances of MR_1 and MR_2 from the price axis. Given that MR_2 starts from a price of £9, then above £9 the only distance being summed is the distance between MR_1 and the price axis. Thus, between £12 and £9 MR is the same as MR_1, i.e.

$$\text{MR} = 12 - 0.3Q$$

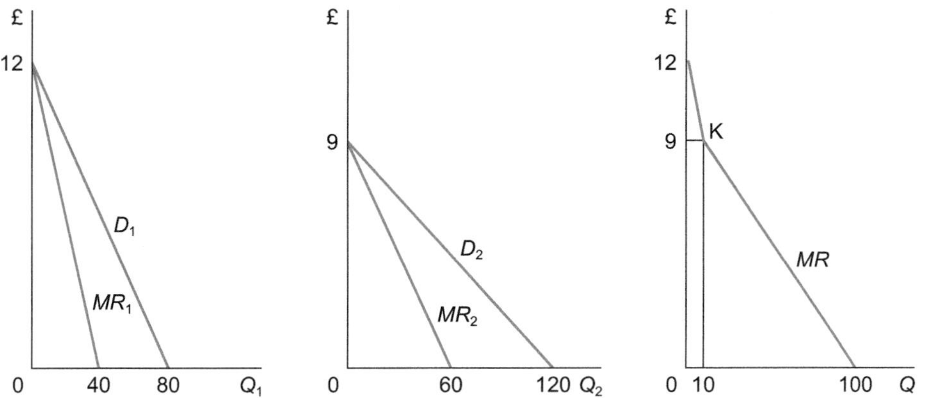

Figure 4.22 Horizontal sum of marginal revenue functions

where Q is aggregate output. If MR = £9 then

$$9 = 12 - 0.3Q$$
$$0.3Q = 3$$
$$Q = 10$$

Thus, the coordinates of the kink K are £9 and 10 units of output.

The proper summation occurs below £9. We are given the schedules

$$\text{MR}_1 = 12 - 0.3Q_1 \quad \text{and} \quad \text{MR}_2 = 9 - 0.15Q_2$$

but if we simply added MR_1 and MR_2 we would be summing vertically instead of horizontally. To be summed horizontally, these marginal revenue functions first have to be re-arranged to obtain their inverses as follows:

$$\text{MR}_1 = 12 - 0.3Q_1$$
$$0.3Q_1 = 12 - \text{MR}_1$$
$$Q_1 = 40 - 3\frac{1}{3}\text{MR}_1 \qquad (1)$$

$$\text{MR}_2 = 9 - 0.15Q_2$$
$$0.15Q_2 = 9 - \text{MR}_2$$
$$Q_2 = 60 - 6\frac{2}{3}\text{MR}_2 \qquad (2)$$

Given that the theory of price discrimination assumes that a firm will adjust thc amount sold in each market until $\text{MR}_1 = \text{MR}_2 = \text{MR}$, then

$$Q = Q_1 + Q_2$$
$$Q = \left(40 - 3\frac{1}{3}\text{MR}\right) + \left(60 - 6\frac{2}{3}\text{MR}\right) \qquad \text{by substituting (1) and (2)}$$
$$Q = 100 - 10\text{MR}$$
$$10\text{MR} = 100 - Q$$
$$\text{MR} = 10 - 0.1\text{Q}$$

This summed MR function will apply above an aggregate output of 10, to the right of K.

PROCEDURE FOR SUMMING FUNCTIONS HORIZONTALLY

1. transform the functions so that quantity is on the left-hand side of the equation sign;
2. sum the functions representing quantities;
3. transform the function back so that quantity returns to the right-hand side;
4. note the quantity range that this summed function applies to, given the intersection points of the functions to be summed on the price axis.

This procedure can also be applied to multiplant monopoly examples where it is necessary to find the horizontally summed marginal cost schedule.

Example 4.22

A monopoly operates two plants whose marginal cost schedules are

$$\text{MC}_1 = 2 + 0.2Q_1 \quad \text{and} \quad \text{MC}_2 = 6 + 0.04Q_2$$

Find the function which describes the horizontal summation of these two functions.

(We shall return to the use of the summed function in determining profit-maximizing price and output levels in Chapter 5.)

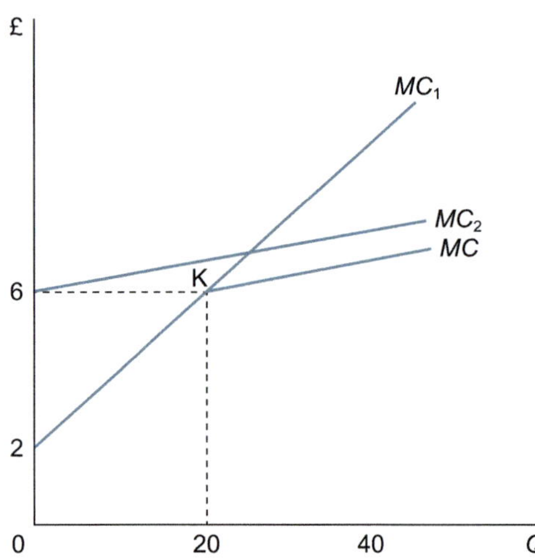

Figure 4.23 Horizontal sum of monopoly's marginal cost functions

Solution

The relevant schedules are illustrated in Figure 4.23. The horizontal sum of MC_1 and MC_2 will be the function MC which is kinked at K. Below £6 only MC_1 is relevant. Therefore, MC is the same as MC_1 from £2 to £6. The corresponding output range can be found by substituting £6 for MC_1. Thus,

$$\begin{aligned} \text{MC}_1 = 6 &= 2 + 0.2Q_1 \\ 4 &= 0.2Q_1 \\ 20 &= Q_1 \end{aligned}$$

Therefore MC = 2 + 0.2Q between $Q = 0$ and $Q = 20$.

Above this output we need to derive the proper sum of the two functions. Given

$$\begin{array}{lrcr} & \text{MC}_1 = 2 + 0.2Q_1 & \text{and} & \text{MC}_2 = 6 + 0.04Q_2 \\ \text{then} & \text{MC}_1 - 2 = 0.2Q_1 & \text{and} & \text{MC}_2 - 6 = 0.04Q_2 \\ & 5\text{MC}_1 - 10 = Q_1 & & 25\text{MC}_2 - 150 = Q_2 \end{array}$$

Summing the functions $Q_1 = 5\text{MC}_1 - 10$ and $Q_2 = 25\text{MC}_2 - 150$ gives total output

$$Q = Q_1 + Q_2 = (5\text{MC}_1 - 10) + (25\text{MC}_2 - 150) \quad (1)$$

A profit-maximizing monopoly will adjust output between two plants until

$$\text{MC}_1 = \text{MC}_2 = \text{MC}$$

Therefore, substituting MC into (2) gives

$$\begin{aligned} Q &= 5\text{MC} - 10 + 25\text{MC} - 150 \\ Q &= 30\text{MC} - 160 \\ 160 + Q &= 30\text{MC} \\ 5\frac{1}{3} + \frac{1}{30}Q &= \text{MC} \end{aligned}$$

This summed MC function applies above the total output level of 20.

In the previous examples the summation of only two linear functions was considered. The method can easily be extended to situations when three or more linear functions are to be summed. However, the inverses of some non-linear functions are not in forms that can be easily summed and so this method is best confined to applications involving linear functions.

QUESTIONS 4.11

Sum the following sets of marginal revenue and marginal cost schedules horizontally to derive functions in the form MR = f(Q) or MC = f(Q) and define the output ranges over which the summed function applies.

1. $\text{MR}_1 = 30 - 0.01Q_1$ and $\text{MR}_2 = 40 - 0.02Q_2$
2. $\text{MR}_1 = 80 - 0.4Q_1$ and $\text{MR}_2 = 71 - 0.5Q_2$
3. $\text{MR}_1 = 48.75 - 0.125Q_1$ and $\text{MR}_2 = 75 - 0.3Q_2$ and $\text{MR}_3 = 120 - 0.15Q_3$

4. $MC_1 = 20 + 0.25Q_1$ and $MC_2 = 34 + 0.1Q_2$
5. $MC_1 = 60 + 0.2Q_1$ and $MC_2 = 48 + 0.4Q_2$
6. $MC_1 = 3 + 0.2Q_1$ and $MC_2 = 1.75 + 0.25Q_2$ and $MC_3 = 4 + 0.2Q_3$

5 Simultaneous linear equations

Learning objectives

After completing this chapter students should be able to:

- solve sets of simultaneous linear equations with two or more variables using the substitution and row operations methods
- relate simultaneous linear equations mathematical solutions to economic analysis, including supply and demand and the basic Keynesian macroeconomic models
- construct and use break-even charts
- recognize when a linear equations system cannot be solved
- derive the reduced form equations for the equilibrium values of dependent variables in basic linear economic models and interpret their meaning
- derive the profit-maximizing solutions to price discrimination and multiplant monopoly problems involving linear functions.

5.1 SYSTEMS OF SIMULTANEOUS LINEAR EQUATIONS

The way to solve single linear equations with one unknown was explained in Chapter 3. We now turn to sets of linear equations with more than one unknown. A simultaneous linear equation system exists when:

(1) there is more than one functional relationship between a set of specified variables, and
(2) all the functional relationships are in a linear form.

The solution to a set of simultaneous equations involves finding values for all the unknown variables.

Where only two variables are involved, a simultaneous equation system can be illustrated on a graph. For example, assume that in a competitive market

the demand schedule is $p = 420 - 0.2q$ (1)

and the supply schedule is $p = 60 + 0.4q$ (2)

DOI: 10.4324/9781003360827-5

If this market is in equilibrium, then the equilibrium price and quantity will be where the demand and supply schedules intersect. As this will correspond to a point which is on both the demand schedule and the supply schedule, the equilibrium values of p and q will be such that both equations (1) and (2) hold. In other words, when the market is in equilibrium, equations (1) and (2) are simultaneously satisfied.

Note that in many of the examples in this chapter schedules for the 'inverse' demand and supply functions are used, i.e. $p = f(q)$ rather than $q = f(p)$. This is because price is normally measured on the vertical axis and we wish to relate the mathematical solutions to graphical analysis.

Simultaneous linear equations systems often involve more than two unknown variables in which case a graphical illustration of the problem may be more complex. It is also possible that a set of simultaneous equations may contain non-linear functions, but these are left until Chapter 7.

5.2 SOLVING SIMULTANEOUS LINEAR EQUATIONS

The basic idea involved in all the different methods of algebraically solving simultaneous linear equation systems is to manipulate the equations until there is a single linear equation with one unknown. This can then be solved using the methods explained in Chapter 3. The value of the variable that has been found can then be substituted back into the other equations to solve for the other unknown values.

Not all sets of simultaneous linear equations have solutions. The general rule is that the number of unknowns must be equal to the number of equations for there to be a unique solution. However, even if this condition is met, one may still come across systems that cannot be solved, e.g. functions which are geometrically parallel and therefore never intersect (see Example 5.2).

We shall first consider four different methods of solving a 2×2 set of simultaneous linear equations, i.e. one in which there are two unknowns and two equations, and then look at how some of these methods can be employed to solve simultaneous linear equation systems with more than two unknowns.

5.3 GRAPHICAL SOLUTION

The graphical solution method can be used when there are only two unknown variables. It will not always give 100% accuracy, but it can be useful for checking that algebraic solutions are not widely inaccurate owing to analytical or computational errors.

Example 5.1

Solve for p and q in the set of simultaneous equations given in Section 5.1.

$$p = 420 - 0.2q \qquad (1)$$
$$p = 60 + 0.4q \qquad (2)$$

Solution

These two functional relationships are plotted in Figure 5.1. Both hold at the intersection point X, where the solution values can be read off the graph as

$$p = 300 \quad \text{and} \quad q = 600$$

A graph can also illustrate why some simultaneous linear equation systems cannot be solved.

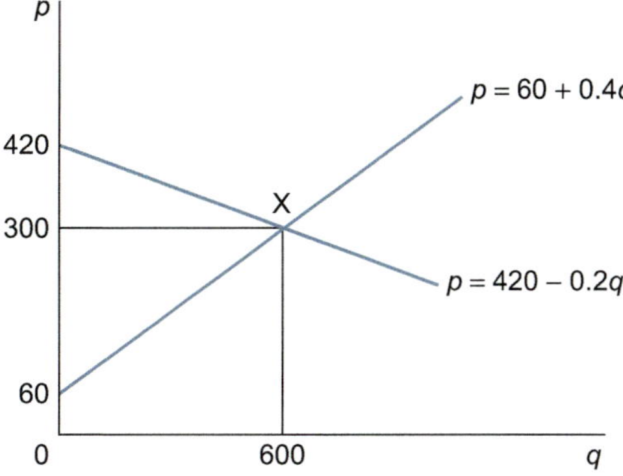

Figure 5.1 Finding market equilibrium by solving simultaneous equations

Example 5.2

Attempt to use graphical analysis to solve for y and x if

$$y = 2 + 2x \quad \text{and} \quad y = 5 + 2x$$

Solution

These two functions are plotted in Figure 5.2. They are obviously parallel lines which never intersect and so there are no values y and x where both functions hold (note that both functions have the same slope coefficients on x). This problem therefore does not have a solution.

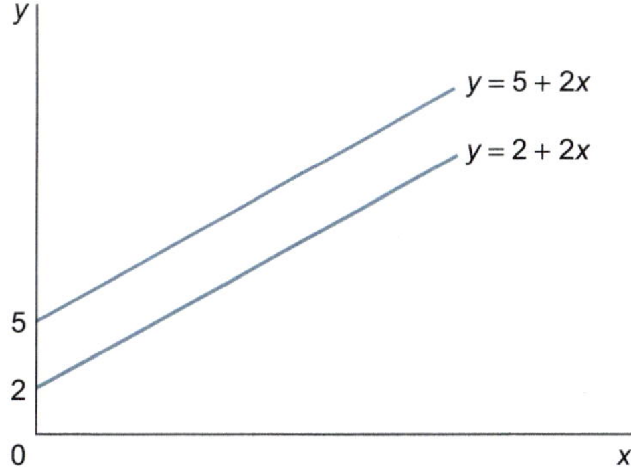

Figure 5.2 System of simultaneous equations with no solution

QUESTIONS 5.1

Solve the following (if a solution exists) using the graphical solution method.

1. Find the equilibrium values of p and q in a competitive market when the inverse demand and supply functions are, respectively,

$$p = 9 - 0.075q \quad \text{and} \quad p = 2 + 0.1q.$$

2. Find x and y when $x = 80 - 0.8y$ and $y = 10 + 0.1x$.
3. Find x and y when $y = -2 + 0.5x$ and $x = 2y - 9$.

5.4 THE EQUATING TO SAME VARIABLE METHOD

The method of equating to the same variable involves rearranging both equations so that the same unknown variable appears by itself on one side of the equality sign. This variable can then be eliminated by setting the other two sides of the equality sign in the two equations equal to each other. The resulting equation in one unknown can then be solved.

Example 5.3

Solve the set of simultaneous equations in Example 5.1 by the equating method.

Solution

In this example no preliminary rearranging of the equations is necessary because a single term in p appears on the left-hand side of both equality signs:

$$p = 420 - 0.2q \qquad (1)$$
$$p = 60 + 0.4q \qquad (2)$$

Then, it must be true that

$$420 - 0.2q = 60 + 0.4q$$

Therefore

$$360 = 0.6q$$
$$600 = q$$

The value of p can be found by substituting this value of q = 600 back into either of the two original equations. Thus,

from (1) $\quad p = 420 - 0.2q = 420 - 0.2(600) = 420 - 120 = 300$

or from (2) $\quad p = 60 + 0.4q = 60 + 0.4(600) = 60 + 240 = 300$

Example 5.4

Assume that a firm can sell as many units of its product at £18 each as it can manufacture in a month. It has to pay £240 fixed costs plus a marginal cost of £14 for each unit produced. How much does it need to produce to break even?

Solution

From the information in the question we can work out that this firm faces:

(a) the total revenue function TR = $18q$ and
(b) the total cost function TC = $240 + 14q$ where q is output.

These functions are plotted in Figure 5.3, which is an example of what is known as a **break-even chart**. This is a basic guide to the profit that can be expected for any given production level.

The break-even point is clearly at B, where the TR and TC schedules intersect and the firm makes neither a profit nor a loss. Since at the break-even point TR = TC, then

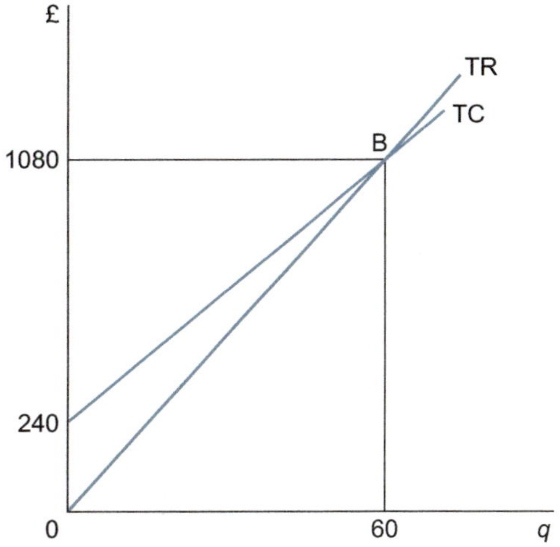

Figure 5.3 Break-even chart

$$\begin{aligned} 18q &= 240+14q \\ 4q &= 240 \\ q &= 60 \end{aligned}$$

Therefore, the output required to break even is 60 units.

Note that in reality a firm's TR and TC functions would not always be linear, and at some point the TR schedule will start to flatten out when the firm has to reduce price to sell more, and TC will get steeper when diminishing marginal productivity causes marginal cost to rise. If this did not happen, then the firm could make infinite profits by indefinitely expanding output.

What happens if you try to use this algebraic method when no solution exists, as in Example 5.2?

Example 5.5

Attempt to use the equating to same variable method to solve for y and x if

$$y=2+2x \quad \text{and} \quad y=5+2x$$

Solution

Eliminating y from the system and equating the other two sides of the equations, we get

$$2+2x=5+2x$$

Subtracting $2x$ from both sides gives $2 = 5$. This is clearly impossible and hence no solution can be found.

QUESTIONS 5.2

1. A competitive market has the demand schedule $p = 610 - 3q$ and the supply schedule $p = 20 + 2q$. Calculate the equilibrium price and quantity.
2. A competitive market has the demand schedule $p = 610 - 3q$ and the supply schedule $p = 50 + 4q$ where p is measured in pounds.
   (a) Find the equilibrium values of p and q.
   (b) What will happen to these values if the government imposes a sales tax of £14 per unit on q?
3. A firm manufactures product x and can sell any amount at a price of £25 a unit. The firm has to pay fixed costs of £200 plus a marginal cost of £20 for each unit produced.
   (a) How much of x must be produced to make a profit?
   (b) If price is cut to £24 what happens to the break-even output?
4. If $y = 16 + 22x$ and $y = -2.5 + 30.8x$, solve for x and y.

5.5 THE SUBSTITUTION METHOD

The substitution method involves rearranging one equation so that one of the unknown variables appears by itself on one side. The other side of the equation can then be substituted into the second equation to eliminate the other unknown.

Example 5.6

Solve the linear simultaneous equation system

$$20x + 6y = 500 \quad (1)$$
$$10x - 2y = 200 \quad (2)$$

Solution

Equation (2) can be rearranged to give

$$\begin{aligned} 10x - 200 &= 2y \\ 5x - 100 &= y \end{aligned} \quad (3)$$

If we substitute the left-hand side of equation (3) for y in equation (1), we get

$$\begin{aligned} 20x + 6y &= 500 \\ 20x + 6(\mathbf{5x - 100}) &= 500 \\ 20x + 30x - 600 &= 500 \\ 50x &= 1{,}100 \\ x &= 22 \end{aligned}$$

To find the value of y we now substitute this value of x into (1) or (2). Thus, in (1)

$$\begin{aligned} 20x + 6y &= 500 \\ 20(22) + 6y &= 500 \\ 440 + 6y &= 500 \\ 6y &= 60 \\ y &= 10 \end{aligned}$$

Example 5.7

Find the equilibrium level of national income in the basic Keynesian macroeconomic model

$$Y = C + I \quad (1)$$
$$C = 40 + 0.5Y \quad (2)$$
$$I = 200 \quad (3)$$

Solution

Substituting the consumption function (2) and the given I value of 200 into (1) gives

$$Y = 40 + 0.5Y + 200$$

Therefore, $0.5Y = 240$

$$Y = 480$$

QUESTIONS 5.3

1. A consumer has a budget of £240 and spends it all on the two goods A and B whose prices are initially £5 and £10 per unit respectively. The price of A then rises to £6 and the price of B falls to £8. What combination of A and B that uses up all the budget is it possible to purchase at both sets of prices?
2. Find the equilibrium value of Y in a basic Keynesian macroeconomic model where

$Y = C + I$	the accounting identity
$C = 20 + 0.6Y$	the consumption function
$I = 60$	exogenously determined investment

3. Solve for x and y when

$$600 = 3x + 0.5y$$
$$52 = 1.5y - 0.2x$$

5.6 THE ROW OPERATIONS METHOD

Row operations entail multiplying or dividing all the terms in one equation by whatever number is necessary to get the coefficient of one of the unknowns equal to the coefficient of that same unknown in another equation. Then, by subtraction of one equation from the other, this unknown can be eliminated.

Alternatively, if two rows have the same absolute value for the coefficient of an unknown but one coefficient is positive and the other is negative, then this unknown can be eliminated by adding the two rows.

Example 5.8

Given the following equations, use row operations to solve for x and y.

$$10x + 3y = 250 \quad (1)$$
$$5x + y = 100 \quad (2)$$

Solution

Multiplying (2) by 3	$15x + 3y = 300$
Subtracting (1)	$10x + 3y = 250$
Gives	$5x = 10$
	$x = 10$

Substituting this value of x back into (1),

$$\begin{aligned} 10(10) + 3y &= 250 \\ 100 + 3y &= 250 \\ 3y &= 150 \\ y &= 50 \end{aligned}$$

Example 5.9

A firm makes two goods A and B which require two inputs K and L. One unit of A requires 6 units of K plus 3 units of L and one unit of B requires 4 units of K plus 5 units of L. The firm has 420 units of K and 300 units of L at its disposal. How much of A and B should it produce if it wishes to exhaust its supplies of K and L totally?

(This question requires you to use the economic information given to set up a mathematical problem in a format that can be used to derive the desired solution. Learning how to set up a problem is just as important as learning how to solve it.)

Solution

The total requirements of input K are 6 for every unit of A and 4 for each unit of B, which can be written as

$$K = 6A + 4B$$

Similarly, the total requirements of input L can be specified as

$$L = 3A + 5B$$

As we know that $K = 420$ and $L = 300$ because all resources are used up, then

$$420 = 6A + 4B \qquad (1)$$
$$300 = 3A + 5B \qquad (2)$$

Now, let's find the solution

$$\begin{array}{ll} \text{Multiplying (2) by2} & 600 = 6A + 10B \\ \text{Subtracting (1)} & \underline{420 - 6A + 4B} \\ & 180 = 6B \\ & 30 = B \end{array}$$

Substituting this value for B into (1) gives

$$\begin{aligned} 420 &= 6A + 4(30) \\ 420 &= 6A + 120 \\ 300 &= 6A \\ 50 &= A \end{aligned}$$

The firm should therefore produce 50 units of A and 30 units of B.

(The method of setting up this problem will be used again when we get to linear programming in Chapter 6.)

QUESTIONS 5.4

1. Solve for x and y if

$$420 = 4x + 5y \quad \text{and} \quad 600 = 2x + 9y$$

2. A firm produces two goods A and B using inputs K and L. Each unit of A requires 2 units of K plus 6 units of L. Each unit of B requires 3 units of K plus 4 units of L. The amounts of K and L available are 120 and 180, respectively. What output levels of A and B will use up all the available K and L?
3. Solve for x and y when

$$160 - 8x - 2y \quad \text{and} \quad 295 = 11x + y$$

5.7 MORE THAN TWO UNKNOWNS

With more than two unknowns, the basic idea is to use one pair of equations to eliminate one unknown and then bring in another equation to eliminate the same variable, repeating the process until a single equation in one unknown is obtained. The exact operations necessary will depend on the format of the particular problem. There are several ways in which row operations can be used to solve most problems and you will only learn to identify which is the quickest method to use through practising examples yourself.

Example 5.10

Solve for x, y and z, given that

$$x + 12y + 3z = 120 \quad (1)$$
$$2x + y + 2z = 80 \quad (2)$$
$$4x + 3y + 6z = 219 \quad (3)$$

Solution

Multiplying (2) by 2 $\quad 4x + 2y + 4z = 160 \quad (4)$

Subtracting (4) from (3) $\quad y + 2z = 59 \quad (5)$

We have now eliminated x from equations (2) and (3) and so the next step is to eliminate x from equation (1) by row operations with one of the other two equations. In this example the easiest way is

Multiplying (1) by 2 $\quad 2x + 24y + 6z = 240$

Subtracting (2) $\quad 2x + y + 2z = 80$

$$23y + 4z = 160 \quad (6)$$

We now have the set of two simultaneous equations (5) and (6) involving two unknowns to solve. Writing these out again, we can now use row operations to solve for y and z.

$$y + 2z = 59 \quad (5)$$
$$23y + 4z = 160 \quad (6)$$

Multiplying (5) by 2 $\quad 2y + 4z = 118$

Subtracting (6) $\quad 23y + 4z = 160$

Gives $\quad -21y = -42$

$$y = 2$$

Substituting this value for y into (5) gives

$$2 + 2z = 59$$
$$2z = 57$$
$$z = 28.5$$

These values for y and z can now be substituted into any of the original equations to find the value of x. Thus, using (1) we get

$$\begin{aligned} x+12(2)+3(28.5) &= 120 \\ x+24+85.5 &= 120 \\ x &= 120-109.5 \\ x &= 10.5 \end{aligned}$$

Therefore, the solutions are $x = 10.5$, $y = 2$ and $z = 28.5$.

Example 5.11

Solve for x, y and z in the following set of simultaneous equations:

$$14.5x+3y+45z=340 \qquad (1)$$
$$25x-6y-32z=82 \qquad (2)$$
$$9x+2y-3z=16 \qquad (3)$$

Solution

Multiplying (1) by 2	$29x+6y+90z=680$	
Adding (2)	$25x-6y-32z=82$	(2)
Gives	$54x \quad +58z=762$	(4)

Having used equations (1) and (2) to eliminate y, we now need to bring in equation (3) to derive a second equation containing only x and z.

Multiplying (3) by 3	$27x+6y-9z = 48$	
Adding (2)	$25x-6y-32z= 82$	(2)
Gives	$52x \quad -41z = 130$	(5)

To eliminate x, we now need to multiply equations (5) and (4) by some values to get a common term that can then be eliminated. For example:

Multiplying (5) by 27	$1{,}404x-1{,}107z = 3{,}510$
Multiplying (4) by 26	$1{,}404x+1{,}508z= 19{,}812$
Subtracting gives	$-2{,}615z =-16{,}302$
	$z = 6.234$

(Note that although final answers are more neatly specified to one or two decimal places, more accuracy will be maintained if the full value of z above is entered when substituting to calculate remaining values of unknown variables.)

Substituting the value of z into (5) gives

$$
\begin{aligned}
52x - 41(6.234) &= 130 \\
52x &= 130 + 255.594 \\
x &= 7.415
\end{aligned}
$$

Substituting for both x and z in (1) gives

$$
\begin{aligned}
14.5(7.415) + 3y + 45(6.234) &= 340 \\
3y &= -48.05 \\
y &= -16.02
\end{aligned}
$$

Thus, solutions to 2 decimal places are

$$x = 7.42, \quad y = -16.02 \text{ and } \quad z = 6.23$$

These examples show how the solution to a 3 × 3 set of simultaneous equations can be solved by row operations. The same method can be used for larger sets but obviously more stages will be required to eliminate the unknown variables one by one until a single equation with one unknown is arrived at.

It is only practical to use the methods of solution for linear equation systems explained here where there are a relatively small number of equations and unknowns. For large systems of equations with more than a handful of unknowns it is more appropriate to use matrix algebra methods and a spreadsheet (see Chapter 16).

QUESTIONS 5.5

1. Solve for x, y and z when

$$
\begin{aligned}
2x + 4y + 2z &= 144 \\
4x + y + 0.5z &= 120 \\
x + 3y + 4z &= 144
\end{aligned}
$$

2. Solve for x, y and z when

$$
\begin{aligned}
12x + 15y + 5z &= 158 \\
4x + 3y + 4z &= 50 \\
5x + 20y + 2z &= 148
\end{aligned}
$$

3. Solve for A, B and C when

$$
\begin{aligned}
32A + 14B + 82C &= 664 \\
11.5A + 8B + 52C &= 349 \\
18A + 26.2B - 62C &= 560.4
\end{aligned}
$$

4. Find the values of x, y and z when

$$
\begin{aligned}
4.5x + 7y + 3z &= 128.5 \\
6x + 18.2y + 12z &= 270.8 \\
3x + 8y + 7z &= 139
\end{aligned}
$$

5. Solve for A, B, C and D when

$$A + 6B + 25C + 17D = 843$$
$$3A + 14B + 60C + 21D = 1{,}286.5$$
$$10A + 3B + 4C + 28D = 1{,}206$$
$$6A + 2B + 12C + 51D = 1{,}096$$

5.8 WHICH METHOD?

There is no hard and fast rule regarding which of the different methods for solving simultaneous equations should be used in different circumstances. The row operations method can be used for most problems, but sometimes it will be quicker to use one of the other methods, particularly in 2 × 2 systems. It may also be quicker to change methods midway. For example, one may find that in a 3 × 3 problem it may be quicker to revert to the substitution method after one of the unknowns has been eliminated by row operations. Only by practising solving problems will you learn how to spot the quickest methods of solving them.

Example 5.12

A firm uses three inputs K, L and R to manufacture its final product. The prices per unit of these inputs are £20, £4 and £2, respectively. If the other two inputs are held fixed, then the marginal product functions, showing the impact on output of an additional unit of input, are

$$MP_K = 20 - 5K$$
$$MP_L = 60 - 2L$$
$$MP_R = 80 - R$$

Using the basic rule for optimal input determination that the last £1 spent on each input should add the same amount to output, find which combination of inputs the firm should use to maximize output if it has a fixed budget of £390.

Solution

The input mix optimization rule requires that

$$\frac{MP_K}{P_K} = \frac{MP_L}{P_L} = \frac{MP_R}{P_R}$$

Therefore, substituting the given marginal product functions, we get

$$\frac{200 - 5K}{20} = \frac{60 - 2L}{4} = \frac{80 - R}{2}$$

Multiplying out two of the three pairwise combinations of equations to get K and R in terms of L gives

$$\begin{aligned}4(200-5K)&=20(60-2L)\\800-20K&=1{,}200-40L\\40L-400&=20K\\2L-20&=K\qquad(1)\end{aligned}$$

$$\begin{aligned}2(60-2L)&=4(80-R)\\120-4L&=320-4R\\4R&=4L+200\\R&=L+50\qquad(2)\end{aligned}$$

The third pairwise combination will not add any new information. Instead, we use the budget constraint

$$20K+4L+2R=390 \qquad (3)$$

Substituting (1) and (2) into (3),

$$\begin{aligned}20(2L-20)+4L+2(L+50)&=390\\40L-400+4L+2L+100&=390\\46L&=690\\L&=15\end{aligned}$$

Substituting this value for L into (1) gives $K=2(15)-20=10$
and substituting this value into (2) gives $R=15+50=65$

Therefore, the optimal input combination is

$$K=10,\quad L=15 \text{ and } \quad R=65$$

Example 5.13

In a closed economy where the usual assumptions of the basic Keynesian macroeconomic model apply,

$$\begin{aligned}C&=£60\text{m}+0.7Y_d\\Y&=C+I+G\\Y_d&=0.6Y\end{aligned}$$

where C is consumption, Y is national income, Y_d is disposable income, I is investment and G is government expenditure. If the values of I and G are exogenously determined as £90 million and £140 million, respectively, what is the equilibrium level of national income?

Solution

Once the given values of I and G are substituted, we have a 3 × 3 set of simultaneous equations with three unknowns:

$$C = 60 + 0.7Y_d \qquad (1)$$
$$Y = C + 90 + 140 = C + 230 \qquad (2)$$
$$Y_d = 0.6Y \qquad (3)$$

This sort of problem is most easily solved by substitution. Substituting (3) into (1) gives

$$C = 60 + 0.7(0.6Y)$$
$$C = 60 + 0.42Y \qquad (4)$$

Substituting (4) into (2) gives

$$Y = (60 + 0.42Y) + 230$$
$$0.58Y = 290$$
$$Y = 500$$

Therefore, the equilibrium value of the national income is £500 million.

Example 5.14

In a competitive market where the supply price (in £) is $p = 3 + 0.25q$ and demand price (in £) is $p = 15 - 0.75q$, the government imposes a per-unit tax of £4. How much of a price rise will this tax mean to consumers? What will be the tax revenue raised?

Solution

The pre-tax equilibrium price and quantity can be found by equating demand and supply price.

$$15 - 0.75q = 3 + 0.25q$$
$$12 = q$$

Substituting this value of q into the supply schedule gives

$$p = 3 + 0.25(12) = 3 + 3 = 6$$

If a per-unit tax is imposed, this would mean that each quantity would be offered for sale by suppliers at the old price plus the amount of the tax. In this case, the tax is £4 and so the supply schedule shifts upwards by £4. Thus, the new supply schedule becomes

$$p = 3 + 0.25q + 4$$
$$= 7 + 0.25q$$

Again, equating demand and supply price

$$15 - 0.75q = 7 + 0.25q$$
$$8 = q$$

Substituting this value of q into the demand schedule

$$\begin{aligned} p &= 15 - 0.75(8) \\ &= 15 - 6 \\ &= 9 \end{aligned}$$

Therefore, consumers see a price rise of £3 from £6 to £9.

(Given that £4 for each unit is given to the government in tax, producers will receive a net price of £5 and so will incur a £1 price reduction.)

$$\begin{aligned} \text{Total tax revenue} &= \text{quantity sold} \times \text{tax per unit} \\ &= 8 \times 4 \\ &= £32 \end{aligned}$$

QUESTIONS 5.6

1. A firm faces the demand schedule $p = 400 - 0.25q$
   the marginal revenue schedule $MR = 400 - 0.5q$
   and the marginal cost schedule $MC = 0.3q$
   What price will maximize profit?
2. A firm buys the three inputs K, L and R at prices per unit of £10, £5 and £3, respectively. The marginal product functions of these three inputs are

   $$\begin{aligned} MP_K &= 150 - 4K \\ MP_L &= 72 - 2L \\ MP_R &= 34 - R \end{aligned}$$

   What input combination will maximize output if the firm's budget is fixed at £285?
3. In a competitive market, the supply schedules is $p = 4 + 0.25q$
   and the demand schedule is $p = 16 - 0.5q$
   What would happen to the price paid by consumers and the quantity sold if
   (a) a per-unit tax of £3 was imposed,
   (b) a proportional sales tax of 20% was imposed?
4. In a Keynesian macroeconomic model of an economy with no foreign trade it is assumed that

   $$\begin{aligned} Y &= C + I + G \\ C &= 0.75Y \\ Y_t &= (1 - t)Y \end{aligned}$$

where the usual notation applies. The tax rate is $t = 0.2$ and Y_t is net after tax income. Exogenously fixed variables are I = £600 million and G = £900 million.

Find the equilibrium value of Y and say whether the government's budget is balanced at this value.

5. In an economy which engages in foreign trade, it is assumed that

$$Y = C + I + G + X - M$$
$$C = 0.9Y_t$$
$$Y_t = (1 - t)Y$$
$$\text{and imports} \quad M = 0.15Y_t$$

Find the equilibrium value of Y given the following values:

$$I = £200\text{m} \quad G = £270\text{m} \quad X = £180\text{m} \quad t = 0.2$$

What is the balance of payments surplus/deficit at this value?

6. In a factor market for labour, a monopsonistic buyer faces

$$\text{the marginal revenue product schedule} \quad \text{MRP}_\text{L} = 244 - 2L$$
$$\text{the supply of labour schedule} \quad w = 20 + 0.4L$$
$$\text{and the marginal cost of labour schedule} \quad \text{MC}_\text{L} = 20 + 0.8L$$

How much labour should it employ, and at what wage, if MRP_L must equal MC_L in order to maximize profit? (*Leave this question for now if you have not yet covered factor supply theory.*)

5.9 COMPARATIVE STATICS AND THE REDUCED FORM OF AN ECONOMIC MODEL

Now that you are familiar with the basic methods for solving simultaneous linear equations, this section will explain how these methods can help you to derive predictions from some economic models. Although no new mathematical methods will be introduced in this section, it is important that you work through the examples to **learn how to set up economic problems in a mathematical format**. This is particularly relevant for those students who can master mathematical methods without too many problems but find it difficult to set up the problem that they need to solve. It is important that you understand the **application of mathematical techniques** to economics, which is the reason why you are studying mathematics as part of your economics course.

Equilibrium and comparative statics

In Section 5.1 we saw how two simultaneous equations representing the supply and demand functions in a competitive market could be solved to determine equilibrium price and quantity. Markets need not always be in equilibrium, however. For example, if

$$\begin{aligned} &\text{Quantity demanded} = q_d = 90 - 0.05p \\ \text{and} \quad &\text{Quantity suppiled} = q_s = -12 + 0.8p \end{aligned}$$

then if price is £100

$$\begin{aligned} q_d &= 90 - 0.05(100) = 90 - 5 = 85 \\ q_s &= -12 + 0.8(100) = -12 + 80 = 68 \end{aligned}$$

and so there would be excess demand equal to

$$q_d - q_s = 85 - 68 = 17$$

In a freely competitive market this situation of excess demand would result in price rising. As price rises, the quantity demanded will fall and the quantity supplied will increase until quantity demanded equals quantity supplied, and the market is in equilibrium.

The time it takes for this adjustment to equilibrium to take place will vary from market to market and the analysis of this dynamic adjustment process between equilibrium situations is considered later in Chapters 14 and 15. Here, we shall just examine how the equilibrium values in an economic model change when certain variables alter. This is known as **comparative static analysis**.

If a market is in equilibrium, it means that quantity supplied equals quantity demanded and so there are no market forces pushing price up or pulling it down. Therefore, price and quantity will remain stable unless something disturbs the equilibrium. One factor that might cause this to happen is a change in the value of an independent variable. In the simple supply and demand model mentioned earlier, both quantity demanded and quantity supplied are determined within the model and so there are no independent variables, but consider the following market model where

$$\begin{aligned} \text{Quantity supplied} &= q_s = -20 + 0.4p \\ \text{Quantity demanded} &= q_d = 160 - 0.5p + 0.1m \\ \text{Average income} &= m \end{aligned}$$

The value of m will just be given, as it will be determined by factors outside this model. It is therefore an independent variable, sometimes known as an exogenous variable. Without knowing the value of m we cannot work out the values for the dependent variables determined within the model (also known as endogenous variables) which are the equilibrium values of p and q.

Once the value of m is known, equilibrium price and quantity can easily be found. For example, if m is £270 then

$$q_d = 160 - 0.5p + 0.1m = 160 - 0.5p + 0.1(270) = 187 - 0.5p$$

In equilibrium $\quad q_s = q_d$

and so

$$\begin{aligned} -20 + 0.4P &= 187 - 0.5P \\ 0.9p &= 207 \\ p &= 230 \end{aligned}$$

Substituting this value for p into the supply function to get equilibrium quantity gives

$$q = -20 + 0.4p = -20 + 0.4(230) = -20 + 92 = 72$$

If factors outside this model cause the value of m to alter then the equilibrium price and quantity will also change. For example, if income rises to £360, then quantity demanded becomes

$$q_d = 160 - 0.5p + 0.1m = 160 - 0.5p + 0.1(360) = 196 - 0.5p$$

and so equating supply and demand to find the new equilibrium price and quantity

$$\begin{aligned} -20 + 0.4p &= 196 - 0.5p \\ 0.9p &= 216 \\ p &= 240 \end{aligned}$$

and so

$$q = -20 + 0.4(240) = 76$$

Thus, this rise of income to £360 causes price to rise to 230 and quantity to fall to 76.

To save having to work out the new equilibrium values in an economic model from first principles every time an exogenous variable changes, it can be useful to derive what is known as the reduced form of an economic model.

Reduced form

The reduced form specifies each of the dependent variables in an economic model as a function of the independent variable(s). This reduced form can then be used to:

- predict what happens to the dependent variables when an independent variable changes
- estimate the parameters of the model from data using regression analysis (which you will learn about in your statistics or econometrics module).

Example 5.15

A per-unit sales tax t is imposed by the government in a competitive market with the

demand function $q = 20 - 1\frac{1}{3}p$ and

supply function $q = -12 + 4p$

Derive reduced form equations for the equilibrium values of p and q in terms of the tax t.

Solution

Firms have to pay the government a per-unit tax of t on each unit they sell. This means that to supply any given quantity firms will require an additional amount t on top of the supply price without the tax, i.e. the supply schedule will shift up vertically by the amount of the tax. To show the effect of this it is easier to work with the inverse demand and supply functions, where price is a function of quantity.

The demand function $q = 20 - 1\frac{1}{3}p$ becomes $p = 15 - 0.75q$ (1)

and the supply function $q = -12 + 4p$ becomes $p = 3 + 0.25q$

After the tax is imposed the inverse supply function becomes

$$p = 3 + 0.25q + t \tag{2}$$

In equilibrium the supply price equals the demand price and so equating (1) and (2)

$$\begin{aligned} 3 + 0.25q + t &= 15 - 0.75q \\ q &= 12 - t \end{aligned} \tag{3}$$

This is the reduced form equation for equilibrium quantity.

From this reduced form we can easily work out that

when $t = 0$ then $q = 12$

when $t = 4$ then $q = 8$

(You can check these solutions are the same as those in Example 5.14. which had the same supply and demand functions.)

In a model with two dependent variables, like this supply and demand model, once the reduced form equation for one dependent variable has been derived then the reduced form equation for the other dependent variable can be derived. This is done by substituting the reduced form for the first variable into one of the functions that make up the model. For example, if the reduced form equation for equilibrium quantity (3) is substituted into the demand function

$$p = 15 - 0.75q$$

it becomes $p = 15 - 0.75(12 - t)$

giving $p = 6 + 0.75t$ (4)

which is the reduced form equation for equilibrium price.

The reduced form equations can also be used to work out the **comparative static effect of a change in *t* on equilibrium quantity or price**, i.e. what happens to these equilibrium values when tax is altered.

In this example the reduced form equation for price (4) tells us that for every one unit increase in t the equilibrium price p increases by 0.75. This is illustrated here for a few values of t:

when $t = 4$ then $p = 6 + 0.75(4) = 6 + 3 = 9$
when $t = 5$ then $p = 6 + 0.75(5) = 6 + 3.75 = 9.75$
when $t = 6$ then $p = 6 + 0.75(6) = 6 + 4.5 = 10.5$

Note that this method can only be used with linear functions. If a dependent variable is a non-linear function of an independent variable, then calculus must be used (see Chapter 10).

Before proceeding any further, students should make sure that they understand an important difference between the supply and demand functions and the reduced form of an economic model. The supply and demand functions give the quantities supplied and demanded for **any** price, which includes prices out of equilibrium. The **reduced form only includes the equilibrium values** of p and q.

Reduced form and comparative static analysis of monopoly

The basic principles for deriving reduced form equations for dependent variables can be applied in various types of economic models and are not confined to supply and demand analysis. The following example shows how the comparative static effect of a per-unit tax on a monopoly can be derived from the reduced form equations.

Example 5.16

A monopoly operates with the marginal cost function MC = 20 + 4q and faces the inverse demand function $p = 400 - 8q$. If a per-unit sales tax t is imposed on its output, derive reduced form equations for the profit-maximizing values of p and q in terms of the tax t and use them to predict the effect of a one-unit increase in the tax on price and quantity. Assume that fixed costs are low enough to allow positive profits to be made.

Solution

The per-unit tax will cause the cost of supplying each unit to rise by amount t and so the monopoly's marginal cost function will change to

$$\text{MC} = 20 + 4q + t$$

For any linear inverse demand function the corresponding marginal revenue function will have the same intercept on the price axis but twice the slope (see Section 9.3 for a proof of this result). Therefore, if

$$p = 400 - 8q \quad \text{then} \quad \text{MR} = 400 - 16q$$

If the monopoly is maximizing profit, then

$$\begin{aligned} \text{MC} &= \text{MR} \\ 20 + 4q + t &= 400 - 16q \\ 20q + t &= 380 \\ q &= 19 - 0.05t \end{aligned}$$

From this reduced form equation for equilibrium q we can see that for every one-unit increase in the sales tax the monopoly's output will fall by 0.05 units.

To find the reduced form equation for equilibrium p we can substitute the reduced form for q into the inverse demand function. Thus,

$$\begin{aligned} p &= 400 - 8q = 400 - 8(19 - 0.05t) \\ &= 400 - 152 + 0.4t = 248 + 0.4t \end{aligned}$$

Thus, the reduced form equation for equilibrium p is

$$p = 248 + 0.4t$$

This tells us that for every one-unit increase in t the monopoly's price will rise by 0.4. So, for example, a £1 tax increase will cause price to rise by 40p.

The effect of a proportional sales tax

In practice sales taxes are often specified as a percentage of the pre-tax price rather than being set as a fixed amount per unit. For example, VAT (value added tax) is levied in the UK at a rate of 20% on most goods and services at the point of sale. To work out the reduced form equations, a proportional tax needs to be specified in decimal format. Thus, a sales tax of 20%, such as VAT, becomes 0.2 in decimal format.

Example 5.17

A proportional sales tax t is imposed in a competitive market where the demand price is given by $p_d = 375 - 2.5q$ and the supply price is given by $p_s = 55 + 4q$. Derive reduced form equations for the equilibrium values of p and q in terms of the tax rate t and use them to predict the effect of an increase in the tax rate on the equilibrium values of p and q.

Solution

To supply any given quantity firms require the original pre-tax supply price p_s plus the proportional tax that is levied at that price. Therefore, the total new price p_s^* that firms will require to supply any given quantity will be

$$p_s^* = p_s(1+t) = (55+4q)(1+t) \tag{1}$$

The supply function therefore swings up as shown in Figure 5.4 (instead of the parallel shift caused by a per-unit tax).

Setting this new supply price function (1) equal to demand price

$$\begin{aligned} p_s^* &= p_d \\ (55+4q)(1+t) &= 375-2.5q \\ 55+55t+4q+4qt &= 375-2.5q \\ 6.5q+4qt &= 320-55t \\ q(6.5+4t) &= 320-55t \\ q &= \frac{320-55t}{6.5+4t} \end{aligned}$$

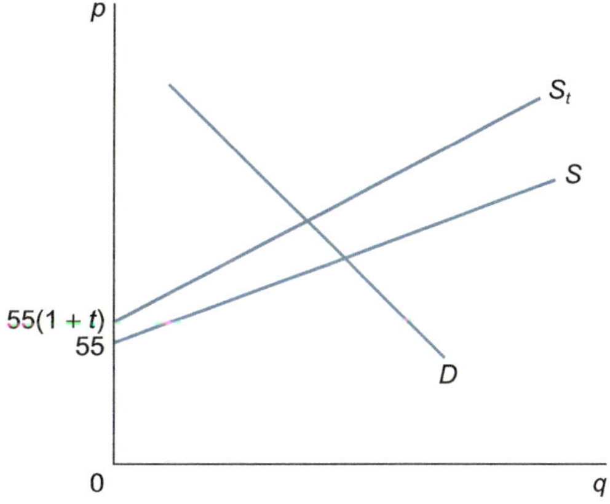

Figure 5.4 The effect of proportional tax on the market equilibrium

This reduced form equation for the equilibrium quantity is a bit more complicated than the one we derived for the per-unit sales tax case. However, we can still use it to work out the predicted value of q for different values of t. Normally we would expect sales taxes to lie between 0% and 100%, giving a value of t in decimal format between 0 and 1.

If $t = 10\% = 0.1$ then $q = \dfrac{320-55(0.1)}{6.5+4(0.1)} = \dfrac{320-5.5}{6.5+0.4} = \dfrac{314.5}{6.9} = 45.58$

If $t = 20\% = 0.2$ then $q = \dfrac{320-55(0.2)}{6.5+4(0.2)} = \dfrac{320-11}{6.5+0.8} = \dfrac{309}{7.3} = 42.33$

If $t = 30\% = 0.3$ then $q = \dfrac{320-55(0.3)}{6.5+4(0.3)} = \dfrac{320-16.5}{6.5+1.2} = \dfrac{303.5}{7.7} = 39.42$

These examples show that as the tax rate increases the value of q falls, as one would expect. However, these equal increments in the tax rate do not bring about equal changes in q because the reduced form equation for equilibrium q is not a simple linear function of t.

Lastly, we can derive the reduced form equation for equilibrium p by substituting the reduced form for q that we have already found into the demand schedule. Thus,

$$p = 375 - 2.5q = 375 - 2.5\left(\frac{320 - 55t}{6.5 + 4t}\right)$$

$$= \frac{2437.5 + 1500t - 800 + 137.5t}{6.5 + 4t}$$

$$= \frac{1637.5 + 1637.5t}{6.5 + 4t}$$

$$= \frac{1637.5(1 + t)}{6.5 + 4t}$$

To check this reduced form equation, we can calculate p for some extreme values of t to see if the prices calculated lie in a reasonable range for this demand schedule.

If $t = 0$ (i.e. no tax) then $p = \frac{1637.5 + 1637.5(0)}{6.5 + 4(0)} = \frac{1637.5}{6.5} = 251.92$

If $t = 100\% = 1$ then $p = \frac{1637.5 + 1637.5}{6.5 + 4} = \frac{3275}{10.5} = 311.81$

These values lie in a range that one would expect for this demand schedule, which intercepts the price axis at 375.

The reduced form of a Keynesian macroeconomic model

Consider the basic Keynesian macroeconomic model used previously in Example 5.7 where

$$Y = C + I \quad (1)$$
$$C = 40 + 0.5Y \quad (2)$$

As the value of investment is exogenously determined we can derive a reduced form equation for the equilibrium value of the dependent variable Y in terms of this independent variable I. Substituting the consumption function (2) into the accounting identity (1) gives

$$Y = 40 + 0.5Y + I$$
$$0.5Y = 40 + I$$
$$Y = 80 + 2I$$

From this reduced form we can directly predict the equilibrium value of Y for any given level of I. For example,

when $I = 200$ then $Y = 80 + 2(200) = 80 + 400 = 480$ (*check with Example 5.7*)
when $I = 300$ then $Y = 80 + 2(300) = 80 + 600 = 680$

From the reduced form equation, we can also see that for every £1 increase in I the value of Y will increase by £2. This ratio of 2 to 1 is the **investment multiplier**.

Reduced forms in models with more than one independent variable

Equilibrium values of dependent variables in an economic model may be determined by more than one independent variable. If this is the case, then all the independent variables will appear in the reduced form equations for these dependent variables.

Consider the following Keynesian macroeconomic model, where the values of investment (I), government expenditure (G) and the tax rate (t) are exogenously determined.

$$Y = C + I + G \quad (1)$$

$$C = 50 + 0.8Y_d \quad (2)$$

And disposable income $$Y_d = (1-t)Y \quad (3)$$

Substituting the function for disposable income (3) into the consumption function (2) gives

$$C = 50 + 0.8Y_d = 50 + 0.8(1-t)Y \quad (4)$$

Substituting (4) into (1) gives

$$\begin{aligned} Y &= 50 + 0.8(1-t)Y + I + G \\ Y(1 - 0.8 + 0.8t) &= 50 + I + G \\ Y &= \frac{50 + I + G}{0.2 + 0.8t} \end{aligned}$$

This reduced form equation tells us that the equilibrium value of Y will be determined by the values of the three exogenous variables I, G and t. For example,

when $I = 180$, $G = 150$ and $t = 0.375$ then

$$Y = \frac{50 + I + G}{0.2 + 0.8t} = \frac{50 + 180 + 150}{0.2 + 0.8(0.375)} = \frac{380}{0.5} = 760$$

The comparative static effect of an increase in one of the three independent variables can only be worked out if the values of the other two are held constant. For example,

if $I = 180$ and $t = 0.375$

then

$$Y = \frac{50 + 180 + G}{0.2 + 0.8(0.375)} = \frac{230 + G}{0.5} = 460 + 2G$$

From this new reduced form equation we can see that (when I is 180 and t is 0.375) for every £1 increase in G there will be a £2 increase in Y, i.e. the government expenditure multiplier is 2.

In Chapter 10, we will return to this form of analysis when we have shown how calculus can be used to derive comparative static effects for economic models with non-linear functions.

QUESTIONS 5.7

1. In a competitive market the supply function is $q_s = -12 + 0.3p$
   and the demand function is $q_d = 80 - 0.2p + 0.1a$
   where a is the price of an alternative substitute good.
   Derive reduced form equations for equilibrium price and quantity in terms of a and use them to predict the values of p and q when a is 160.
2. A per-unit tax t is imposed on all items sold in a competitive market where

   $$q_s = -10 + 0.5p \quad \text{and} \quad q_d = 200 - 2p$$

   Derive reduced form equations for equilibrium price and quantity in terms of tax t and use them to predict the values of p and q when t is 5.
3. A monopoly faces the marginal cost function MC = $12 + 6q$
   and the demand function $q = 75 - 0.5p$
   If a per-unit tax t is imposed on its output, derive reduced form equations for the profit-maximizing values of p and q in terms of the tax t and use them to predict these values when t is 5.
4. In a Keynesian macroeconomic model the following functional relations hold

   $$Y = C + I + G$$
   $$C = 20 + 0.75Y_d$$
   $$Y_d = (1 - t)Y$$

   and Y_d denotes disposable income.
   (a) If the values of investment and government expenditure (I and G) are exogenously fixed at 50 and 30, respectively, derive a reduced form equation for equilibrium Y in terms of t and use it to predict Y when the tax rate t is 20%.
   (b) Explain what will happen to this reduced form equation and the equilibrium level of Y if G changes to 40.
5. A proportional sales tax v is imposed in a competitive market where

   $$p_d = 800 - 4q \quad \text{and} \quad p_s = 50 + 5q$$

   Derive reduced form equations for the equilibrium values of p and q in terms of the tax rate v and use them to predict p and q when v is 15%.

5.10 PRICE DISCRIMINATION

In Section 4.10 we examined how linear functions could be summed 'horizontally'. We shall now use this method to help tackle some problems involving price discrimination and, in the following section, multiplant firm and cartel pricing. It is assumed that the main economic principles underpinning these models will be explained in your economics course and so only the methods of calculating prices and output are explained here.

In **third-degree price discrimination**, firms charge different prices in separate markets. The theory of price discrimination says that to maximize profits firms should:

(a) split total sales between the different markets so that the marginal revenue from the last unit sold in each market is the same, and
(b) decide on the total sales level by finding the output level where the aggregate marginal revenue function (derived by horizontally summing the marginal revenue schedules from each individual market) intersects the firm's marginal cost function.

It is usually assumed that the firm practising price discrimination is a monopoly. All the examples in this section assume that the firm faces linear demand schedules in each of the separate markets. We shall also make use of the rule that the marginal revenue schedule corresponding to a linear demand schedule will have the same intercept on the price axis but twice the slope. The method of solution is best explained with some examples.

Example 5.18

A monopoly can sell in two separate markets at different prices (in £) and faces the marginal cost schedule

$$\text{MC} = 1.75 + 0.05q$$

The two demand schedules are

$$p_1 = 12 - 0.15q_1 \quad \text{and} \quad p_2 = 9 - 0.075q_2$$

What price should it charge and how much should it sell in each market to maximize profit?

Solution

It helps to draw a sketch diagram when tackling this type of problem so that you can relate the different quantities to the economic model. Note that the demand schedules in this example, illustrated in Figure 5.5, are the same as those in Example 4.20 in the

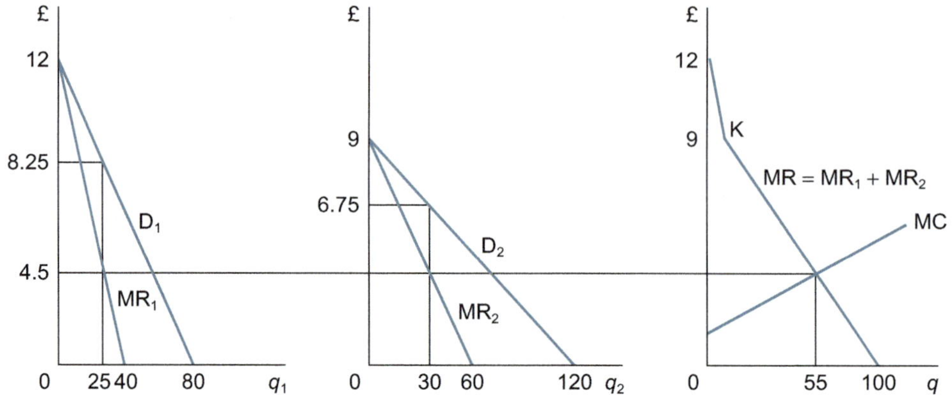

Figure 5.5 Profit maximization by a monopoly operating in two markets

previous chapter when the marginal revenue summation process was explained in more detail.

First, the relevant MR schedules and their inverse functions are derived from the demand schedules. Given

$$p_1 = 12 - 0.15q_1 \qquad\qquad p_2 = 9 - 0.075q_2$$

$$\text{then} \quad MR_1 = 12 - 0.3q_1 \qquad\qquad MR_2 = 9 - 0.15q_2$$

$$\text{and so} \quad q_1 = 40 - \frac{MR_1}{0.3} \quad (1) \qquad\qquad q_2 = 60 - \frac{MR_2}{0.15} \quad (2)$$

For profit maximization,

$$MR_1 = MR_2 = MR \quad (3)$$

and

$$q = q_1 + q_2 \quad (4)$$

Therefore, substituting (1), (2) and (3) into (4)

$$q = \left(40 - \frac{MR}{0.3}\right) + \left(60 - \frac{MR}{0.15}\right) = \frac{12 - MR + 18 - 2MR}{0.3}$$

$$\frac{30 - 3MR}{0.3} = 100 - 10MR$$

and also

$$MR = 10 - 0.1q \quad (5)$$

This function does not apply above £9 as only MR_1 applies above this price. In this example, Figure 5.5 shows that MC will cut MR in the section below the kink K.

The aggregate profit-maximizing output is found where

$$\text{MR} = \text{MC}$$

Thus, using the MC function given in the question and the aggregated marginal revenue function (5) derived earlier we get

$$\begin{aligned} 10 - 0.1q &= 1.75 + 0.05q \\ 8.25 &= 0.15q \\ 55 &= q \end{aligned}$$

$$\begin{aligned} &\text{Therefore,} \quad \text{MR} = 10 - 0.1(55) = 10 - 5.5 = 4.5 \\ &\text{and so} \qquad \text{MR}_1 = 4.5 \qquad \text{MR}_2 = 4.5 \end{aligned}$$

To determine the prices and output levels in each market, we now just substitute these MR values into the inverse marginal revenue functions (1) and (2) derived earlier. Thus,

$$q_1 = 40 - \frac{\text{MR}_1}{0.3} = 40 - \frac{4.5}{0.3} = 40 - 15 = 25$$

$$q_2 = 60 - \frac{\text{MR}_2}{0.15} = 60 - \frac{4.5}{0.15} = 60 - 30 = 30$$

You can check these output figures to ensure that $q_1 + q_2 = q$.

Relating these calculations to Figure 5.5, what we have done is found the intersection point of MR and MC to determine the profit-maximizing levels of q and MR. Then a horizontal line is drawn across to see where this level of marginal revenue cuts MR_1 and MR_2. This enables us to read off q_1 and q_2 and the corresponding prices p_1 and p_2. These prices can be determined by simply substituting the previous values of q_1 and q_2 into the two demand schedules specified in the question. Thus,

$$\begin{aligned} p_1 &= 12 - 0.15q_1 = 12 - 0.15(25) = 12 - 3.75 = £8.25 \\ p_2 &= 9 - 0.075q_2 = 9 - 0.075(30) = 9 - 2.25 = £6.75 \end{aligned}$$

Finally, refer back to the sketch diagram to ensure that the relative magnitudes of the answer correspond to those read off the graph. It is easy to get mixed up in the various stages of the calculation in this type of problem. From Figure 5.5, we can see that p_1 should be greater than p_2 which checks out with the previous answers.

Not all price discrimination models involve the horizontal summation of marginal revenue schedules. In **first-degree (perfect) price discrimination**, each individual unit is sold at a different price. Because the prices of other units do not have to be reduced for a firm to increase sales, the marginal revenue from each unit is the price it sells for. Therefore, the marginal revenue schedule is the same as the demand schedule, instead of lying below it.

In **second-degree price discrimination**, a firm breaks the market up into a series of price bands. In a two-part pricing scheme, this might mean that the first few units are sold at a previously determined price and then a price is chosen for the remaining units that will maximize profits, given the first price and the marginal cost schedule.

The following example explains how the relevant prices and quantities can be calculated under these different forms of price discrimination.

Example 5.19

A monopoly faces the demand schedule $p = 16 - 0.064q$ and the marginal cost schedule $MC = 2.2 + 0.019q$. It has decided that the first 60 units will be sold at a price of £12.16 each. Given this constraint, what price for the remaining units will maximize profits? How will the corresponding total output compare with output when

(a) the firm can only set a single price?
(b) perfect price discrimination takes place?

Solution

We can check that the price of £12.16 for the first 60 units corresponds to point A on the demand schedule in Figure 5.6 since

$$p = 16 - 0.064q = 16 - 0.064(60) = 16 - 3.84 = £12.16$$

If the firm wishes to sell more output, it will not have to reduce the price of these first 60 units. It therefore effectively faces the marginal revenue schedule MR′. This is constructed by assuming that the zero on the quantity axis is moved 60 units to the right to point B. MR′ is then drawn in the usual way with the same 'intercept' on the price axis (effectively point A) but twice the slope of the demand schedule. The firm should then employ the usual rule for profit maximization, which is to produce the output level at which marginal revenue MR′ equals marginal cost.

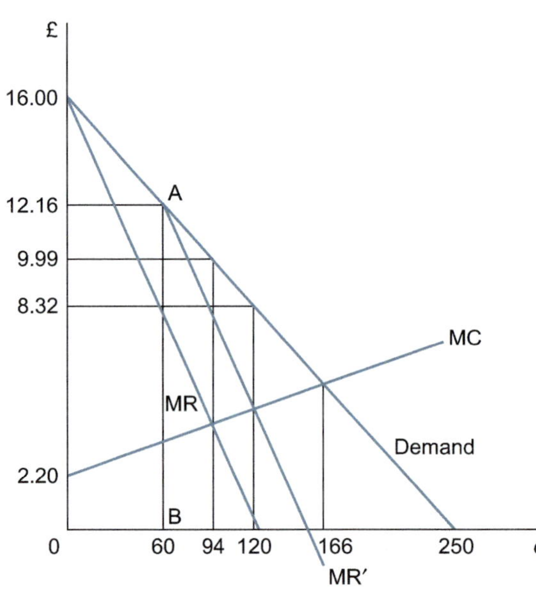

Figure 5.6 Profit maximization by a monopoly

To derive a function for MR′, define $q' = q - 60$, i.e. q' measures output from point B on the quantity axis. The demand schedule over this output range has the same slope as the original demand schedule (−0.064) but the new 'intercept' value of £12.16. It is therefore described by the function

$$p = 12.16 - 0.064q'$$

and therefore

$$MR' = 12.16 - 0.128q' \tag{1}$$

using the rule that a marginal revenue function has twice the slope of a linear demand schedule.

Substituting the original definition of output for q' into (1)

$$\begin{aligned} MR' &= 12.16 - 0.128(q - 60) \\ &= 12.16 - 0.128q + 7.68 \\ &= 19.84 - 0.128q \end{aligned} \tag{2}$$

Profit maximization, subject to the given price constraint on the first 60 units, requires

$$MR' = MC$$

Therefore, equating (2) and the given MC function

$$\begin{aligned} 19.84 - 0.128q &= 2.2 + 0.019q \\ 17.64 &= 0.147q \\ 120 &= q \end{aligned}$$

This is total output. The amount sold at the second (lower) price will be

$$q' = q - 60 = 120 - 60 = 60$$

The price for these units will be

$$p = 12.16 - 0.064q' = 12.16 - 0.064(60) = 12.16 - 3.84 = £8.32$$

The total output level of 120 units under this second-degree price discrimination policy can be compared with

(a) Single-price profit maximization:

given the demand schedule $p = 16 - 0.064q$
then $MR = 16 - 0.128q$

Single-price profit maximization occurs when

$$\begin{aligned} MR &= MC \\ 16 - 0.128q &= 2.2 + 0.019q \\ 13.8 &= 0.147q \\ 93.877 &= q \text{ (marked as 94 on graph)} \end{aligned}$$

Output is therefore lower than the 120 units produced under a two-part pricing scheme. This is what one would expect given that price discrimination allows a firm to sell extra output without reducing the price of all previously sold units and hence shifts the relevant marginal revenue schedule to the right.

The profit-maximizing single price can be found from the demand schedule as

$$p = 16 - 0.064q = 16 - 0.064(93.877) = £9.99$$

This is higher than the £8.32 price for the second batch of output under the second-degree price discrimination example earlier.

(b) Perfect price discrimination:

If all units are sold at different prices and the prices of the first units sold do not have to be reduced to sell more, then marginal revenue is the same as the demand schedule, i.e.,

$$\text{MR} = 16 - 0.064q$$

The profit-maximizing output is determined where

$$\begin{aligned} \text{MR} &= \text{MC} \\ 16 - 0.064q &= 2.2 + 0.019q \\ 13.8 &= 0.083q \\ 166.265 &= q \text{ (shown as 166 on graph)} \end{aligned}$$

This is greater than the two-part pricing discrimination output, which is what is expected. The greater the number of different segments a market can be broken up into, the higher will be the profits that can be extracted and the output that can be sold.

Note that in the previous example, and in all the others in this section, we shall assume that total costs, which are not actually specified, are low enough to allow an overall profit to be made.

QUESTIONS 5.8

1. A price-discriminating monopoly sells in two markets whose demand functions are

$$q_1 = 160 - 10p_1 \quad \text{and} \quad q_2 = 240 - 20p_2$$

and it faces the marginal cost schedule MC = 1.2 + 0.02q where $q = q_1 + q_2$. How much should it sell in each market and at what prices to maximize profits?

2. A monopoly faces the marginal cost schedule MC = 1.1 + 0.01q and can price-discriminate between the two markets where

$$p_1 = 10 - 0.1q_1 \quad \text{and} \quad p_2 = 6 - 0.04q_2$$

How much should it sell in each market to maximize profit, and at what prices?

3. A price-discriminating monopoly sells in two markets whose demand schedules are

$$p_1 = 12.5 - 0.0625q_1 \quad p_2 = 7.2 - 0.002q_2$$

and faces the horizontal marginal cost schedule MC = 5. What price and output should it choose for each market?

4. A monopoly faces the horizontal marginal cost schedule MC = 42 and can operate a two-part pricing scheme in the market with the demand schedule

$$p = 180 - 0.6q$$

If the first 100 units are sold at a price of £120 each, what price should be charged for the remaining units in order to maximize profit?

5. A monopoly sells in a market where $p = 12 - 0.06q$
and has the marginal cost schedule MC = 3 + 0.04q.
If it can operate second-degree price discrimination, what price should it sell the remaining units for if it has already decided to sell the first 50 units for a price of £9?

6. A price-discriminating monopoly sells in two markets whose demand functions are

$$q_1 = 120 - 6p_1 \quad \text{and} \quad q_2 = 110 - 8p_2$$

Calculate the profit-maximizing price and sales levels for each market if its marginal cost function is MC = 2.26 + 0.02q.7.

7. A monopoly has the demand function $q = 1050 - 5p$
and the marginal cost schedule MC = 20 + 0.8q
   (a) If it can practise first-degree price discrimination how much should it sell?
   (b) If it can practise second-degree price discrimination and has already made the decision to sell the first 100 units at a price of £190, what price should it charge for the rest of the units it sells?

5.11 MULTIPLANT MONOPOLY

The theory of multiplant monopoly is analogous to the model of third-degree price discrimination explained earlier, except that it is marginal cost schedules that are summed rather than marginal revenue schedules. The basic principles of the multiplant model are:

(a) The firm should adjust production so that the marginal cost of the last unit produced in each plant is equal to the marginal cost of the last unit produced by the other plant(s).
(b) Total output is determined where the aggregate marginal cost schedule (derived by horizontally summing the marginal cost schedules in each individual plant) intersects the firm's marginal revenue schedule.

The firm is usually assumed to be a monopoly so that the demand and marginal revenue schedules can be clearly defined. The multiplant monopoly model can also be used to determine price and output levels for the different (single-plant) firms in a cartel where perfect collusion takes place. This is a less likely scenario, however, as perfect collusion within cartels is beset with many problems, as you should know from your economics course, and is also usually illegal.

Example 5.20

A firm operates two plants whose marginal cost schedules are

$$\text{MC}_1 = 2 + 0.2q_1 \quad \text{and} \quad \text{MC}_2 = 6 + 0.04q_2$$

It is a monopoly seller in a market where the demand function is

$$q = 660 - 10p,$$

where q is aggregate output and all costs and prices are measured in £.

How much should the firm produce in each plant, and at what price should total output be sold, if it wishes to maximize profits?

Solution

We need to derive the horizontally summed marginal cost schedule MC, find where it intersects MR, and then see which output levels this marginal cost value corresponds to in each plant. Price is read off the demand schedule at the aggregate output level. (You will note that the marginal cost schedules to be summed are the same as those in Example 4.21 and Figure 4.23.)

From the demand function we can derive the inverse demand function

$$p = 66 - 0.1q$$

and we know that the marginal revenue schedule will have the same intercept and twice the slope. Thus,

$$\mathrm{MR} = 66 - 0.2q \tag{1}$$

To set MC = MR and solve for q we need to derive MC as a function of q. To do this, we first derive the inverse functions of the individual plant MC schedules, as shown here:

$$\begin{aligned} \mathrm{MC}_1 &= 2 + 0.2q_1 & \mathrm{MC}_2 &= 6 + 0.04q_2 \\ \mathrm{MC}_1 - 2 &= 0.2q_1 & \mathrm{MC}_2 - 6 &= 0.04q_2 \\ 5\mathrm{MC}_1 - 10 &= q_1 & 25\mathrm{MC}_2 - 150 &= q_2 \end{aligned}$$

Given that $q = q_1 + q_2$ by definition and $\mathrm{MC} = \mathrm{MC}_1 = \mathrm{MC}_2$ for profit maximization, then by substituting the previous functions for q_1 and q_2 in terms of MC we get

$$\begin{aligned} q &= (5\mathrm{MC} - 10) + (25\mathrm{MC} - 150) \\ q &= 30\mathrm{MC} - 160 \\ q + 160 &= 30\mathrm{MC} \\ \frac{q+160}{30} &= \mathrm{MC} \end{aligned} \tag{2}$$

Setting MC = MR, from (1) and (2) we now get

$$\begin{aligned} \frac{q+160}{30} &= 66 - 0.2q \\ q + 160 &= 1{,}980 - 6q \\ 7q &= 1{,}820 \\ q &= 260 \end{aligned}$$

Substituting this aggregate output level into (2) gives

$$\mathrm{MC} = \frac{q+160}{30} = \frac{260+160}{30} = \frac{420}{30} = 14$$

Therefore, $\mathrm{MC}_1 = \mathrm{MC}_1 = \mathrm{MC}_2 = \mathrm{MC} = 14$

and so

$$\begin{aligned} q_1 &= 5(14) - 10 = 70 - 10 = 60 \\ q_2 &= 25(14) - 150 = 350 - 150 = 200 \end{aligned}$$

We can easily check that these output levels for the individual plants correspond to the aggregate output of 260 calculated earlier, since

$$q_1 + q_2 = 60 + 200 = 260 = q$$

To find the price at which this aggregate output is sold, simply substitute this value of q into the demand schedule. Therefore,

$$p = 66 - 0.1q = 66 - 0.1(260) = 66 - 26 = £40$$

The basic principles explained earlier can also be applied to more complex problems where there are more than two plants.

Example 5.21

A firm operates four plants whose marginal cost schedules are

$$\begin{aligned} \text{MC}_1 &= 20 + q_1 & \text{MC}_3 &= 40 + q_3 \\ \text{MC}_2 &= 40 + 0.5q_2 & \text{MC}_4 &= 60 + 0.5q_4 \end{aligned}$$

and it is a monopoly seller in a market where $p = 580 - 0.3q$

How much should it produce in each plant and at what price should its output be sold if it wishes to maximize profit?

Solution

First we find the inverses of the marginal cost functions. Thus,

$$\begin{array}{cccc} \text{MC}_1 = 20 + q_1 & \text{MC}_2 = 40 + 0.5q_2 & \text{MC}_3 = 40 + q_3 & \text{MC}_4 = 60 + 0.5q_4 \\ q_1 = \text{MC}_1 - 20 & q_2 = 2\text{MC}_2 - 80 & q_3 = \text{MC}_3 - 40 & q_4 = 2\text{MC}_4 - 120 \end{array}$$

Given that

$$q = q_1 + q_2 + q_3 + q_4$$

and for profit maximization

$$\text{MC} = \text{MC}_1 = \text{MC}_2 = \text{MC}_3 = \text{MC}_4$$

then, by summing all the inverses of the individual MC functions and substituting MC, we can write

$$\begin{aligned} q &= (\text{MC} - 20) + (2\text{MC} - 80) + (\text{MC} - 40) + (2\text{MC} - 20) \\ q &= 6\text{MC} - 260 \\ \frac{q + 260}{6} &= \text{MC} \end{aligned} \qquad (1)$$

Since

$$\begin{aligned} p &= 580 - 0.3q \\ \text{MR} &= 580 - 0.6q \end{aligned} \qquad (2)$$

To maximize profits MC = MR, so we set (1) and (2) equal

$$\begin{aligned} \frac{q+260}{6} &= 580-0.6q \\ q+260 &= 3{,}480-3.6q \\ 4.6q &= 3{,}220 \\ q &= 700 \end{aligned}$$

For this aggregate output level the marginal cost is

$$\text{MC} = \frac{q+260}{6} = \frac{700+260}{6} = \frac{960}{6} = 160$$

Substituting this value of MC into the individual inverse marginal cost functions to find plant output levels gives

$$\begin{aligned} q_1 &= \text{MC}_1 - 20 = 160 - 20 = 140 \\ q_2 &= 2\text{MC}_2 - 80 = 320 - 80 = 240 \\ q_3 &= \text{MC}_3 - 40 = 160 - 40 = 120 \\ q_3 &= \text{MC}_3 - 40 = 160 - 40 = 120 \end{aligned}$$

These total to 700, which checks out with the answer for q earlier.

The price to sell at is found by substituting the total output of 700 units into the demand schedule given in the question. Thus,

$$p = 580 - 0.3q = 580 - 0.3(700) = 580 - 210 = £370$$

Note that we did not draw a sketch diagram for the previous example to check whether or not the MR schedule cuts the aggregated MC schedule at a level where output by all four plants is positive, i.e. where the value of MC is above the intercept on the vertical axis for each individual MC schedule. However, as all four output levels were calculated as positive numbers, we know that this must be the case.

In this type of question, if the usual mathematical method throws up a negative quantity for output by one or more plants (or a negative sales figure in a price discrimination model), then this means that output in this plant (or plants) should be zero. The question should then be reworked with the marginal cost schedule for any such plants (or the MR schedule from any such markets) excluded from the aggregated MC schedule.

Price discrimination with multiplant monopoly

It is possible to apply the principles of both price discrimination and multiplant monopoly at the same time, if all the necessary conditions hold.

Example 5.22

A multiplant monopoly operates two plants whose marginal cost schedules are

$$\mathrm{MC}_1 = 42.5 + 0.5q_1 \quad \text{and} \quad \mathrm{MC}_2 = 130 + 2q_2$$

It also sells its product in two separable markets, A and B, whose demand schedules are

$$p_\mathrm{A} = 360 - q_\mathrm{A} \quad \text{and} \quad p_\mathrm{B} = 280 - 0.4q_\mathrm{B}$$

Calculate how much it should produce in each plant, how much it should sell and charge in each market.

Solution

First derive the aggregate MC function by the usual method. Given

$$\begin{aligned} &\mathrm{MC}_1 = 42.5 + 0.5q_1 \qquad \mathrm{MC}_2 = 130 + 2q_2 \\ \text{then} \quad &q_1 = 2\mathrm{MC}_1 - 85 \qquad q_2 = 0.5\mathrm{MC}_2 - 65 \end{aligned}$$

To maximize profits, output is adjusted between the plants so that

$$\begin{aligned} \mathrm{MC}_1 &= \mathrm{MC}_2 = \mathrm{MC} \\ \text{Therefore} \quad q &= q_1 + q_2 = (2\mathrm{MC} - 85) + (0.5\mathrm{MC} - 65) \qquad (1) \\ q &= 2.5\mathrm{MC} - 150 \\ 60 + 0.4q &= \mathrm{MC} \end{aligned}$$

Next, derive the aggregate MR function for the two markets. Given

$$\begin{aligned} &p_\mathrm{A} = 360 - q_\mathrm{A} \text{ and} \qquad p_\mathrm{B} = 280 - 04q_\mathrm{B} \\ \text{then} \quad &\mathrm{MR}_\mathrm{A} = 360 - 2q_\mathrm{A} \qquad \mathrm{MR}_\mathrm{B} = 280 - 0.8q_\mathrm{B} \\ &q_\mathrm{A} = 180 - 0.5\mathrm{MR}_\mathrm{A} \qquad q_\mathrm{B} = 350 - 1.25\mathrm{MR}_\mathrm{B} \end{aligned}$$

To maximize profits, sales are adjusted so that $\mathrm{MR}_\mathrm{A} = \mathrm{MR}_\mathrm{B} = \mathrm{MR}$

$$\begin{aligned} \text{Therefore} \quad q &= q_\mathrm{A} + q_\mathrm{B} = (180 - 0.5\mathrm{MR}) + (350 - 1.25\mathrm{MR}) \\ q &= 530 - 1.75\mathrm{MR} \qquad (2) \\ \mathrm{MR} &= \frac{530 - q}{1.75} \end{aligned}$$

To maximize profits MC = MR. Therefore, equating (1) and (2)

$$\begin{aligned}60 + 0.4q &= \frac{530 - q}{1.75}\\ 105 + 0.7q &= 530 - q\\ 1.7q &= 425\\ q &= 250\end{aligned}$$

$$\begin{aligned}&\text{Thus, MC} &&= 60 + 0.4q = 60 + 0.4(250) = 60 + 100 = 160\\ &\text{and also MR} &&= \text{MC} = 160\end{aligned}$$

To find production levels in the two plants, substitute this value of MC into the inverse MC_1 and MC_2 functions earlier. Thus,

$$\begin{aligned}q_1 &= 2\text{MC} - 85 = 2(160) - 85 = 320 - 85 = 235\\ q_2 &= 0.5\text{MC} - 65 = 0.5(160) - 65 = 80 - 65 = 15\end{aligned}$$

To find sales levels in each market, substitute this value of MR into the inverse MR functions earlier. Thus,

$$\begin{aligned}q_A &= 180 - 0.5\text{MR} = 180 - 0.5(160) = 180 - 80 = 100\\ q_B &= 350 - 1.25\text{MR} = 350 - 1.25(160) = 350 - 200 = 150\end{aligned}$$

A quick check shows that the two production levels and the two sales levels both add to 250, which is what is expected.

Finally, the prices charged in the two markets A and B are found by substituting the previous values of q_A and q_B into the demand schedules. Thus,

$$\begin{aligned}p_A &= 360 - q_A = 360 - 100 = £260\\ p_B &= 280 - 0.4q_B = 280 - 0.4(150) = 280 - 60 = £220\end{aligned}$$

QUESTIONS 5.9

1. A monopoly operates two plants whose marginal cost schedules are

$$MC_1 = 2 + 0.1q_1 \quad \text{and} \quad MC_2 = 4 + 0.08q_2$$

and sells in a market where the demand function is $q = 1160 - 20p$.
How much should it produce in each plant and at what price should its product be sold?

2. A multiplant monopoly sells in a market where the demand schedule is

$$p = 253.4 - 0.025q$$

and produces in two plants whose marginal cost schedules are

$$\mathrm{MC}_1 = 20 + 0.0625q_1 \quad \text{and} \quad \mathrm{MC}_2 = 50 + 0.1q_2$$

How should it split output between the two plants in order to maximize profit? What price should it sell at?

3. A firm operates two plants whose marginal cost schedules are

$$\mathrm{MC}_1 = 22.5 + 0.25q_1 \quad \text{and} \quad \mathrm{MC}_2 = 15 + 0.25q_2$$

It is also a monopoly which can price-discriminate between two markets, A and B, whose demand schedules are

$$p_\mathrm{A} = 600 - 0.125q_\mathrm{A} \quad \text{and} \quad p_\mathrm{B} = 850 - 0.1q_\mathrm{B}$$

If it wishes to maximize profits, how much should it produce in each plant, how much should it sell in each market, and what prices should it sell at?

4. A multiplant monopoly produces using two plants with the marginal cost schedules

$$\mathrm{MC}_1 = 8 + 0.2q_1 \quad \text{and} \quad \mathrm{MC}_2 = 10 + 0.05q_2$$

It can also price-discriminate between three markets whose demand schedules are

$$p_\mathrm{A} = 150 - 0.1875q_\mathrm{A} \quad \text{and} \quad p_\mathrm{B} = 80 - 0.15q_\mathrm{B} \quad \text{and} \quad p_\mathrm{C} = 80 - 0.1q_\mathrm{C}$$

In order to maximize profits, how much should it produce in each plant, how much should it sell in each market, and what prices should it sell at?

5. A monopoly operates three plants with marginal cost schedules

$$\mathrm{MC}_1 = 0.1 + 0.02q_1 \quad \text{and} \quad \mathrm{MC}_2 = 0.3 + 0.004q_2 \quad \text{and} \quad \mathrm{MC}_3 = 0.2 + 0.008q_3$$

How much should it make in each plant to maximize profit if its market demand function is

$$q = 140 - 5p$$

and what price will the total output be sold at?

6 Linear programming

Learning objectives

After completing this chapter students should be able to:

- set up and illustrate linear programming problems
- use graphical analysis to solve constrained maximization problems
- use graphical analysis to solve constrained minimization problems.

Although basically an extension of the linear algebra covered in Chapter 5, the technique of linear programming involves special features which distinguish it from other linear algebra applications. When all relevant functions are linear, this technique enables one to:

- calculate the profit-maximizing output mix of a multi-product firm subject to restrictions on input availability, or
- calculate the input mix that will minimize costs subject to minimum quality standards being met.

This makes it an extremely useful tool for managerial decision-making.

However, it should be noted that, from a pure economic theory viewpoint, linear programming cannot make any general predictions about price or output for a large number of firms. Its usefulness lies in the realm of managerial (or business) economics where economic techniques can help an individual firm to make efficient decisions.

6.1 CONSTRAINED MAXIMIZATION

A resource allocation problem that a firm may encounter is how to decide on the product mix which will maximize profits when it has limited amounts of the various inputs

DOI: 10.4324/9781003360827-6

required for the different products that it makes. The firm's objective is to maximize profit and so profit is what is known as the 'objective function'. It tries to optimize this function subject to the constraint of limited input availability. This is why it is known as a 'constrained optimization' problem.

When both the objective function and the constraints can be expressed in a linear form then the technique of linear programming can be used to try to find a solution. (Constrained optimization of non-linear functions is explained in Chapter 12.) We shall restrict the analysis here to objective functions which have only two variables, e.g. when only two goods contribute to a firm's profit. This enables us to use graphical analysis to help find a solution, as explained in the following example.

Example 6.1

A firm manufactures two goods A and B using three inputs K, L and R. The firm has at its disposal 150 units of K, 120 units of L and 40 units of R. The net profit contributed by each unit sold is £4 for A and £1 for B (note that these are per-unit profits and not prices).

Each unit of A produced requires 3 units of K, 4 units of L plus 2 units of R.

Each unit of B produced requires 5 units of K, 3 units of L and none of R.

What combination of A and B should the firm manufacture to maximize profits given these constraints on input availability?

Solution

From the per-unit profit figures of £4 for A and £1 for B we can see that the linear objective function for profit which the firm wishes to maximize will be

$$\pi = 4A + B$$

where A and B represent the quantities of goods A and B that are produced.

The total amount of input K required is 3 for each unit of A plus 5 for each unit of B and we know that only 150 units of K are available, so the constraint on input K is

$$3A + 5B \leq 150 \quad (1)$$

Similarly, for L

$$4A + 3B \leq 120 \quad (2)$$

and for R

$$2A \leq 40 \quad (3)$$

The weak inequality sign ≤ is used in these constraints because total usage of the inputs can be less than or equal to the maximum amount of each input available.

As the firm cannot produce negative quantities of the two goods, we can also add the two non-negativity constraints on the solutions for the optimum values of A and B, i.e.

$$A \geq 0 \tag{4}$$

and

$$B \geq 0 \tag{5}$$

Now turn to the graph in Figure 6.1 which measures A and B on its axes. The first step in the graphical solution of a linear programming problem is to mark out what is known as the '**feasible area**'. This will contain all the values of A and B that satisfy all the previous constraints (1) to (5). This is done **by eliminating the areas which could not possibly contain the solution**.

We can easily see that the non-negativity constraints (4) and (5) mean that the solution must lie on, or above, the A axis and on, or to the right of, the B axis.

To mark out the other constraints we consider in turn what would happen if the firm entirely used up its quota of each of the inputs K, L and R.

If all the available K was used up, then in constraint (1) an equality sign would replace the ≤ sign and it would become the function

$$3A + 5B = 150 \tag{6}$$

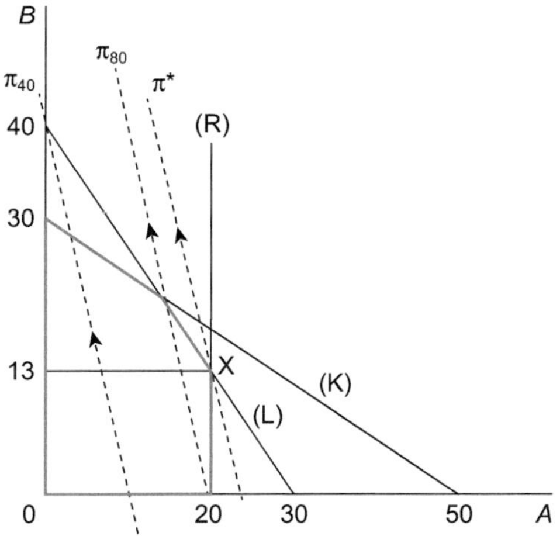

Figure 6.1 Linear programming: constrained maximization

This linear constraint can easily be marked out by joining its intercepts on the two axes. When $A = 0$ then $B = 30$ and when $B = 0$ then $A = 50$. Thus, the constraint will be the straight line marked (K). This is rather like a budget constraint. If all the available K is used then the firm's production mix will correspond to a point somewhere on the constraint line (K). It is also possible to use less than the total amount available, in which case the firm would produce a combination of A and B below this constraint. Points above this constraint are not feasible, though, as they correspond to more than 150 units of K.

In a similar fashion we can deduce that all points above the constraint line (L) are not feasible because when all the available L is used up then

$$4A+3B=120 \quad (7)$$

The constraint on R is shown by the vertical line (R) since when all available R is used up then

$$2A=40 \quad (8)$$

Points to the right of this line will not be feasible.

Having marked out the individual constraints, we can now delineate the area which contains combinations of A and B which satisfy all five constraints. This is shown by the blue lines in Figure 6.1.

We know that the firm's objective function is $\pi = 4A + B$. But as we do not yet know what the profit is, how can we draw in this function? To overcome this problem, first make up a value for profit, which when divided by the two per-unit profit values (£4 and £1) will give numbers within the range shown on the graph. For example, if we suppose profit is £40, then we can draw in the broken line π_{40} corresponding to the function

$$40=4A+B$$

If we had chosen a figure for profit of more than £40 then we would have obtained a line parallel to this one, but further away from the origin; e.g. the line π_{80} corresponds to the function $80 = 4A + B$.

If the firm is seeking to maximize profit then it needs to **find the furthest profit line from the origin that passes through or just touches the feasible area**. All profit lines will have the same slope and so, using π_{40} as a guideline, we can see that the highest feasible profit line is π^* which just touches the edge of the feasible area at X. The optimum values of A and B can then simply be read off the graph, giving $A = 20$ and $B = 13$ (approximately).

A more accurate answer may be obtained algebraically, once the graph has been used to determine the optimum point, since the solution to a linear programming problem will nearly always be at the intersection of two or more constraints. (Exceptionally, the objective function may be parallel to a constraint, see Example 6.3.)

The graph in Figure 6.1 tells us that the solution to this problem is where the constraints (L) and (R) intersect. Thus, we have the two simultaneous equations

$$4A + 3B = 120 \quad (7)$$
$$2A = 40 \quad (8)$$

which can easily be solved to find the optimum values of A and B.

$$\begin{aligned} &\text{From}(10) & A &= 20 \\ &\text{Substituting in}(9) & 4(20) + 3B &= 120 \\ & & 3B &= 40 \\ & & B &= 13.33 \quad (\text{to 2dp}) \end{aligned}$$

Thus, maximum profit is

$$\begin{aligned} \pi &= 4A + B = 4(20) + 13.33 \\ &= 80 + 13.33 = £93.33 \end{aligned}$$

The optimum combination X is on the constraints for L and R, but below the constraint for K. Therefore, as the K constraint does not 'bite', there must be some spare capacity, or what is often called 'slack', for K. When the firm produces 20 of A and 13.33 of B, then its usage of K is

$$\begin{aligned} 3A + 5B &= 3(20) + 5(13.33) \\ &= 60 + 66.67 = 126.67 \end{aligned}$$

The amount of K available is 150 units; therefore the slack is

$$150 - 126.67 = 23.33 \text{ units of K}$$

Now that the different steps involved in solving a linear programming problem have been explained let us work through another problem.

Example 6.2

A firm produces two goods A and B, which each contribute a net profit of £1 per unit sold. It uses two inputs K and L. The input requirements are:

3 units of K plus 2 units of L for each unit of A
2 units of K plus 3 units of L for each unit of B.

If the firm has 600 units of K and 600 units of L at its disposal, how much of A and B should it produce to maximize profit?

Solution

Using the same method as in the previous example we can see that the constraints are:

$$\begin{array}{lll} \text{for input K} & 3A+2B \leq 600 & (1) \\ \text{for input L} & 2A+3B \leq 600 & (2) \\ \text{non-negativity} & A \geq 0 \text{ and } B \geq 0 & \end{array}$$

The feasible area is therefore as marked out by the blue lines in Figure 6.2.

As profit is £1 per unit for both A and B, the objective function is

$$\pi = A + B$$

If we suppose profit is £200, then

$$200 = A + B$$

This function corresponds to the line π_{200} which can be used as a guideline for the slope of the objective function. The line parallel to π_{200} that is furthest away from the origin but still within the feasible area will represent the maximum profit. This is the line π^* through point M. The optimum values of A and B can thus be read off the graph as 120 of each.

Alternatively, once we know that the optimum combination of A and B is at the intersection of the constraints (K) and (L), the values of A and B can be found from the simultaneous equations

$$3A + 2B = 600 \quad (1)$$
$$2A + 3B = 600 \quad (2)$$

From (1) $\quad 2B = 600 - 3A$

$$B = 300 - 1.5A \quad (3)$$

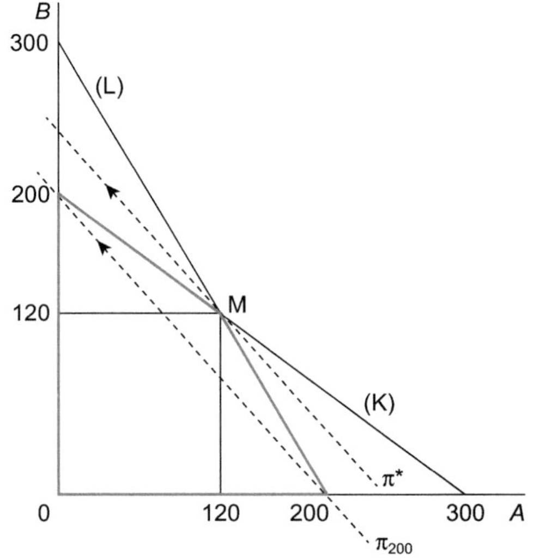

Figure 6.2 Linear programming: constrained profit maximization

Substituting (3) into (2)

$$\begin{aligned} 2A+3(300-1.5A)&=600 \\ 2A+900-4.5A&=600 \\ 300&=2.5A \\ 120&=A \end{aligned}$$

Substituting this value of A into (3)

$$B=300-1.5(120)=120$$

As both A and B equal 120 then maximum profit is

$$\pi^*=120+120=£240$$

The optimum combination at M is where both constraints (K) and (L) bite. There is therefore no slack for either K or L.

It is possible that the objective function will have the same slope as one of the constraints. In this case there will not be one optimum combination of the inputs as all points along the section of this constraint that forms part of the boundary of the feasible area will correspond to the same value of the objective function.

Example 6.3

A firm produces two goods x and y which require inputs of raw material (R), labour (L) and components (K) in the following quantities:

- 1 unit of x requires 12 kg of R, 10 hours of L and 15 units of K
- 1 unit of y requires 21 kg of R, 10 hours of L and 6 units of K

Both x and y add £200 per unit sold to the firm's profits. The firm can use up to a total of 252 kg of R, 150 hours of L and 180 units of K. What production mix of x and y will maximize profits?

Solution

The constraints can be written as

$$12x+21y\le 252 \qquad \text{(R)}$$
$$10x+10y\le 150 \qquad \text{(L)}$$
$$15x+6y\le 180 \qquad \text{(K)}$$
$$x\ge 0, \qquad y\ge 0$$

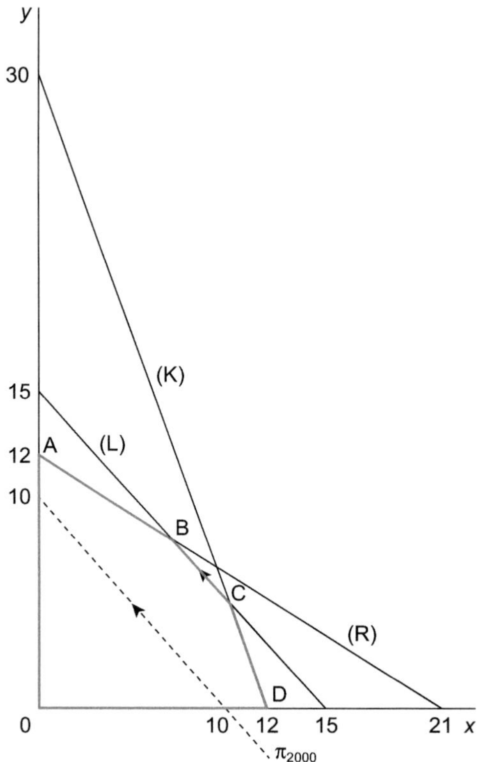

Figure 6.3 Linear programming: constrained profit maximization

These are shown in Figure 6.3 where the feasible area is marked out by the shape ABCD0. The objective function is

$$\pi = 200x + 200y$$

To find the slope of this objective function, assume profit is £2,000. This could be achieved by producing 10 of x and none of y, or 10 of y and no x, and is therefore shown by the broken line π_{2000}. This line is parallel to the constraint (L). Therefore, if we slide out the objective function π to find the maximum value of profit within the feasible area we can see that it coincides with the boundary of the feasible area along the stretch BC.

What this means is that both points B and C, and anywhere along the portion of the constraint line (L) between these points, will give the same (maximum) profit figure.

At B the constraints (R) and (L) intersect. Therefore, these two resources are used up completely and so

$$12x + 21y = 252 \quad (1)$$

$$10x + 10y = 150 \quad (2)$$

$$x = 15 - y \quad (3)$$

Substituting (3) into (1)

$$\begin{aligned} 12(15-y)+21y &= 252 \\ 180-12y+21y &= 252 \\ 9y &= 72 \\ y &= 8 \end{aligned}$$

Substituting this value of y into (3) $\quad x = 15 - 8 = 7$

Thus, profit at B is

$$\pi = 200x + 200y = 200(7) + 200(8) = £1,400 + £1,600 = £3,000$$

At C the constraints (L) and (K) intersect, giving the simultaneous equations

$$10x + 10y = 150 \qquad (2)$$
$$15x + 6y = 180 \qquad (4)$$

Using (3) again to substitute for x in (4),

$$\begin{aligned} 15(15-y)+6y &= 180 \\ 225-15y+6y &= 180 \\ 45 &= 9y \\ 5 &= y \end{aligned}$$

Substituting this value of y into (3)

$$x = 15 - 5 = 10$$

Thus, profit at C is

$$\pi = 200x + 200y = 200(10) + 200(5) = 2,000 + 1,000 = £3,000$$

which, as expected, is the same as the profit achieved at B. Profit will also be £3,000 for any other point on the line BC. This example therefore illustrates how a linear programming problem may not have a unique solution if the objective function has the same slope as one of the constraints that bounds the feasible area.

You should also note that the solution to a linear programming problem may be on one of the axes, where a non-negativity constraint operates. The following example illustrates such a case.

Example 6.4

A company uses inputs K and L to manufacture goods A and B. It has available 200 units of K and 180 units of L and the input requirements are

10 units of K plus 30 units of L for each unit of A,
25 units of K plus 15 units of L for each unit of B.

If the per-unit profit is £80 for A and £30 for B, what combination of A and B should it produce to maximize profit and how much of K and L will be used in doing this?

Solution

The resource constraints are

$$10A + 25B \leq 200 \qquad \text{(K)}$$

$$30A + 15B \leq 180 \qquad \text{(L)}$$

$$A \geq 0 \qquad B \geq 0$$

The corresponding feasible area ZXY0 is marked out in Figure 6.4.

The objective function is

$$\pi = 80A + 30B$$

To find the slope of the objective function, assume total profit is £240. This could be obtained by selling 8 of B or 3 of A, and so the broken line π_{240} in Figure 6.4 illustrates the combinations of A and B that would yield this level of profit. The maximum profit mix is obtained when a line parallel to π_{240} is drawn as far from the origin as possible but still within the feasible area. This will be line π^* through point Y.

Therefore, profit is maximized at Y, where no B is produced and 6 units of A are produced.

Maximum profit = 6 × £80 = £480.

In this example only the constraint (L) bites and so there will be slack in the (K) constraint. The total requirement of K to produce 6 units of A will be 60. There are 200 units of K available and so 140 remain unused. All 180 units of L are used up.

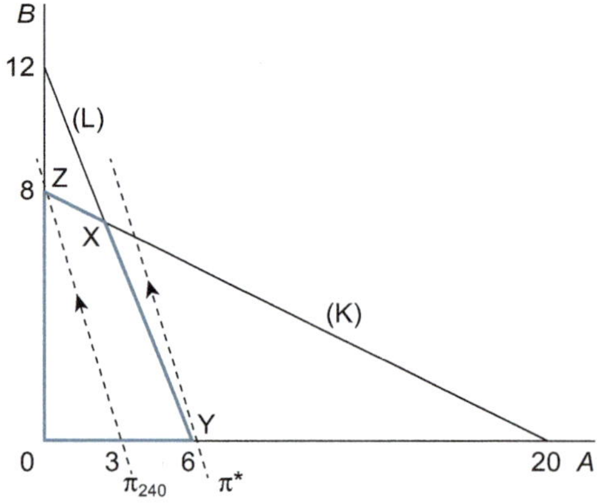

Figure 6.4 Linear programming: profit maximizing amounts of K and L

QUESTIONS 6.1

1. A firm manufactures products A and B using the two inputs X and Y in the following quantities:

   1 tonne of A requires 80 units of X plus 148 units of Y,
   1 tonne of B requires 200 units of X plus 120 units of Y.

   The profit per tonne of A is £20, and the per-tonne profit of B is £30. If the firm has at its disposal 1,600 units of X and 1,800 units of Y, what combination of A and B should it manufacture in order to maximize profit? (Fractions of a tonne may be produced.) Should the firm change its production mix if per-tonne profits alter to (a) £25 each for both A and B, or (b) £30 for A and £20 for B?
2. A firm produces the goods A and B using the four inputs W, X, Y and Z in the following quantities:

   1 unit of A requires 9 units of W, 30 of X, 20 of Y and 20 of Z,
   1 unit of B requires 13 units of W, 55 of X, 28 of Y and 20 of Z.

   The firm has available 468 units of W, 1,980 units of X, 1,120 units of Y and 800 units of Z. What production mix will maximize its total profit if each unit of A adds £60 to profit and each unit of B adds £75?
3. A firm sells two versions of a device for cutting and drilling. Version A is sold direct to the public in DIY stores, yielding a profit per unit of £50, and version B is sold to other firms for industrial use, yielding a per-unit profit of £20. Each day the firm is able to use 400 hours of labour, 750 kg of raw material and 240 metres of packaging material. These inputs are required to produce A and B in the following quantities: one version A device requires 20 hours of labour, 50 kg of raw material and 20 metres of packaging, whilst one of version B only requires 20 hours of labour plus 30 kg of raw material. How many of each version should be produced each day in order to maximize profit?
4. A firm uses three inputs X, Y and Z to manufacture two goods A and B. The requirements per tonne are as follows.

   For A: 5 loads of X, 4 containers of Y and 6 hours of Z,
   For B: 5 loads of X, 6 containers of Y and 2 hours of Z.

   Each tonne of A brings in £400 profit and each tonne of B brings in £300. What combination of A and B should the firm produce to maximize profit if it has at its disposal 150 loads of X, 240 containers of Y and 150 hours of Z?
5. A firm makes two food products A and B and the contribution to profit is £2 per unit of A and £3 per unit of B. There are three stages in the production process: cleaning, mixing and tinning. The number of hours of each process required for each product and the total number of hours available for each

process are given in Table 6.1. Given these constraints what combination of A and B should the firm produce to maximize profit?

6. A firm manufactures two compounds A and B using two raw materials R and Q, in addition to labour and a mixing additive. Input requirements per tonne are:

   For A: 1 container of R, 3 sacks of Q, 4 hours labour and 2 tins of mixing additive,

   For B: 2 containers of R, 5 sacks of Q and 3 hours labour, but no mixing additive.

   Both A and B add £200 per tonne to the firm's profits and it has at its disposal 60 containers of R, 150 sacks of Q, 120 hours of labour and 50 tins of mixing additive.

   What combination of A and B should it produce to maximize profits, assuming that fractions of a tonne can be manufactured? What will these profits be? What surplus amounts of the inputs will there be?

7. A firm manufactures two products A and B which sell for respectively £900 and £2,000 each. It uses the four processes cutting, drilling, finishing and assembly and the requirements per unit of output are:

   For A: 5 hours cutting, 18 hours drilling, 9 hours finishing and 10 hours assembly,

   For B: 15 hours cutting, 7 hours drilling, 15 hours finishing and 10 hours assembly.

   How can this firm maximize its weekly sales revenue if the capacity of its factory is limited to 390 hours cutting, 630 hours drilling, 450 hours finishing and 400 hours assembly per week?

8. If a firm is faced with the constraints described in Question 2, in Questions 5.4, what combination of A and B will maximize profit if A contributes £30 per unit to profit and B contributes £10?

9. Show that more than one solution exists if one tries to maximize the objective function

   $$\pi = 4A + 4B$$

   subject to the constraints

   $$20A + 20B \leq 60$$
   $$20A + 80B \leq 120$$
   $$A \geq 0 \text{ and } \quad B \geq 0$$

10. A firm has £120,000 to invest. It can buy shares in company X which cost £2 each and give an expected annual return of 6%, or shares in company Y which cost £4 each and give an expected annual return of 8%. It is advised

not to put more than 60% of its total investments into any one type of share. What investment portfolio will maximize the expected return? (You may answer this question with or without a diagram.)

Table 6.1 Hours required at stages of production

	Hours of		
	Cleaning	Mixing	Tinning
1 unit of A requires	3	6	2
1 unit of B requires	6	2	1.5
Total hours available	210	120	60

6.2 CONSTRAINED MINIMIZATION

Another problem a firm might be faced with is how to minimize the cost of producing a good subject to constraints regarding its quality. If the objective function and the constraints are all linear functions, then the method used for constrained minimization is analogous to that used in the maximization problems. The main differences in constrained minimization problems are that:

- the feasible area is usually **above** the constraint lines
- one needs to find the objective function line that is **nearest** to the origin within the feasible area.

The following examples show how this method operates.

Example 6.5

A firm manufactures a medicinal product containing three ingredients X, Y and Z. Each unit produced must contain at least 100 g of X, 30 g of Y and 75 g of Z. The product is made by mixing the inputs A and B which contain different mixtures of ingredients X, Y and Z, and come in containers costing respectively £3 and £6 each.

These inputs A and B contain X, Y and Z in the following quantities:

1 container of A contains 50 g of X, 10 g of Y and 15 g of Z
1 container of B contains 20 g of X, 10 g of Y and 50 g of Z

What mix of A and B will minimize the cost per unit of the product subject to the above quality constraints? (It does not matter if these minimum requirements are exceeded and all other production costs can be ignored.)

Solution

Total usage of X will be 50 g for each container of A plus 20 g for each container of B. Total usage must be at least 100 g. This quality constraint for X can be written as

$$50A+20B\geq 100$$

Note that this constraint has the greater than or equal to weak inequality sign ≥ instead of the ≤ weak inequality sign used in the maximization problems in the previous section. The quality constraints on Y and Z can also be written as

$$10A+10B\geq 30$$
$$15A+50B\geq 75$$

As negative amounts of the inputs A and B are not feasible there are also the two non-negativity constraints

$$A\geq 0 \text{ and } \quad B\geq 0$$

If the quality constraint for X is only just met then

$$50A+20B=100 \tag{X}$$

The line representing this function is drawn as (X) in Figure 6.5. Any combination of A and B above this line will more than satisfy the quality constraint for X. Any combination of A and B below this line will not satisfy this constraint and will therefore not be feasible.

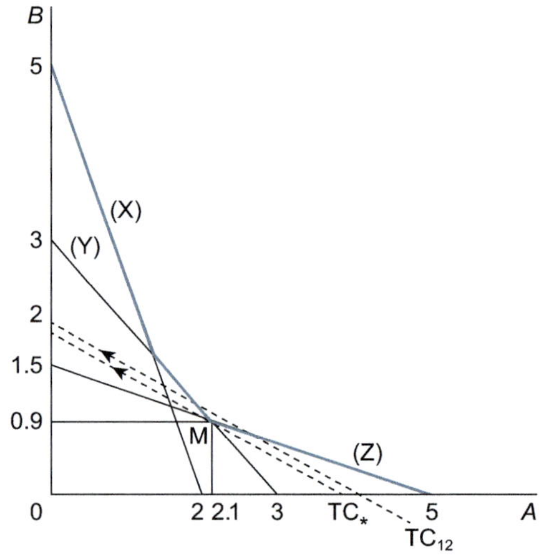

Figure 6.5 Linear programming: constrained minimization

In a similar fashion the constraints for Y and Z are shown by the lines representing the functions

$$10A+10B=30 \quad \text{(Y)}$$

and

$$15A+50B=75 \quad \text{(Z)}$$

Taking all the constraints into account, the feasible area is marked out by the blue lines in Figure 6.5, or at least its lower bounds are. As these are minimum constraints, then theoretically there are no upper limits to the amounts of A and B that could be used to make a unit of the final product.

The objective function is total cost (TC) per unit which the firm is seeking to minimize. Given the prices of A and B of £3 and £6 respectively, then

$$\text{TC}=3A+6B$$

To obtain a guideline for the slope of the TC function, assume any value for TC that is easily divisible by the two prices of £3 and £6. For example, if TC is assumed to be £12 then the line TC_{12} can be drawn representing the function

$$12=3A+6B$$

This line has a slope of −0.5. One now needs to ask the question 'can a line with this slope be drawn closer to the origin (thus representing a smaller value for TC) but still going through the feasible area?' In this case, the answer is 'yes'. The line TC* through M represents the lowest cost method of combining A and B that still satisfies the three quality constraints. The optimum amounts of A and B can now be read off the graph at M as approximately 2.1 and 0.9 respectively.

More accurate answers can be obtained algebraically. The optimum combination M is where the quality constraints for Y and Z intersect. These correspond to the linear equations

$$10A+10B=30 \quad (1)$$
$$15A+50B=75 \quad (2)$$

Dividing (2) by 5 we get

$$3A+10B=15$$

Subtracting (1)

$$
\begin{aligned}
10A+10B&=30\\
-7A&=-15\\
A&=\frac{15}{7}=2\frac{1}{7}
\end{aligned}
$$

Substituting this value for A into (1)

$$10\left(\frac{15}{7}\right)+10B=\frac{150}{7}+10B=30 \qquad (3)$$

Multiplying (3) by 7

$$\begin{aligned}150+70B&=210\\ 70B&=60\\ B&=\frac{6}{7}\end{aligned}$$

Thus, the firm should use $2\frac{1}{7}$ containers of A plus $\frac{6}{7}$ of a container of B for every unit of the final product it makes. As long as large quantities of the product are made, the firm does not have to worry about unused fractions of containers. It just needs to use containers A and B in the ratio $2\frac{1}{7}$ to $\frac{6}{7}$ which is the same as the ratio 2.5 to 1.

The constraint on X does not bite and so there is some slack. In a minimization problem, slack means overabundance. The total amount of X contained in a unit of the final product will be

$$50A+20B=50\left(\frac{15}{7}\right)+20\left(\frac{6}{7}\right)=\frac{50\times 15}{7}+\frac{20\times 6}{7}=\frac{750+120}{7}=\frac{970}{7}=138.57\text{ g}$$

This exceeds the minimum requirement of 100 g of X by 38.57 g.

Example 6.6

A firm makes a product that has minimum input requirements for the four ingredients W, X, Y and Z. These cannot be manufactured individually and can only be supplied as part of the composite inputs A and B.

1 litre of A includes 20 g of W, 5 g of X, 5 g of Y and 20 g of Z
1 litre of B includes 90 g of W, 7 g of X and 4 g of Y but no Z

One drum of the final product must contain at least 7,200 g of W, 1,400 g of X, 1,000 g of Y and 1,200 g of Z. (The volume of the drum is fixed and not related to the volume of inputs A and B as evaporation occurs during the production process.) If a litre of A costs £9 and a litre of B costs £16 how many litres of A and B should the firm use to minimize the cost of a drum of the final product? Assume that all other costs can be ignored.

Solution

The minimum input requirements can be written as

$$20A+90B\geq 7{,}200 \qquad \text{(W)}$$
$$5A+7B\geq 1{,}400 \qquad \text{(X)}$$

$$5A + 4B \geq 1{,}000 \qquad \text{(Y)}$$
$$20A \geq 1{,}200 \qquad \text{(Z)}$$

plus the non-negativity conditions $A \geq 0$ and $B \geq 0$. These constraints are shown in Figure 6.6.

If only the minimum 7,200 g of W is included in the final product, then

$$20A + 90B = 7{,}200$$

If no B was used then 7,200/20 = 360 litres of A is needed to satisfy this constraint.

If no A was used then 7,200/90 = 80 litres of B would be needed.

Thus, the values where the linear constraint (W) hits the A and B axes are 360 and 80 respectively. Combinations of A and B below this line do not satisfy the minimum amount of W requirement. The other constraints, for X, Y and Z, are constructed in a similar fashion and the feasible area is marked out by the blue lines in Figure 6.6.

To find a guideline for the slope of the objective function, we need to suggest a suitable cost figure that is divisible by both 9 and 16, so assume that the total cost (TC) of A and B is £1,440, giving the budget constraint

$$1{,}440 = 9A + 16B$$

This particular budget constraint is shown by the broken line TC_{1440} and does not go through the feasible area. Therefore, total cost must be greater than £1,440. An increased budget will mean a budget line further from the origin but still with the same slope as TC_{1440}. The budget line with this slope that is closest to the origin and that also passes through the feasible area is TC*.

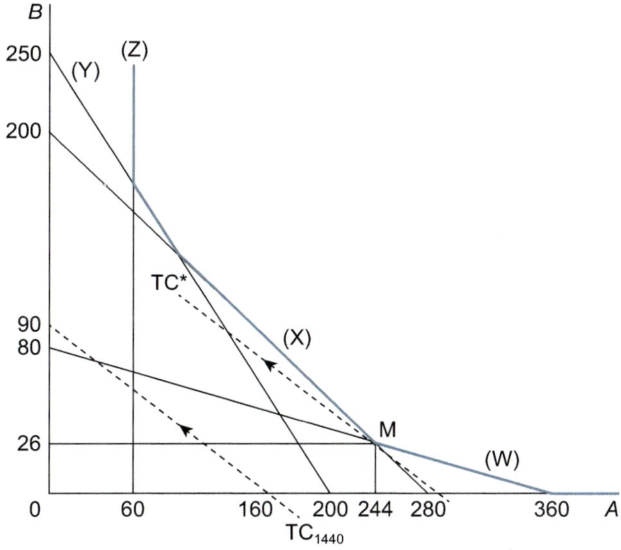

Figure 6.6 Linear programming: constrained cost minimization

Minimum TC is therefore achieved by using the combination of A and B corresponding to point M. Approximate values read off the graph at M are 244 litres of A and 26 litres of B.

More accurate answers can be obtained algebraically as we know that M is at the intersection of the constraints for W and X. This means that the minimum requirements for W and X are only just met and so

$$20A + 90B = 7{,}200 \quad \text{for W} \qquad (1)$$
$$5A + 7B = 1{,}400 \quad \text{for X} \qquad (2)$$

Multiplying (2) by 4 gives

$$20A + 28B = 5{,}600$$
$$20A + 90B = 7{,}200$$
$$\text{Subtracting (1)} \qquad -62B = -1{,}600$$
$$B = \frac{1{,}600}{62} = 25.8 \qquad \text{(to 1 dp)}$$

Substituting this value for B into (1) gives

$$20A + 90(25.8) = 7{,}200$$
$$20A + 2{,}322 = 7{,}200$$
$$A = \frac{4{,}878}{20} = 243.9$$

Therefore, the firm should use 243.9 litres of A and 25.8 litres of B for each drum of the final product.

The total input cost will be

$$243.9 \times £9 + 25.8 \times £16 = £2{,}195.10 + £412.80 = £2{,}607.90$$

QUESTIONS 6.2

1. Find the minimum value of the function $C = 40A + 20B$ subject to the constraints

$$10A + 40B \geq 40$$
$$30A + 20B \geq 60$$
$$10A \geq 10$$
$$A \geq 0, B \geq 0$$

   Will there be slack in any of the constraints at the optimum combination of A and B? If so, what is this excess capacity?

2. A firm manufactures a product that, per litre, must contain at least 18 g of chemical X and 10 g of chemical Y. The rest of the product is water whose

costs can be ignored. The two inputs A and B contain X and Y in the following quantities:

1 unit of A contains 6 g of X and 5 g of Y,
1 unit of B contains 9 g of X and 2 g of Y.

The per-unit costs of A and B are £2 and £6 respectively. What combination of A and B will give the cheapest way of producing a litre of the final product?

3. A firm mixes the two inputs Q and R to make a vitamin supplement in liquid form. The inputs Q and R contain the four vitamins A, B, C and D in the following amounts:

   6 mg of A, 50 mg of B, 35 mg of C and 12 mg of D per unit of Q,
   30 mg of A, 25 mg of B, 30 mg of C and 20 mg of D per unit of R.

   The inputs Q and R cost respectively 5p and 12p per unit. Each centilitre of the final product must contain at least 60 mg of A, 100 mg of B, 105 mg of C and 60 mg of D. What is the cheapest way of making the final product? Which vitamins will exceed the minimum requirements per centilitre using this method?

4. A delivery firm has two types of van, A and B, and carries three types of load, X, Y and Z. Each van is capable of carrying a mixed load, but only in certain proportions, given the special size and weight of the different loads. When fully loaded,

   type A can carry 20 of X, 15 of Y and 15 of Z,
   type B can carry 10 of X, 60 of Y and 15 of Z.

   A typical daily delivery schedule requires the firm to carry 200 loads of X, 450 loads of Y and 225 loads of Z. Each van is only loaded for deliveries once a day. The smaller van, A, costs £50 a day to run and the larger van, B, costs £100 a day. How many of each type of van should the firm use to minimize total running costs? Would there be space in the vans for any more of any of the loads X, Y or Z should more orders be placed?

5. A firm uses two inputs R and T which cost £40 each per tonne. They both contain the chemical compounds G and H in the following quantities:

   1 tonne of R contains 6 kg of G and 3 kg of H,
   1 tonne of T contains 15 kg of G and 4 kg of H.

   The final product must contain at least 180 kg of G and 60 kg of H per batch. How many tonnes of R and T should the firm use to minimize the cost of a batch of the final product? Will the amount of G or H it contains exceed the minimum requirement?

6. An aircraft manufacturer fitting out the interior of a plane can use two fitments A and B, which contain components X, Y and Z in the following quantities:

   1 unit of A contains 3 units of X, 4 units of Y plus 2 units of Z,
   1 unit of B contains 6 units of X, 5 units of Y plus 8 units of Z.

   The aircraft design is such that there must be at least 540 units of X, 600 units of Y and 480 units of Z in total in the plane. If each unit of A weighs 4 kg and each unit of B weighs 6 kg, what combination of A and B will minimize the total weight of these fitments in the plane?

6.3 MIXED CONSTRAINTS

Some linear programming problems may contain both 'less than or equal to' and 'greater than or equal to' constraints. It is also possible to have equality constraints, i.e. where one variable must exactly equal a specified quantity. For example, a medicinal product may have to contain a specific amount of some ingredient.

Example 6.7

Minimize the objective function $C = 12A + 8B$ subject to the constraints

$$10A + 40B \geq 40 \quad (1)$$
$$12A + 16B \leq 48 \quad (2)$$
$$A = 1.5 \quad (3)$$

Solution

The constraints are marked out in Figure 6.7. Constraint (1) means that the feasible area must be above the line

$$10A + 40B = 40$$

Constraint (2) means that the feasible area must be below the line

$$12A + 16B = 48$$

Constraint (3) means that the feasible area must be along the vertical line through $A = 1.5$. The only section of the graph that satisfies all three of these constraints is the blue section LM of the vertical line through $A = 1.5$.

If C is assumed to be 24 then the line C_{24} representing the function

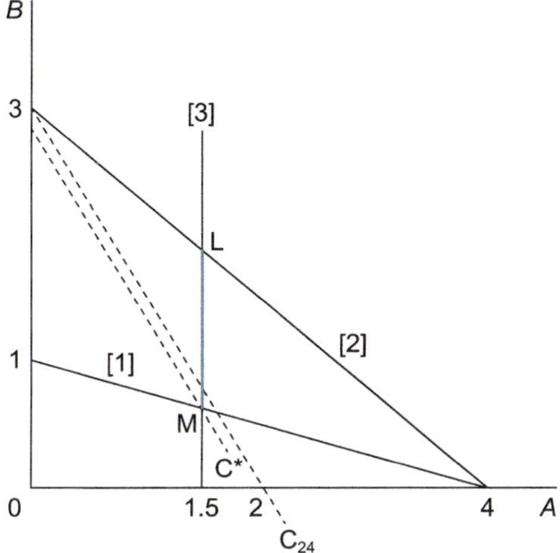

Figure 6.7 Linear programming: mixed constraints

$$24 = 12A + 8B$$

can be drawn in and has a slope of −1.5. To minimize *C*, one needs to find the closest line to the origin that has this slope and also passes through the feasible area. This will be the line *C** through M.

The optimum value of A is therefore obviously 1.5.

The optimum value of B occurs at the intersection of the two lines

$$10A + 40B = 40$$

$$\text{Thus,}\quad 10(1.5) + 40B = 40$$

$$15 + 40B = 40$$

$$40B = 25$$

$$B = 0.625$$

QUESTIONS 6.3

1. A firm makes two goods A and B using the three inputs X, Y and Z in the following quantities:

   20 units of X, 8 units of Y and 20 units of Z per unit of A,
   20 units of X, 20 units of Y and 14 units of Z per unit of B.

   The per-unit profit of A is £1,500, and for B the figure is £1,000. Input availability is restricted to 60 units of X, 40 units of Y and 70 units of Z. The firm has already committed itself to a contract to supply one customer with 1 unit of B. What combination of A and B should it produce to maximize total profit?

2. A company produces two industrial compounds X and Y that are mixed in a final product. They both contain one common input, R. The amount of R in 1 tonne of X is 8 litres and the amount of R in 1 tonne of Y is 12 litres. A load of the final product must contain at least 240 litres of R to ensure that its quality level is met. No R is lost in the production process of combining X and Y.

   The total cost of a tonne of X is £30 and the total cost of a tonne of Y is £15. If the firm has already signed a contract to buy 7.5 tonnes of X per week, what mix of X and Y should the firm use to minimize the cost of a load of the final product?
3. A firm manufactures two goods A and B which require the two inputs K and L in the following amounts:

   1 unit of A requires 6 units of K and 4 of L,
   1 unit of B requires 8 units of K and 10 of L.

   The firm has at its disposal 96 units of K and 100 of L. The per-unit profit of A is £600 and for B the figure is £300. The firm is under contract to produce a minimum of 6 units of B. How many units of A should it make to maximize profit?

6.4 MORE THAN TWO VARIABLES

When the objective function in a linear programming problem contains more than two variables then it cannot be solved by graphical analysis. An advanced mathematical technique known as the *simplex method* can be used for these problems. This is based on the principle that the optimum value of the objective function will usually be at the intersection of two or more constraints.

It is an iterative method that can be very time consuming to use manually and for most practical purposes it is best to use a computer program package to do the necessary calculations. If you have access to a linear programming computer package then you may try to use it now that you understand the basic principles of linear programming. The way that data are entered will depend on the computer package you use and you will need to consult the relevant instructions.

7 Quadratic equations

Learning objectives

After completing this chapter students should be able to:

- Use factorization to solve quadratic equations with one unknown variable
- Use the quadratic equation solution formula
- Identify quadratic equations that cannot be solved
- Set up and solve economic problems that involve quadratic functions
- Construct a spreadsheet to plot quadratic and higher order polynomial functions.

7.1 SOLVING QUADRATIC EQUATIONS

A quadratic equation is one that can be written in the form

$$ax^2 + bx + c = 0$$

where x is an unknown variable and a, b and c are constant parameters with $a \neq 0$. For example,

$$6x^2 + 2.5x + 7 = 0$$

A quadratic equation that includes terms in both x and x^2 cannot be rearranged to get a single term in x, so we cannot use the method used to solve linear equations.

There are three possible methods one might try to use to solve for the unknown in a quadratic equation:

(i) by plotting a graph
(ii) by factorization
(iii) using the quadratic formula.

DOI: 10.4324/9781003360827-7

In the next three sections we shall see how each can be used to tackle the following question. Let us start from setting up an economic example.

If a monopoly faces the linear demand schedule

$$p = 85 - 2q \tag{1}$$

at what output will its total revenue be 200?

It is not immediately obvious that this question involves a quadratic equation. We first need to use economic analysis to set up the mathematical problem to be solved. By definition we know that total revenue will be

$$\text{TR} = pq \tag{2}$$

So, substituting the function for p from (1) into (2), we get

$$\text{TR} = (85 - 2q)q = 85q - 2q^2$$

This is a quadratic function that tells us the value of TR for any given output. What the question asks is 'at what value of q will this function be equal to 200'? The mathematical problem is therefore to solve the following quadratic equation

$$200 = 85q - 2q^2 \tag{3}$$

All three solution methods require all terms to be brought together on one side of the equality sign, leaving a zero on the other side. It is also necessary to put the terms in the order given in the previous definition of a quadratic equation, i.e.

unknown squared (q^2), unknown (q), constant.

Thus, (3) can be rewritten as

$$2q^2 - 85q + 200 = 0$$

We will use this quadratic equation to demonstrate each of the three methods in the following sections.

Before we run through these methods, however, you should note that an equation involving terms in x^2 and a constant, but not x, can usually be solved by a simpler method. For example, suppose that

$$5x^2 - 80 = 0$$

then

$$5x^2 = 80$$

$$x^2 = 16$$

$$x = 4 \quad \text{or} \quad x = -4$$

(Note that both $x = 4$ and $x = -4$ will give 16 when squared.)

7.2 GRAPHICAL SOLUTION

Drawing a graph of a quadratic function can be a long-winded and not very accurate process that involves separately plotting each individual value of the variable within the range that is being considered. It is therefore usually not a very practical method for solving a quadratic equation. The graphical method can be useful, however, not so much for finding an approximate value for the solution, but for explaining why certain quadratic equations do not have a solution whilst others have two solutions. Only a rough sketch diagram is necessary for this purpose.

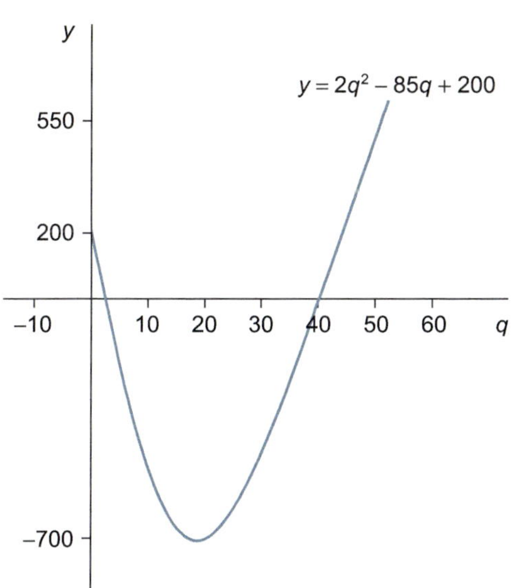

Figure 7.1 Quadratic function with two horizontal intercepts

Example 7.1

Show graphically that a solution does exist for the quadratic equation

$$2q^2 - 85q + 200 = 0$$

Solution

We first need to define a function

$$y = 2q^2 - 85q + 200$$

If the graph of this function cuts the q axis then $y = 0$, and we have a solution to the quadratic equation specified in the question. Next, we calculate a few values of the function to get an approximate idea of its shape.

When $q = 0$, then $y = 200$
When $q = 1$, then $y = 2 - 85 + 200 = 117$

and so the graph initially falls.

When $q = 3$, then $y = 18 - 255 + 200 = -37$

and so it must cut the q axis as y has gone from a positive to a negative value.

When $q = 50$, then $y = 5{,}000 - 4{,}250 + 200 = 950$

and so the value of y rises again and must cut the q axis a second time.

These values indicate that the graph is a U-shape, as shown in Figure 7.1. This cuts the horizontal axis twice and so there are two values of q for which y is zero, which means that there are two solutions to the question. It is difficult to read values off a graph like this with 100% accuracy, but precise values of these solutions, 2.5 and 40, can be found by the other two methods explained in the following sections or by computation of y for different values of q. (See spreadsheet solution method in the 'Plotting quadratic functions with a spreadsheet' section.)

If we slightly change the problem in Example 7.1 we can see why there may not always be a solution to a quadratic equation.

Example 7.2

Find out if there is an output level at which total revenue is 1,500 for the function

$$\text{TR} = 85q - 2q^2$$

Solution

The quadratic equation to be solved is

$$1{,}500 = 85q - 2q^2$$

which can be rewritten as

$$2q^2 - 85 + 1{,}500 = 0$$

If we now specify the function

$$y = 2q^2 - 85 + 1{,}500$$

and calculate a few values, we can see that it falls and then rises again but never cuts the q axis, as Figure 7.2 shows.

When $q = 0$, then $y = 1{,}500$
When $q = 10$, then $y = 850$
When $q = 20$, then $y = 600$
When $q = 25$, then $y = 625$

There are therefore no solutions to this quadratic equation, i.e. there is no output at which total revenue will be 1,500.

Although one would never try to plot the whole graph of a quadratic function manually, one may of course get a computer plot.

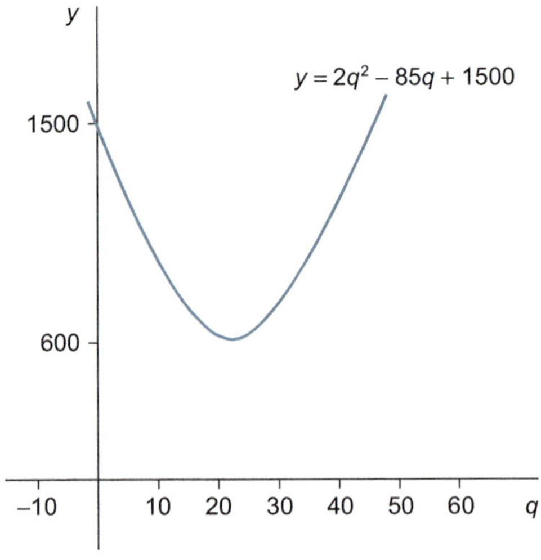

Figure 7.2 Quadratic function with no horizontal intercept

Plotting quadratic functions with a spreadsheet

An Excel spreadsheet for calculating different values of the function y in Example 7.1 can be constructed by following the instructions in Table 7.1. Rather than building in formulae that are specific to this example, this spreadsheet is constructed in a format that can be used to plot any function in the form $y = aq^2 + bq + c$ once the parameters a, b and c are entered in the relevant cells. The range for q has been chosen to ensure that it includes the values when y is zero, which is what we are interested in finding.

If you construct this spreadsheet, you should get the series of values shown in Table 7.2. (Note that this has been amended from the format specified earlier in order to get three columns of values to fit on one page.) The q values which correspond to a y value of zero can now be read off, giving the solutions 2.5 and 40.

You may use this spreadsheet to plot a graph of the function $y = 2q^2 - 85q + 200$. Assuming that you have q and y in single columns, then you just use the Chart Options commands to obtain a plot with q measured on the X axis and y as variable A on the vertical axis. (If you do not know how to use Chart Options, refer back to Example 4.17.) To make the chart clearer to read, enlarge it a bit by dragging the corner. The legend box for y can also be cut out to allow the chart area to be enlarged. This should give you a plot similar to Figure 7.3, which clearly shows how this function cuts the horizontal axis twice.

Table 7.1 Setting up a spreadsheet for Example 7.2

CELL	Enter	Explanation
A1	Ex.7.2	Label to remind you what example this is
B1	QUADRATIC SOLUTION TO y = aq^2+bq+c = 0	Title of spreadsheet (Note that this label is not an actual Excel formula.)
B2	a =	These are labels that tell you that the actual parameter values will go in the cells next to them. Right justify these labels.
D2	b =	
F2	c =	
C2	2	These are the actual parameter values for this example.
E2	–85	
G2	200	
A4	q	Column heading label
B4	y	Column heading label
A5	0	Initial value for q
A6	=A5+0.5	Calculates a 0.5 unit increment in q
A7 to A90	*Copy formula from cell* A6 *down column* A	Calculates a series of values of q in 0.5 unit increments
B5	=$C$2*A5^2+$E$2*A5+$G$2	This formula calculates the value of the function corresponding to the value of q in cell A5 and the parameter values in cells C2, E2 and G2. Note that the $ sign is used so that these cell references do not change when this function is copied down the y column.
B6 to B92	*Copy formula from cell* B5 *down column* B	Calculates values for y in each row corresponding to values of q in column A.

Table 7.2 Excel output based on Table 7.1

	A	B	C	D	E	F	G	H
1	Ex 7.2	QUADRATIC SOLUTION TO y = aq^2+bq+c = 0						
2		a =	2	b =	-85	c =	200	
3	q	y	q	y	q	y	q	y
4	0	200	11	−493	22	−702	33	−427
5	0.5	158	11.5	−513	22.5	−700	33.5	−403
6	1	117	12	−532	23	−697	34	−378
7	1.5	77	12.5	−550	23.5	−693	34.5	−352
8	2	38	13	−567	24	−688	35	−325
9	**2.5**	0	13.5	−583	24.5	−682	35.5	−297
10	3	−37	14	−598	25	−675	36	−268
11	3.5	−73	14.5	−612	25.5	−667	36.5	−238
12	4	−108	15	−625	26	−658	37	−207
13	4.5	−142	15.5	−637	26.5	−648	37.5	−175
14	5	−175	16	−648	27	−637	38	−142
15	5.5	−207	16.5	−658	27.5	−625	38.5	−108
16	6	−238	17	−667	28	−612	39	−73
17	6.5	−268	17.5	−675	28.5	−598	39.5	−37
18	7	−297	18	−682	29	−583	40	0
19	7.5	−325	18.5	−688	29.5	−567	40.5	38
20	8	−352	19	−693	30	−550	41	77
21	8.5	−378	19.5	−697	30.5	−532	41.5	117
22	9	−403	20	−700	31	−513	42	158
23	9.5	−427	20.5	−702	31.5	−493	42.5	200
24	10	−450	21	−703	32	−472	43	243
25	10.5	−472	21.5	−703	32.5	−450	43.5	287

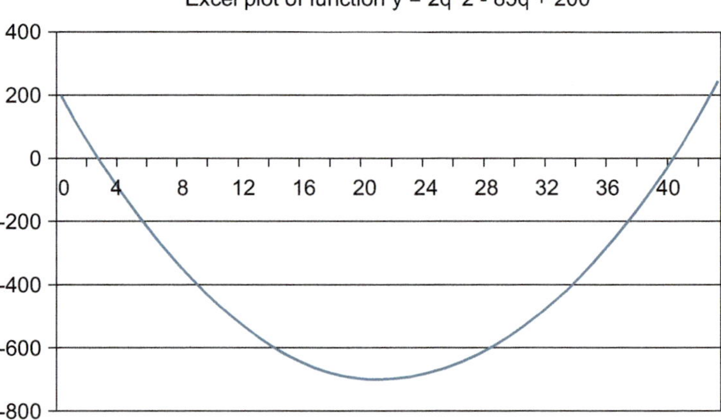

Figure 7.3 Plot of a quadratic function using a spreadsheet

This spreadsheet can easily be amended to calculate values and plot graphs of other quadratic functions by entering different values for the parameters a, b and c in cells C2, E2 and G2. For example, to calculate values for the function from Example 7.2

$$y = 2q^2 - 85q + 1{,}500$$

the value in cell G2 should be changed to 1,500. A computer plot of this function should produce the shape shown in Figure 7.2, confirming again that this function will not cut the horizontal axis and that there is no solution to the quadratic equation

$$0 = 2q^2 - 85q + 1{,}500$$

7.3 FACTORIZATION

In Chapter 3 factorization was explained, i.e. how some expressions can be broken down into terms which when multiplied together give the original expression. For example,

$$a^2 - 2ab + b^2 = (a - b)(a - b)$$

If a quadratic function which has been rearranged to equal zero can be factorized in this way, then one or the other of the two factors must equal zero when the quadratic equation is set to zero.

(Remember that if $A \times B = 0$ then either A or B, or both, must be zero.)

Example 7.3

Solve by factorization the quadratic equation

$$2q^2 - 85q + 200 = 0$$

Solution

This expression can be factorized as

$$(2q - 5)(q - 40) = 2q^2 - 85q + 200$$

Therefore, $\quad (2q - 5)(q - 40) = 0$

This will hold when the expressions in either of the brackets equal zero, i.e.

$$2q - 5 = 0 \quad \text{or} \quad q - 40 = 0$$

giving solutions $\quad q = 2.5 \quad \text{or} \quad q = 40$

As expected, these are the same solutions as those found by the graphical method.

It may be the case that mathematically a quadratic equation has one or more solutions with a negative value that will not apply in an economic problem. One cannot have a negative output, for example.

Example 7.4

Solve by factorization the quadratic equation

$$2x^2 - 6x - 20 = 0$$

Solution

By factorization $(2x - 10)(x + 2) = 0$

Therefore, $2x - 10 = 0$ or $x + 2 = 0$

$x = 5$ or $x = -2$

If x represented output or another variable that cannot be measured in negative quantities, then $x = 5$ would be the only answer we would use.

If a quadratic equation cannot be factorized, then the formula method in Section 7.5 must be used. However, the formula method can also be used when an equation can be factorized. Therefore, if you cannot quickly see a way of factorizing, then you should use the formula method. Factorization is only useful as a short-cut way of solving certain quadratic equations. It defeats the objective of the exercise if you spend half an hour trying to find a way of factorizing an expression when it would be quicker to use the formula.

It should also go without saying that quadratic equations for which no solutions exist cannot be factorized. For example, it is not possible to factorize the equation

$$2q^2 - 85q + 1{,}500 = 0$$

which we have already shown to have no solution.

QUESTIONS 7.1

1. Solve for x in the equation $x^2 - 5x + 6 = 0$.
2. Find the output at which total revenue is £600 if a firm's demand schedule is

   $$p = 70 - q$$

3. Find when average cost will be 40 if a firm faces the average cost function

   $$\text{AC} = 40q^{-1} + 10q$$

4. Is there a positive solution for x when

   $$0 = 12x^2 + 90x - 48?$$

5. At what output level will TC = £150 if firm faces the total cost schedule

   $$\text{TC} = 6 - 2q + 2q^2? \quad (\text{assume } q > 2 \text{ so TC} > 0)$$

7.4 THE QUADRATIC FORMULA

Any quadratic equation expressed in the form

$$ax^2 + bx + c = 0$$

where a, b and c are given parameters and for which a solution exists can be solved for x by using the quadratic formula

$$x = \frac{-b \pm \sqrt{b^2 - 4ac}}{2a}$$

(The sign ± means + or −.) The proof for this formula is rather complex, but there is no real need for you to understand how the formula is derived. You just need to know that it works.

Example 7.5

Use the quadratic formula to solve the quadratic equation

$$2q^2 - 85q + 200 = 0$$

Solution

In the quadratic formula applied to this example $a = 2$, $b = -85$ and $c = 200$ and, of course, $x = q$. Note that the minus signs for any negative coefficients must be included. One also needs to take special care to remember to use the rules for arithmetic operations using negative numbers. Substituting these values for a, b and c into the formula we get

$$q = \frac{-(-85) \pm \sqrt{(-85)^2 - 4 \times 2 \times 200}}{2 \times 2}$$

$$= \frac{85 \pm \sqrt{7{,}225 - 1{,}600}}{4} = \frac{85 \pm \sqrt{5{,}625}}{4}$$

$$= \frac{85 \pm 75}{4} = \frac{160}{4} \text{ or } \frac{10}{4} = 40 \text{ or } 2.5$$

These are, of course, the same as the solutions found by factorization in Example 7.3.

What happens if you try to use the quadratic formula when no solution exists? We can find out by applying the formula to the quadratic equation in Example 7.2, where a sketch graph showed that there was no solution.

Example 7.6

Use the quadratic formula to try to solve

$$2q^2 - 85q + 1{,}500 = 0$$

Solution

In this example $a = 2$, $b = -85$ and $c = 1{,}500$. Therefore,

$$q = \frac{-(-85) \pm \sqrt{(-85)^2 - 4 \times 2 \times 1{,}500}}{2 \times 2}$$

$$= \frac{85 \pm \sqrt{7{,}225 - 12{,}000}}{4} = \frac{85 \pm \sqrt{-4{,}775}}{4}$$

We are now stuck! We cannot find the square root of a negative number. In other words, no solution exists. It will always be the case that the quadratic formula will require the square root of a negative number if no solution exists.

QUESTIONS 7.2

(Use the quadratic formula to solve these problems.)

1. Solve for x if $0 = x^2 + 2.5x - 125$.
2. A firm faces the demand schedule $q = 400 - 2p - p^2$. What price does it need to charge to sell 100 units?
3. If a firm's demand function is $q = 100 - p$, what quantities need to be sold to bring in a total revenue of
   (a) £100 (b) £1,000 (c) £10,000?
   (Give answers to 2 decimal places, where they exist.)
4. At what output levels will average cost be 50 if a firm faces the average cost function $\text{AC} = 100q^{-1} + 4q$?

7.5 QUADRATIC SIMULTANEOUS EQUATIONS

If one or more equations in a simultaneous equation system are quadratic, then it may be possible to eliminate all but one unknown and to reduce the problem to a single quadratic equation. If this can be solved, then the other unknowns can be found by substitution.

Example 7.7

Find the equilibrium values of p and q in a competitive market with the following functions:

Demand: $q = 200p^{-1}$

Supply: $q = -15 + 0.5p$

Solution

In equilibrium, demanded quantity equals quantity supplied. Therefore,

$$200p^{-1} = -15 + 0.5p$$

Multiplying through by p,

$$\begin{aligned} 200 &= -15p + 0.5p^2 \\ 0 &= 0.5p^2 - 15p - 200 \\ 0 &= p^2 30p - 400 \\ 0 &= (p-40)(p+10) \end{aligned}$$

Therefore, $p - 40 = 0$ or $p + 10 = 0$

$p = 40$ or $p = -10$

We can ignore the second solution as negative prices are unlikely. Thus, the equilibrium price is 40.

Substituting this value into the supply function gives equilibrium quantity

$$q = -15 + 0.5p = -15 + 0.5(40) = 5$$

You should now be able to link the different mathematical techniques you have learned so far to tackle more complex problems. If you have covered the theory of perfect competition in your economics course, then you should be able to follow the analysis in the following example.

Example 7.8

An industry is made up of 100 identical firms, all with the cost schedules

$$\text{AC} = 40q^{-1} + 0.4q^2, \quad \text{TC} = 40 + 0.4q^3, \quad \text{MC} = 1.2q^2$$

They sell in a market where the demand function is

$$Q = 875 - 12.5p$$

where Q is industry output (and q is an individual firm's output).

(i) What will be the short-run price, industry output and profit for each firm?
(ii) What will happen to price, industry output and the number of firms in the long run? (Assume new entrants have the same cost structure.)

Solution

(i) The industry supply schedule is the horizontal sum of the individual firms' marginal cost schedules. Given the marginal cost function

$$\text{MC} = 1.2q^2$$

$$\left(\frac{MC}{1.2}\right)^{0.5} = q$$

There are 100 firms, and so the amount supplied by the whole industry is

$$Q = 100q = 100\left(\frac{MC}{1.2}\right)^{0.5} \tag{1}$$

In perfect competition MC corresponds to the price at which any given quantity will be supplied, and so (1) can be rewritten as

$$Q = 100\left(\frac{p}{1.2}\right)^{0.5}$$

Therefore,

$$0.01Q = \left(\frac{p}{1.2}\right)^{0.5}$$

$$(0.01Q)^2 = \frac{p}{1.2}$$

$$0.0001Q^2 = \frac{p}{1.2}$$

$$0.00012Q^2 = p \tag{2}$$

The function (2) is the industry supply schedule.

The given demand function is $Q = 875 - 12.5p$
Thus, the inverse demand function is $p = 70 - 0.08Q$ (3)

In equilibrium, demand price equals supply price. Thus, equating (3) and (2) we get

$$70 - 0.08Q = 0.00012Q^2$$

$$0 = 0.00012Q^2 + 0.08Q - 70 \tag{4}$$

Using the quadratic formula to solve (4)

$$\begin{aligned} Q &= \frac{-0.08 \pm \sqrt{0.0064 + 0.0336}}{0.00024} = \frac{-0.08 \pm \sqrt{0.04}}{0.00024} \\ &= \frac{-0.08 + 0.2}{0.00024} \text{ or } \frac{-0.08 - 0.2}{0.00024} \\ &= \frac{0.12}{0.00024} \text{ or } \frac{-0.28}{0.00024} \\ &= 500 \quad \text{(ignoring the negative answer)} \end{aligned}$$

Substituting this value of Q into the demand schedule (3) gives

$$p = 70 - 0.08(500) = 70 - 40 = £30$$

Each of the 100 firms produces the same amount q. Therefore, given the total industry output of 500, the amount each individual firm produces will be

$$q = \frac{Q}{100} = \frac{500}{100} = 5$$

Each firm's profit will be

$$\begin{aligned} \text{TR} - \text{TC} &= pq - (40 + 0.4q^3) \\ &= 30(5) - (40 + 50) \\ &= 150 - 90 = £60 \end{aligned}$$

(ii) If existing firms are making a profit then in the long run new entrants will be attracted into the industry. This will shift the supply schedule to the right and price will be driven down until each firm is only just breaking even, when price equals the lowest value on the firm's U-shaped average cost schedule.
How do we find when AC is at its minimum point? The MC and AC functions are given in the question. From cost theory you should know that MC always cuts AC at its minimum point. Therefore,

$$\begin{aligned} \text{MC} &= \text{AC} \\ 1.2q^2 &= 40q^{-1} + 0.4q^2 \\ 0.8q^2 &= 40q^{-1} \\ q^3 &= 50 \\ q &= 3.684 \quad (\text{to 3 dp}) \end{aligned}$$

Given that $\quad \text{AC} = 40q^{-1} + 0.4q^2$

When $q = 3.684$, then $\quad \text{AC} = 40(3.684)^{-1} + 0.4(3.684)^2 = 16.2865$

Therefore, $\quad p = £16.29$ (to the nearest penny)

The old supply schedule does not apply now because of the increased number of firms in the industry. Therefore, substituting this price, that corresponds to the minimum point on AC, into the demand schedule (3) to get industry output Q gives

$$\begin{aligned} 16.29 &= 70 - 0.08Q \\ 0.08Q &= 53.71 \\ Q &= 671.375 \end{aligned}$$

We already know that each firm produces 3.684 units in long-run equilibrium at the minimum point on AC. Therefore, the new number of firms in the industry is

$$\frac{Q}{q} = \frac{671.375}{3.684} = 182.24 = 182 \text{ firms}$$

Given that there were originally 100 firms, the number of new entrants is therefore 82.

Note that the fraction is rounded down to the nearest whole number. Any extra firms would bring price below the break-even level.

QUESTIONS 7.3

1. If $y = 255 - x - x^2$ and $y = 180 + \frac{2}{3}x^2 - 21x$, find x and y.
2. Find x and y given the functions

$$2y + 4x^2 + 10x - 36 = 0 \quad \text{and} \quad 4y - 10x^2 + 24x = 24$$

3. A monopoly faces the marginal cost function MC = $0.5q^2$ and the marginal revenue function MR = $200 - 4q$. What output will maximize its profits?
4. A price-discriminating monopoly sells in two markets with demand schedules

$$p_1 = 200 - 2q_1 \quad \text{and} \quad p_2 = 120 - 5q_2$$

Total output $q = q_1 + q_2$ and marginal cost is MC = $40 + 0.5q^2$.
How much should it sell in each market, and at what price, to maximize profits?

5. A firm's marginal cost schedule is MC = $2.3 + 0.00012q^2$ and it sells its output in two separate markets with demand schedules

$$p_1 = 25 - 0.125q_1 \quad \text{and} \quad p_2 = 12 - 0.05q_2$$

What prices and quantities will maximize profits if this firm is a price-discriminating monopoly?

7.6 POLYNOMIALS

Quadratic equations are a special case of polynomial equations. The general format of a polynomial function is

$$y = a_0 + a_1x + a_2x^2 + a_3x^3 + \ldots + a_nx^n$$

where n is any non-negative integer. Quadratic equations contain polynomials where $n = 2$. Linear equations contain polynomials where $n = 1$.

When n is greater than 2, the solution of a polynomial equation by algebraic means becomes complex and time consuming. For practical purposes, however, you can use a spreadsheet to find a solution by the iterative method. This means calculating values of a function for different values of the unknown variable until a solution or a good approximation to it is found. As a spreadsheet can quickly perform the necessary calculations, it is an ideal tool for the calculation of polynomial solutions.

The format of the spreadsheet required will depend on the problem tackled. Next are some examples of how this sort of problem can be approached.

Example 7.9

A firm's total costs (TC) are given by the function

$$TC = 420 + 32.5q - 6.25q^2 + 0.8q^3$$

where q is output level and TC is measured in pounds. If the firm's management is given a budget of £43,000, what output can it produce?

Solution

The basic approach is to construct a spreadsheet that will calculate TC for different values of q. We can then experiment with different ranges for the values of q until we find which value of q will correspond to the given value for TC.

The method for constructing the spreadsheet is similar to that used for quadratic equations as set out previously in Table 7.1. This time the spreadsheet calculates the cubic TC function that corresponds to the parameters entered. The instructions for doing this are set out in Table 7.3.

Table 7.3 Setting up a spreadsheet for Example 7.9

CELL	Enter	Explanation
A1	Ex.7.9	Label to remind you what example this is
B2	CUBIC POLYNOMIAL SOLUTION TO	Title of spreadsheet (Note that this is not an actual Excel formula.)
B3	TC =a + bq + cq^2 + dq^3	
F2	Parameter	Labels that tell you that the parameter values will be shown below
F3	Values	
E4	a =	These are labels that tell you that the parameter values will go in the cells next to them. Right justify these cells.
E5	b =	
E6	c =	
E7	d =	
F4	420	These are the actual parameter values for *a*, *b*, *c* and d, respectively, for this example.
F5	32.5	
F6	–6.25	
F7	0.8	
A3	Q	Column heading labels
B3	TC	
C3	MC	
A4	0	Initial value for *q*
A5	=A4+1	Calculates a one-unit increment in *q*
A6 to A45	*Copy formula from cell A5 down column A*	Calculates a series of values of *q* in one-unit increments
B4	=F$4+F$5*A4+F$6*A4^2+F$7*A4^3	Formula to calculate value of TC corresponding to value of *q* in cell A4 and parameter values in cells F4, F5, F6 and F7. Note the $ sign used to anchor row references for when this formula is copied down row B.

(Continued)

Table 7.3 *(Continued)*

CELL	Enter	Explanation
B5 to B45	*Copy formula from cell* B4 *down column* B	Calculates values for TC in each row corresponding to values of q in column A.
C5	=B5-B4	Calculates values MC as the change in TC from a one-unit increment in q.
C6 to C45	*Copy formula from cell* C5 *down column* C	Calculates MC of a unit of q corresponding to increment in TC shown in column B.
B4 to C45	*Highlight these columns and format to 2 decimal places*	TC and MC are both monetary values measured in £ so use numerical format 0.00

Although AC and MC may initially fall as a firm's output increases, its TC function should never fall. It would therefore be useful to have a check that this cubic TC function always increases as q increases and MC is never negative. To do this, the spreadsheet also includes a third column where values of MC are calculated. (See Section 9.4 for further analysis of cubic functions with this property.)

The values calculated by this spreadsheet are shown in Table 7.4, which tells us that when q is 40, TC will be 42,920. Thus, if output is constrained to whole units, 40 is the maximum output that the firm's management can produce for a budget of £43,000.

This spreadsheet also confirms that MC declines in value then increases but is never negative. This is what we would expect. Save your spreadsheet for use with other examples.

Table 7.4 Excel output based on Table 7.3

	A	B	C	D	E	F
1	Ex 7.9	CUBIC POLYNOMIAL SOLUTION TO				
2		TC =a + bq + cq^2 + dq^3				Parameter
3	Q	TC	MC			Values
4	0	420.00			a =	420
5	1	447.05	27.05		b =	32.5
6	2	466.40	19.35		c =	−6.25
7	3	482.85	16.45		d =	0.8
8	4	501.20	18.35			
9	5	526.25	25.05			
10	6	562.80	36.55			
11	7	615.65	52.85			
12	8	689.60	73.95			
13	9	789.45	99.85			
14	10	920.00	130.55			
15	11	1086.05	166.05			
16	12	1292.40	206.35			
17	13	1543.85	251.45			
18	14	1845.20	301.35			

(Continued)

Table 7.4 *(Continued)*

	A	B	C	D	E	F
19	15	2201.25	356.05			
20	16	2616.80	415.55			
21	17	3096.65	479.85			
22	18	3645.60	548.95			
23	19	4268.45	622.85			
24	20	4970.00	701.55			
25	21	5755.05	785.05			
26	22	6628.40	873.35			
27	23	7594.85	966.45			
28	24	8659.20	1064.35			
29	25	9826.25	1167.05			
30	26	11100.80	1274.55			
31	27	12487.65	1386.85			
32	28	13991.60	1503.95			
33	29	15617.45	1625.85			
34	30	17370.00	1752.55			
35	31	19254.05	1884.05			
36	32	21274.40	2020.35			
37	33	23435.85	2161.45			
38	34	25743.20	2307.35			
39	35	28201.25	2458.05			
40	36	30814.80	2613.55			
41	37	33588.65	2773.85			
42	38	36527.60	2938.95			
43	39	39636.45	3108.85			
44	40	42920.00	3283.55			
45	41	46383.05	3463.05			

This example was constructed for a range of values of q that contained the answer we were seeking. If you had no idea where the solution to this cubic polynomial lay then you could get a 'ball park' estimate by producing a range of values in jumps of 10 in the column headed q by entering the formula =A4+10 in cell A5 and then copying it down the column for a few dozen rows. This would tell you that when $q = 31$, TC = £19,254.05, and when $q = 41$, TC = £46,383.05. Therefore, TC = £43,000 must lie somewhere between these values of q. Once you have a rough idea of where the solution value for q will lie, you can change the q column so that values increase by one-unit increments, or smaller if necessary, until the actual solution is pinpointed.

To solve other cubic polynomials, one simply enters the corresponding parameters into the spreadsheet set up for Example 7.9 and adjusts the range of the independent variable (q) until the solution is found.

Table 7.5 Using a spreadsheet for Example 7.10

	A	B	C	D	E	F
1	Ex 7.10	CUBIC POLYNOMIAL SOLUTION TO				
2		TC =a + bq + cq^2 + dq^3				Parameter
3	Q	TC	MC			Values
4	0	880.00				880
5	1	939.00	59.00		b =	72
6	2	978.00	39.00		c =	–14.5
7	3	1006.00	28.00		d =	1.5
8	4	1032.00	26.00			
9	5	1065.00	33.00			
10	6	1114.00	49.00			
11	7	1188.00	74.00			
12	8	1296.00	108.00			
13	9	1447.00	151.00			
14	10	1650.00	203.00			
15	11	1914.00	264.00			
16	12	2248.00	334.00			
17	13	2661.00	413.00			
18	14	3162.00	501.00			
19	15	3760.00	598.00			
20	16	4464.00	704.00			
21	17	5283.00	819.00			
22	18	6226.00	943.00			
23	19	7302.00	1076.00			
24	20	8520.00	1218.00			
25	21	9889.00	1369.00			
26	22	11418.00	1529.00			

Example 7.10

If TC = $880 + 72q - 14.5q^2 + 1.5q^3$, at what value of q will TC = £9,889?

Solution

Entering the new values for a, b, c and d into the spreadsheet constructed for Example 7.9 and adjusting the range of q, one should get a spreadsheet similar to Table 7.5. This shows that $q = 21$ when TC = £9,889.

This spreadsheet can also be adjusted to cope with more complex polynomials. Its crucial part is the formula in cell B4. This needs to be amended to calculate the value of the new polynomial function if more terms are added. Note, however, that large **polynomial equations may have several solutions**. In particular, if there are both positive and negative coefficients, a polynomial function may equal zero at more than two values of the independent variable. You can usually deduce the number of solutions from the format of the equation, or get a plot from your spreadsheet to see how many times the function crosses the horizontal axis. On the other hand, no solutions may exist, in which case a graph will not cut the axis; e.g. although there will be a negative solution, there is no positive value of x which will satisfy the equation

$$0 = 8 + 32x + 6x^2 + 0.9x^3$$

Example 7.11

Assuming $x < 1{,}000$, is there a positive value of x that is a solution to the function

$$0 = -770{,}077.6 + 262x - 74x^2 + 12x^3 + 2x^4 - 0.05x^5\,?$$

Solution

To solve this equation we need to calculate values of the polynomial function

$$y = -770{,}077.6 + 262x - 74x^2 + 12x^3 + 2x^4 - 0.05x^5$$

Table 7.6 Setting up a spreadsheet for Example 7.11
(Only shows changes needed to adapt Table 7.3 for this example.)

CELL	Enter	Explanation
A1	Ex.7.11	Label for example number
B2	y =a + bx + cx^2 + dq^3 + ex^4 + fx^5	Title of new formula.
E8	e =	Additional labels for the two extra parameter values. (Right justify)
E9	f =	
A3	x	New labels for column headings.
B3	y	
Column C	*Highlight and hit Edit-Clear-All*	This column can be cleared as MC not calculated for this example.
F4	−770077.6	These are the actual parameter values for *a*, *b*, *c*, *d*, *e* and *f*, respectively, for this example.
F5	262	
F6	−74	
F7	12	
F8	2	
F9	−0.05	
A4	30	Initial value for *x* (ball-park range found.)
Rows 15 onward	*Highlight-Edit-Delete*	Delete extra rows as only need range of *x* from 30 to 40 for this example.
B4	=F$4+F$5*A4+F$6*A4^2+F$7*A4^3+ F$8*A4^4+F$9*A4^5	Calculates the value of *y* function corresponding to value of *x* in cell A4 and the parameter values in cells F4 to F9. Note $ sign used to anchor row references.
B5 to B14	*Copy formula from cell B4 down column B*	Calculates values for *y* in each row corresponding to values of *x* in column A.

Table 7.7 Excel output based on Table 7.6

	A	B	C	D	E	F
1	Ex 7.11	CUBIC POLYNOMIAL SOLUTION TO				
2		y =a + bx + cx^2 + dq^3 + ex^4 + fx^5				Parameter
3	x	y				Values
4	30	–99817.60			a =	–770077.6
5	31	–59993.15			b =	262
6	32	–24823.20			c =	–74
7	33	4298.75			d =	12
8	34	25835.20			e =	2
9	35	38098.65			f =	–0.05
10	36	39245.60				
11	37	27270.55				
12	38	0.00				
13	39	–44913.55				
14	40	–109997.60				

and find the value(s) of x where this function equals zero. To do this, call up the spreadsheet created for Example 7.9 and then follow the instructions in Table 7.6 to add two new terms so that it will be able to calculate values for polynomials in the format

$$y = a + bx + cx^2 + dx^3 + ex^4 + fx^5$$

Once the basic spreadsheet has been created, the ball-park method explained earlier can be used to narrow down the possible solution range to between 30 and 40. This should give you a spreadsheet that looks like Table 7.7. This clearly shows that y is zero when x is 38 and so this is the solution. (If you try increasing the range of x you will see that there are no other solutions in the range $0 < x < 1{,}000$.)

QUESTIONS 7.4

(You will need to use a spreadsheet to tackle these questions.)

1. How many units of q can be produced for £60,000 if the total cost function is

$$\text{TC} = 86 + 152q - 12q^2 + 0.6q^3 ?$$

2. What output can be produced for £160,000 if

$$\text{TC} = 130 + 62q - 3.5q^2 + 0.15q^3 ?$$

3. Solve for x when

$$0 = -1{,}340 + 14x + 2x^2 - 1.5x^3 + 0.2x^4 + 0.005x^5 - 0.0002x^6$$

8 Financial mathematics – series, time and investment

Learning objectives

After completing this chapter students should be able to:

- calculate the final sum, the initial sum, the time period and the interest rate for an investment
- derive the annual equivalent rate for part year investments
- calculate the net present value and internal rate of return for an investment, constructing relevant spreadsheets when required
- use the appropriate investment appraisal method to decide whether an investment project is worthwhile
- find the sum of finite and infinite geometric series
- calculate the value of an annuity and estimate the income that a given size annuity could provide
- analyze the feasibility of different patterns of income withdrawal from a draw-down pension scheme
- calculate monthly repayments and the APR for a loan
- calculate the amount accumulated in different forms of savings schemes
- solve problems involving the growth and decline over discrete time periods of different economic variables, including the depletion of natural resources.

8.1 DISCRETE AND CONTINUOUS GROWTH

In economics we come across many variables that grow, or decline, over time. A sum of money invested in a deposit account will grow as interest accumulates on it. The amount of oil left in an oil field will decline as production continues over the years. This chapter explains how mathematics can help answer certain problems concerned with these variables that change over time. The main area of application is finance, including methods of appraising different forms of investment. Other applications

DOI: 10.4324/9781003360827-8

include the management of natural resources, where the implications of different depletion rates are also analyzed.

The interest earned on money invested in a deposit account is normally paid at set regular intervals. Calculations of the return are therefore usually made with respect to specific time intervals. For example, Figure 8.1(a) shows the amount of money in a deposit account at any given moment in time assuming an initial deposit of £1,000 and interest credited at the end of each year at a rate of 10%. There is not a continuous relationship between time and the total sum in the deposit account. Instead, there is a 'jump' at the end of each year when the interest on the account is paid. This is an example of a 'discrete' function. Between the occasions when interest is added there is no change in the value of the account.

A **discrete function** can therefore be defined as one where the value of the dependent variable is known for specific values of the independent variable but does not continuously change between these values. Hence one gets a series of values rather than a continuum. For example, some public sector salaries are based on scales with series of increments. A hypothetical scale linking completed years of service to salary might be:

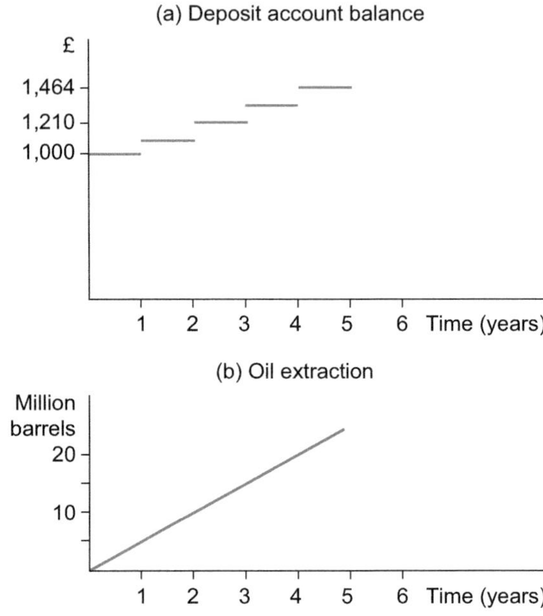

Figure 8.1 Discrete and continuous growth over time

$$0\text{ yrs} = £28{,}000,\ \ 1\text{ yrs} = £31{,}000,\ \ 2\text{ yrs} = £35{,}000,\ \ 3\text{ yrs} = £40{,}000$$

The relationship between salary and years of service is a discrete function. At any moment in time one knows what the salary level will be but there is not a continuous relationship between time and salary level.

An example of a continuous function is illustrated in Figure 8.1(b). This shows the cumulative total amount of oil extracted from an oil field when there is a steady 5 million barrels per year extraction rate. There is a continuous smooth function showing the relationship between the amount of oil extracted and the time elapsed because the oil flows continuously through pipelines throughout the year.

In this chapter we analyze a number of discrete-variable problems. Algebraic formulae are developed to solve some applications of discrete functions and methods of solution using spreadsheets are explained where appropriate, for investment appraisal analysis in particular. The analysis of continuous growth requires the use of the exponential function and will be explained later, in Chapter 15.

8.2 INTEREST

Time is money. If you borrow money you have to pay interest on it. If you invest money in a deposit account, you expect to earn interest on it. From an investor's point of view, the interest rate can be looked on as the 'opportunity cost of capital'. If a sum of money is tied up in a project for a year then the investor loses the interest that could have been earned by investing the money elsewhere, perhaps by putting it in a deposit account.

Simple interest is the interest that accrues on a given sum in a set period of time. It is *not* reinvested along with the original capital. The amount of interest earned on a given investment each time period will be the same (if interest rates do not change) as the total amount of capital invested remains unaltered.

Example 8.1

An investor puts £20,000 into a deposit account and has the annual interest paid directly into a separate current account and then spends it. The deposit account pays 2.5% interest. How much interest is earned in the fifth year?

Solution

The interest paid each year will remain constant at 2.5% of the original investment of £20,000. Thus, in year 5 the interest will be

$$0.025 \times £20{,}000 = £500$$

Most investment decisions, however, need to take into account the fact that any interest earned can be reinvested and so compound interest, explained later, is more relevant. The calculation of simple interest is such a basic arithmetic exercise that the only mistake you are likely to make is to transform a percentage figure into a decimal fraction incorrectly.

Example 8.2

How much interest will be earned on £400 invested for a year at an annual interest rate of 0.5%?

Solution

To convert any percentage figure to a decimal fraction you must divide it by 100. Therefore

$$0.5\% = \frac{0.5}{100} = 0.005$$

and so 0.5% of $£400 = 0.005 \times £400 = £2$

If you can remember that 1% = 0.01 then you should be able to transform any interest rate specified in percentage terms into a decimal fraction in your head. Try to do this for the following interest rates:

(i) 1.5%
(ii) 30%
(iii) 0.075%
(iv) 1.02%
(v) 0.6%

Now check your answers with a calculator. If you got any wrong you really ought to go back and revise Section 2.5 before proceeding. Converting decimal fractions back to percentage interest rates is, of course, simply a matter of multiplying by 100; e.g. 0.02 = 2%, 0.4 = 40%, 1.25 = 125%, 0.008 = 0.8%.

Compound interest is interest which is added to the original investment every time it accrues. The interest added in one time period will itself earn interest in the following time period. The total value of an investment will therefore grow over time.

Example 8.3

If £600 is invested for 3 years at 8% interest compounded annually at the end of each year, what will the final value of the investment be?

Solution

	£
Initial sum invested	600.00
Interest at end of year 1 = 0.08 × 600	48.00
Total sum invested for year 2	648.00
Interest at end of year 2 = 0.08 × 648	51.84
Total sum invested for year 3	699.84
Interest at end of year 3 = 0.08 × 699.84	55.99
Final value of investment	755.83

Example 8.4

If £5,000 is invested at an annual rate of interest of 2% and interest earned is reinvested, how much will the investment be worth after 2 years?

Solution

Initial sum invested	5,000
Year 1 interest = 0.02 × 5,000	100
Sum invested for year 2	5,100
Year 2 interest = 0.02 × 5,100	102
Final value of investment	£5,202

These examples only involved the calculation of interest for a few years and did not take too long to solve from first principles. To work out the final sum of an investment after longer time periods one could construct a spreadsheet, but an even quicker method is to use the formula explained next.

Calculating the final value of an investment

Consider an investment at compound interest where:

A is the initial sum invested,
F is the final value of the investment,
i is the interest rate per time period (as a decimal fraction),
n is the number of time periods.

The value of the investment at the end of each year will be $(1 + i)$ times the sum invested at the start of the year. For instance, the £648 at the start of year 2 is 1.08 times the initial investment of £600 in Example 8.3. The value of the investment at the start of year 3 is 1.08 times the value at the start of year 2, and so on. Thus, for any investment

Value after 1 year $= A(1 + i)$
Value after 2 years $= A(1 + i)(1 + i) = A(1 + i)^2$
Value after 3 years $= A(1 + i)^2(1 + i) = A(1 + i)^3$, etc.

We can see that each value is A multiplied by $(1 + i)$ to the power of the number of years that the sum is invested. Thus, after n years the initial sum A is multiplied by $(1 + i)^n$.

The **formula for the final value *F* of an investment** of an amount A invested for n time periods at interest rate i is therefore

$$F = A(1+i)^n$$

Let us rework Examples 8.3 and 8.4 using this formula just to check that we get the same answers.

Example 8.3 (reworked)

If £600 is invested for 3 years at 8% then the known values for the formula will be

$$A = £600 \quad n = 3 \quad i = 8\% = 0.08$$

Thus, the final sum will be

$$F = A(1+i)^n = 600(1/08)^3 = 600(1.259712) = £755.83$$

Example 8.4 (reworked)

£5,000 invested for 2 years at 2% means that

$$A = £5,000 \qquad n = 2 \qquad i = 2\% = 0.02$$
$$F = A(1+i)^n = 5,000(1.02)^2 = 5,000(1.0404) = £5,202$$

Having satisfied ourselves that the formula works we can now tackle some more difficult problems.

Example 8.5

If £4,000 is invested for 10 years at an interest rate of 3.25% per annum what will the final value of the investment be?

Solution

$$A = £4{,}000 \qquad n = 10 \qquad i = 3.25\% = 0.0325$$
$$F = A(1+i)^n = 4{,}000(1.0325)^{10} = 4{,}000(1.3768943) = £5{,}507.58$$

Sometimes a compound interest problem may be specified in a rather different format, but the method of solution is still the same.

Example 8.6

You estimate that you will need £8,000 in 3 years' time to buy a new car, assuming a reasonable trade-in price for your old car. You have £7,000 in a fixed interest building society account earning 4.5%. Will this generate enough to buy the car?

Solution

You need to work out the final value of your savings to see whether it will be greater than £8,000. Using the usual notation,

$$A = £7{,}000 \qquad n = 3 \qquad i = 0.045$$
$$F = A(1+i)^n = 7{,}000(1.045)^3 = 7{,}000(1.141166) = £7{,}988.16$$

So, the answer is 'almost'. You will have to find another £12 to get to £8,000, but perhaps you can get the dealer to knock this off the price.

Changes in interest rates

What if interest rates are expected to change before the end of the investment period? The final sum can be calculated by slightly adjusting the usual formula.

Example 8.7

Interest rates are expected to be 4% for the next 2 years and then fall to 2% for the following 3 years. How much will £2,000 be worth if it is invested for 5 years?

Solution

After 2 years the final value of the investment will be

$$F = A(1+i)^n = 2{,}000(1.04)^2 = 2{,}000(1.0816) = £2{,}163.20$$

If this sum is then invested for a further 3 years at the new interest rate of 2% then the final sum is

$$F = A(1+i)^n = 2{,}163.20(1.02)^3 = 2{,}163.20(1.061208) = £2{,}299.85$$

This could have been worked out in one calculation by finding

$$F = 2{,}000(1.04)^2(1.02)^3 = £2{,}299.85$$

Therefore, the formula for the final sum F that an initial sum A will accrue to after n time periods at interest rate i_n and q time periods at interest rate i_q will be

$$F = A\left(1+i_n\right)^n\left(1+i_q\right)^q$$

If more than two interest rates are involved then the formula can be adapted along the same lines.

Example 8.8

What will £20,000 invested for 10 years be worth if the expected rate of interest is 12% for the first 3 years, 9% for the next 2 years and 8% thereafter?

Solution

$$F = 20{,}000(1.12)^3(1.09)^2(1.08)^5 = £49{,}051.90$$

QUESTIONS 8.1

1. If £4,000 is invested at 5% interest for 3 years what will the final sum be?
2. How much will £200 invested at 12% be worth at the end of 4 years?
3. A parent invests £6,000 for a 7-year-old child in a fixed interest scheme which guarantees 8% interest. How much will the child have at the age of 21?
4. If £525 is invested in a deposit account that pays 6% interest for 6 years, what will the final sum be?
5. What will £24,000 invested at 11% be worth at the end of 5 years?
6. Interest rates are expected to be 10% for the next 3 years and then to fall to 8% for the following 3 years. How much will an investment of £3,000 be worth at the end of 6 years?

8.3 PART YEAR INVESTMENT AND THE ANNUAL EQUIVALENT RATE

If the duration of an investment is less than a year the usual final sum formula does not always apply. It is usually the custom to specify interest rates on an equivalent annual basis for part year investments, but two different types of annual interest rates can be used:

(a) the nominal annual interest rate, and
(b) the annual equivalent rate (AER).

The ways that these annual interest rates relate to part year investments differ and they are used in different circumstances. Although it is the AER that most individual savers and investors will encounter, we shall first look at part year investment based on the less common nominal interest rates approach.

Nominal annual interest rates

For large institutional investors on the money markets, on certain forms of investments such as government bonds, a nominal annual interest rate may be quoted for part year investments. To find the interest that will actually be paid, this nominal annual rate is multiplied by the fraction of the year that it is quoted for.

Example 8.9

What interest is payable on a £100,000 investment for 6 months at a nominal annual interest rate of 6%?

Solution

Six months is 0.5 of one year and so the interest rate that applies is

$$0.5 \times (\text{nominal annual rate}) = 0.5 \times 6\% = 3\%$$

Therefore, interest earned is

$$3\% \text{ of } £100{,}000 = £3{,}000$$

and the final sum is

$$F = (1.03)100{,}000 = £103{,}000$$

If this nominal annual interest rate of 6% applied to a 3-month investment, then the actual interest payable would be a quarter of 6% which is 1.5%. If it applied to an investment for one month, then the interest payable would be 6% divided by 12 which gives 0.5%.

The calculation of part year interest payments on this basis could, in theory, give investors a total annual return that is greater than the nominal interest rate if they could keep reinvesting through the year at the same part year interest rate. The total final value of the investment can be calculated with reference to these new time periods using the $F = A(1 + i)^n$ formula as long as the interest rate i and the number of time periods n refer to the same time periods.

For example, if £100,000 can be invested for four successive 3-month periods at a nominal annual interest rate of 6% then, letting i represent the effective quarterly interest rate and n represent the number of 3-month periods

$$A = £100{,}000 \qquad n = 4 \qquad i = 0.25 \times 6\% = 1.5\% = 0.015$$
$$F = A(1+i)^n = 100{,}000(1.015)^4 = £106{,}136.35$$

This final sum gives a 6.13635% return on the initial £100,000 sum invested.

The more frequently that interest based on the nominal annual rate is paid the greater will be the total annual return when all the interest is compounded. For example, if a nominal annual interest rate of 6% is paid monthly at 0.5% a month and £100,000 is invested for 12 months then

$$A = £100{,}000 \qquad n = 12 \qquad i = 0.5\% = 0.005$$
$$F = A(1+i)^n = 100{,}000(1.005)^{12} = £106{,}167.78$$

This new final sum is greater than that achieved from quarterly interest payment and is equivalent to an annual rate of 6.17%.

The annual equivalent rate (AER) and annual percentage rate (APR)

Individual investors are usually quoted an annual equivalent rate (AER), which is a more accurate reflection of the interest earned on part year investments. For example, interest on the money you may have in a building society account will normally be worked out on a daily basis, although you will only be told the AER and the interest may only be credited on your account once a year. For loan repayments the annual equivalent rate is usually referred to as the annual percentage rate (APR). If you take out a bank loan you will usually be quoted an APR even though you will be asked to make monthly repayments.

The nominal interest rate examples earlier have already demonstrated that the AER is not simply 12 times the monthly interest rate. To determine the relationship between part year interest rates and their true AER, consider another example.

Example 8.10

If interest is credited monthly at a monthly rate of 0.9% how much will £100 invested for 12 months accumulate to?

Solution

Using the standard investment formula where the time period n is measured in months:

$$A = £100 \qquad n = 12 \qquad i = 0.9\% = 0.009$$
$$F = A(1+i)^n = 100(1.009)^{12} = 100(1.1135) = £111.35$$

This final sum of £111.35 after investing £100 for one year corresponds to an annual rate of interest of 11.35%. This is greater than 12 times the monthly rate of 0.9%, since

$$12 \times 0.9\% = 10.8\%$$

The calculations in this example tell us that the ratio of the final sum to the initial sum invested is $(1.009)^{12}$. Using the same principle, the **corresponding AER for any given monthly rate of interest i_m** can be found using the formula

$$\text{AER} = (1 + i_m)^{12} - 1$$

Because $(1 + i_m)^{12}$ gives the ratio of the final sum F to the initial amount A the '−1' has to be added to the formula in order to get the proportional increase in F over A.

The APR on a loan is effectively the same thing as the AER and so the same formula applies to convert a monthly loan rate to the annual rate. In practice the charges that lenders include and the profit margin they take into account when setting lending rates mean that there are differences between their AER and the APR rates. However, the same mathematical principles will apply regardless of whether one is converting a monthly deposit account rate to an AER or a monthly lending rate to an APR.

Example 8.11

If the monthly rate of interest on a loan is 1.75% what is the corresponding APR?

Solution

$$\begin{aligned} i_m &= 1.75\% = 0.0175 \\ \text{APR} &= (1 + i_m)^{12} - 1 = (1.0175)^{12} - 1 \\ &= 1.2314393 - 1 = 0.2314393 = 23.14\% \end{aligned}$$

If you have a credit card you can use this formula to check that the correct APR is quoted for the monthly interest charges on outstanding balances that are levied. For example, if interest is charged at 1.527% per month on a credit card and the bank supplying it says that this corresponds to an APR of 19.9%, then this can be checked as follows:

$$\text{APR} = (1 + i_m)^{12} - 1 = (1.01527)^{12} - 1 = 1.19944 - 1 = 0.19944 = 19.9\%$$

The calculation of monthly loan repayments from a given APR will be explained later, in Section 8.11.

Daily interest rates

Savers may put money into deposit accounts or make withdrawals at any time throughout the year and so different amounts will remain in their accounts for different periods of time. The interest on these accounts is therefore usually calculated on a daily basis. However, only the AER is widely publicized as this is much more useful to savers to help them make comparisons between different possible investment opportunities. The relationship between the daily interest rate i_d on a deposit account and the AER can be formulated as

$$\text{AER} = (1 + i_d)^{365} - 1$$

For example, if a building society tells you that it will pay you an AER of 6% on a savings account, what it actually will do is credit interest at a rate of 0.015954% a day. We can check this out using the formula

$$\begin{aligned}\text{AER} &= (1 + i_d)^{365} - 1 = (1.00015954)^{365} - 1 \\ &= 1.06 - 1 = 0.66 = 6\%\end{aligned}$$

To derive a formula for the daily interest rate i_d that corresponds to a given AER, we just need to derive the inverse of AER formula. Thus,

$$\begin{aligned}\text{AER} &= (1 + i_d)^{365} - 1 \\ \text{AER} + 1 &= (1 + i_d)^{365} \\ \sqrt[365]{\text{AER}+1} &= 1 + i_d \\ \left(\sqrt[365]{\text{AER}+1}\right) - 1 &= i_d\end{aligned}$$

Example 8.12

A building society account pays interest on a daily basis at an AER of 4.5%. If you deposited £2,750 in such an account how much would you get back if you closed the account 254 days later?

Solution

Given AER = 4.5% = 0.045 then, using the formula derived earlier, the daily interest rate is

$$\begin{aligned}i_d &= \left(\sqrt[365]{\text{AER}+1}\right) - 1 = \left(\sqrt[365]{0.045 + 1}\right) - 1 = 1.0001206 - 1 \\ &= 0.0001206 = 0.01206\%\end{aligned}$$

The final sum accumulated when £2,750 is invested at this daily rate for 254 days will therefore be

$$F = 2750(1 + 0.0001206)^{254} = 2750(1.0311045) = £2,835.54$$

Interest rates on Treasury Bills

A government Treasury Bill, like other forms of bond, guarantees the owner a fixed sum of money payable at a fixed date in the future. So, for example, a 3-month Treasury Bill for £100,000 is effectively a promise from the government that it will pay £100,000 to the owner on a date 3 months from when it was issued. The prices that the institutional investors who trade in these bills will pay for them will reflect the returns that can be made on other similar investments.

Suppose that investors are currently willing to pay £95,000 for 12-month £100,000 Treasury Bills when they are issued. This would mean that they consider an annual return of £5,000 on their £95,000 investment to be acceptable. An annual return of £5,000 on a £95,000 investment is equivalent to an interest rate of

$$i = \frac{5{,}000}{95{,}000} = 0.0526316 = 5.26\%$$

However, in the financial press the interest rates quoted relate to a nominal annual rate of return based on the final sum paid out when the Treasury Bill matures. Thus, in the previous example, the 12-month Treasury Bill rate quoted would be 5%, because this is the discount the price of £95,000 yields on the final maturity sum of £100,000. This is why they are called discount rates.

Although the previous example considered a 12-month Treasury Bill so that the equivalent annual rate of return could be easily compared, in practice UK government Treasury Bills are normally issued for shorter periods. Also, the nominal annual rates are usually quoted using fractions, such as $4\frac{5}{16}\%$, rather than in decimal format.

Example 8.13

If an annual discount rate of $4\frac{7}{8}\%$ is quoted for 3-month Treasury Bills, what would it cost to buy a tranche of these bills with redemption value of £100,000? What would be the annual equivalent rate of return on the sum paid for them?

Solution

A nominal annual rate of $4\frac{7}{8}\%$ corresponds to a 3-month rate of

$$\frac{4\frac{7}{8}}{4} = \frac{39}{8} \times \frac{1}{4} = 1\tfrac{7}{32} = 1.21875\%$$

As this rate is actually the discount on the maturity sum then the cost of 3-month Treasury Bills with redemption value of £100,000 would be

$$£100{,}000(1 - 0.0121875) = £98{,}781.25$$

and the amount of the discount is £1,218.75.

Therefore, the rate of return on the sum of £98,781.25 invested for 3 months is

$$\frac{1,218.75}{98,781.25} = 0.012338 = 1.2338\%$$

If this investment could be compounded for four 3-month periods at this quarterly rate of 1.2338% then the annual equivalent rate would be

$$\text{AER} = (1.012338)^4 - 1 = 1.050273 - 1 = 0.050273 = 5.0273\%$$

It should be noted, however, that effective interest rates on different issues of government bonds can fluctuate substantially and reinvestment at exactly the same rate would be unlikely.

QUESTIONS 8.2

1. If £40,000 is invested at a monthly rate of 1% what will it be worth after 9 months? What is the corresponding AER?
2. A sum of £450,000 is invested at a monthly interest rate of 0.6%. What will the final sum be after 18 months? What is the corresponding AER?
3. Which is the better investment for someone wishing to invest a sum of money for 2 years:
   (a) an account which pays 0.9% monthly, or
   (b) an account which pays 11% annually?
4. If £1,600 is invested at a quarterly rate of interest of 4.5% what will the final sum be after 18 months? What is the corresponding AER?
5. How much interest is earned on £50,000 invested by an institutional investor in money markets for 3 months at a nominal annual interest rate of 5%? If money can be reinvested each quarter at the same rate, what is the AER?
6. If a credit card company charges 1.48% a month on any outstanding balance, what APR is it charging?
7. A building society pays an AER of 5.5% on an investment account, calculated on a daily basis. What daily rate of interest will it pay?
8. If 3-month government Treasury Bills are offered at an annual discount rate of $4\frac{7}{16}$%, what would it cost to buy bills with redemption value of £500,000? What would the AER be for this investment?

8.4 TIME PERIODS, INITIAL AMOUNTS AND INTEREST RATES

The formula for the final sum of an investment contains the four variables F, A, i and n. So far we have only calculated F for given values of A, i and n. However, if the values of any three of the variables in this equation are given then one can usually calculate the fourth.

Initial amount

A formula to calculate A, when values for F, i and n are given, can be derived as follows. Since the final sum formula is

$$F = A(1+i)^n$$

then dividing through by $(1 + i)^n$, we get

$$\frac{F}{(1+i)^n} = A$$

This **initial sum formula** can also be written as

$$A = F(1+i)^{-n}$$

Example 8.14

How much money needs to be invested now in order to accumulate a final sum of £12,000 in 4 years' time at an annual rate of interest of 3%?

Solution

Using the formula derived earlier, the initial amount is

$$A = F(1+i)^{-n} = 12{,}000(1.03)^{-4} = \frac{12{,}000}{1.1255} = £10{,}661.84$$

What we have actually done in the previous example is find the sum of money that is equivalent to £12,000 in 4 years' time if interest rates remain at 3%. Theoretically, an investor should be indifferent between (a) £10,661.84 now and (b) £12,000 in 4 years' time. The £10,661.84 is therefore known as the 'present value' (PV) of the £12,000 in 4 years' time. We shall come back to this concept later when methods of appraising different types of investment projects are explained.

Time period

Calculating the time period is trickier than the calculation of the initial amount. From the final sum formula

$$F = A(1+i)^n$$

$$\text{Then} \quad \frac{F}{A} = (1+i)^n$$

If the values of F, A and i are given and one is trying to find n, this means that one has to work out to what power $(1 + i)$ has to be raised to equal F/A. One way of doing this is by using logarithms.

Example 8.15

For how many years must £1,000 be invested at 10% in order to accumulate £1,600?

Solution

$$A = £1{,}000 \quad F = £1{,}600 \quad i = 10\% = 0.1$$

Substituting these values into the formula

$$\frac{F}{A} = (1+i)^n$$

$$\text{we get} \quad \frac{1{,}600}{1{,}000} = (1+0.1)^n \qquad (1)$$

$$1.6 = (1.1)^n$$

If equation (1) is specified in logarithms then

$$\log 1.6 = \log(1.1)^n$$

$$\text{since } \log x^n = n \log x$$

$$\log 1.6 = n \log 1.1 \qquad (2)$$

$$\text{taking logs we get } 0.20412 = n \times 0.0413927$$

$$n = \frac{0.20412}{0.0413927} = 4.93 \text{ years}$$

If investments must be made for whole years, then the answer is 5 years.

This answer can be checked using the final sum formula

$$F = A(1+i)^n = 1{,}000(1.1)^5 = £1{,}610.51$$

If the £1,000 is invested for a full 5 years, then it accumulates to just over £1,600.

A general formula to solve for n can be derived as follows from the final sum formula:

$$F = A(1+i)^n$$

$$\frac{F}{A} = (1+i)^n$$

Taking logs

$$\log\left(\frac{F}{A}\right) = n \log(1+i)$$

Therefore, the **time period formula** is

$$\frac{\log(F/A)}{\log(1+i)} = n \qquad (3)$$

Example 8.16

How many years will £2,000 invested at 5% take to accumulate to £3,000?

Solution

Given values are $A = 2{,}000 \quad F = 3{,}000 \quad i = 5\% = 0.05$

Using these in the time period formula derived earlier gives

$$n = \frac{\log(F/A)}{\log(1+i)} = \frac{\log 1.5}{\log 1.05} = \frac{0.1760913}{0.0211893} = 8.34 \text{ year}$$

Example 8.17

How long will any sum of money take to double its value if it is invested at 12.5%?

Solution

Let the initial sum be A. Therefore, the final sum is

$$F = 2A \quad \text{and} \quad i = 12.5\% = 0.125$$

Substituting these values for F and i into the final sum formula gives

$$F = A(1+i)^n$$

$$2A = A(1.125)^n$$

$$2 = (1.125)^n$$

Taking logs $\log 2 = n \log 1.125$

$$n = \frac{\log 2}{\log 1.125} = \frac{0.30103}{0.0511525} = 5.9 \text{ years}$$

Interest rates

A method of calculating the interest rate on an investment is explained in the following example.

Example 8.18

If £4,000 invested for 10 years is projected to accumulate to £6,000, what interest rate is used to derive this forecast?

Solution

$$A = 4{,}000 \quad F = 6{,}000 \quad n = 10$$

Substituting these values into the final sum formula

$$F = A(1+i)^n$$

$$\text{Gives} \quad 6{,}000 = 4{,}000(1+i)^{10}$$

$$1.5 = (1+i)^{10}$$

$$1+i = \sqrt[10]{1.5} = 1.0413797$$

$$i = 0.0414 = 4.14\%$$

A general formula for calculating the interest rate can be derived in a similar fashion, starting with the familiar final sum formula.

$$F = A(1+i)^n$$

$$\frac{F}{A} = (1+i)^n$$

$$\sqrt[n]{\frac{F}{A}} = (1+i)$$

$$\sqrt[n]{\frac{F}{A}} - 1 = i$$

This interest rate formula can also be written as

$$i = \left(\frac{F}{A}\right)^{1/n} - 1$$

Example 8.19

At what interest rate will £3,000 accumulate to £10,000 after 15 years?

Solution

Using the interest rate formula earlier,

$$i = \sqrt[n]{\left(\frac{F}{A}\right)} - 1 = \sqrt[15]{\left(\frac{10{,}000}{3{,}000}\right)} - 1 = \sqrt[15]{(3.3333)} - 1$$

$$= 1.083574 - 1 = 0.083574 = 8.36\%$$

Example 8.20

An initial investment of £50,000 increases to £56,711.25 after 2 years. What interest rate has been applied?

Solution

$$A = 5{,}000 \quad F = 56{,}711.25 \quad n = 2$$

Therefore,

$$\frac{F}{A} = \frac{56,711.25}{50,000} = 1.134225$$

Substituting these values into the interest rate formula gives

$$i = \sqrt[n]{\left(\frac{F}{A}\right)} - 1 = \sqrt[2]{(1.13455)} - 1 = 1.065 - 1 = 0.065 = 6.5\%$$

QUESTIONS 8.3

1. How much needs to be invested now in order to accumulate £10,000 in 6 years' time if the interest rate is 8%?
2. What sum invested now will be worth £500 in 3 years' time if it earns interest at 12%?
3. Do you need to invest more than £10,000 now if you wish to have £16,400 in 15 years' time and you have a deposit account paying 4% annual interest?
4. You need to have £7,500 on 1 January next year. How much do you need to invest at 1.3% per month if your investment is made on 1 June?
5. How much do you need to invest now in order to earn £25,000 in 10 years' time if the interest rate is
   (a) 10% (b) 8% (c) 6.5%?
6. How many complete years must £2,400 be invested at 5% in order to accumulate a minimum of £3,000?
7. For how long must £5,000 be kept in a deposit account paying 8% interest before it accumulates to £7,500?
8. If it can earn 9.5% interest, how long would any given sum of money take to treble its value?
9. If one needs to have a final sum of £20,000 how many years must one wait if £12,500 is invested at 9%?
10. How long will £70,000 take to accumulate to £100,000 if it is invested at 11%?
11. What annual rate of interest would cause
    (a) £6,000 to accumulate to £10,000 in 5 years?
    (b) £50,000 to accumulate to £60,000 in 2 years?
    (c) £3,000 to accumulate to £4,000 in 4 years?
    (d) £3,000 to accumulate to £8,000 in 10 years?
    (e) £600 to accumulate to £900 in 5 years?
12. What monthly rate of interest must be paid on a sum of £2,800 if it is to accumulate to £3,000 after 8 months?
13. Would you prefer (a) £5,000 now or (b) £8,000 in 4 years' time if money can be borrowed or lent at 11%?

8.5 INVESTMENT APPRAISAL: NET PRESENT VALUE

Assume that you have £10,000 to invest and that someone offers you the following proposal: pay £10,000 now and get £11,000 back in 12 months' time. Assume that the returns on this investment are guaranteed and there are no other costs involved. What would you do? Perhaps you would compare this return of 10% with the rate of interest your money could earn in a deposit account, say 4%. In a simple example like this the comparison of rates of return, known as the internal rate of return (IRR) method, is perhaps the most intuitively obvious method of judging the proposal.

This is not the preferred method for investment appraisal, however. The net present value (NPV) method has several advantages over the IRR method of comparing the project rate of return with the market interest rate. These advantages are explained more fully in the following section, but first it is necessary to understand what the NPV method involves.

We have already come across the concept of present value (PV) in Section 8.4. If a certain sum of money will be paid to you at some given time in the future, its PV is the amount of money that would accumulate to this sum if it was invested now at the ruling rate of interest. The process of finding the present value is often referred to as **discounting**.

Example 8.21

What is the present value of £1,500 payable in 3 years' time if the relevant interest rate is 4%?

Solution

Using the initial amount investment formula

$$F = £1,500 \quad i = 0.04 \quad n = 3$$

$$A = \frac{F}{(1+i)^n} = \frac{1,500}{(1.04)^3} = \frac{1,500}{1.124864} = £1,333.49$$

An investor would be indifferent between £1,333.49 now and £1,500 in 3 years' time. Thus £1,333.49 is the PV of £1,500 in 3 years' time at 4% interest.

In all the examples in this chapter it is assumed that future returns are assured with 100% certainty. In reality, investors may place greater importance on earlier returns just because the future is thought to be riskier. If some form of measure of the degree of risk can be estimated then more advanced mathematical methods exist which can be used to adjust the investment appraisal methods explained in this chapter. However, here we just assume that estimated future returns, and costs, are correct. An investor has to try to make the most rational decision based on whatever information is available.

The **net present value** (NPV) of an investment project is defined as the PV of the future returns minus the cost of setting up the project.

Example 8.22

An investment project involves an initial outlay of £600 now and a return of £1,000 in 5 years' time. Money can be invested elsewhere at 3%. What is the NPV?

Solution

The PV can be found using the initial amount formula as

$$A = F(1+i)^{-n} = 1000(1.03)^{-5} = £862.61$$

Therefore, $\text{NPV} = £862.61 - £600 = £262.61$

This project is clearly worthwhile. The £1,000 in 5 years' time is equivalent to £862.61 now and so the outlay required of only £600 makes it a bargain. In other words, one is being asked to pay £600 for something worth £862.61.

Another way of looking at the situation is to consider what alternative sum could be earned by the investor's £600. If £862.61 was invested for 5 years at 3% it would accumulate to £1,000. Therefore, the lesser sum of £600 must obviously accumulate to a smaller sum, which will be

$$F = A(1+i)^n = 600(1.03)^5 = 600(1.159274) = £695.56$$

The investor thus has the choice of

(a) putting £600 into this investment project and securing £1,000 in 5 years' time, or
(b) investing £600 at 3%, accumulating £695.56 in 5 years.

Option (a) is clearly the winner.

If the outlay is less than the PV of the future return, an investment must be a profitable venture. Therefore, the **criterion for deciding whether or not an investment project is worthwhile** is

$$\text{NPV} > 0$$

As well as deciding whether specific projects are profitable or not, an investor may have to decide how to allocate limited capital resources to competing investment projects. The **rule for choosing between projects** is that they should be ranked according to their NPV. If only one out of a set of possible projects can be undertaken then the one with the largest NPV should be chosen, as long as its NPV is positive.

Example 8.23

An investor can put money into any one of the following three ventures:

Project A costs £2,000 now and pays back £3,000 in 4 years
Project B costs £2,000 now and pays back £4,000 in 6 years
Project C costs £3,000 now and pays back £4,800 in 5 years

The current interest rate is 10%. Which project should be chosen?

Solution

$$\begin{aligned}
\text{NPV of project A} &= 3{,}000(1.1)^{-4} - 2{,}000 \\
&= 2{,}049.04 - 2{,}000 = £49.04 \\
\text{NPV of project B} &= 4{,}000(1.1)^{-6} - 2{,}000 \\
&= 2{,}257.90 - 2{,}000 = £257.90 \\
\text{NPV of project C} &= 4{,}800(1.1)^{-5} - 3{,}000 \\
&= 2{,}980.42 - 3{,}000 = -£19.58
\end{aligned}$$

Project B has the largest NPV and is therefore the best investment. Project C has a negative NPV and so would not be worthwhile even if there was no competition.

The investment examples considered so far have only involved a single return payment at some given time in the future. However, most real investment projects involve a stream of returns occurring over several time periods. The same principle for calculating NPV is used to assess these projects, but with the initial outlay being subtracted from the sum of the PVs of the different future returns.

Example 8.24

An investment proposal involves an initial payment of £40,000 now and then returns of £10,000, £30,000 and £20,000 respectively in 1, 2 and 3 years' time. If money can be invested elsewhere at 10% is this a worthwhile investment?

Solution

$$\begin{aligned}
\text{PV of £10,000 in 1 year's time} &= \frac{£10{,}000}{1.1} = £9{,}090.91 \\
\text{PV of £30,000 in 2 years' time} &= \frac{£30{,}000}{1.1^2} = £24{,}793.39 \\
\text{PV of £20,000 in 3 years' time} &= \frac{£20{,}000}{1.1^3} = £15{,}026.30 \\
\text{Total PV of future returns} & \qquad £48{,}910.60 \\
\textit{less}\ \text{initial outlay} & \qquad -£40{,}000
\end{aligned}$$

This NPV is greater than zero and so the project is worthwhile. At an interest rate of 10% one would need to invest a total of £48,910.60 to get back the projected returns and so £40,000 is clearly a bargain price.

The further into the future an expected return occurs the greater will be the discounting factor. This is obvious in Example 8.25, where the returns are the same each time period and the PV of each successive year's return is smaller than that of the previous year because it is multiplied by $(1 + i)^{-1}$.

Example 8.25

An investment project requires an initial outlay of £7,500 and will pay back £2,000 at the end of each of the next 5 years. Is it worthwhile if capital can be invested elsewhere at 12%?

Solution

$$\text{PV of £2,000 in 1 year's time} = \frac{£2,000}{1.12} = £1,785.71$$

$$\text{PV of £2,000 in 2 years' time} = \frac{£2,000}{1.12^2} = £1,594.39$$

$$\text{PV of £2,000 in 3 years' time} = \frac{£2,000}{1.12^3} = £1,423.56$$

$$\text{PV of £2,000 in 4 years' time} = \frac{£2,000}{1.12^4} = £1,271.04$$

$$\text{PV of £2,000 in 5 years' time} = \frac{£2,000}{1.12^5} = £1,134.85$$

Total PV of future returns	£7,209.55
less initial outlay	–£7,500.00
NPV of the project	–£ 290.45

The NPV < 0 and so this is not a worthwhile investment.

Investment appraisal using a spreadsheet

From the previous examples one can see that the mathematics involved in calculating the NPV of a project can be quite time consuming. For this type of problems a spreadsheet program can be a great help. Although Excel has a built-in NPV formula, this does not take the initial outlay into account and so care has to be taken when using it. We shall therefore construct a spreadsheet to calculate NPV from first principles.

To derive an algebraic formula for calculating NPV, assume that R_j is the net return in year j, i is the given rate of interest, n is the number of time periods in which returns occur and C is the initial cost of the project. Thus,

$$\text{NPV} = \frac{R_1}{1+i} + \frac{R_2}{(1+i)^2} + \cdots + \frac{R_n}{(1+i)^n} - C$$

Using the $\sum$ notation this can be written as

$$\text{NPV} = \sum_{j=1}^{n} \frac{R_j}{(1+i)^j} - C$$

If the initial outlay C in time period 0 is treated as a negative return (i.e. $R_0 = -C$) the formula can be more neatly stated as

$$\text{NPV} = \sum_{j=0}^{n} \frac{R_j}{(1+i)^j}$$

As $(1 + i)^0 = 1$, there will be no discounting of the initial outlay in the first term

$$\frac{R_0}{(1+i)^0}$$

The following example shows how a spreadsheet program based on this formula can be used to work out the NPV of a project. The answer obtained is then compared with the solution using the Excel built-in NPV function.

Example 8.26

An investment project requires an initial outlay of £25,000 with the following expected returns:

£5,000 at the end of year 1,
£6,000 at the end of year 2,
£10,000 at the end of year 3,
£10,000 at the end of year 4,
£10,000 at the end of year 5.

Is this a viable investment if money can be invested elsewhere at 15%?

Solution

Following the instructions for creating an Excel spreadsheet set out in Table 8.1, should give you a spreadsheet with the values shown in Table 8.2. This calculates the PVs of the returns in each year separately, including the outlay in year 0. It then sums the PVs, giving a total NPV of £1,149.15 which is positive and hence means that the project is a viable investment opportunity.

This can be compared with the answer obtained using the Excel built-in NPV formula. Because this formula always treats the number in the first cell of the range as the return at the end of year 1, the computed answer of £26,149.15 is the total PV of the returns in years 1 to 5 only. To get the overall NPV of the project, one has to subtract the initial outlay. (As the outlay amount was entered as a negative quantity this means it is added.) This adjusted Excel NPV figure is the same as the NPV calculated from first principles. Having an answer computed by two separate methods is a useful check.

The spreadsheet created for the previous example can be used to work out the NPV for other projects. The initial cost and returns need to be entered in cells B4 to B9 and the new interest rate goes in cell D2. Obviously, if there are more (or fewer) years when returns occur then rows will need to be added (or deleted or left blank) and the cell ranges in the NPV sum formulae adjusted accordingly.

As investment appraisal involves the comparison of different projects, as well as the assessment of the financial viability of individual projects, a spreadsheet can be adapted to work

Table 8.1 Setting up a spreadsheet for Example 8.26

CELL	Enter	Explanation
A1	Ex.8.26	Label to remind you what example this is
A3	YEAR	Column heading label
B3	RETURN	Column heading label
C3	PV	Column heading label
C1	Interest rate =	Label to tell you interest rate goes in next cell.
D1	15%	Value of interest rate. (NB Excel automatically treats this % format as 0.15 in any calculations.)
A4 to A9	*Enter numbers* 0 *to* 5	These are the time periods.
B4	−25000	Initial outlay (negative because it is a cost)
B5	5000	Returns at end of years 1 to 5
B6	6000	
B7	10000	
B8	10000	
B9	10000	
C4	=B4/(1+$D$1)^A4	Formula calculates PV corresponding to return in cell B4, time period in cell A4 and interest rate in cell D1. Note the $ to anchor cell D1.
C5 to C9	*Copy cell C4 formula down column C*	Calculates PV for return in each time period. Format to 2 dp as monetary values
B11	NPV =	Label to tell you NPV goes in next cell.
C11	=SUM(C4:C9)	Calculates NPV of project by summing PVs for each year in cells C4–C9, which includes the negative return of the initial outlay.
B13	Excel NPV	Label tells you Excel NPV goes in next cell.
B14	less cost =	Label tells you outlay goes in next cell.
C13	=NPV(D1,B5:B9)	The Excel NPV formula will calculate NPV based only on the interest rate in D1 and the 5 years of future returns in cells B5 to B9.
C14	=C13+B4	Adjusts the Excel computed NPV in C13 by subtracting initial outlay in B4. (This was entered as a negative number so it is added.)

Table 8.2 Excel output based on Table 8.1

	A	B	C	D
1	Ex 8.26		Interest rate=	15%
2				
3	YEAR	RETURN	PV	
4	0	−25000	−25000	
5	1	5000	4347.83	
6	2	6000	4536.86	
7	3	10000	6575.16	
8	4	10000	5717.53	
9	5	10000	4971.77	
10				
11		NPV =	1149.15	
12				
13		Excel NPV	£26,149.15	
14		less cost =	£1,149.15	

out the NPV for more than one project. The following example shows how the spreadsheet created for Example 8.26 can be extended so that two projects can be compared.

Example 8.27

An investor has to choose between two projects A and B whose outlay and returns are set out in Table 8.3. Which is the better investment if the going rate of interest is 10%?

Table 8.3 Outlays and returns of the two projects

(All values in £)	Project A	Project B
Initial outlay	30,000	30,000
Return in 1 year's time	6,000	8,000
Return in 2 years' time	10,000	8,000
Return in 3 years' time	10,000	8,000
Return in 4 years' time	10,000	8,000
Return in 5 years' time	8,000	8,000

Solution

Call up the worksheet which you created for Example 8.26 and make the changes shown in Table 8.4. This should give you a spreadsheet that looks similar to Table 8.5. The computed NPV for project B is £326.29 compared with £3,029.66 for project A. Therefore, although both projects are financially viable, the better investment is project A because it has the greater NPV.

Table 8.4 Setting up a spreadsheet for Example 8.27

CELL	Enter	Explanation
A1	Ex.8.27	New example label
B3	PROJECT A	Changed column heading label
C3	PV A	Changed column heading label
D1	10%	New value of interest rate
D3	PROJECT B	New column heading label for project B returns
E3	PV B	New column heading label for project B PVs
B4	−30000	Project A initial outlay
B5	6000	Project A returns at end of years 1 to 5
B6	10000	
B7	10000	
B8	10000	
B9	8000	
D4	−30000	Project B initial outlay
D5	8000	Project B returns at end of years 1 to 5
D6	8000	
D7	8000	
D8	8000	
D9	8000	
E4	=D4/(1+$D$1)^A4	Formula calculates PV for project B corresponding to return in cell D4.
E5 to E9	*Copy cell* E4 *formula down column* E	Calculates PV for project B for return in each time period.
E11	=SUM(E4:E9)	Calculates NPV of B by summing PVs for each time period.
E13	=NPV(D1,D5:D9)	Excel NPV formula applied to project B.
E14	=E13+D4	Adjusts the Excel NPV for project B.
C12	=B3	Writes 'PROJECT A' under relevant NPV.
E12	=D3	Writes 'PROJECT B' under relevant NPV.

Table 8.5 Excel output based on Table 8.4

	A	B	C	D	E
1	Ex 8.27		Interest rate =	10%	
2					
3	YEAR	PROJECT A	PV A	PROJECT B	PV B
4	0	–30000	–30000.00	–30000	–30000.00
5	1	6000	5454.55	8000	7272.73
6	2	10000	8264.46	8000	6611.57
7	3	10000	7513.15	8000	6010.52
8	4	10000	6830.13	8000	5464.11
9	5	8000	4967.37	8000	4967.37
10					
11		NPV =	3029.66		326.29
12			PROJECT A		PROJECT B
13		Excel NPV	£33,029.66		£30,326.29
14		less cost =	£3,029.66		£326.29

If you do not have access to a spreadsheet program, then you can still use a calculator to work out the NPV of different projects from first principles or you can use a financial calculator with an NPV function.

QUESTIONS 8.4

1. The following investment projects all involve an outlay now and a single return at some point in the future. Calculate the NPV and say whether or not each is a worthwhile investment:
   (a) £1,100 outlay, £1,500 return after 3 years, interest rate 8%,
   (b) £750 outlay, £1,000 return after 5 years, interest rate 9%,
   (c) £10,000 outlay, £12,000 return after 3 years, interest rate 8%,
   (d) £50,000 outlay, £75,000 return after 3 years, interest rate 14%,
   (e) £50,000 outlay, £100,000 return after 5 years, interest rate 14%,
   (f) £5,000 outlay, £7,000 return after 3 years, interest rate 6%,
   (g) £5,000 outlay, £7,750 return after 5 years, interest rate 6%,
   (h) £5,000 outlay, £8,500 return after 6 years, interest rate 6%.
2. An investor has to choose between the following three projects:
   A requires an outlay of £35,000 and returns £60,000 after 4 years,
   B requires an outlay of £40,000 and returns £75,000 after 5 years,
   C requires an outlay of £25,000 and returns £50,000 after 6 years.
   Which project would you advise this investor to put money into if the cost of capital is 10%?

Table 8.6 Returns from the three projects

Year	Project A	Project B	Project C
1	15,000	5,000	20,000
2	15,000	10,000	15,000
3	15,000	20,000	10,000
4	15,000	25,000	5,000

All values are given in £.

3. A firm has a choice between three investment projects, all of which involve an initial outlay of £36,000. The returns at the end of each of the next 4 years are given in Table 8.6. If the cost of borrowing capital is 15%, say (a) whether each project is viable or not, and (b) which is the best investment.
4. Is the project with the returns shown in Table 8.7 worthwhile if it requires an initial outlay now of £100,000 and money can be invested elsewhere at 6%?

Table 8.7 Investment project returns

Return at end of year	
Year 1	£10,000
Year 2	£12,000
Year 3	£15,000
Year 4	£18,000
Year 5	£20,000
Year 6	£20,000
Year 7	£20,000
Year 8	£15,000

5. Would you put £40,000 into a project which pays back nothing in the first year but then brings annual net returns of £12,000 from the end of year 2 until the end of year 6, assuming an interest rate of 8%?
6. A project requires an initial outlay of £20,000 and will pay back

   £1,000 at the end of years 1 and 2,
   £2,000 at the end of years 3 and 4,
   £5,000 at the end of years 5, 6, 7, 8, 9 and 10.

   Is this a worthwhile investment if the rate of interest is (a) 9%, (b) 10%?
7. Which of the three projects shown in Table 8.8 is the best investment if the interest rate is 20%?

Table 8.8 Outlays and returns of the three projects

	Project A	Project B	Project C
Outlay now	85,000	40,000	40,000
Return after year 1	20,000	15,000	10,000
Return after year 2	24,000	20,000	12,000
Return after year 3	30,000	25,000	12,000
Return after year 4	30,000	0	12,000
Return after year 5	25,000	0	15,000
Return after year 6	20,000	0	15,000

All values are given in £.

8.6 THE INTERNAL RATE OF RETURN

The internal rate of return (IRR) method of investment appraisal involves finding the rate of return (r) on a project and comparing it with the market rate of interest (i). If $r > i$ then the project is viable. Alternative projects can be ranked according to the value of the different rates of return.

Example 8.28

Find the IRR for the three projects in Table 8.9, decide whether they are viable if the market rate of interest is 7%, and then rank them in order of profitability according to the IRR method.

Table 8.9 Initial outlays and returns

	Project A	Project B	Project C
Initial outlay	£5,000	£4,000	£8,000
Return after 1 year	£5,750	£4,300	£8,500

Solution

In this simple example it is obvious from basic arithmetic that

$$\text{IRR for A} = r_A = \frac{750}{5{,}000} = 0.15 = 15\%$$

$$\text{IRR for B} = r_B = \frac{300}{4{,}000} = 0.075 = 7.5\%$$

$$\text{IRR for C} = r_C = \frac{500}{8{,}000} = 0.0625 = 6.25\%$$

Only projects A and B are viable, with an IRR of more than the market rate of interest of 7%. C is not viable. A is preferred to B because $r_A > r_B$.

From this example, one can see that the IRR is the rate of interest which, if applied to the initial outlay, gives the return in year 1. Put another way, r is the rate of interest at which the PV of the future return equals the initial outlay, thus making the NPV of the whole project zero. This principle can be used to help calculate the IRR for more complex problems.

Example 8.29

Use the IRR method to evaluate the following project given a market rate of interest of 11%.

Initial outlay	£75,000
Return at end of year 1	£15,000
Return at end of year 2	£20,000
Return at end of year 3	£20,000
Return at end of year 4	£25,000
Return at end of year 5	£25,000
Return at end of year 6	£12,000

Solution

To obtain the IRR one needs to find the value of r for which

$$0 = -75,000 + 15,000(1+r)^{-1} + 20,000(1+r)^{-2} + 20,000(1+r)^{-3} + 25,000(1+r)^{-4} + 25,000(1+r)^{-5} + 12,000(1+r)^{-6}$$

However, the algebraic method of solution of this sort of polynomial is far too complex and time consuming to consider using here. The most practical method is to use a spreadsheet. Excel has a built-in IRR formula which can immediately calculate r. You could also find r by using a spreadsheet to calculate the project NPV for a range of interest rates and then identifying the interest rate at which NPV is zero. Instructions for constructing a spreadsheet to solve this problem by both methods are shown in Table 8.10.

Table 8.10 Setting up a spreadsheet for Example 8.29

CELL	Enter	Explanation
A1	Ex.8.29	Label to remind you what example this is
A3	YEAR	Column heading label
B3	RETURN	Column heading for project returns
D2	Interest	Column heading label for the range of interest rates for
D3	rate	which NPV will be computed
E3	NPV	Column heading label
A4 to A10	*Enter numbers* 0 *to* 6	These are the time periods for this example.
B4	–75000	Initial outlay (negative because it is a cost)
B5	15000	Project returns at end of years 1 to 6
B6	20000	
B7	20000	
B8	25000	
B9	25000	
B10	12000	
D4	4%	Interest rate to start range used
D5	=D4+0.01	Calculates a 1% rise in interest rate.
D6 to D20	*Copy cell* D5 *formula down column* D	Calculates a series of interest rates with increments of 1%.
E4	=NPV(D4,B$5:B$10)+B$4	Calculates project NPV corresponding to interest rate in D4 using Excel NPV formula less outlay in B4. Note the $ to anchor rows.
E5 to E20	*Copy cell E4 formula down column E*	Calculates NPV corresponding to interest rates in column D.
A12	IRR =	Label to tell you IRR calculated in next cell.
B12	= IRR(B4:B10)	Excel IRR formula calculates IRR of project returns in cells B4 to B10, which includes the negative return of the initial outlay.

The resulting spreadsheet should look like Table 8.11. This shows that the rate of interest that corresponds to an NPV of zero will lie somewhere between 14% and 15%, which checks out with the precise value for the IRR of 14.14% computed in cell B12. The market rate of interest given in the question is 11% and so, as the calculated IRR of 14.14% exceeds this, the project is worthwhile.

Table 8.11 Excel output based on Table 8.10

	A	B	C	D	E
1	Ex.8.29				
2				Interest	
3	YEAR	RETURN		Rate	NPV
4	0	−75000		4%	27096.19
5	1	15000		5%	23813.36
6	2	20000		6%	20686.58
7	3	20000		7%	17706.57
8	4	25000		8%	14864.67
9	5	25000		9%	12152.86
10	6	12000		10%	9563.64
11				11%	7090.04
12	IRR =	14.14%		12%	4725.54
13				13%	2464.06
14				14%	299.94
15				15%	−1772.14
16				16%	−3757.14
17				17%	−5659.71
18				18%	−7484.20
19				19%	−9234.68
20				20%	−10914.99

Deficiencies of the IRR method

Although the IRR method may appear to be the most obvious and easily understood criterion for deciding on investment projects, and is still frequently used, it has several deficiencies which can make it less useful than the NPV method.

First, the IRR ignores the total value of the profit, as illustrated in the following example.

Example 8.30

Project A involves an initial outlay of £18,000 and a return in 1 year's time of £20,000. Project B involves an initial outlay of £2,000 and a return in 1 year's time of £2,500. The interest rate is 6%. Which would be the better investment?

Solution

The IRR method ranks B as the better investment opportunity, since

$$r_A = \frac{20,000}{18,000} - 1 = 1.11 - 1 = 0.11 = 11\%$$

$$r_B = \frac{2,500}{2,000} - 1 = 1.25 - 1 = 0.25 = 25\%$$

The NPV method, however, would rank A as the better investment since

$$\text{NPV}_A = -18,000 + \frac{20,000}{1.06} = -18,000 + 18,867.92 = £867.92$$

$$\text{NPV}_B = -2,000 + \frac{2,500}{1.06} = -2,000 + 2,358.49 = £358.49$$

If the firm has a straightforward choice between A and B, then A is clearly the better investment. (The possibility of the firm using its initial £18,000 for investing in nine separate projects all with the same returns as B is ruled out.)

Assume that the firm has up to £18,000 at its disposal. If it puts this all into project A, then at the end of the year its total assets will be £20,000.

If it puts £2,000 into project B, then it can also invest the remaining £16,000 elsewhere at the going rate of interest of 6%. Its total assets will therefore be

	Return project B	£2,500
plus	£16,000 invested at 6% = 16,000 × 1.06	£16960
	Total assets	£19,460

This shows that the firm is in a worse financial position at the end of the year than if it had chosen project A. Thus, the NPV recommendation of project A is the correct one.

The second advantage that the NPV investment appraisal method has over the IRR method is that it can easily cope with forecasts of variable interest rates. The IRR method just involves comparing a project's IRR with one given market interest rate and so it could not be applied to cases such as in Example 8.31.

Example 8.31

An investment project involves an initial outlay of £25,000 and net annual returns as follows:

£6,000 at the end of year 1,
£8,000 at the end of year 2,
£8,000 at the end of year 3,
£10,000 at the end of year 4,
£6,000 at the end of year 5.

The interest rate is currently 15% but it is forecast to fall to 12% next year and 10% the following year. It will then rise by 1 percentage point each year. Is the project worthwhile?

Solution

Because of the variation in interest rates, the Excel NPV formula cannot be used and we have to adjust the basic discounting formula to allow for the different discount rates that apply to each year. Thus,

$$\text{NPV} = -25{,}000 + \frac{6{,}000}{1.15} + \frac{8{,}000}{1.15 \times 1.12} + \frac{8{,}000}{1.15 \times 1.12 \times 1.1}$$

$$+ \frac{10{,}000}{1.15 \times 1.12 \times 1.1 \times 1.11} + \frac{6{,}000}{1.15 \times 1.12 \times 1.1 \times 1.11 \times 1.12}$$

$$= -25{,}000 + \frac{6{,}000}{1.15} + \frac{8{,}000}{1.288} + \frac{8{,}000}{1.4168} + \frac{10{,}000}{1.572648} + \frac{6{,}000}{1.7613657}$$

$$= -25{,}000 + 5{,}217.39 + 6{,}211.18 + 5{,}646.53 + 6{,}358.70 + 3{,}406.45$$

$$= -25{,}000 + 26{,}840.25 = £1{,}840.25$$

This NPV is positive and so the investment is worthwhile.

Although this is not a straightforward NPV calculation, a spreadsheet can be constructed to do the calculations. One suggested format for solving Example 8.31 is

Table 8.12 Setting up a spreadsheet for Example 8.31

	A	B	C	D	E	F
1	Ex 8.31	NPV	WITH	VARIABLE	INTEREST	RATES
2						
3	YEAR	i	DISCOUNT	FACTOR	RETURN	PV
4	0	0	=1/(1 + B4)	1	−25000	=D4*E4
5	1	0.15	=1/(1 + B5)	=D4*C5	6000	=D5*E5
6	2	0.12	=1/(1 + B6)	=D5*C6	8000	=D6*E6
7	3	0.1	=1/(1 + B7)	=D6*C7	8000	=D7*E7
8	4	0.11	=1/(1 + B8)	=D7*C8	10000	=D8*E8
9	5	0.12	=1/(1 + B9)	=D8*C9	6000	=D9*E9
10						
11				TOTAL	NPV	=SUM(F4.F9)

Table 8.13 Excel output based on Table 8.12

	A	B	C	D	E	F
1	Ex 8.31	NPV	WITH	VARIABLE	INTEREST	RATES
2						
3	YEAR	i	DISCOUNT	FACTOR	RETURN	PV
4	0	0	1	1	−25000	−25000.00
5	1	0.15	0.8695652	0.869565	6000	5217.39
6	2	0.12	0.8928571	0.776398	8000	6211.18
7	3	0.1	0.9090909	0.705816	8000	5646.53
8	4	0.11	0.9009009	0.63587	10000	6358.70
9	5	0.12	0.8928571	0.567741	6000	3406.45
10						
11				TOTAL	NPV =	1840.25

shown in Table 8.12, which shows the formulae to enter in relevant cells. This should produce the figures shown in Table 8.13, which confirm that NPV is £1,840.25.

A third drawback of the IRR method is that there may not be one unique solution for r when there are several negative terms in the polynomial to be solved. This point was made in Chapter 7 when the solution of polynomial equations was discussed. Apart from the initial outlay, negative returns may occur if further investment is required, or if a company has to pay to dismantle a project or to return the site to an environmentally acceptable state at the end of its useful life. However, investment projects with multiple solutions for the IRR are unusual and you are unlikely to come across them.

QUESTIONS 8.5

1. Calculate the IRR for the projects in Table 8.14 and then say whether or not the IRR ranking is consistent with the NPV ranking for these projects if the market rate of interest is 15%.

Table 8.14 Initial outlays and returns

	Project A	Project B	Project C	Project D
Outlay now	£20,000	£6,000	£25,000	£10,000
Return after 1 year	£24,000	£8,500	£30,000	£12,000

2. Two projects A and B each involve an initial outlay of £40,000 and guarantee the returns (in £) given in Table 8.15. The market rate of interest is 18%. Which is the better investment according to (a) the IRR criterion, (b) the NPV criterion?

Table 8.15 Investment returns from the two projects

	Project A	Project B
End of year 1	15,000	10,000
End of year 2	20,000	12,000
End of year 3	25,000	12,000
End of year 4	0	12,000
End of year 5	0	15,000
End of year 6	0	15,000

All values are given in £.

3. Using a spreadsheet, find the IRR and show that the NPV of the following project is zero when the discount rate used is approximately equal to this IRR.

Outlay now:	£25,000		
Annual returns:	(1) £4,000;	(2) £6,000;	(3) £7,500;
	(4) £7,500;	(5) £10,000;	(6) £10,000.

8.7 GEOMETRIC SERIES AND ANNUITIES

An annuity is a financial product that will give a fixed return each time period. Customers normally pay an up-front lump sum in exchange for this income stream.

For example, someone might pay a fixed sum now to secure a guaranteed payment of £9,000 a year for the next 20 years.

As final salary based pensions schemes disappear, and it becomes more common for workers to move between employers throughout their working life, more and more people need to buy an annuity to provide a regular income in old age. Throughout their working life they save money in a pension scheme to accumulate a 'pension pot', normally with tax relief on contributions, and then they can use this to buy an annuity when they retire.

Pension annuities are normally guaranteed for life. The calculations that a pension provider will use to determine what annual income can be bought for a given sum will depend on the life expectancy of the purchaser, which will provide an estimate of how many years the annuity pension will be paid out for, and also on the market interest rate. However, to explain the basic principles involved, in the examples in this chapter we will assume that an annuity is taken out for a fixed time period, e.g. 20 years.

As an annuity is a stream of payments over a period of time, we can use the present value formula to help calculate the price of an annuity. The present value of a fixed return of £a per year for the next n years when the interest rate is i will be

$$\text{PV} = \frac{a}{1+i} + \frac{a}{(1+i)^2} + \cdots + \frac{a}{(1+i)^n}$$

This can also be written as

$$\text{PV} = a(1+i)^{-1} + a(1+i)^{-2} + \cdots + a(1+i)^{-n}$$

This sequence of terms is an example of what is known as a 'geometric series'. We shall derive the formula for the sum of a geometric series because it can be used to determine the price of an annuity and it will also help us to analyze other financial products such as drawdown pensions.

Geometric series

A geometric series is a sequence of terms where each successive term is the previous term multiplied by a **common ratio**. The series starts with a given initial term and may contain any number of terms.

Example 8.32

If the given initial term is 24 and the common ratio is 5, what is the corresponding geometric series? (Find up to six terms.)

Solution

The series will be

	24	24×5	24×5^2	24×5^3	24×5^4	24×5^5
or	24	120	600	3,000	15,000	75,000

Example 8.33

A firm's sales revenue is initially £40,000 and then grows by 20% each successive year. What is the pattern of sales revenue over 5 years?

Solution

Each year's sales are 120% of the previous year's. The time profile of sales revenue is therefore a geometric series with an initial term of £40,000 and a common ratio of 1.2. Thus (in £) the series is

$$40{,}000 \quad 40{,}000\times1.2 \quad 40{,}000\times1.2^2 \quad 40{,}000\times1.2^3 \quad 40{,}000\times1.2^4$$

If we use the algebraic notation a for the initial term, k for the common ratio and n for the number of terms, then the general form of a geometric series will be

$$a, ak, ak^2, \ldots, ak^{n-1}$$

Note that the last (nth) term is ak^{n-1} and not ak^n, because multiplication by k does not start until the second term.

Sum of a geometric series

The sum of a geometric series can be found by simply adding all the terms together. This is easy enough to do using a calculator for the previous examples. More complex series are more difficult to sum in this way, however, and so we need to derive a formula for summing them.

The general format for the sum of a geometric series with n terms will be

$$\text{GP}_n = a + ak + ak^2 + \cdots + ak^{n-1} \qquad (1)$$

Multiplying each term by k gives

$$k\text{GP}_n = \quad ak + ak^2 + \cdots + ak^{n-1} + ak^n$$

Subtracting (1) $\quad \text{GP}_n = a + ak + ak^2 + \cdots + ak^{n-1}$

gives $\quad (k-1)\text{GP}_n = -a + ak^n$

Therefore,

$$\text{GP}_n = \frac{-a+ak^n}{k-1} = \frac{-a\left(1-k^n\right)}{k-1} = \frac{(-1)a\left(1-k^n\right)}{(-1)(1-k)} = \frac{a\left(1-k^n\right)}{1-k}$$

Thus, **the formula for the sum of a geometric series** is

$$\text{GP}_n = \frac{a\left(1-k^n\right)}{1-k}$$

The following examples illustrate how this formula can be used to sum some simple numerical sequences of numbers.

Example 8.34

Use the geometric series sum formula to sum the geometric series

$$15 \quad 45 \quad 135 \quad 405 \quad 1,215 \quad 3,645$$

Solution

In this geometric series with six terms, each number except the first is 3 times the previous one. Thus,

$$a=15, \quad k=3, \quad n=6$$

Substituting these values into the geometric series sum formula we get

$$\text{GP}_n = \frac{a\left(1-k^n\right)}{1-k} = \frac{15\left(1-3^6\right)}{-2} = \frac{15(-728)}{-2} = 15 \times 364 = 5,460$$

You can check that this formula gives the same answer as that found using a calculator. In fact, in simple examples like this using the calculator may be the quicker method, but in other more complex cases the formula will provide the quickest method of solution.

Example 8.35

A firm expects its sales to grow by 12% per month. If its January sales figure is £9,200 per month what will its expected total annual sales be?

Solution

The firm's total annual sales will be the sum of the geometric series

$$9,200 + 9,200(1.12) + 9,200(1.12)^2 + \cdots + 9,200(1.12)^{11}$$

Thus a = 9,200, k = 1.12 and n = 12, and so the sum is

$$\text{GP}_n = \frac{9,200\left(1-1.12^{12}\right)}{1-1.12} = \frac{9,200(1-3.895976)}{1-1.12} = \frac{-26,642.979}{-0.12} = £222,024.83$$

Annuity prices

The formula for the sum of a geometric series can be used in present value (PV) calculations for a constant stream of returns such as the regular payments made to someone who has bought an annuity. However, one has to be very careful not to get the algebraic terminology mixed up when the initial payback figure includes the common ratio, as in the following example.

Example 8.36

An annuity will pay £8,000 at the end of each year for 5 successive years, the first payment being 12 months from the initial purchase date. What is the maximum price any rational investor would pay for such an annuity if the opportunity cost of capital is 10%?

Solution

The maximum annuity purchase price will be the PV of the stream of returns, using 10% as the discount rate. Therefore (in £)

$$\text{PV} = \frac{8{,}000}{1.1} + \frac{8{,}000}{1.1^2} + \frac{8{,}000}{1.1^3} + \frac{8{,}000}{1.1^4} + \frac{8{,}000}{1.1^5}$$

This is a geometric series with five terms. The first term a is $\frac{8{,}0000}{1.1}$ (*not* 8,000). The common ratio k is $\frac{1}{1.1}$. Therefore,

$$\text{PV} = \frac{a\left(1-k^n\right)}{1-k} = \frac{\frac{8{,}000}{1.1}\left[1-\left(\frac{1}{1.1}\right)^5\right]}{1-\frac{1}{1.1}}$$

$$= \frac{8{,}000(1-0.6209211)}{1.1\left(1-\frac{1}{1.1}\right)} = \frac{8{,}000(0.3790789)}{1.1-1}$$

$$= \frac{3{,}032.6312}{0.1} = £30{,}326.31 \text{ annuity price}$$

In this example some of the terms cancelled out. The same terms will cancel in any annuity PV calculations and so a simplified general formula for the PV of an annuity can be derived. Assuming an annual payment of R for n years and an interest rate of i, the annuity value is

$$\text{PV} = \frac{R}{1+i} + \frac{R}{(1+i)^2} + \cdots + \frac{R}{(1+i)^n}$$

In this geometric series the initial term is

$$a = \frac{R}{1+i}$$

And the common ratio is

$$k = \frac{1}{1+i}$$

Therefore,

$$\text{PV} = \frac{a(1-k^n)}{1-k} = \frac{\frac{R}{1+i}\left[1-\left(\frac{1}{1+i}\right)^n\right]}{1-\frac{1}{1+i}} = \frac{\frac{R}{1+i}\left[1-\left(\frac{1}{1+i}\right)^n\right]}{\frac{1+i-1}{1+i}}$$

$$= \frac{R\left[1-\frac{1}{(1+i)^n}\right]}{1+i-1} = \frac{R\left[1-(1+i)^{-n}\right]}{i}$$

Thus, the **formula for the PV of an annuity** is

$$\text{PV} = \frac{R\left[1-(1+i)^{-n}\right]}{i}$$

We can use this formula to check the answer to Example 8.36.

Given that $R = 8{,}000$, $i = 0.1$ and $n = 5$, then

$$\text{PV} = \frac{R\left[1-(1+i)^{-n}\right]}{i} = \frac{8{,}000\left[1-(1.1)^{-5}\right]}{0.1} = £30{,}326.31$$

This is the same answer as that derived from first principles.

Example 8.37

An annuity will pay £2,000 a year for the next 5 years, with the first payment in 12 months' time. Capital can be invested elsewhere at an interest rate of 4%. Is £8,000 a reasonable price to pay for this annuity?

Solution

For this annuity (in £)

$$\text{PV} = 2{,}000(1.04)^{-1} + 2{,}000(1.04)^{-2} + 2{,}000(1.04)^{-3} + 2{,}000(1.04)^{-4} + 2{,}000(1.04)^{-5}$$

In this example, the annual payment $R = 2{,}000$, $i = 0.04$ and $n = 5$. Therefore,

$$\text{PV} = \frac{R\left[1=(1+i)^{-n}\right]}{i} = \frac{2{,}000\left[1-(1.04)^{-5}\right]}{0.4} = \frac{2{,}000(1-0.8219271)}{0.04}$$

$$= £8{,}903.64$$

The PV of this annuity is greater than its purchase price of £8,000 and so it is clearly a worthwhile investment.

Example 8.38

What would you pay for an annuity that promises to pay £450 a year for the next 10 years given an interest rate of 8%?

Solution

Given R = £450, i = 8% = 0.08 and n = 10, the present value of the future returns is

$$\text{PV} = \frac{R\left[1 = (1+i)^{-n}\right]}{i} = \frac{450\left[1-(1.08)^{-10}\right]}{0.08} = \frac{450(1-0.4631935)}{0.08}$$
$$= £3{,}019.54$$

Thus, any price less than £3,019.54 would make this annuity a worthwhile purchase.

> **QUESTIONS 8.6**
>
> 1. In the geometric series that follows, (i) identify the common ratio, (ii) say what the sixth term will be and (iii) calculate the sum of each series up to ten terms using the formula for summation of a geometric series.
>   (a) 8, 20, 50, …
>   (b) 0.5, 1.5, 4.5, …
>   (c) 2, 2.8, 3.92, …
>   (d) 60, 48, 38.4, …
>   (e) 2.4, 1.8, 1.35, …
> 2. A firm starts producing a new product. It sells 420 units in January and then sales increase by 10% each month. What will total demand be in the last 6 months of the year?
> 3. What would be the maximum price you would pay for the following annuities if money can be invested elsewhere at 8%?
>   (a) Annuity A pays £200 a year for the next 8 years.
>   (b) Annuity B pays £900 a year for the next 4 years.
>   (c) Annuity C pays £6,000 a year for the next 12 years.
> 4. Would you pay £3,500 for an annuity which guarantees to pay you £750 annually for the next 7 years if you can invest money elsewhere at 9%?
> 5. What would be a reasonable price to pay for a pension plan which guarantees to pay £200 a month for the next 2 years if you can earn 1.2% a month on your bank deposit account?

8.8 PERPETUAL ANNUITIES

Some forms of annuities are called 'perpetual annuities' which promise a fixed annual monetary return indefinitely. For example, a bond that pays a fixed 6% return every year on a nominal price of £100 is effectively a perpetual annuity of £6. The present value of such an annuity at a rate of interest i would be

$$\frac{6}{1+i}+\frac{6}{(1+i)^2}+\cdots+\frac{6}{(1+i)^n}+\cdots$$

As n continues to infinity, each successive term gets smaller and smaller, but the sum of this sequence continues to grow as n gets bigger. You cannot sum such an infinite series of numbers without using the formula for the sum of an infinite geometric series.

The PV of the stream of returns from a perpetual annuity is an infinite geometric progression. Whether or not one can find the sum of an infinite geometric progression depends on whether the progression is convergent or divergent. Before looking at the formal mathematical conditions for convergence or divergence, these concepts are illustrated with some simple examples.

When you were at school you may have come across the teaser about the frog jumping across a pond, which goes something like this: 'A frog is sitting on a lily leaf in the middle of a circular pond. The pond is 10 metres in radius and the frog jumps 5 metres with its first jump. Its second jump is 2.5 m, its third jump 1.25 m and so on. How many jumps will it take for the frog to reach the edge of the pond? Assume that each time it jumps it lands on a leaf.'

The answer is, of course, 'never'. Each time the frog manages to jump half of the remaining distance to the edge of the pond. The total distance the frog travels in n jumps is given by the sum of the geometric series

$$5+5(0.5)+5(0.5)^2+\cdots+5(0.5)^{n-1}$$

As n gets larger the sum of this series continues to increase but never actually reaches 10 metres. Only if an infinite number of jumps can be made will the total distance travelled be 10 metres. Thus, in this example we have a geometric series which converges on 10 metres.

Geometric series may also be divergent. For example, the sequence

40 60 90 135 ...etc.

can be written as the geometric series

$$40 \quad 40(1.5) \quad 40(1.5)^2 \quad 40(1.5)^3 \quad \ldots \quad 40(1.5)^n$$

It is intuitively obvious that each successive term is larger than the previous one. Therefore, as the number of terms approaches infinity the sum of the series will also become infinitely large. Also, the last term $40(1.5)^n$ will itself become infinitely large. There is, thus, no set quantity towards which the sum of the series converges.

As you will probably have already guessed by now, the way to distinguish a convergent and a divergent geometric series is to look at the value of the common ratio k, or rather its absolute value as it is possible to have a negative common ratio.

If $|k| > 1$ then successive terms become larger and larger and the series **diverges**.
If $|k| < 1$ then successive terms become smaller and the series **converges**.

To find the sum of a convergent geometric series (such as the case of a perpetual annuity) let us look again at the general formula for the sum of a geometric series:

$$\mathrm{GP}_n = \frac{a\left(1-k^n\right)}{1-k}$$

This can be rewritten as

$$\mathrm{GP}_n = \frac{a}{1-k} - \left(\frac{a}{1-k}\right)k^n \qquad (1)$$

If $-1 < k < 1$ then $k^n \to 0$ as $n \to \infty$ (i.e. the value of k^n approaches zero as n approaches infinity) and so the second term in (1) will disappear and the sum to infinity will be

$$\mathrm{GP}_n = \frac{a}{1-k} \qquad (2)$$

We can now use formula (2) for the frog example. The total distance jumped is

$$\sum_{n-0}^{\infty} 5(0.5)^n$$

In this geometric series $k = 0.5$ and $a = 5$. The sum for an infinite number of terms will, thus, be

$$\frac{a}{1-k} = \frac{5}{1-0.5} = \frac{5}{0.5} = 10 \text{ metres}$$

The PV of a perpetual annuity can also be found using this formula although care must be taken to include the discounting factor in the initial term, as explained in the following example.

Example 8.39

What is the PV of an annuity which will pay £6 a year *ad infinitum*, with the first payment due in 12 months' time? Assume that capital can be invested elsewhere at 5%.

Solution

$$\mathrm{PV} = \frac{6}{1.05} + \frac{6}{1.05^2} + \cdots + \frac{6}{1.05^n} \qquad \text{where } n \to \infty$$

In this geometric series $a = \dfrac{6}{1.05}$ and $k = \dfrac{1}{1.05}$

This is clearly convergent as $|k| < 1$. The sum to infinity is therefore

$$\text{NPV} = \frac{a}{1-k} = \frac{\frac{6}{1.05}}{1-\frac{1}{1.05}} = \frac{6}{1.05\left(1-\frac{1}{1.05}\right)} = \frac{6}{1.05-1} = \frac{6}{0.05} = £120$$

A simplified formula for the PV of a perpetual annuity can be derived as certain terms will always cancel out, as in Example 8.39.

Assume that an annuity pays a fixed return R each year, starting in 12 months' time, and the opportunity cost of capital is i. For this annuity

$$\text{PV} = R(1+i)^{-1} + R(1+i)^{-2} + \ldots + R(1+i)^{-n} \quad \text{where } n \to \infty$$

In this infinite geometric series the initial value is $a = R(1+i)^{-1}$ and the constant ratio $k = (1+i)^{-1}$. Therefore, using the formula for the sum of an infinite converging geometric series

$$\text{PV} = \frac{a}{1-k} = \frac{R(1+i)^{-1}}{1-(1+i)^{-1}} = \frac{R}{(1+i)\left[1-(1+i)^{-1}\right]} = \frac{R}{1+i-1} = \frac{R}{i}$$

Thus, the **formula for the PV of a perpetual annuity** is

$$\text{PV} = \frac{R}{i}$$

Reworking Example 8.39 using this formula we get

$$\text{PV} = \frac{6}{0.05} = £120$$

which is identical to the answer derived from first principles, although the formula obviously makes the calculations much easier.

Example 8.40

An investment opportunity involves an initial outlay of £175,000 and gives a £5,000 annual return, starting in 12 months' time and continuing indefinitely. Capital can be invested elsewhere at 2.5%. Is this worth considering?

Solution

The PV of the annual income stream can be calculated using the formula for the PV of a perpetual annuity as

$$\text{PV} = \frac{R}{i} = \frac{5{,}000}{0.025} = £200{,}000$$

This is greater than the initial outlay of £175,000 and so this investment is clearly an attractive proposition.

Example 8.41

What would be the maximum price you would pay for a perpetual annuity that will pay £520 per annum, starting in 12 months' time, given an interest rate of 3.25%?

Solution

$$\text{PV} = \frac{R}{i} = \frac{520}{0.0325} = £16{,}000$$

This is the maximum price a rational investor would pay for this annuity.

QUESTIONS 8.7

1. Identify which of the following geometric series are convergent and then calculate the sum to which these series converge as the number of terms approaches infinity:
   (a) 4, 6, 9, …
   (b) 120, 96, 76.8, …
   (c) 0.8, −1.2, 1.8, …
   (d) 36, 12, 4, …
   (e) 500, 500(0.48), $500(0.48)^2$, …
   (f) 850, $850(1.2)^{-1}$, $850(1.2)^{-2}$, …
2. What is the maximum price you would pay for a perpetual annuity that will commence annual payments of £400 in 12 months' time if the market rate of interest is 13%?
3. Is it worth paying £40,000 for a perpetual annuity of £1,500 per annum, commencing payments in 12 months' time, if money can be invested elsewhere at 3%?
4. What would the price of an annuity paying £12,000 per annum be (starting in 12 months' time) if the market rate of interest is
   (a) 5%, (b) 10%, (c) 15%, (d) 20%?
5. A government bond guarantees an annual payment of £140 in perpetuity. What will it be priced at, given a market rate of interest of 4%?

8.9 PENSION POTS, ANNUITY INCOME AND DRAWDOWN PENSIONS

In Section 8.7 we saw how the price of an annuity guaranteeing a fixed annual income could be calculated. However, although many people will rely on an annuity for much of their pension income when they retire, they may put the question the other way around and ask how much income they can get for a given size 'pension pot'. (A pension pot is just the total amount accumulated in an individual's pension fund.) Alternatively, instead of buying an annuity when they retire, they may decide just to spend a certain amount from their pension pot each year, and will wish to know how much they can draw out of their pension pot in this way. This section addresses these questions and provides some answers. The analysis of savings schemes that may help accumulate a pension pot is considered later in Section 8.12.

Pension pots and annuity income

The formula for the price of an annuity can be used to derive a formula that will show how big an annuity can be bought for a given sum, as the following example explains.

Example 8.42

If a woman retiring at the age of 67 has saved £350,000 in her pension pot and interest rates are 4%, how big an annual pension could she expect to receive if she bought a 20-year annuity?

(Note: in practice pension annuities guarantee to pay an income for life, but to explain the calculations we shall just assume that the annuity provider sets the price based on 20 years average life expectancy for a 67-year-old woman.)

Solution

From the question we know that $i = 4\% = 0.04$ and $n = 20$. We do not know the annual payment R but we do know that PV = £350,000 because this is what is spent on the annuity. If we reverse the sides of the annuity price formula this becomes

$$\frac{R\left[1-(1+i)^{-n}\right]}{i} = PV$$

Multiplying both sides by i and dividing by $[1 - (1 + i)^{-n}]$ we get

$$R = \frac{iPV}{\left[1-(1+i)^{-n}\right]}$$

So, substituting in the given values for this example, annual pension income is

$$R = \frac{0.04 \times 350{,}000}{\left[1-(1+0.04)^{-20}\right]} = \frac{14{,}000}{[1-0.4563869]} = £25{,}753.61$$

Thus, for any similar questions we can use this **formula for annual pension income from an annuity**

$$R = \frac{iPV}{\left[1-(1+i)^{-n}\right]}$$

Example 8.43

What annual income could be expected if someone is prepared to pay £175,000 for a 20-year annuity when they retire and interest rates are expected to remain at 2.4%?

Solution

The known values are

$$i = 2.4\% = 0.024, \quad n = 20, \quad \text{PV} = 175{,}000$$

so the annuity income formula gives annual income as

$$R = \frac{iPV}{\left[1-(1+i)^{-n}\right]} = R = \frac{0.024 \times 175{,}000}{\left[1-(1+0.024)^{-20}\right]} = £11{,}119.98$$

Drawdown pensions

Although some private pension schemes may require individuals to purchase an annuity when they retire, there are other ways that people can try to ensure that they have an adequate income in their retirement. One approach is to save money into a pension pot and then to drawdown and use a proportion of the total pension pot during each year of retirement. Because, in the UK, income tax relief is given on contributions to pension schemes there may be certain restrictions on how a pension pot may be used. For example, people in the UK are allowed to withdraw money from approved pension schemes as a lump sum, rather than being forced to buy an annuity when they retire, and this drawdown pension method has become more widely used.

Although this approach allows individuals more freedom of choice regarding when they take money out of their pension pot to fund their retirement, there is an obvious possible downside, as individuals may be left without a pension income if they use up their pension pot too fast. On the other hand, if individuals use up their pension pot so slowly that some remains when they die then the remainder will be added to their estate that they will leave as inheritance to their heirs. This contrasts with annuity-based pensions which, although they normally pay out each year until you die, will leave no balance for your heirs to inherit.

In practice the interest earned and actual payments from a drawdown pension may be affected by tax, and a person may possibly receive other pensions or state benefits if they completely exhaust their pension fund. However, here we just consider the basic mathematics of drawing down a regular income from a fixed sized pension pot.

Drawdown pensions and the depletion of pension pots

To work out how fast a pension pot will be depleted, assume that income is taken in one lump sum each year; the pension pot is invested in a fixed interest account and that:

G is the total pension pot that an individual has saved when they retire,
M is the amount the individual takes out of the pension pot as income each year, and
i is the annual interest rate earned on the invested pension pot, which is assumed to be constant.

If the individual takes their income M at the start of each year, then the amount that will be invested at interest rate i at the start of their first year of retirement will be $G - M$. Thus, the value of the remaining funds in the pension pot at the end of their first year of retirement, after this balance of $G - M$ has earned one year's interest, will be

$$G_1 = (G - M)(1+i) = G(1+i) - M(1+i)$$

After the second year's income M is deducted, the amount invested for the second year of retirement will be the above balance less M which is

$$G(1+i) - M(1+i) - M$$

Thus, the value of the remaining funds in the pension pot at the end of the second year of retirement, after a year's interest has been earned on this balance, will be

$$G_2 = [G(1+i) - M(1+i) - M](1+i) = G(1+i)^2 - M(1+i)^2 - M(1+i)$$

If a further amount M is then taken out at the start of year 3 then at the end of the third year of retirement the pension pot remaining will be

$$\begin{aligned} G_3 &= \left[G(1+i)^2 - M(1+i)^2 - M(1+i) - M\right](1+i) \\ &= G(1+i)^3 - M(1+i)^3 - M(1+i)^2 - M(1+i) \end{aligned}$$

From these calculations for the first few years, it can be seen that a pattern is starting to emerge in the formulae. Assuming that the fund does not become exhausted then, applying the same calculation method, the amount left in the pension pot at the end of year n will be

$$\begin{aligned} G_n &= G(1+i)^n - M(1+i)^n - M(1+i)^{n-1} - M(1+i)^{n-2} - \ldots - M(1+i) \\ &= G(1+i)^n - M\left[(1+i)^n + (1+i)^{n-1} + (1+i)^{n-2} + \ldots + (1+i)\right] \end{aligned}$$

If the sequence of terms within the square brackets is written in reverse order, then it can be recognized as a geometric series of n terms with a common ratio of $(1 + i)$ and an initial term also equal to $(1 + i)$. Thus, we can use the standard formula to find the sum of this geometric series, where $a = k = (1 + i)$, which gives

$$\mathrm{GP}_n = \frac{a\left(1-k^n\right)}{1-k} = \frac{(1+i)\left(1-(1+i)^n\right)}{1-(1+i)} = \frac{(1+i)\left(1-(1+i)^n\right)}{-i}$$

Substituting this geometric series sum into the function

$$G_n = G(1+i)^n - M\left[(1+i)^n + (1+i)^{n-1} + (1+i)^{n-2} + \ldots + (1+i)\right]$$

derived earlier for the amount left in a pension pot at the end of year n, gives the general **formula showing the amount left in a pension pot after *n* years** as

$$G_n = G(1+i)^n - M\left[\frac{(1+i)\left(1-(1+i)^n\right)}{-i}\right]$$

Example 8.44

A person retires with a pension pot of £200,000 which is invested at an annual rate of return of 4%. How much will be left in the pension pot after 20 years if this person withdraws an annual income of £12,000 at the start of each year?

Solution

In this example the relevant values are:

$$\text{Initial pension pot } G = 200{,}000, \; i = 4\% = 0.04, \; n = 20, \; M = 12{,}000$$

Thus, using the formula for the amount left in a pension pot after n years we get

$$\begin{aligned} G_{20} &= G(1+i)^n - M\left[\frac{(1+i)\left(1-(1+i)^n\right)}{-i}\right] \\ &= 200{,}000(1.04)^{20} - 12{,}000\left[\frac{(1.04)\left(1-(1.04)^{20}\right)}{-0.04}\right] \\ &= 200{,}000(2.191123143) - 12{,}000\left[\frac{(1.04)(1-(1-2.191123143))}{-0.04}\right] \\ &= 438{,}224.6286 - 12{,}000[29.77807858] = £66{,}594.21 \end{aligned}$$

Sometimes the amount in a pension pot will not be enough to fund an individual's desired pension income, as in the next example.

Example 8.45

If a person wishes to take an annual income of £18,000 for 20 years after they retire with a pension pot of £250,000, will this be feasible if the pension pot is invested in an account paying 3% interest per annum?

Solution

Relevant given values are:

$$\text{Initial pension pot } G = 250{,}000, \; i = 3\% = 0.03, \; n = 20, \; M = 18{,}000$$

Thus, using this formula the amount left in a pension pot after 20 years will be

$$G_{20} = 250{,}000(1.03)^{20} - 18{,}000\frac{(1.03)\left(1-(1.03)^{20}\right)}{-0.03}$$
$$= 451{,}527.81 - 498{,}176.74 = -£46{,}648.93$$

This negative amount tells us that the pension fund will have been exhausted before the end of 20 years of retirement, so this individual may need to revise their retirement plans.

In this example we can work out approximately when the pension fund will become exhausted by working, backwards, as the shortfall of £46,648.93 divided by £18,000 = 2.6, so this will be soon after retirement year 17. However, this rough calculation ignores any interest earned.

To get a more precise answer it is necessary to use a spreadsheet to work out by iteration how long a given income can be taken from a fixed pension scheme, as the formula cannot easily be adapted to calculate n when other values are given. The easiest way to do this is to enter values for G, M, i and n in the first 4 columns on a spreadsheet and then enter the drawdown pension final value formula in the fifth column with reference to the values in the first 4 columns. All values and formulae can then be copied down the spreadsheet page, except for n, which is set at gradually increasing values until the final value calculated in column 5 is close to zero, and the corresponding value of n will then show when the pension pot will be exhausted.

Your spreadsheet should give you a set of values similar to those shown in Table 8.16. The amount remaining will equal zero somewhere between years 17 and 18, but if pension can only be taken in one lump sum at the start of the year then the last time the full £18,000 pension income can be taken is year 17, with only £9,752.07 left for year 18 income.

Table 8.16 Pension pot balance over years

Pension Pot G	Income per year M	Interest annual i	Years taken n	Amount remaining Gn
250,000	18,000	3%	15	£44,667.99
250,000	18,000	3%	16	£27,468.03
250,000	18,000	3%	17	£9,752.07
250,000	18,000	3%	18	–£8,495.37
250,000	18,000	3%	19	–£27,290.23
250,000	18,000	3%	20	–£46,648.93

In practice most people will prefer to receive their pension income on a monthly basis rather than in one lump sum at the start of the year. The formula for the amount left in a pension pot can easily be adapted for monthly payments if we redefine terms as:

G is the total pension pot that an individual has saved when they retire,
M is the amount the individual takes out of the pension pot as income in each time period,
i is the interest rate per time period earned on the invested pension pot.

If M denotes the monthly income taken then i will be the monthly interest rate and, of course, the number of time periods n will measure the number of months for which this income is taken.

Thus, the following formula can be redefined as showing the amount left in a drawdown pension fund after n time periods when an amount M is withdrawn at the start of each time period and the fund is invested at an interest rate of i per time period.

$$G_n = G(1+i)^n - M\left[\frac{(1+i)\left(1-(1+i)^n\right)}{-i}\right]$$

Example 8.46

If a person retires with a pension pot of £200,000 which is invested at an annual rate of return of 4% and withdraws an annual income of £12,000, taken in equal instalments of £1,000 at the start of each month, how much will be left in the pension pot after 20 years?

Solution

Note that the amounts are similar to those in Example 8.44, apart from the monthly income, so we first have to adjust other parameters to monthly values.

When the annual rate is 4% then the monthly interest rate is

$$i_m = \left(\sqrt[12]{1.04}\right) - 1 = 1.0032737 - 1 = 0.0032737 = 0.32737\%$$

and the number of time periods is $n = 20 \times 12 = 240$

Thus, the relevant values are now:

$$\text{Initial pension pot } G = 200{,}000,\ i = 0.0032737,\ n = 240,\ M = 1{,}000$$

Thus, using the formula for the amount left in a pension pot after n time periods gives

$$\begin{aligned} G_{240} &= 200{,}000(1.0032737)^{240} - 1{,}000\left[\frac{(1.0032737)\left(1-(1.0032737)^{240}\right)}{-0.0032737}\right] \\ &= 438{,}224.6286 - 1{,}000[365.03] \\ &= £73{,}191.78 \end{aligned}$$

This amount remaining in the pension pot after 20 years is slightly higher than the amount found in Example 8.44 because funds in the pension pot are, on average, invested for 6 months longer when income is taken monthly and thus earn more interest. For example, instead of the invested amount in year 1 being reduced from the initial £200,000 by £12,000 to £188,000 from the start of the year, only 1 month's income is taken, so £199,000 earns interest in January. The amount then invested at the start of February in year 1 of retirement is £198,000 plus the interest earned in January, and so on through the whole investment period.

QUESTIONS 8.8

1. What annual income could someone expect if the market interest rate is 4%, they spend all their pension pot on a 20-year annuity and the amount saved in their personal pension fund is: (a) £80,000, (b) £200,000, (c) £450,000?
2. If the interest rate increased to 7% would this give a better annual income to the person buying the annuities in the previous question?
3. During the stock market collapse in 2008 a person on the verge of retirement saw the value of their personal pension fund fall from £320,000 to £240,000. Interest rates also fell from 5% to 2%. If they were planning to buy a 20-year annuity when they retired, what difference would these changes have made to the annual pension they would have expected to receive?
4. Will a pension pot of £400,000 be sufficient to provide an annual drawdown pension of £25,000 for the next 20 years, taken at the start of each year, if it is invested at an annual interest rate of 3%?
5. A person on the verge of retiring hopes that their pension pot of £150,000, which is invested in a bank account paying 2.5% per annum, will allow them to drawdown an annual pension of £10,000 for the next 20 years. What would you advise them if they consulted you about this pension plan?
6. Explain briefly what impact a rise in interest rates will have on drawdown pensions in general and then show what impact a rise in the annual interest rate from 2% to 3% will have on an individual who has a pension pot of £325,000 and plans to take an annual drawdown pension of £20,000 for the next 20 years.
7. Will a pension pot of £450,000 be sufficient to provide a drawdown pension of £2,500 per month for the next 20 years if it is invested at an annual interest rate of 4%?
8. Can a monthly drawdown pension of £1,750 over the next 20 years be achieved if a pension pot of £260,000 is invested at an AER of 5%?
9. An individual has saved £150,000 in their pension pot, which is invested at an annual interest rate of 3.5%, and hopes to take a drawdown pension monthly income of £1,000. What would you advise them regarding the feasibility of this pension plan if they hope to draw this pension for: (a) the next 20 years, (b) the next 15 years?
10. Construct a spreadsheet to find out how many years an annual pension of £20,000 can be drawn down at the start of each year from a pension pot of £200,000 invested at 5%.
11. For how many full years can an annual pension of £25,000 be drawn down at the start of each year from an initial pension pot of £385,000 invested in a bank account paying 4% interest per annum?

8.10 DRAWDOWN PENSION INCOME

The drawdown pension formula can be adapted to find the maximum income that can be taken from a given pension pot over a given time period. If the pension pot is exhausted exactly then G_n, the amount remaining at the start of time period n, will be zero. Thus, using the drawdown pension income formula

$$G_n = 0 = G(1+i)^n - M\left[\frac{(1+i)\left(1-(1+i)^n\right)}{-i}\right]$$

Then, by adding the second term on the right-hand side to both sides, we get

$$M\left[\frac{(1+i)\left(1-(1+i)^n\right)}{-i}\right] = G(1+i)^n$$

Dividing both sides by the term in square brackets, so that the right-hand side of this equation is multiplied by the inverse of this term, gives the **formula for the drawdown pension maximum income** that can be taken from an initial pension pot of size M as

$$M = \frac{-iG(1+i)^n}{(1+i)\left(1-(1+i)^n\right)}$$

This formula can be applied to drawdown pensions taken annually, monthly or for any other time period unit as long as the amount withdrawn, the interest rate i and the number of time periods n all refer to the same unit of time. However, it should be remembered that individuals cannot predict in advance how long they will live to collect their pension. Thus, these calculations are just based on whatever 'guesstimated' time period the individual uses to make their pension decision.

Example 8.47

What is the maximum annual income that can be taken at the start of each year from an initial pension pot of £150,000 invested at 4% per annum if someone retiring plans to exhaust their pension pot after 20 years?

Solution

The relevant values are: $G = 150{,}000,\ i = 4\% = 0.04,\ n = 20$

Thus, the formula for the maximum income that can be taken gives

$$M = \frac{-iG(1+i)^n}{(1+i)\left(1-(1+i)^n\right)} = \frac{(-0.04)150{,}000(1.04)^{20}}{(1.04)\left(1-(1.04)^{20}\right)} = £10{,}612.75$$

Example 8.48

What maximum income can be taken at the start of each month from an initial pension pot of £300,000 if a retired person plans to exhaust their pension pot after 20 years and it is invested in an account paying an AER of 3.5%?

Solution

Initial pension pot $G = 300{,}000$; number of time periods $n = 20 \times 12 = 240$; when the annual rate is 3.5% then the monthly interest rate is

$$i = \left(\sqrt[12]{1.035}\right) - 1 = 1.002871 - 1 = 0.002871 = 0.2871\%$$

Thus, the formula for the maximum monthly income that can be taken gives

$$M = \frac{-iG(1+i)^n}{(1+i)\left(1-(1+i)^n\right)} = \frac{(-0.002871)300{,}000(1.002871)^{240}}{(1.002871)\left(1-(1.002871)^{240}\right)} = £1{,}726.47$$

QUESTIONS 8.9

1. What maximum annual income over a 20-year period can be drawn down from a pension pot of £180,000 invested in an account paying 3% per annum?
2. If an individual has saved £200,000 in their pension pot and plans to draw-down a pension income at the start of each year for the next 20 years, what is the maximum income they can get if the balance remaining in their pension pot earns 5% a year interest?
3. Assuming that pension funds are invested at 4.5% per annum, compare the maximum annual pensions that can be drawn down for the following two cases:
   (a) Individual A has accumulated a pension pot of £475,000 and plans to retire early and to drawdown this pension for 35 years.
   (b) Individual B has only accumulated £65,000 but retires late and only plans to drawdown their pension for 15 years.
4. What maximum monthly income can be drawn down over a 20-year period from a pension pot of £250,000 invested in an account paying 5% per annum?
5. If an individual has saved £420,000 in their pension pot and plans to draw-down a pension income at the start of each month for 20 years, what is the maximum monthly pension income they can get if their pension pot earns interest at an AER of 2.5%?
6. Assuming that a pension pot of £50,000 can be invested to earn interest at an AER of 4%, what is the maximum monthly income that can be drawn down over a 25-year period?

8.11 LOAN REPAYMENTS AND MORTGAGES

If someone takes out a loan now, to be paid off in regular equal instalments over a given time period, how can these payments be calculated? The following example shows how the formula for calculating the PV of an annuity can be adapted to help answer this type of problem. As most loans are paid off monthly then the monthly rate of interest is used and time periods must refer to the number of months involved.

Loan repayments are usually set at a fixed amount per month. Therefore, from a lender's viewpoint, monthly repayments can be viewed as a monthly annuity which pays R per month for the period of the loan. If the lender is willing to exchange a loan of amount L now for this regular stream of payments then this must be the value of the PV of this 'annuity'. Therefore, to find the level of monthly payment R for a given size loan L we can adapt the formula for the PV of an annuity as follows. Since

$$\text{PV} = \frac{R\left[1-(1+i)^{-n}\right]}{i} = L$$

Then $$R\left[1-(1+i)^{-n}\right] = iL$$

This gives the general **formula for calculating loan repayments**

$$R = \frac{iL}{1-(1+i)^{-n}}$$

Example 8.49

If a £2,000 loan is taken out now to be paid back over the next 12 months at a monthly interest rate of 2% what will the monthly payments be?

Solution

The known values for this example are $L = 2{,}000$, $i = 2\% = 0.02$ and $n = 12$. Substituting these into the loan repayment formula gives monthly repayments as

$$R = \frac{iL}{1-(1+i)^{-n}} = \frac{0.02 \times 2{,}000}{1-(1.02)^{-12}} = \frac{40}{1-0.7884934} = £189.12$$

Example 8.50

What will be the monthly repayments on a loan of £6,000 taken out for 5 years at a monthly interest rate of 0.7%?

Solution

Given values are: $L = £6{,}000, \quad i = 0.7\% = 0.007, \quad n = 5 \times 12 = 60$

Using the loan repayment formula, the monthly repayments will be

$$R = \frac{iL}{1-(1+i)^{-n}} = \frac{0.007 \times 6{,}000}{1-(1.007)^{-60}} = \frac{42}{1-0.658008} = £122.81$$

Mortgages, monthly payments and the APR

In the previous examples the monthly interest rate was given. However, in practice banks and building societies usually quote customers the APR for loans and mortgages rather than the monthly rate of interest. If only the APR for a loan is quoted, then it will be necessary to calculate the equivalent monthly interest rate before working out monthly repayments.

Example 8.51

If a loan of £4,200 is taken out over a period of 3 years at an APR of 6.8% what will the monthly repayments be?

Solution

First, we need to convert the APR of 6.8% to its equivalent monthly rate i_m. To do this, a formula for i_m can be derived from the APR formula from Section 8.3. Thus, given that

$$\text{APR} = (1+i_m)^{12} - 1$$

then $$\text{APR} + 1 = (1+i_m)^{12}$$

Taking the 12th root of both sides of this equation gives

$$\sqrt[12]{(1+APR)} = 1 + i_m$$

and so

$$\sqrt[12]{(1+APR)} + -1 = i_m$$

Substituting the value APR = 6.8% = 0.068 into this formula gives the monthly interest rate as

$$i_m = \sqrt[12]{(1.068)} - 1 = 1.0054974 - 1 = 0.0054974 = 0.54974\%$$

Note that the monthly interest rate should **not** be rounded down and all decimal places should be retained until the final answer is obtained, otherwise rounding will lead to an inaccurate final answer. In fact, there is no real need to write it as a percentage, as the decimal format is used to complete the exercise.

The values to be entered into the loan repayment formula are therefore

$$L = 4{,}200, \quad i = 0.0054974, \quad n = 3 \times 12 = 36$$

and so the monthly repayments will be

$$R = \frac{iL}{1-(1+i)^{-n}} = \frac{0.0054974 \times 4,200}{1-(1.0054974)^{-36}} = \frac{23.1}{1-0.820815} = £128.92$$

Example 8.52

What will be the monthly payments on a repayment mortgage of £225,000 taken out over 25 years if the interest rate is fixed at 4.5% APR?

Solution

First, find the monthly interest rate

$$i_m = \sqrt[12]{(1.045)} - 1 = 1.003674809 - 1 = 0.003674809$$

and so the values to be entered into the loan repayment formula are

$$L = 225,000, \quad i = 0.003674809, \quad n = 25 \times 12 = 300$$

Therefore, the monthly repayments will be

$$R = \frac{iL}{1-(1+i)^{-n}} = \frac{0.003674809 \times 225,000}{1-(1.003674809)^{-300}} = £1,239.13$$

Calculating the maximum loan available for a given repayment

The loan repayment formula can be used to derive a formula for the size of loan that any given repayments would correspond to. Starting from the loan repayment formula

$$R = \frac{iL}{1-(1+i)^{-n}}$$

if both sides are divided by i and multiplied by $1 - (1 + i)^{-n}$ this gives the **loan size formula**

$$\frac{R\left[1-(1+i)^{-n}\right]}{i} = L$$

In fact, as you may have noticed, this is the same as the annuity price formula that we started with to derive the loan repayment formula. This is not surprising, as the loan provider is effectively receiving regular 'annuity' payments from the borrower. In practice the interest rates applied to lenders and borrowers may differ, as higher rates will normally be charged to individual borrowers to reflect the higher risk, plus the need for the lender to make a profit, but the basic principle used to calculate the loan size is the same as that used to calculate an annuity price.

Example 8.53

Suppose that a potential house buyer can only afford to make monthly mortgage payments of £800. What is the maximum mortgage loan they could obtain assuming that they wish to take out a 25-year repayment mortgage at an APR of 5.6%?

Solution

First, we need to convert the APR of 5.6% to its equivalent monthly rate. Thus,

$$i_m = \sqrt[12]{(1.056)} - 1 = 1.00455101 - 1 = 0.00455101$$

The values to be entered into the loan size formula are therefore

$$R = 800, \quad i = 0.00455101, \quad n = 25 \times 12 = 300$$

Giving maximum loan size

$$L = \frac{R\left[1-(1+i)^{-n}\right]}{i} = \frac{800\left[1-(1+0.00455101)^{-300}\right]}{0.00455101} = £130,767.45$$

Example 8.54

Someone wants a loan to buy a car but can only afford a monthly repayment of £200. What is the maximum loan that a lender would approve to be paid back over 5 years at an APR of 12%?

Solution

Converting the APR of 12% to its equivalent monthly rate gives

$$i_m = \sqrt[12]{(1.12)} - 1 = 1.0094888 - 1 = 0.0094888$$

The values to be entered into the loan size formula are therefore

$$R = 200, \quad i = 0.0094888, \quad n = 5 \times 12 = 60$$

Giving maximum loan size

$$L = \frac{R\left[1-(1+i)^{-n}\right]}{i} = \frac{200\left[1-(1+0.0094888)^{-60}\right]}{0.0094888} = £9,117.56$$

Calculating the interest rate on a loan

From an individual consumer's viewpoint, you may be more interested in finding out the interest rate you have to pay on a loan. All lenders now have to quote their APR by law, but you may still wish to check this.

Example 8.55

A car dealer offers a £12,000 car for a £4,000 deposit now followed by 24 monthly payments of £400. What is the APR on this effective loan of £8,000?

Solution

As in the earlier examples, treat the stream of repayments as an annuity for the lender. Referring again to the formula for loan repayments

$$R = \frac{iL}{1-(1+i)^{-n}}$$

we can see that even if we know the other values (L = 8,000, R = 400, n = 24) this still leaves us with the awkward equation

$$400 = \frac{i \times 8{,}000}{1-(1+i)^{-24}}$$

to solve for i (the monthly interest rate) which can then be used to calculate the APR.

The quickest way to solve this is to use a spreadsheet which calculates the repayment values that correspond to a range of monthly interest rates which, in turn, will correspond to specific values for the APR. Instructions for constructing an appropriate format are shown in Table 8.17, which should give the actual spreadsheet values shown in

Table 8.17 Setting up a spreadsheet for Example 8.55

CELL	Enter	Explanation
A1	Ex.8.55	Label to remind you what example this is
B1	LOAN =	Label to tell you loan value goes in next cell
C1	8000	Loan value for this example
D1	n MONTHS=	Label to tell you number of months for repayment goes in next cell
E1	24	Number of months for this example
A3	APR	Column heading labels
B3	MONTHLY i	
C3	REPAYMENT	
A4	15.00%	Start of (guessed) interest rate for APR range
A5	=A4+0.0025	Gives increment of 0.25%
A6 to A24	*Copy cell A5 formula down column A*	Gives a range of values for APR in 0.25% increments. (Format to 2 dp.)
B4	=(1+A4)^(1/12)–1	Formula calculates monthly interest rate corresponding to APR in cell A4
B5 to B24	*Copy cell B4 formula down column B*	Calculates monthly interest rates corresponding to APR in column A
C4	=B4*C$1/(1-(1+B4)^-E$1)	Formula calculates repayment corresponding to total loan in cell C1, number of months in cell E1 and monthly interest rate in cell B4, which is determined by APR in column A
C5 to C24	*Copy cell C4 formula down column C*	Calculates repayment corresponding to different APR values

Table 8.18 Excel output based on Table 8.17

	A	B	C	D	E
1	Ex 8.55	LOAN =	8000	n MONTHS=	24
2					
3	APR	MONTHLY i	REPAYMENT		
4	15.00%	1.17%	384.32		
5	15.25%	1.19%	385.15		
6	15.50%	1.21%	385.98		
7	15.75%	1.23%	386.81		
8	16.00%	1.24%	387.64		
9	16.25%	1.26%	388.47		
10	16.50%	1.28%	389.30		
11	16.75%	1.30%	390.13		
12	17.00%	1.32%	390.95		
13	17.25%	1.33%	391.78		
14	17.50%	1.35%	392.61		
15	17.75%	1.37%	393.43		
16	18.00%	1.39%	394.26		
17	18.25%	1.41%	395.08		
18	18.50%	1.42%	395.90		
19	18.75%	1.44%	396.73		
20	19.00%	1.46%	397.55		
21	19.25%	1.48%	398.37		
22	19.50%	1.50%	399.19		
23	19.75%	1.51%	400.01	*<< Solution*	
24	20.00%	1.53%	400.83		

Table 8.18. Once a repayment equal (or very close) to £400 has been identified then the corresponding monthly interest rate and APR can be read off. Near the bottom of Table 8.18 we can see that a £400.01 repayment corresponds to a 1.51% monthly interest rate and a 19.75% APR, which is the solution to this problem.

This spreadsheet format can be used to solve other similar problems. You only need to change the total loan figure in cell C1 and the time period in cell E2 to compute a new set of repayment figures for a range of monthly interest rates and you may have to extend the interest rate range or change the initial trial value in cell A4.

Example 8.56

A loan company will require 36 monthly payments of £438.25 in return for a loan of £12,500. What APR is it charging?

Solution

Using the spreadsheet constructed for Example 8.55, enter the new values for the loan and time period in cells C1 and E1. In row 12 you should then be able to read off the values:

APR	MONTHLY i	REPAYMENT
17.00%	1.32%	438.25

The APR this company charges is therefore 17%.

Although lenders are always supposed to quote the APR they charge borrowers, in recent years these comparisons have become a bit more complicated because of different charges and deals offered in the UK mortgage market. For example, one building society may offer a lower APR on mortgage loans than another building society but will charge an up-front fee to arrange the mortgage. The 'APR Equivalent' rate on such deals will normally be provided, but not always, and so the following example explains a method of calculating it.

Example 8.57

Which of the following two mortgage deals is the best for someone wishing to take out a mortgage loan of £195,000 repayable over 25 years?

(a) Lender A will offer this mortgage at an APR of 3.8%.
(b) Lender B will offer this mortgage at an APR of 3.65% but will also require an up-front arrangement fee of £1,400.

Solution

We know the APR for lender A is 3.8% so we need to calculate the true APR for lender B taking into account the initial fee payment and then compare APR rates. To do this we first calculate the monthly payments R for the loan from lender B.

$$\text{Lender B's monthly interest rate} \quad i_m = \sqrt[12]{(1+0.0365)} - 1 = 0.002991938$$
$$\text{and} \quad L = 195{,}000, \qquad n = 25 \times 12 = 300$$

Therefore, monthly repayments will be

$$R = \frac{iL}{1-(1+i)^{-n}} = \frac{0.002991938 \times 195{,}000}{1-(1+0.002991938)^{-300}} = £985.69$$

However, although this monthly mortgage repayment has been calculated on the basis of a loan of £195,000, the actual loan is effectively only for £193,600 if we deduct the £1,400 special fee at the outset. We therefore have to work out what interest rate would correspond to monthly payments of £985.69 over 25 years for a loan of £193,600.

The spreadsheet method can be used to find this interest rate. Using a similar spreadsheet to that used for Example 8.55, we start by entering the fixed values of £193,600 for the loan and 300 for the number of months. Then we construct columns for (i) the APR, (ii) the corresponding monthly interest rate and (iii) the monthly repayments that would be required given the monthly interest rate in the previous column.

Starting at a value of, say, 3.6%, if the APR is then increased down the column (i) in increments of 0.01%, the repayments column will show increasing monthly payment values. When the repayment value equals (or is very close to) the £985.69 payment that lender B requires we have found our answer. In this particular example, a spreadsheet will show that monthly payments will equal £985.69 when the APR is 3.72%.

Thus, the true APR of lender B is 3.72%, which is below the APR of 3.8% charged by lender A and so lender B offers the better deal, although the true APR of 3.72% is higher than the headline APR of 3.65%.

QUESTIONS 8.10

1. What will be the monthly repayments on a loan of £6,500 taken out over 5 years at a monthly interest rate of 1.2%?
2. You wish to buy a car priced at £6,000 by putting down a cash deposit of £2,000 and borrowing £4,000, the loan being paid back in monthly instalments over 2 years. How much will you have to budget to pay out of your salary if the monthly interest rate is 1.4%?
3. A loan company will lend you £5,000, repayable over the next 3 years in monthly payments. What will these payments be if the APR on the loan is 24.6%?
4. What will be the monthly payments on a repayment mortgage of £275,000 taken out over 20 years at an APR of 3.9%?
5. If a borrower can only afford to repay £1,200 a month, what is the maximum 25-year repayment mortgage loan they could get from a building society that charges 4.2% APR?
6. If a borrower can repay £950 a month, what is the maximum 20-year repayment mortgage loan they could get from a building society that charges 6% APR?
7. A loan of £1,000 is taken out. What APR is being charged if the monthly payments are:
   (a) £33.00 over 36 months? (b) £22.00 over 48 months?
8. A car dealer has on offer a special '0% finance' deal on the advertised price of £12,475 for a particular model. This requires an initial deposit of £2,995 followed by 24 monthly payments of £395.00. If you could get the price reduced to £10,000 if you paid cash and can earn 4% per annum on money invested in a building society, which method would you use to purchase this car?
9. If government legislation requires lenders to clearly specify that they charge high interest rates when the APR that they charge borrowers exceeds 50%, would a private loan dealer who lends £1,000 to be repaid over a year at £25 a week have to make this declaration?
10. Which of the following two mortgage deals is the best one for a 25-year mortgage loan of £230,000?
    Building Society A will offer the loan at an APR of 5%.
    Building Society B will offer the loan at an APR of 4.9% but also requires an up-front arrangement fee of £2,500.

8.12 SAVINGS SCHEMES

There are basically two forms of savings schemes, depending on whether payments into the scheme are made:

(i) at the start of each time period, or
(ii) at the end of each time period.

The way of calculating the final sum accumulated will depend on which of these methods applies and so there are two different formulae for savings schemes when regular fixed amounts are paid in, both based on the geometric series sum formula.

Saving schemes that individual savers enter into normally fall into the first category. For example, an individual may start to make payments on the first day of each month and save regularly for a given number of years. Thus, the last payment in would be on the first day of the last month before the scheme ended.

Certain savings schemes used by companies, however, require payments in at the end of each time period. Sometimes companies will put a certain amount of money each year into a scheme to ensure that they will have enough funds to replace machinery when it wears out, and this is often known as a **sinking fund**. For example, a firm may spend £800,000 on a specialist drilling machine that is expected to last for 5 years and may choose to put a certain sum of money into a special sinking fund so that at the end of 5 years it will have enough money to purchase a replacement machine. In this sort of scheme the last payment occurs on the last day of the last time period.

The rest of this section will focus on type (i) savings schemes.

Savings schemes type (i): payments at start of time period

Payments into any savings scheme will accumulate interest from the date they are invested. Early payments at the start will be invested for longer and so will accumulate more interest than later payments. In fact, each individual payment into the savings scheme will accumulate to a different final amount because each will be invested for a different length of time.

If interest at a rate of i per time period is earned then, assuming that the same amount is paid in each time period, the final sum accumulated by a payment made in any given time period will be greater than the final sum accumulated by a payment into the scheme in the following time period by a ratio of $(1 + i)$. The following examples show how this enables us to use the geometric sum formula to calculate the total amount accumulated in the savings scheme.

Example 8.58

A 5-year saving scheme requires investors to pay in £5,000 now followed by £5,000 at 12-monthly intervals. Interest is credited at 4% at the end of each year. What will the final sum be at the end of the fifth year?

Solution

At the end of year 5, the £5,000 invested at the start of year 5 will be worth 5,000(1.04) as it will only earn interest for 1 year.

The £5,000 invested at the start of year 4 will be worth $5{,}000(1.04)^2$ as it will earn interest in both years 4 and 5.

In a similar fashion we can calculate the final values of other payments into the savings scheme until we get to the £5,000 invested at the start of year 1, which will be invested for the full 5 years and will be worth $5{,}000(1.04)^5$.

Thus, showing the values of payments in reverse chronological order, the final sum will be

$$5{,}000(1.04)+5{,}000(1.04)^2+5{,}000(1.04)^3+5{,}000(1.04)^4+5{,}000(1.04)^5$$

This is a geometric series with

Initial term	$a=5{,}000(1.04)=5{,}200$	[*Not just* £5,000]	
Common ratio	$k=1.04$	and	number of time periods $n=5$

The total amount accumulated will therefore be this geometric series sum:

$$\text{GP}_n=\frac{a\left(1-k^n\right)}{1-k}=\frac{5{,}000(1.04)\left(1-1.04^5\right)}{1-1.04}$$

$$=\frac{5{,}200(1-12166529)}{-0.04}=\frac{-1{,}1265951}{-0.04}=£28{,}164.88$$

Thus, for any regular annual savings scheme where the amount saved is S and the interest rate is i, the final sum accumulated after n time periods will be the geometric series with the first term $a = S(1 + i)$ and common ratio $k = 1 + i$.
The final sum will therefore be

$$\text{GP}_n=\frac{a\left(1-k^n\right)}{1-k}=\frac{S(1+i)\left(1-(1+i)^n\right)}{1-(1+i)}=\frac{S(1+i)\left(1-(1+i)^n\right)}{-i}$$

This gives the **formula for the sum of a savings scheme with payments at the start of each time period**:

$$\text{Final sum}=\frac{S(1+i)\left(1-(1+i)^n\right)}{-i}$$

Example 8.59

A 10-year saving scheme requires investors to pay in £750 at the start of each year. If interest is credited at 4.5% at the end of each year, what will the final sum be?

Solution

In this example the regular amount saved $S = 750$, $i = 0.045$, $n = 10$.

So, using the savings scheme formula

$$\text{Final sum} = \frac{S(1+i)\left(1-(1+i)^n\right)}{-i} = \frac{750(1.045)\left(1-1.045^{10}\right)}{-0.045} = £9,630.88$$

What we have just found is the final sum that will be accumulated from a regular annual saving scheme, but what if we wanted to calculate the annual amount that a saver would need to invest to generate a given final sum?

Example 8.60

Assume that a wealthy parent wants to put an annual amount into a saving scheme that will cover the £30,000 student debt that their child expects to have when they leave university in 3 years' time. How much per year would they have to pay into a savings scheme that paid 3.5% interest at the end of each year?

Solution

As we already know the final sum, we have to adapt the geometric sum formula and work backwards to find the unknown amount to save each time period, S.

In the geometric series showing the sum of values of payments into the savings scheme, the initial term will be the present value of the last payment 1 year ago = $S(1 + i) = S(1.035)$ and we know the final sum will be £30,000. Thus, putting these values into the savings scheme formula:

$$\text{Final sum } F = \frac{S(1+i)\left(1-(1+i)^n\right)}{-i} = \frac{S(1.035)\left(1-1.035^{3}\right)}{-0.035} = £30,000$$

We can then cross multiply to find the amount needed to be saved each year as

$$S = \frac{-0.035(30,000)}{(1.035)\left(1-1.035^{3}\right)} = £9,331.43$$

In a similar fashion we can cross multiply from the savings scheme final sum formula to get a **general formula for the amount that needs to be saved to accumulate a given final sum** as

$$S = \frac{-iF}{(1+i)\left(1-(1+i)^n\right)}$$

Monthly savings schemes

In the previous examples only one savings deposit per year was made, but it is more usual for people to pay into savings schemes monthly with a standing order from their bank. As the interest rate quoted for a savings account will normally be the AER, then to work out the final value of a monthly savings scheme the equivalent monthly interest rate must be used.

Example 8.61

If a savings account pays an AER of 5%, credited monthly, how much would be accumulated if you put £100 into this savings account on the first day of every month for 5 years?

Solution

First, find the monthly interest rate $i_m = \left(\sqrt[12]{\text{AER}+1}\right) - 1 = \left(\sqrt[12]{1.05}\right) - 1 = 0.00407412$

Given the amount saved per month $S = 100$ and $n = 5 \times 12 = 60$ months then, using the savings scheme formula, the final sum accumulated will be

$$F = \frac{S(1+i)\left(1-(1+i)^n\right)}{-i} = \frac{100(1.00407412)\left(1-1.00407412^{60}\right)}{-0.004074125} = £6{,}809.00$$

(Note: it is easy to do a 'ball park' check for this sort of problem to ensure that answers are approximately in the right range. If £100 a month was saved for 5 years without earning interest then the total saved would be 100 × 12 × 5 = £6,000, so the answer we expect when interest is earned should be a bit higher than this, which it is.)

Example 8.62

How much needs to be put into a savings account on the first day of every month to accumulate a final sum of £25,000 after 8 years if the account earns interest at an AER of 5%?

Solution

The monthly interest rate $i_m = \left(\sqrt[12]{\text{AER}+1}\right) - 1 = \left(\sqrt[12]{1.05}\right) - 1 = 0.00407412.$

Other given values are $F = 25{,}000$ and $n = 8 \times 12 = 96$.

Thus, the amount that needs to be saved per month is

$$S = \frac{-iF}{(1+i)\left(1-(1+i)^n\right)} = \frac{-(0.00407412)\times 25{,}000}{(1+0.00407412)\left(1-(1+0.00407412)^{96}\right)}$$

$$= \frac{-101.8530946}{(1.00407412)(1-1.477455444)} = £212.46$$

QUESTIONS 8.11

1. If £3,500 is put into a savings scheme on 1 January each year how much will be saved by the end of 8 years if the annual interest rate is 4%?
2. How much would you accumulate in 10 years if your savings earned interest at 3% per annum and you invested £1,500 at the start of each year?

3. If you wanted to have £25,000 in 5 years' time, what annual amount would need to be saved at the start of each year if your savings account pays 3.5% annual interest?
4. How much would you accumulate in 5 years' time in a savings account paying an AER of 5% if you invested £250 a month via a direct debit?
5. If you wanted to have £36,000 in 8 years' time, what monthly amount would need to be saved if an AER of 4.5% is paid on your savings account?
6. If you put £250 a month into a savings account for 40 years, what annual pension could you expect to receive if you used the total amount saved to buy a 20-year annuity, assuming the interest rate stayed at an AER of 4% throughout the whole period?

8.13 SINKING FUND SAVINGS SCHEMES

This section analyzes type (ii) savings schemes, where payments are made in at the end of each time period. These are also known as sinking fund schemes because they are sometimes used by firms who pay money into a 'sinking fund' on a regular basis to try to ensure that sufficient funds will be available to replace plant or machinery when it gets to the end of its operational life.

In this form of savings scheme the last payment does not earn any interest because it is made on the day the scheme ends. Similarly, all other payments will earn interest for one less time period than in type (i) savings schemes where payments are made at the start of period payment. The following example explains how a formula can be derived for the final sum accumulated.

Example 8.63

A sinking fund savings scheme requires a firm to pay in £8,000 at the end of each year for the next 6 years. Interest is accrued on the balance in the fund at the end of each year at a rate of 4%. What will the final sum be at the end of the sixth year?

Solution

At the end of year 6 the amounts of £8,000 invested at the end of each year will have accumulated to different final amounts as follows:

End of year 6 payment, final value = £8,000 as no interest earned.
End of year 5 payment, final value = £8,000(1.04) as 4% interest earned for 1 year.
End of year 4 payment, final value = £8,000$(1.04)^2$ as 4% interest earned for 2 years.
End of year 3 payment, final value = £8,000$(1.04)^3$ as 4% interest earned for 3 years.
End of year 2 payment, final value = £8,000$(1.04)^4$ as 4% interest earned for 4 years.
End of year 1 payment, final value = £8,000$(1.04)^5$ as 4% interest earned for 5 years.

Thus, the total final sum in the savings scheme can be found by adding these final values up:

$$\text{Final sum} = 8{,}000 + 8{,}000(1.04) + 8{,}000(1.04)^2 + \ldots + 8{,}000(1.04)^5$$

This is a geometric series with the first term $a = 8{,}000$, ratio $k = 1.04$ and $n = 6$.

The sum will therefore be

$$\text{GP}_n = \frac{a\left(1-k^n\right)}{1-k} = \frac{8{,}000\left(1-1.04^6\right)}{1-1.04}$$

$$= \frac{8{,}000(1-1.265319)}{-0.04} = \frac{-2122.556}{-0.04} = £53{,}063.80$$

Thus, for any regular annual savings scheme where the annual amount saved, S, is paid in **at the end** of each time period (such as a sinking fund) and the annual interest rate is i then the final sum accumulated after n years will be the geometric series with the first term $a = S$ and common ratio $k = 1 + i$. The final sum will therefore be

$$\text{GP}_n = \frac{a\left(1-k^n\right)}{1-k} = \frac{S\left(1-(1+i)^n\right)}{1-(1+i)} = \frac{S\left(1-(1+i)^n\right)}{-i}$$

Thus, the **formula for the sum of a sinking fund savings scheme with payments in at the end of each time period** is:

$$\text{Final sum} = \frac{S\left(1-(1+i)^n\right)}{-i}$$

Example 8.64

If £250 is paid into a savings scheme at the end of each year for 10 years how much will be accumulated if the scheme pays an annual interest rate of 5%?

Solution

In this example the amount saved $S = 250$, $i = 5\% = 0.05$, $n = 10$.

Using the formula for a savings scheme with payments at the end of each time period:

$$\text{Final sum} = \frac{S\left(1-(1+i)^n\right)}{-i} = \frac{250\left(1-(1.5)^{10}\right)}{-0.05} = £3{,}144.47$$

The same formula will apply if payments into the savings scheme are made monthly although, of course, the monthly interest rate must be used and n will be measured in months.

Example 8.65

If a savings account pays an AER of 5%, how much would be accumulated if you put £100 into this account on the last day of every month for 5 years?

Solution

First find the monthly interest rate $i_m = \left(\sqrt[12]{\text{AER}+1}\right) - 1 = \left(\sqrt[12]{1.05}\right) - 1 = 0.00407412$ and $n = 5 \times 12 = 60$ months. Thus,

$$\text{Final sum} = \frac{S\left(1-(1+i)^n\right)}{-i} = \frac{100\left(1-(1.05)^{60}\right)}{-0.05} = £6{,}781.37$$

It should be noted that this final amount is less than the final amount of £6,809 calculated in Example 8.61 where the same amounts, time period and interest rate applied but payments were made in at the start of each time period. This is what we would expect, as each payment in is made a month later and so over the whole period one month's interest on one payment in is lost. We can prove this as follows:

$$\text{The difference between the two totals is } £6{,}809.00 - £6{,}781.27 = £27.63$$

If one monthly payment of £100 was invested for 5 years at the monthly interest rate $i_m = 0.00407412$ (corresponding to 5% AER), then the total amount accumulated would be

$$F = A(1+i)^n = 100(1+0.00407412)^{60} = £127.63$$

Subtracting the £100 invested gives the net interest accumulated of £27.63, which is the difference between the totals from the two different methods as given earlier.

Thus, in general, when comparing type (i) and (ii) savings schemes where the same amount S is saved over the same number of time periods n at the same interest rate i, then the savings scheme type (i) where payments are made at the start of each time period will accumulate a final amount greater than the type (ii) savings scheme where payments are made at the end of each time period. The difference between the two schemes' total amounts accumulated will be

$$S(1+i)^n - S$$

Example 8.66

A bank deposit account pays an annual interest rate of 3.5%. Compare the outcomes of the two savings schemes (a) and (b) using this account, where

(a) requires £2,000 to be paid in at the end of each year for 8 years
(b) requires £2,000 to be paid in at the start of each year for 8 years.

Solution

In both schemes: the amount saved $S = 2{,}000$, $i = 3.5\% = 0.035$, $n = 8$.

If we use the formula for a savings scheme with payments in at the end of each time period, then for scheme (a):

$$\text{Final sum} = \frac{S\left(1-(1+i)^n\right)}{-i} = \frac{2{,}000\left(1-(1.035)^8\right)}{-0.035} = £18{,}103.37$$

Using the formula for the difference between amounts accumulated in the two types of schemes gives the difference as

$$S(1+i)^n - S = 2{,}000(1.035)^8 - 2{,}000 = 2633.62 - 2{,}000 = £633.62$$

Thus, the amount accumulated for scheme (b) will be the amount accumulated in scheme (a) plus this difference, giving the final sum of scheme (b) as:

$$\text{Final sum accumulated} = £18{,}103.37 + £633.62 = £18{,}736.99$$

To check this answer, you can calculate the answer for (b) using the type (i) savings scheme method.

QUESTIONS 8.12

1. A company estimates that it will have to pay £45,000 in 8 years' time to replace some specialist machinery that it owns. Will it accumulate enough to pay for this machinery if it invests £5,000 at the end of each year in a sinking fund account paying 3% annual interest?
2. How much will be accumulated after 4 years if £85 is invested at the end of each month in an account paying 1.75% AER interest?
3. If an individual pays £125 via a standing order at the end of each month into a savings account earning interest at 3.5% AER how much will they have accumulated at the end of 10 years?
4. A delivery firm wanting to ensure that it will be able to buy a new delivery van when its current one wears out decides to pay £400 at the end of each month into an account earning 4.5% AER interest. How much will it have available to buy a new van after 6 years?
5. If interest on a bank account can be earned at an annual rate of 2.5% what would be the difference between the final amounts accumulated if:
   (a) £7,500 is paid into this account at the end of each month for the next 4 years, and
   (b) £7,500 is paid into this account at the start of each month for the next 4 years?

6. An individual decides to save £120 a month for 6 years in a savings account that credits interest each month at an AER of 4.25%. Assuming payments start next month, what difference would it make to the final sum accumulated if this individual's standing order pays into this account on the first day of each month instead of the last day of each month?

8.14 OTHER APPLICATIONS OF GROWTH AND DECLINE

Rather than introducing new mathematical methods, this section considers how the techniques already explained in this chapter to solve problems concerned with finance and investment can be adapted to some non-financial problems. Note that growth and decline are still treated as discrete processes in all these examples.

Example 8.67

There are limited world reserves of mineral M. The current level of extraction is 45 million tonnes a year, with all mined material being used up by manufacturing industry. This extraction level is expected to increase at 3% per annum. Total estimated reserves are 1,200 million tonnes. When will they be expected to run out?

Solution

The pattern of consumption will be (in millions of tonnes) the geometric series

$$45, 45(1.03), 45(1.03)^2, \ldots, 45(1.03)^{n-1}$$

where n is the number of years that mining continues. The initial term is $a = 45$ and the common ratio $k = 1.03$. The sum of this geometric series is therefore

$$\frac{a\left(1-k^n\right)}{1-k} = \frac{45\left(1-1.03^n\right)}{1-1.03}$$

which must sum up to 1,200 if all reserves are used up. Therefore,

$$\begin{aligned} 1,200 &= \frac{45\left(1-1.03^n\right)}{-0.03} \\ -0.8 &= 1-1.03^n \\ 1.03^n &= 1.8 \end{aligned}$$

Putting this in logarithmic form we get

$$n \log 1.03 = \log 1.8$$

$$n = \frac{\log 1.8}{\log 1.03} = \frac{0.2552725}{0.0128372} = 19.885$$

Therefore, mineral M is expected to run out within 20 years.

Example 8.68

A developing country currently produces 3,600 million units of food per annum and this production level is expected to increase by 4% a year. Its population is currently 2.5 million and expected to grow by 6% per annum. The minimum recommended average intake of food is 1,200 units per person per year. Assuming no changes in production or population growth rates, no imports and exports and no foreign aid, when will food production fall below the subsistence level?

Solution

The minimum level of demand for food is 1,200 × population.

Initial demand is therefore 1,200 × 2.5 million = 3,000 million units.

The rate of growth of the population is 6% and so total demand for food after n years will be $3{,}000(1.06)^n$ million units.

Initial production is 3,600 million units of food. The rate of growth of production is 4% and so total production after n years will be $3{,}600(1.04)^n$ million units.

The subsistence level is reached in year n when

$$\text{food demand} = \text{food production}$$

$$3{,}000(1.06)^n = 3{,}600(1.04)^n$$

$$\left(\frac{1.06}{104}\right)^n = \frac{3{,}600}{3{,}000}$$

$$(1.0192307)^n = 1.2$$

putting this in log form,

$$n \log 1.01923 = \log 1.2$$

$$n = \frac{\log 1.2}{\log 1.01923} = \frac{0.079182}{0.0082722} = 9.572$$

Therefore, food production will fall below the subsistence level after 9 years.

Example 8.69

Estimated reserves of an oil field are 84 million barrels. What annual growth in the rate of extraction will exhaust the oil in 12 years given that this year's production is 6 million barrels?

Solution

Total oil extraction will be the sum of the geometric series with 12 terms:

$$6+6(1+r)+6(1+r)^2+\ldots+6(1+r)^{11}$$

where r is the growth in the extraction rate. The oil field will be exhausted when this totals to 84. Therefore, given the initial term $a = 6$ and the common ratio $k = 1 + r$, and employing the formula for the sum of a geometric series

$$84 = \text{GP}_n = \frac{a\left(1-k^n\right)}{1-k} = \frac{6\left[1-(1+r)^{12}\right]}{1-(1+r)}$$

$$84 = \frac{6\left[1-(1+r)^{12}\right]}{-r}$$

The easiest way to solve for r in this equation is to set up a spreadsheet to calculate different values of GP_n for different values of r and then see which one is closest to 84.

A spreadsheet which does this is shown in Table 8.19. To construct this spreadsheet yourself, you should now be able to enter the labels, given parameter values and the range of interest rates without any difficulty. The crucial calculation is the formula that calculates the total amount of extraction corresponding to the given initial extraction rate, the time period and the interest rate. This is achieved by entering the formula =C$2*(1−(1+A6)^C$3)/−A6 in cell B6 and then copying it down the column.

Table 8.19 Using a spreadsheet for Example 8.69

	A	B	C	D
1	Ex 8.69	OIL	RESERVES	
2	INITIAL	EXTRACTION =	6	m barrels
3		TIME PERIOD =	12	years
4	GROWTH	TOTAL		
5	r	EXTRACTION		
6	2.00%	80.473		
7	2.05%	80.699		
8	2.10%	80.927		
9	2.15%	81.155		
10	2.20%	81.384		
11	2.25%	81.613		
12	2.30%	81.844		
13	2.35%	82.075		
14	2.40%	82.307		
15	2.45%	82.540		
16	2.50%	82.773		
17	2.55%	83.008		
18	2.60%	83.243		
19	2.65%	83.479		
20	2.70%	83.715		
21	2.75%	83.953	<< *solution*	

	A	B	C	D
22	2.80%	84.191		
23	2.85%	84.430		
24	2.90%	84.670		
25	2.95%	84.911		
26	3.00%	85.152		

You can now read off the growth rate that corresponds to the total extraction amount which is closest to 84, which is 2.75% giving total extraction of 83.953. If you wanted to get a more precise answer you could make the growth rate increments smaller close to this approximate solution. This should show that a growth rate of 2.76% in the annual level of extraction will exhaust the oil reserves in 12 years.

Example 8.70

In a water authority's area the current river flows allow a maximum extraction rate of 100 million gallons of water per day and current usage is 25 million gallons per day. When will a crisis point be reached if water consumption grows by 4% per annum? What rate of growth would allow current supply sources to be sufficient for the next 100 years?

Solution

Consumption rate in n years' time (in millions of gallons) will be $25(1.04)^n$. The crisis point will be reached when consumption equals the maximum extraction rate which will be when

$$100 = 25(1.04)^n$$
$$4 = 1.04^n$$

Putting this in log form this gives

$$\log 4 = n \log 1.04$$
$$n = \frac{\log 4}{\log 1.04} = \frac{0.060206}{0.017033} = 35.346$$

Therefore, a growth rate of 4% can be sustained for 35 years.

If current water supplies are to last another 100 years at growth rate r, then the current maximum supply rate of 100 million gallons per day will equal demand when

$$100 = 25(1+r)^{100}$$
$$4 = (1+r)^{100}$$
$$\sqrt[100]{4} = 1+r$$
$$1.0139595 = 1+r$$
$$0.0139595 = r$$

Therefore, current water supplies will be sufficient for the next 100 years with a growth rate of just under 1.4%.

Example 8.71

A retailer has to order stock of a particular summer seasonal product in one batch at the start of the season. The first week's sales are expected to be 200 units. Past years' sales suggest that demand will then grow by 5% a week for the next 14 weeks and then fall by 10% a week for the remaining 10 weeks of the season. How much stock needs to be ordered to meet the anticipated sales for the whole 25-week season?

Solution

This problem involves the summing of two separate geometric series. Sales over the first 15 weeks are expected to be

$$200+200(1.05)+200(1.05)^2+\ldots+200(1.05)^{14}$$

In this geometric series $a = 200$, $k = 1.05$ and $n = 15$, and so its sum will be

$$\frac{a\left(1-k^n\right)}{1-k}=\frac{200\left(1-1.05^{15}\right)}{1-1.05}=\frac{200(1-2.0789282)}{-0.05}=4,315.7128$$

Thus, to the nearest whole unit over the first 15 weeks, sales will be 4,316. In the 15th week (which is part of the first geometric series) the sales will be

$$200(1.05)^{14}=395.98632=396 \text{ (to nearest whole unit)}$$

Over the remaining 10 weeks the total sales will therefore be

$$396(0.9)+396(0.9)^2+\ldots+396(0.9)^{10}$$

In this geometric series $a = 396(0.9)$, $k = 0.9$ and $n = 10$. Its sum will therefore be

$$\frac{a\left(1-k^n\right)}{1-k}=\frac{396(0.9)\left(1-0.9^{10}\right)}{1-0.9}=\frac{356.4(0.6513216)}{0.1}=2,321.31$$

Thus, to the nearest whole unit, expected total sales over the whole season will be

$$4,316+2,321=6,637 \text{ units}$$

Example 8.72

Average annual income in a developing country is \$4,200, and average annual expenditure on food is \$2,800. If average income rises at 3% per annum and income elasticity of demand for food is 0.8, when will average expenditure on food reach \$3,400?

Solution

For every 1% rise in average income Y, there will be a 0.8% rise in food consumption F, given an income elasticity of demand of 0.8. Therefore a 3% growth in Y will mean a 2.4% growth in F. The mathematical problem then becomes 'how long will it take $2,800 to grow to $3,400 at a growth rate of 2.4%?' and so we need to solve for n in the equation

$$3{,}400 = 2{,}800(1.024)^n$$

$$\frac{3{,}400}{2{,}800} = 1.024^n$$

$$1.2142857 = 1.024^n$$

Putting this in log form

$$\log 1.2142857 = n \log 1.024$$

$$n = \frac{\log 1.2142857}{\log 1.024} = \frac{0.084309}{0.0103} = 8.1865$$

Therefore, average food expenditure will reach $3,400 after approximately 8.19 years.

QUESTIONS 8.13

1. Total reserves of mineral Z are 140 million tonnes. Current annual consumption is 18 million tonnes. If consumption is expected to grow by 4.5% a year, how long will these reserves last?
2. A developing country's gross national product (GNP) is forecast to grow at 3% per annum and its population is expected to expand at 1.5% per annum. GNP is currently $12,000 million and the population is 15 million, giving a GNP per capita of $800. When will GNP per capita reach $1,000?
3. What rate of growth of consumption will allow the current reserves of a natural resource to last for the next 50 years if this year's consumption is 8 million tonnes and total reserves are 1,220 million tonnes?
4. World annual usage of mineral M is declining by 5% a year. The current rate of extraction is 65 million tonnes per year. Total reserves in existence amount to 1,500 million tonnes. Will they last forever if the 5% per annum decline persists?
5. The estimated reserves of a resource R are 1,650 million tonnes. Current annual consumption is 80 million tonnes. What percentage reduction in the annual consumption rate will ensure that the resource never runs out, assuming that consumption falls each year by the same percentage?

8A APPENDIX: ASSET VALUATION

This appendix provides an introduction to some basic methods used for the valuation of financial assets, or securities. These methods will be particularly useful for students who go on to specialize in financial economics, perhaps with the intention of becoming a financial or investment analyst.

Securities are tradable financial assets. Here we will just focus on bonds (debt securities) and shares (equity securities), which are the two most common types of financial assets that you will encounter in introductory finance courses.

Bonds can be issued by either governments or firms and they are a form of debt that must be repaid. In return for purchasing a bond, an investor is guaranteed to receive a fixed payment in each period, usually a year, and then a fixed 'nominal' value at the end of the investment period when the bond is redeemed. However, the actual prices at which bonds are traded may differ from their nominal value. Asset valuation methods can help determine what the correct price of a bond should be, which will help potential investors decide if current market prices offer good investment opportunities or not.

Shares are used by firms to raise funds. When a public company issues and sells new shares it is effectively selling a stake in the ownership of the company. If the company makes profits the board of directors will decide how much of these will be paid to shareholders and then a 'dividend' is declared, which is the amount paid out per share. Shareholders may also make a capital gain, or loss, if the share price rises, or falls. Expected dividends will influence this price and, unlike the fixed payments on bonds, dividend payments can fluctuate substantially.

8A.1 VALUATION OF BONDS

When buying a bond an investor is lending funds to the bond issuer. As with any loan, the lender will require the original sum of money plus interest to be paid back. The rate of interest earned on bonds is typically expressed as a percentage of the original sum per year and its value depends on many factors, including inflation, length of the loan, how liquid the asset is (i.e. if it can be easily sold or exchanged) and the perceived risk associated with the issuer. Government bonds are normally considered to be relatively risk-free assets and the effective interest rate on government bonds is often used as a benchmark for the cost of capital in financial markets. The way that government bonds and their yields are valued can therefore have a significant impact on the interest rates that businesses and households pay to borrow money. The basic techniques explained in this section to determine the price and yield of bonds also apply to the valuation of corporate bonds issued by companies to raise funds.

Although actual market values may fluctuate, all bonds have two fixed values:

(i) **Nominal value**: This is the value that will be paid back when the term of the bond expires. It is also called the **face value**.

(ii) **Coupon payments**: These are the payments that are paid to the bond holder, usually at annual or 6-monthly intervals.

The fixed value coupon payment can be expressed as a percentage of a bond's nominal value. For example, a bond with a nominal value of £100 that pays interest at a nominal rate of 4% per year will pay a single annual coupon payment of £4.

To determine the value of a bond we need to calculate the present value of the stream of payments received by a bond holder, which will consist of:

(i) the stream of regular coupon payments with a fixed value, and
(ii) the final payment of the bond's face value upon its maturity.

In the following examples, the bond price that is determined is considered its 'correct' price based on the valuation method used. This will help a potential investor determine whether the actual market price is over- or under-valued.

Example 8A.1

Determine the price of a 3-year government bond that has £100 face value and provides annual coupon payments of £4 at the end of each year. Assume that the market interest rate is 6% and does not change over the asset's life.

Solution

To find the price of this bond, P_B, we need to calculate the present value of the stream of coupon payments plus the redemption value, giving

$$P_B = \frac{4}{1+0.06} + \frac{4}{(1+0.06)^2} + \frac{4}{(1+0.06)^3} + \frac{100}{(1+0.06)^3} = 94.65$$

Thus, the price an investor would agree to pay for this bond should not be higher than £94.65. Note that the denominator $(1 + 0.06)^3$ has the same power in the two final elements of the previous equation because both the last coupon payment and the redemption payment take place upon the bond's maturity at the end of year 3. This example illustrates the fact that the market price of a bond will usually differ from its nominal value.

Using the same principles as in the previous example, we can derive a more general **formula for the price of a bond** as:

$$P_B = \frac{C}{1+i} + \frac{C}{(1+i)^2} + \ldots \frac{C}{(1+i)^t} + \frac{RV}{(1+i)^t}$$

where C is coupon payment, RV redemption (face) value, i the market interest rate and t represent the number of periods to maturity.

Example 8A.2

Consider again the bond from Example 8A.1 and find its price if the market interest rate is expected to be 8% over the 3-year period.

Solution

To find the new price we need to discount all the received payments at 8%, and so

$$P_B = \frac{4}{1+0.08} + \frac{4}{(1+0.08)^2} + \frac{4}{(1+0.08)^3} + \frac{100}{(1+0.08)^3} = 89.69$$

The new bond price of £89.69 is lower than when the interest rate was 6%. Thus, the effect of a higher interest rate is to lower the price of the bond.

The assumption that the market interest rate does not change over a number of years is unrealistic. In reality investors may see market interest rates change every day or even many times in a day. The present value formula used to determine the bond price can easily accommodate such interest rate changes, using the method explained earlier in Example 8.31. However, to explain the method of doing this and to keep the calculations manageable, we will consider a case when interest rates only change at the end of each year.

Example 8A.3

Consider a 5-year bond with a redemption value of £100 and annual coupon payments of £2. It is expected that the relevant market rate of interest will be 3% in year 1, 4% in year 2, 5% in year 3, 6% in year 4 and 7% in year 5. What price should an investor pay for it?

Solution

Adapting the present value formula to account for changes in the interest rate gives

$$P_B = \frac{2}{(1.03)} + \frac{2}{(1.03)(1.04)} + \frac{2}{(1.03)(1.04)(1.05)} + \frac{2}{(1.03)(1.04)(1.05)(1.06)}$$

$$+ \frac{2+100}{(1.03)(1.04)(1.05)(1.06)(1.07)} = 87.22$$

Thus, the price an investor should pay for this bond today should not exceed £87.22.

These examples assumed that investors bought bonds at their start date and then held onto them until the maturity date. However, in reality investors often sell and buy bonds in the financial market before their maturity and as a consequence the same bond may have several owners during its lifetime (provided the terms of bond issue allow that). We also assumed coupon payments were made only once a year, while it is a common practice to make coupon payments twice a year (or at other frequencies). The principle of finding the price of a bond is still the same in these situations – one has to find the present value of the stream of all future payments.

Another factor that has to be taken into account is what happens to the valuation if a bond purchase is between coupon dates. Should a seller of a bond receive a higher price that reflects the 'accrued interest' if they hold it for part of the time interval between coupon payments but then sell it to a new investor who only has to wait a shortened time interval to get the next coupon payment? There are two ways in which prices are quoted in the bond markets. The **clean price**, which does not include any accrued interest, is used in the British and American bond markets. It is the present value of the stream of the future payments, excluding any interest that would be payable on the next coupon payment. In contrast, most European markets quote the **dirty price**. This is the actual price an investor pays to purchase a bond and it includes the present value of all future payments, including interest accruing on the next coupon payment. On the coupon payment date the dirty and clean prices are equal. After that date the dirty price will increasingly exceed the clean price until the day when the next coupon payment is due and the two prices are equal again.

Bond yield

As explained earlier, estimating the correct value for a bond can become complex when the realities and future uncertainties of changing interest rates are taken into account. However, given that bonds are regularly traded on bond markets, actual market prices are readily available and a potential investor may be more likely to be interested in finding the overall rate of return on an investment in bonds. This rate of return is often described as **yield** and is used to assess the attractiveness of the bond and to compare with other investment opportunities, within or outside the bond market. There are three basic measures of the yield:

(i) The current yield

This is the most basic measure of yield. Current yield, Y_C, is calculated as the percentage ratio of the coupon payment, C, to the current bond price, P_B. Thus,

$$Y_C = \frac{C}{P_B} \times 100$$

The current yield is very easy to find because an investor considering a purchase of a bond will know the current market price and value of coupon payments. For example, the bond from Example 8A.3 pays the coupon of £2 and has the price of £87.22 (assuming the market price reflects this valuation) and so its current yield is

$$Y_C = \frac{2}{87.22} \times 100 = 0.0229 = 2.29\%$$

The current yield does not, however, take into account the difference between the current market price of the bond and its redemption value upon maturity. This means that it ignores potential capital gains or losses and so it is a rather imprecise measure of the true bond yield.

(ii) Simple yield to maturity

A more reliable yield measure, which takes into account capital gains and losses, is the simple yield to maturity, Y_S. This is calculated using the formula:

$$Y_S = \left[\frac{C}{P_B} + \frac{(100 - P_B)/T}{P_B}\right] \times 100$$

where T is the number of years to the bond's maturity. As you can see, the simple yield to maturity is the current yield plus the capital gain or loss (i.e. the £100 redemption value minus current price) per time period as proportion of the current market price P_B. This measure assumes that any capital gains or losses are spread evenly over the bond's life.

Example 8A.4

Find the simple yield to maturity of the bond from Example 8A.3 which pays the coupon of £2, has the current market price of £87.22 and can be redeemed in 5 years' time.

Solution

Substituting $C = 2$, $P_B = 87.22$ and $T = 5$ into the previous formula gives

$$Y_S = \left[\frac{2}{87.22} + \frac{(100-87.22)/5}{87.22}\right] \times 100 = (0.0229 + 0.0293) \times 100 = 0.0522 = 5.22\%$$

This shows that the simple yield to maturity is the current yield plus the capital gains or losses expressed as an annual rate of return. In this example the bond holder will earn 2.29% a year from the coupon payments and 2.93% a year from capital gains.

(iii) Yield to maturity (YTM)

Although the simple yield to maturity gives a better picture of a bond's rate of return than the current yield, it is still not a full measure because it does not take into account the fact that coupon payments may be reinvested and earn additional interest. A better measure is the yield to maturity (YTM), which takes into account all coupon payments, capital gains and losses, and the number of periods remaining until maturity and also assumes that coupon payments are immediately reinvested. This YTM measure helps investors to compare a given bond's return with the rates of return on other bonds, or with other investment opportunities.

To calculate the yield to maturity we need to find the value of YTM that satisfies the equation:

$$P_B = \frac{C}{1+YTM} + \frac{C}{(1+YTM)^2} + \ldots \frac{C}{(1+YTM)^t} + \frac{RV}{(1+YTM)^t}$$

As before, P_B is the bond actual market price, i.e. the dirty price, C is the coupon payment, RV is the redemption value at maturity and YTM represents the yield to maturity, which is the unknown variable. As you may have noticed, this method of calculation of YTM is similar to that used to calculate an investment's internal rate of return (IRR).

The calculations required to find YTM are relatively simple when only two time periods are involved but get more complex as the number of periods increases and a polynomial equation of a higher order needs to be solved. However, as we did in Section 8.6 for calculating IRR, we can use a spreadsheet to do these complex calculations and find the YTM, as the following example explains.

Example 8A.5

Consider the bond from Example 8A.3 again and find its yield to maturity (YTM). The bond pays a coupon of £2 per year, has the current (dirty) price of £87.22, can be redeemed in 5 years' time and its redemption value is £100.

Solution

Finding the yield to maturity requires solving the following equation for YTM:

$$87.22 = \frac{2}{1+YTM} + \frac{2}{(1+YTM)^2} + \frac{2}{(1+YTM)^3} + \frac{2}{(1+YTM)^4} + \frac{2}{(1+YTM)^5} + \frac{100}{(1+YTM)^5}$$

This is too complex to solve manually, but Table 8A.1 sets out the required steps to find YTM using the Excel YIELD function and Table 8A.2 shows the spreadsheet solution. The bond's yield to maturity is 4.9476%. This means that the stream of payments from an investment in the bond purchased at £87.22 and held for 5 years represent an

Table 8A.1 Setting up a spreadsheet for Example 8A.5

CELL	Enter	Explanation
A1	Ex.8A.5	Label to remind you what example this is
A2	Settlement date	The day on which an investor buys the bond
A3	Maturity date	The day on which the bond will be redeemed
A4	Percent coupon	The bond's annual coupon rate
A5	Price	The bond's dirty price (on settlement date)
A6	Redemption value	The amount at which the bond will be redeemed
A7	Frequency	The number of coupon payments per year
A8	YIELD TO MATURITY	Label tells you YTM will be calculated in next cell
B2	01/01/2016	Enter the settlement date; in this example we assume it is 1 January 2016
B3	01/01/2021	It is a 5-year bond so it will reach maturity on 1 January 2021
B4	2%	The payable coupons are £2 per annum which is 2% of £100 face value
B5	87.22	The bond's dirty price
B6	100	The bond's redemption/face value is £100
B7	1	The coupon payments are made once a year
B8	=YIELD(B2,B3,B4,B5,B6,B7)	Calculates YTM for the information in cells B2–B7. You can also use the function wizard by clicking on the *fx* symbol and searching for YIELD.

Table 8A.2 Excel output based on Table 8.A1

	A	B
1	Ex 8A.5	
2	Settlement date	01/01/2016
3	Maturity date	01/01/2021
4	Percent coupon	2%
5	Price	87.22
6	Redemption value	100
7	Frequency	1
8	YIELD TO MATURITY	4.9476%

average return of 4.9476% a year, assuming that coupon payments are reinvested at the same rate as the yield to maturity. In other words, the investor is investing £87.22 at the annual rate of 4.9476% for the period of 5 years.

This YTM rate can be compared with other potential investments. Alternatively, one could compare the total value of the investment at the end of this period, which will be

$$£87.22 \times (1+0.049476)^5 = £111.04$$

Note: instead of using the YIELD function, the YTM can also be found using the IRR function if the bond price is entered as a negative outlay in time period zero, and coupon values are entered as returns for each year, apart from the final time period where the redemption value is added to the coupon.

8A.2 VALUATION OF SHARES

There are different types of shares issued by companies, but we will just focus on **ordinary shares**, which are the most common type. Shares represent a form of ownership and give their holder the right to a share of the company's profits, voting rights, the power to change management and priority when buying new shares. Shareholders receive profits through dividend distributions, which in turn depend on the company's profits.

Companies normally receive investors' funds directly when they sell a 'new issue' of shares. However, after the initial issue, shares are traded on the stock market and their prices will rise if investors expect the firm to obtain high profits and consequently pay a higher dividend. As you can imagine, it can be difficult to predict a firm's future performance, and the further into the future a forecast is for then the larger the uncertainty and room for error. This means that share valuation models can get very complex when they attempt to include additional factors in order to account for this uncertainty. However, in this section we shall just explore some basic share valuation methods, which will be extended into more complex share valuation models if you go on to study more advanced finance courses.

In contrast to bonds, shares do not have a redemption date and are not usually bought back by a company. If we consider the special case when an investor wants to hold on to a share indefinitely and not to sell it at any time in future, and furthermore assume that the same dividend D is paid on each share each year, then the returns on this share can be treated as an infinite stream of amount D per year. Thus, the maximum price the investor will be willing to pay for a share will be the present value of all the future dividend payments. This is analogous to a perpetual annuity (see Section 8.8) and so, if i is the relevant discount rate, the price of this equity, P_E, can be easily calculated by adapting the perpetual annuity formula to give

$$P_E = \frac{D}{i}$$

For example, if a share pays a constant dividend of £3 each year and applies a 12% discount rate, then the price of this share should be £3/0.12 = £25. This represents the present value of all the dividend payments in the infinite future.

The assumption that the dividend is exactly the same every year is, however, rather unrealistic as market conditions change and firms are subject to various business and economic cycles. This means that firms' profits vary from year to year and so do share dividends. A firm may even choose not to pay a dividend at all in some years if it has not managed to achieve good earnings or if it wants to re-invest its profits. Therefore, we need a method of pricing shares that can accommodate changes in dividends from year to year, and also the fact that most shareholders will consider selling their shares at some point in the future. A method that does this, as before based on the present value of all the future payments a shareholder is expected to receive, is the more flexible **dividend pricing approach**, expressed by the formula:

$$P_E = \frac{D_1}{1+i_1} + \frac{D_2}{(1+i_2)^2} + \ldots \frac{D_t}{(1+i_t)^t} + \frac{P_{E(t)}}{(1+i_t)^t}$$

Here P_E is the current price of the share, which depends on the values of dividends paid in each year, D_1, D_2, …, D_t, and the expected price of the share in year t, $P_{E(t)}$, which is when the investor is considering selling their share. All the future payments are discounted at the interest rates i_t. (Note that this formula allows the discount rate i_t to change each year to better reflect likely conditions.) It has to be stressed, however, that a share's value estimated with this formula is only as reliable as the expectations of the future dividend payments and value of the share in t years' time and is therefore still subject to a large degree of uncertainty.

Example 8A.6

Estimate the current value of a share which promises to pay a dividend of £1 in 1 year's time, £3 in 2 years' time, £5 in 3 years' time and £2 in 4 years' time. The investor expects that the price of this share will be £30 at the end of year 4. The central bank's long-term forecast shows that the relevant interest rate should remain constant at 4% over the next 4 years.

Solution

We have all the information required to calculate the current share price using the dividend pricing formula. Substituting the given values, we get

$$P_E = \frac{1}{1.04} + \frac{3}{1.04^2} + \frac{5}{1.04^3} + \frac{2}{1.04^4} + \frac{30}{1.04^4} = 35.53$$

Thus, a current share price of £35.53 would reflect the discounted value of the dividend payments over the next 4 years plus the discounted expected value of the share at the end of year 4.

The Gordon growth model

In the previous example we used predictions of dividends that varied from one year to another. However, in practice it is difficult enough to forecast firms' profits, and it is even more difficult to predict what proportion of these profits will be distributed as dividends. A way around this problem may be to assume that dividends grow every year at some constant growth rate g. This approach is referred to as the **Gordon growth model**, named after Myron J. Gordon who suggested it in the early 1960s. This model equates the current share price to the present value of all future dividend payments and therefore is also referred to as the **dividend discount model** (DDM). The assumption of constant dividend growth allows us to express all future dividends in terms of the dividend in the current year and the expected dividend growth rate g. Thus,

$$\begin{aligned}
&\text{Dividend in current year} = D_0 \\
&\text{Dividend in year1, } D_1 \quad = D_0(1+g) \\
&\text{Dividend in year2, } D_2 \quad = D_0(1+g)^2 \\
&\text{Dividend in year3, } D_3 \quad = D_0(1+g)^3 \\
&\quad \vdots \qquad\quad \vdots \quad \vdots \qquad\quad \vdots \\
&\text{Dividend in year } t,\ D_t \quad = D_0(1+g)^t
\end{aligned}$$

As shares do not have a redemption date, the present value of a share therefore becomes the discounted stream of all future dividend payments to infinity, which is

$$P_E = \frac{D_0(1+g)}{1+i} + \frac{D_0(1+g)^2}{(1+i)^2} + \ldots + \frac{D_0(1+g)^t}{(1+i)^t} + \ldots \tag{1}$$

The Gordon growth model assumes that the discount rate remains unchanged over time. Although this is a strong assumption, it enables the previous formula to be significantly simplified and simplicity is the main appeal of this model.

A closer look at the formula in (1) reveals that it is an infinite geometric series with

$$\text{initial term } a = \frac{D_0(1+g)}{1+i} \quad \text{and} \quad \text{common ratio } k = \frac{1+g}{1+i} \tag{2}$$

From Section 8.8 we know that the sum of an infinite geometric series simplifies to:

$$a + ak + ak^2 + ak^3 + \ldots = \frac{a}{1-k} \tag{3}$$

Therefore, substituting a and k values from (2) into (3) gives

$$P_E = \frac{a}{1-k} = \frac{\frac{D_0(1+g)}{1+i}}{1-\frac{1+g}{1+i}} = \frac{\frac{D_0(1+g)}{1+i}}{\frac{1+i}{1+i}-\frac{1+g}{1+i}} = \frac{\frac{D_0(1+g)}{1+i}}{\frac{i-g}{1+i}} = \frac{D_0(1+g)}{i-g}$$

and so the **Gordon growth model formula for the price of a share** is

$$P_E = \frac{D_0(1+g)}{i-g}$$

Thus, the Gordon growth model offers a simple way of calculating what the current price of a share should be, assuming we know the current year's dividend, D_0, the expected growth rate of the dividend, g, and the applicable discount rate, i.

One restriction of this model is that i must be greater than g. If the denominator $(i - g)$ in the previous formula took a non-positive value then the result would have no economic sense because the calculated share price would be negative if $i < g$, or infinite if $i = g$.

As with any model, the Gordon growth model represents a simplified approach to calculating the value of a share. Its result is based on the two key assumptions that both the rate of the dividend and the market interest rate remain constant over time. As the future is never certain and markets can be volatile, in reality it is very difficult to predict whether a firm will pay a dividend several years from now or what the market interest rate will be. Furthermore, if the dividend paid by a firm represents a high percentage of its profits then, according to the Gordon growth model, the present value of its shares will increase. However, this ignores the fact that there will be less profit left for the firm to invest in its business. This may result in smaller profits in the future and hence actual dividend payments may become lower. In other words, the shareholders may receive a greater share of a cake which is becoming smaller.

Example 8A.7

An investor is considering a purchase of shares in a high-tech company which is known for its innovative products. The current market price is £45.95 per share. Recently the firm paid a dividend of £2.40 per share. The firm's future prospects look optimistic: it has several new products being developed which should be introduced to the market in the coming years and ensure handsome profits. Market analysts forecast that the dividends paid to shareholders will grow at the rate of 7% each year. Use the Gordon growth model to determine the value of the firm's shares to the investor if the required rate of return is 12%. Should the investor purchase this equity at the current market price?

Solution

From this example we know that $D_0 = 2.40$, $g = 0.07$ and $i = 0.12$. We substitute these values into the Gordon growth model formula to find the share's price:

$$P_E = \frac{D_0(1+g)}{i-g} = \frac{2.4(1+0.07)}{0.12-0.07} = 51.36$$

According to the Gordon growth model, the price the investor should pay for this equity should not exceed £51.36. Thus, the current market price of £45.95 appears to offer a good investment opportunity.

Example 8A.8

The market trades equity in a public listed company at £6.25 a share. The company announces that it will pay a dividend of £0.14 in the coming year. This announcement attracts your attention as an investor and you decide to ask your financial analyst for advice. They expect the dividend to grow at a constant rate of 5% a year into the foreseeable future. Is £6.25 a share a good price for this equity if you require the rate of return to be at least 7.5% to cover your costs of obtaining capital?

Solution

The Gordon growth model can be used to calculate the maximum price to be paid for this equity. If the dividend paid in the coming year is going to be £0.14, then $D_1 = 0.14$. Recall that the model assumes that the dividend grows at the same rate each year and therefore $D_1 = D_0(1 + g)$. Substituting this value into the Gordon growth model formula, we get

$$P_E = \frac{D_0(1+g)}{i-g} = \frac{D_1}{i-g}$$

If the values $D_1 = 0.14$, $g = 0.05$ and $i = 0.075$ are inserted then

$$P_E = \frac{0.14}{0.075-0.05} = 5.6$$

Assuming consistent dividend growth at 5% a year, the present value of all future dividend payments is equal to £5.60 which is less than the price of £6.25 at which the share is traded. Thus, the share is overvalued by the market and, according to the Gordon growth model, you should not purchase it.

The share valuation methods presented in this Appendix are relatively basic and, as you may have already realized, they are far from perfect. The biggest challenge facing stock market analysts are uncertainty and difficulty in predicting the future. Several alternative methods of share valuation have been proposed, attempting to take into account various types of information relating to the firm itself or general market and economic conditions, and you may go on to study these in more advanced finance courses.

QUESTIONS 8A

1. A government bond promises the coupon payment of £3.50 a year and can be redeemed in 3 years' time at the redemption value of £100. What should be this bond's market price if the relevant market rate of interest is 3.25%?
2. A 5-year bond has the face value of £100 and offers annual coupon payments of £2. What is the maximum price an investor should agree to pay if the expected rate of interest is forecast to stay at 4.5%?
3. A 3-year bond pays the coupon of £2.50 a year and has the face value of £100. How much would you agree to pay for it if you expected the interest rate to be 3% in year 1, 3.5% in year 2 and 3.8% in year 3?
4. A bond has the face value of £100 and promises coupon payments of £1.75 a year. Its current market price is £90 and it will be redeemed in 10 years' time. Calculate:
   (i) the current yield,
   (ii) the current yield to maturity,
   (iii) the yield to maturity (YTM) using the YIELD function in a spreadsheet.
5. Use the dividend pricing approach to find the maximum price an investor should pay for a share which is expected to pay the dividend of £4.50 in 1 year's time, £4 in 2 years' time and £5 in 3 years' time. The investor expects to re-sell the share after 3 years at the price of £50. Assume that the rate of interest required by the market is 6% and will remain unchanged.
6. As an investor you are considering a purchase of equity in a commodity trading firm whose shares trade currently at £40 each. Recently the company paid the dividend of £0.70 a share. The commodity markets are forecasted to be on the rise and analysts expect that the dividends paid by this particular firm will grow at 10% a year. Use the Gordon growth model to assess whether the current market price provides a worthwhile investment if the market rate of interest is 12%.
7. A firm is expected to pay the dividend of £1.60 a share in a year from now. This dividend is expected to grow each year at a constant rate of 3%. The current rate of interest puts the cost of capital at 5%. What is the value of this equity according to the Gordon growth model?

9 Introduction to calculus

Learning objectives

After completing this chapter students should be able to:

- differentiate functions with one unknown variable
- find the slope of a function using differentiation
- derive marginal revenue and marginal cost functions using differentiation and relate them to the slopes of the corresponding total revenue and cost functions
- calculate point elasticity for non-linear demand functions
- use calculus to find the sales tax that will maximize tax yield
- derive the Keynesian multiplier using differentiation.

9.1 DIFFERENTIAL CALCULUS

This chapter introduces some of the basic techniques of calculus and their application to economic problems. We shall be concerned here with what is known as the 'differential calculus'.

Differentiation is a method used to find the slope of a function at any point. Although this is a useful tool in itself, it also forms the basis for some very powerful techniques for solving optimization problems, which are explained in this and the following chapters. The basic technique of differentiation is quite straightforward and easy to apply. Consider the simple function

$$y = 6x^2$$

To derive an expression for the slope of this function for any value of x, **the basic rules of differentiation** require you to:

(a) multiply the whole term by the value of the power of x, and
(b) deduct 1 from the power of x.

DOI: 10.4324/9781003360827-9

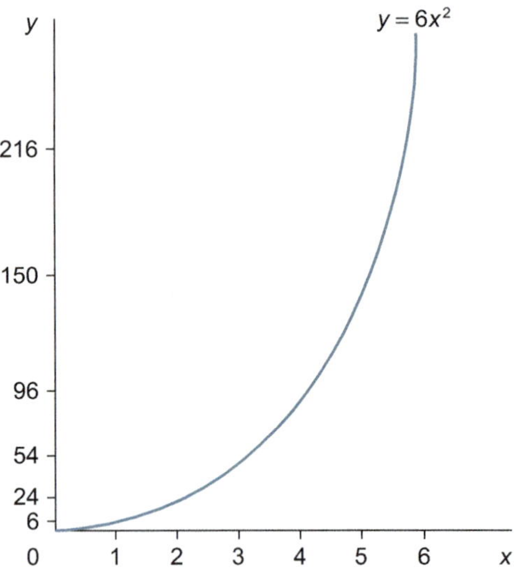

Figure 9.1 Variable slope of a quadratic function

In this example there is a term in x^2 and so the power of x is reduced from 2 to 1. Using the previous rule the expression for the slope of this function therefore becomes

$$2 \times 6x^{2-1} = 12x$$

This is known as the derivative of y with respect to x, and is usually written as $\frac{dy}{dx}$, which is read as 'dy by dx'.

We can check that this is approximately correct by looking at the graph of the function $y = 6x^2$ in Figure 9.1. Any term in x^2 will rise at an ever increasing rate as x becomes larger. In other words, the slope of this function must increase as x increases. The slope is the derivative of the function with respect to x, which we have just worked out to be $12x$. As x increases the term $12x$ will also increase and so we can confirm that the formula derived for the slope of this function does behave in the expected fashion.

To determine the actual value of the slope of the function $y = 6x^2$ for any given value of x, one simply enters the given value of x into the formula

$$\text{slope} = 12x$$

When $x = 4$, then slope $= 48$; when $x = 5$, then slope $= 60$, etc.

Example 9.1

What is the slope of the function $y = 4x^2$ when x is 8?

Solution

By differentiating y we get

$$\text{Slope} = \frac{dy}{dx} = 2 \times 4x^{2-1} = 8x$$

When $x = 8$, then slope $= 8 \times 8 = 64$.

Example 9.2

Find a formula that gives the slope of the function $y = 6x^3$ for any value of x.

Solution

$$\text{Slope} = \frac{dy}{dx} = 3 \times 4x^{3-1} = 18x^2 \text{ for any value of } x$$

Example 9.3

What is the slope of the function $y = 45x^4$ when $x = 10$?

Solution

$$\text{Slope} = \frac{dy}{dx} = 180x^3$$

When $x = 10$, then slope = 180 × 1,000 = 180,000

QUESTIONS 9.1

1. Derive an expression for the slope of the function $y = 12x^3$.
2. What is the slope of the function $y = 6x^4$ when $x = 2$?
3. What is the slope of the function $y = 0.2x^4$ when $x = 3$?
4. Derive an expression for the slope of the function $y = 52x^3$.

9.2 RULES FOR DIFFERENTIATION

The rule for differentiation can be formally stated as:

If $y = ax^n$ where a and n are given parameters then $\frac{dy}{dx} = nax^{n-1}$.

When there are several terms in x added together or subtracted in a function then this rule for differentiation is applied to each term individually. (The special rules for differentiating functions where terms are multiplied or divided are explained in Chapter 13.)

Example 9.4

Differentiate the function $y = 3x^2 + 10x^3 - 0.2x^4$.

Solution

$$\frac{dy}{dx} = 2 \times 3x^{2-1} + 3 \times 10x^{3-1} - 4 \times 0.2x^{4-1} = 6x + 30x^2 - 0.8x^3$$

Example 9.5

Find the slope of the function $y = 6x^2 - 0.5x^3$ when $x = 10$.

Solution

$$\text{Slope} = \frac{dy}{dx} = 12x - 1.5x^2$$

When $x = 10$, slope $= 12 \times 10 - 1.5 \times 100 = 120 - 150 = -30$

Example 9.6

Derive an expression for the slope of function y at any value of x.

$$y = 4x^2 + 2x^3 - x^4 + 0.1x^5$$

Solution

$$\text{Slope} = \frac{dy}{dx} = 8x + 6x^2 - 4x^3 + 0.5x^4$$

In using the formula for differentiation, one has to remember that $x^1 = x$ and $x^0 = 1$.

Example 9.7

Differentiate the function $y = 8x$.

Solution

$$y = 8x = 8x^1$$

$$\frac{dy}{dx} = 1 \times 8x^{1-1} = 8x^0 = 8 \times 1 = 8$$

Example 9.8

Derive an expression for the slope of the function $y = 30x - 0.5x^2$ for any value of x.

Solution

$$\text{Slope} = \frac{dy}{dx} = 30x^0 - 2(0.5)x = 30 - x$$

Example 9.9

Differentiate the function $y = 14x$.

Solution

$$\frac{dy}{dx} = 14x^{1-1} = 14x^0 = 14$$

This example illustrates the point that the derivative of any term in x (to the power of 1) is simply the value of the parameter that x is multiplied by.

We can also relate this result to the slope of the original function $y = 14x$, which will be an upward sloping straight line. The slope of any straight line is constant along its entire length, which is what this derivative shows, and it is positive, which means the straight line slopes up from left to right.

Any constant terms always disappear when a function is differentiated. To understand why, consider a function with one constant such as the function $y = 5$. This could be written as $y = 5x^0$. Differentiating this function gives

$$\frac{dy}{dx} = 0(5x^{-1}) = 0$$

This result means that the slope of this function is zero, which is what we would expect given that the graph of the function $y = 5$ will be a horizontal straight line.

Example 9.10

Differentiate the function $y = 20 + 4x - 0.5x^2 + 0.01x^3$.

Solution

$$\frac{dy}{dx} = 4 - x + 0.03x^2$$

Example 9.11

Derive an expression for the slope of the function $y = 6 + 3x - 0.1x^2$.

Solution

$$\text{Slope} = \frac{dy}{dx} = 3 - 0.2x$$

Even when the power of x in a function is negative or not a whole number, the same rules for differentiation still apply.

Example 9.12

What is the slope of the function $y = 4x^{0.5}$ when $x = 4$?

Solution

$$\text{slope} = \frac{dy}{dx} = 0.5 \times 4x^{0.5-1} = 2x^{-0.5}$$

$$\text{When } x = 4, \text{ the slope} = 2 \times 4^{-0.5} = 2 \times \left(\frac{1}{2}\right) = 1$$

Example 9.13

Differentiate the function $y = x^{-1} + x^{0.5}$.

Solution

$$\frac{dy}{dx} = -1 \times x^{-1-1} + 0.5x^{0.5-1} = -x^{-2} + 0.5x^{-0.5}$$

What differentiation actually does is look at the effect of an infinitely small change in the independent variable x on the dependent variable y in a function $y = f(x)$. This may seem a strange concept, but to help to understand it better, consider the following example which shows how a function can be differentiated from first principles.

Example 9.14

Differentiate the function $y = 6x + 2x^2$ from first principles.

Solution

Assume that x is increased by the small amount Δx (Δ is the Greek letter 'delta' which usually signifies a change in a variable). This will produce a small change Δy in y.

The new value of y (i.e. $y + \Delta y$) can be found by substituting the new value of x (i.e. $x + \Delta x$) into the original function

$$y = 6x + 2x^2 \qquad (1)$$

Thus,

$$y + \Delta y = 6(x + \Delta x) + 2(x + \Delta x)^2$$

$$y + \Delta y = 6x + 6\Delta x + 2x^2 + 4x\Delta x + 2(\Delta x)^2$$

Subtracting (1)

$$y = 6x + 2x^2$$

gives

$$\Delta y = 6\Delta x + 4x\Delta x + 2(\Delta x)^2$$

Dividing through by Δx gives

$$\frac{\Delta y}{\Delta x} = 6 + 4x + 2\Delta x$$

If Δx becomes infinitely small, then the last term disappears and

$$\frac{\Delta y}{\Delta x} = 6 + 4x \qquad (2)$$

By definition, $\frac{dy}{dx}$ is the effect of an infinitely small change in x on y.

Thus, from (2) $\frac{dy}{dx} = 6 + 4x$

This is the same result for dy/dx that would be obtained using the basic rules for differentiation. It is obviously quicker to use these rules than to differentiate from first principles. However, Example 9.14 should help you to better understand the underlying rationale for the rules of differentiation and their application to economics in the next sections.

QUESTIONS 9.2

1. Differentiate the function $y = x^3 + 60x$.
2. What is the slope of the function $y = 12 + 0.5x^4$ when $x = 5$?
3. Derive a formula for the slope of the function $y = 4 + 4x^{-1} - 4x$.
4. What is the slope of the function $y = 4x^{0.5}$ when $x = 4$?
5. Differentiate the function $y = 25 - 0.1x^{-2} + 2x^{0.3}$.

9.3 MARGINAL REVENUE AND TOTAL REVENUE

Up to this point we have been using the usual algebraic notation for a single variable function, assuming that y is dependent on x. Changing the notation so that we can look at some economic applications does not alter the rule for differentiation as long as functions are specified in a form where one variable is dependent on another.

In introductory economics texts, marginal revenue (MR) is sometimes defined as the increase in total revenue (TR) received from sales caused by an increase in output by 1 unit. This is not a precise definition though. It only gives an approximate value for MR and it will vary if the units that output is measured in are changed. A more precise definition of MR is that it is the rate of change of TR relative to increases in output.

In Figure 9.2 the rate of change of TR between points B and A is

$$\frac{\Delta TR}{\Delta q} = \frac{AC}{BC} = \text{the slope of the line AB}$$

which is an approximate value for MR over this output range.

Now, suppose that the distance between B and A gets smaller. As point B moves along TR towards A the slope of the line AB gets closer to the value of the slope of TT′, which is the tangent to TR at A. (A tangent

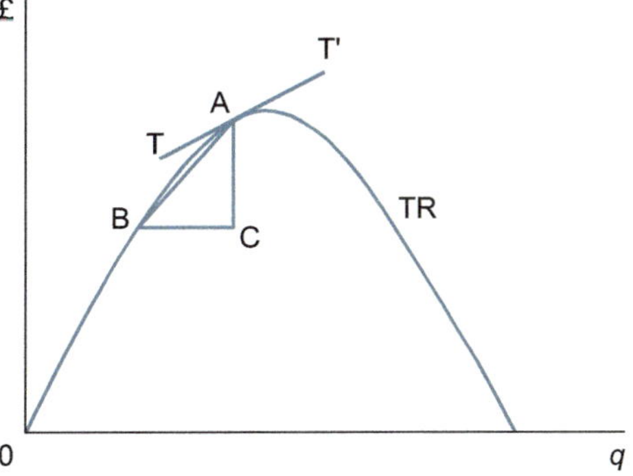

Figure 9.2 Rate of change of total revenue between two points

to a curve at any point is a straight line having the slope at that point.) Thus, for a very small change in output, MR will be almost equal to the slope of TR at A. If the change becomes infinitesimally small, then the slope of AB will exactly equal the slope of TT′. Therefore, MR will be equal to the slope of the TR function at any given output.

We know that the slope of a function can be found by differentiation and so it must be the case that

$$\text{MR} = \frac{\text{dTR}}{\text{d}q}$$

Example 9.15

Given that TR = $80q - 2q^2$, derive a function for MR.

Solution

$$\text{MR} = \frac{\text{dTR}}{\text{d}q} = 80 - 4q$$

This result helps explain some of the properties of the relationship between TR and MR. The linear demand schedule D in Figure 9.3 represents the function

$$p = 80 - 2q \tag{1}$$

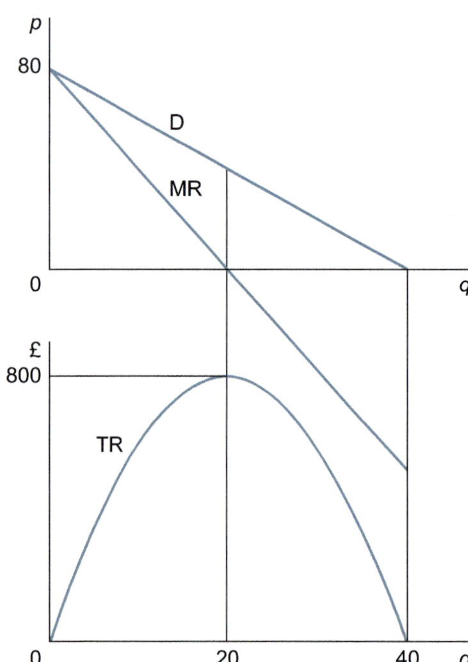

Figure 9.3 Demand and revenue functions

We know that by definition TR = pq. Therefore, substituting (1) for p,

$$\text{TR} = (80 - 2q)q = 80q - 2q^2$$

which is the same as the TR function in Example 9.15. This TR function is plotted in the lower section of Figure 9.3 and the function for MR, already derived, is plotted in the top section.

You can see that when TR is rising, MR is positive, as one would expect. As the rate of increase of TR gets smaller so does the value of MR. When TR is at its maximum, MR is zero, and when TR is falling, MR is negative.

With the function for MR derived earlier it is very straightforward to find the exact value of the output at which TR is a maximum. The TR function is horizontal with slope zero at its maximum point and so MR is also zero for this output. Thus, when TR is at its maximum

$$\begin{aligned} \text{MR} = 80 - 4q &= 0 \\ 80 &= 4q \\ 20 &= q \end{aligned}$$

One can also see that the MR function has the same intercept on the vertical axis as this straight line demand schedule, but twice its slope. We can show that this result holds for any linear downward-sloping demand schedule.

Given the function $\quad p = a - bq$

Then $\quad \text{TR} = pq = (a - bq)q = aq - bq^2$

$$\text{MR} = \frac{\text{dTR}}{\text{d}q} = a - 2bq$$

Thus, both the demand schedule and the MR function have a as the intercept on the vertical axis, and the slope of MR is $2b$ which is obviously twice the demand schedule's slope.

It should be noted that this result does not hold for non-linear demand schedules. If a demand schedule is non-linear then it is best to derive the slope of the MR function from first principles.

Example 9.16

Derive the MR function for the non-linear demand schedule $p = 80 - q^{0.5}$.

Solution

$$\text{TR} = pq = \left(80 - q^{0.5}\right)q = 80q - q^{1.5}$$

$$\text{MR} = \frac{\text{dTR}}{\text{d}q} = 80 - 1.5q^{0.5}$$

In this non-linear case the intercept on the price axis is still 80 but the slope of MR is 1.5 times the slope of the demand function. (Note, though, that the slope is not simply the coefficient of the term in q and the original demand function will have a slope of $-0.5q^{-0.5}$, which will change as q changes, so derive the slope of MR yourself and compare.)

For those of you who are still not convinced that the idea of looking at an 'infinitesimally small' change can help find the rate of change of a function at a point, Example 9.17 shows how a spreadsheet can be used to calculate rates of change for very small increments. This example is for illustrative purposes only. The main reason for using calculus in the first place is to enable the immediate calculation of rates of change at any point of a function.

Example 9.17

For the total revenue function

$$\text{TR} = 500q - 2q^2$$

find the value of MR when $q = 80$ (i) using calculus, and (ii) using a spreadsheet that calculates increments in q above the given value of 80 that get progressively smaller. Compare the two answers.

Solution

(i) $$\text{MR} = \frac{\text{dTR}}{\text{d}q} = 500 - 4q$$

Thus, when $q = 80$

$$\text{MR} = 500 - 4(80) = 500 - 320 = 180$$

(ii) The spreadsheet shown in Table 9.2 can be constructed by following the instructions in Table 9.1. This spreadsheet shows that as increments in q (relative to the initial given value of 80) become smaller and smaller, the value of MR (i.e. $\Delta\text{TR}/\Delta q$) approaches 180. This is consistent with the answer obtained by calculus in (i).

Table 9.1 Setting up a spreadsheet for Example 9.17

CELL	Enter	Explanation
A1 to B4 and A6 to E6	*Enter labels as shown in* Table 9.2.	*Labels to indicate where initial values go plus column heading labels*
D2	TR = 500q -2q^2	*Label to remind you what function is used.*
D3	80	Given initial value for q.
D4	=500*D3-2*D3^2	Calculates TR corresponding to given q value.
B7	10	Initial size of increment in q.
B8	=B7/10	Calculates an increment in q that is only 10% of the value of the one in cell above.
B9 to B13	*Copy cell* B8 *formula down column* B.	Calculates a series of increments in q that get smaller and smaller each time.
A7	=B7+D$3	Calculates new value of q by adding the increment in cell A7 to the given value of 80.
A8 to A13	*Copy cell* A7 *formula down column* A.	Calculates a series of values of q that increase by smaller and smaller increments each time.
C7	=500*A7-2*A7^2	Calculates TR corresponding to value of q in cell A7.
C8 to C13	*Copy cell* C7 *formula down column* C.	Calculates a series of values of TR corresponding to values of q in column A.
D7	=C7-D$4	Calculates the change in TR relative to the initial given value in cell D4.
D8 to D13	*Copy cell* D7 *formula down column* D.	Calculates series of changes in TR corresponding to increments in q in column B.
E7	=D7/B7	Calculates ΔTR / Δq

CELL	Enter	Explanation
E8 to E13	*Copy cell E7 formula down column E.*	Calculates values of ΔTR / Δ*q* corresponding to decreasing increments in *q* and TR.
A7 to E13	*Widen columns and increase number of decimal places as necessary.*	The point of this example is to show how the value of ΔTR / Δ*q* converges on dTR/d*q* so all the decimal places need to be shown.

Table 9.2 Excel output based on Table 9.1

	A	B	C	D	E
1	Ex 9.17	DIFFERENTIATION OF TR FUNCTION			
2		GIVEN FUNCTION		TR = 500q − 2q^2	
3			INITIAL q VALUE =	80	
4			INITIAL TR VALUE =	27200	
5					Marginal Revenue
6	q	Delta q	TR	Delta TR	(DeltaTR)/(Delta q)
7	90	10	28800	1600	160
8	81	1	27378	178	178
9	80.1	0.1	27217.98	17.98	179.8
10	80.01	0.01	27201.7998	1.7998	179.98
11	80.001	0.001	27200.18	0.179998	179.998
12	80.0001	0.0001	27200.018	0.01799998	179.9998
13	80.00001	0.00001	27200.0018	0.001800000	179.9999802

QUESTIONS 9.3

1. Given the demand schedule $p = 120 - 3q$ derive a function for MR and find the output at which TR is a maximum.
2. For the demand schedule $p = 40 - 0.5q$ find the value of MR when $q = 15$.
3. Find the output at which MR is zero when $p = 720 - 4q^{0.5}$ describes the demand schedule.
4. A firm knows that the demand function for its output is $q = 800 - 2p$. What price should it charge to maximize sales revenue?

9.4 MARGINAL COST AND TOTAL COST

Just as MR can be shown to be the rate of change of the TR function, so marginal cost (MC) is the rate of change of the total cost (TC) function. In fact, in nearly all situations where one is dealing with the concept of a marginal increase, the marginal function is equal to the rate of change of the original function, i.e. to derive the marginal function one just differentiates the original function.

Example 9.18

Given $\text{TC} = 6 + 4q^2$ derive the MC function.

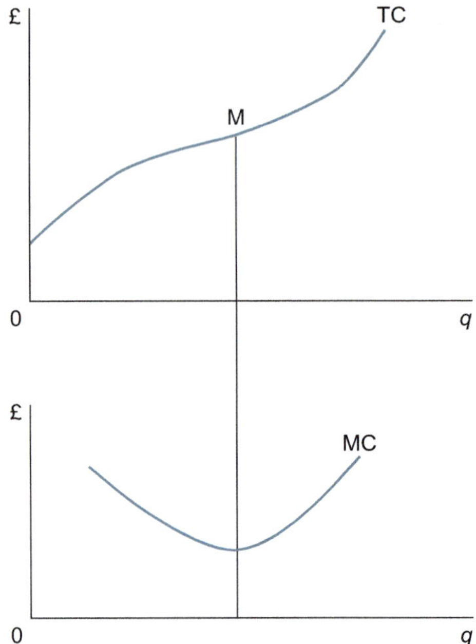

Figure 9.4 Total cost and marginal cost

Solution

$$\mathrm{MC} = \frac{\mathrm{dTC}}{\mathrm{d}q} = 8q$$

This example is somewhat unrealistic in that it assumes an MC function that is a straight line. This is because the TC function is given as a simple quadratic function, whereas one normally expects a TC function to have a shape similar to that shown in Figure 9.4. This represents a cubic function with certain properties to ensure that:

(a) the rate of change of TC first falls and then rises, and
(b) TC never actually falls as output increases, i.e. MC is never negative. (Although it is quite common to find economies of scale causing *average* costs to fall, no firm is going to find the *total* cost of production falling when output increases.)

The flattest, or least steep, point on this TC schedule is at M, which corresponds to the minimum value of MC.

A cubic total cost function has the previous properties if

$$\mathrm{TC} = aq^3 + bq^2 + cq + d$$

where a, b, c and d are parameters such that

$$a, c, d > 0, b < 0 \quad \text{and} \quad b^2 < 3\,ac.$$

This applies to the TC functions in the following examples.

Example 9.19

If TC = $2.5q^3 - 13q^2 + 50q + 12$ derive the MC function.

Solution

$$\mathrm{MC} = \frac{\mathrm{dTC}}{\mathrm{d}q} = 7.5q^2 - 26q + 50$$

Example 9.20

When will average variable cost be at its minimum value for the below TC function?

$$TC = 40 + 82q - 6q^2 + 0.2q^3$$

Solution

The theory of costs tells us that MC will cut the minimum point of both the average cost (AC) and the average variable cost (AVC) functions. We therefore need to derive the MC and AVC functions and find where they intersect.

It is obvious from this TC function that total fixed costs (TFC) = 40 and total variable costs (TVC) = $82q - 6q^2 + 0.2q^3$. Therefore,

$$AVC = \frac{TVC}{q} = 82 - 6q + 0.2q^2$$

and

$$MC = \frac{dTC}{dq} = 82 - 12q + 0.6q^2$$

Setting $\quad MC = AVC$

$$82 - 12q + 0.6q^2 = 82 - 6q + 0.2q^2$$
$$0.4q^2 = 6q$$
$$q = \frac{6}{0.4} = 15$$

So, the minimum point of AVC corresponds to an output of 15.

(When you have covered the analysis of maximization and minimization in the next chapter, come back to this example and see if you can think of another way of solving it.)

QUESTIONS 9.4

1. If TC = 65 + $q^{1.5}$ what is MC when q = 25?
2. Derive an expression for MC if TC = $4q^3 - 20q^2 + 60q + 40$.
3. If TC = $0.5q^3 - 3q^2 + 25q + 20$ derive functions for:
   (a) MC, (b) AC, (c) the slope of AC.
4. What is special about MC if TC = $25 + 0.8q$?

9.5 PROFIT MAXIMIZATION

We are now ready to see how calculus can help determine the output where a firm will maximize profits, as the following examples illustrate. At this stage we shall just use the MC = MR rule for profit maximization, which you will probably have learned in your economics course (if not, this will be explained in Section 10.5). The second condition (MC cuts MR from below) will be dealt with in the next chapter.

Example 9.21

A monopoly faces the demand schedule $p = 460 - 2q$
and the total cost schedule $\mathrm{TC} = 20 + 0.5q^2$.

How much should it sell to maximize profit and what will this maximum profit be? (All costs and prices are in £.)

Solution

To find the output where MC = MR we first need to derive the MC and MR functions.

$$\text{Given} \quad \mathrm{TC} = 20 + 0.5q^2$$

$$\text{then} \quad \mathrm{MC} = \frac{\mathrm{dTC}}{\mathrm{d}q} = q \qquad (1)$$

$$\text{As} \quad \mathrm{TR} = pq = (460 - 2q)q = 460q - 2q^2 \qquad (2)$$

$$\text{then} \quad \mathrm{MR} = \frac{\mathrm{dTR}}{\mathrm{d}q} = 460 - 4q$$

To maximize profit MR = MC. Therefore, equating (1) and (2),

$$\begin{aligned} 460 - 4q &= q \\ 460 &= 5q \\ 92 &= q \end{aligned}$$

The actual maximum profit when the output is 92 will be

$$\begin{aligned} \mathrm{TR} - \mathrm{TC} &= (460q - 2q^2) - (20 + 0.5q^2) = 460q - 2q^2 - 20 - 0.5q^2 \\ &= 460q - 2.5q^2 - 20 = 460(92) - 2.5(8{,}464) - 20 \\ &= 42{,}320 - 21{,}160 - 20 = £21{,}140 \end{aligned}$$

Example 9.22

A firm faces the demand schedule $p = 184 - 4q$
and the TC function $\mathrm{TC} = q^3 - 21q^2 + 160q + 40$.

What output will maximize profit?

Solution

$$\begin{aligned}
&\text{Given} && \text{TR} = pq = (184 - 4q)q = 184q - 4q^2 \\
&\text{then} && \text{MR} = \frac{\text{dTR}}{\text{d}q} = 184 - 8q \\
&\text{From the TC function} && \text{MC} = \frac{\text{dTC}}{\text{d}q} = 3q^2 - 42q + 160
\end{aligned}$$

To maximize profits MC = MR. Therefore,

$$3q^2 - 42q + 160 = 184 - 8q$$
$$3q^2 - 34q - 24 = 0$$
$$(q - 12)(3q + 2) = 0$$

$$q - 12 = 0 \quad \text{or} \quad 3q + 2 = 0$$
$$q = 12 \quad \text{or} \quad q = -\frac{2}{3}$$

One cannot produce a negative quantity and so the firm must produce 12 units of output in order to maximize profits.

QUESTIONS 9.5

1. A monopoly faces the following TR and TC schedules:

   $$\text{TR} = 300q - 2q^2$$
   $$\text{TC} = 12q^3 - 44q^2 + 60q + 30$$

   What output should it sell to maximize profit?

2. A firm faces the demand schedule $p = 190 - 0.6q$ and thc total cost function $\text{TC} = 40 + 30q + 0.4q^2$.
   (a) What output will maximize profit?
   (b) What output will maximize total revenue?
   (c) What will the output be if the firm makes a profit of £4,760?
3. Find the output where profit is maximized for a firm with the total revenue and total cost functions

   $$\text{TR} = 52q - q^2 \quad \text{TC} = \frac{q^3}{3} - 2.5q^2 + 34q + 4$$

9.6 RE-SPECIFYING FUNCTIONS

Many of the examples considered so far have included a demand schedule in the format

$$p = a + bq$$

Although, as was explained in Chapter 4, economic theory normally defines a demand function in the format q = f(p), with q being the dependent variable rather than p. However, because the usual convention is to have p on the vertical axis in supply and demand graphical analysis, and also because cost functions have q as the independent variable, it usually helps to work with the inverse demand function p = f(q). The following examples show how to derive the relationship between MR and q by finding the inverse demand function.

Example 9.23

Derive the MR function for the demand function $q = 400 - 0.1p$.

Solution

Given
$$\begin{aligned} q &= 400 - 0.1p \\ 10q &= 4{,}000 - p \\ p &= 4{,}000 - 10q \end{aligned}$$

Using this inverse demand function, we can now derive

$$\begin{aligned} \text{TR} &= pq = (4{,}000 - 10q)q = 4{,}000q - 10q^2 \\ \text{MR} &= \frac{\text{dTR}}{\text{d}q} = 4{,}000 - 20q \end{aligned}$$

Example 9.24

A firm faces the demand schedule $q = 200 - 4p$
and the cost schedule $\text{TC} = 0.1q^3 - 0.5q^2 + 2q + 8$.

What price will maximize profit?

Solution

The demand function $q = 200 - 4p$
has the inverse demand function $p = 50 - 0.25q$

This is a linear demand schedule and so MR has the same intercept and twice the slope. Thus,

$$\text{MR} = 50 - 0.5q$$

From the TC function we can get

$$\text{MC} = \frac{\text{dTC}}{\text{d}q} = 0.3q^2 - q + 2$$

To maximize profits MC = MR. Therefore, equating the MR and MC functions

$$0.3q^2 - q + 2 = 50 - 0.5q$$
$$0.3q^2 - 0.5q - 48 = 0$$

Using the quadratic equation formula to solve

$$q = \frac{-(-0.5) \pm \sqrt{(-0.5)^2 - 4 \times 0.3 \times (-48)}}{2 \times 0.3}$$
$$= \frac{0.5 \pm \sqrt{0.25 + 57.6}}{0.6} = \frac{0.5 \pm \sqrt{57.85}}{0.6}$$

Disregarding the negative solution as output cannot be negative

$$q = \frac{0.5 + 7.6}{0.6} = \frac{8.1}{0.6} = 13.5$$

Substituting this output into the demand function gives the price than maximizes TR as

$$p = 50 - 0.25q = 50 - 3.375 = 46.625$$

QUESTIONS 9.6

1. Given the demand function $q = 150 - 3p$ derive a function for MR.
2. A firm faces the demand function $q = 40 - p^{0.5}$ (where $p^{0.5} \geq 0$, $q \leq 40$) and the cost schedule $\text{TC} = q^3 - 2.5q^2 + 50q + 16$. What price should it charge to maximize profit?
3. Derive the MR function for the demand function $q = (60 - 2.5p)^{0.5}$.

9.7 POINT ELASTICITY OF DEMAND

Price elasticity of demand is defined as

$$e = \frac{\text{percentage change in quantity}}{\text{percentage change in price}}$$

However, looking at the changes in price and quantity between points A and B on the demand schedule D in Figure 9.5, you may ask 'percentage of what?' Clearly the change in quantity Δq is a much larger percentage of q_1 than of the larger quantity q_2. Although arc elasticity gives an approximate 'average' measure, a more precise measure can be obtained by finding the elasticity of demand at a single point on the demand schedule.

In Chapter 4 some simple examples of point elasticity based on linear demand schedules were considered. With the aid of calculus, we can now also derive point elasticity for non-linear demand schedules.

If the movement along D from A to B in Figure 9.5 is very small then we can assume

$$p_1 = p_2 = p \quad \text{and} \quad q_1 = q_2 = q \quad \text{and} \quad \text{so}$$

$$e = \frac{\frac{\Delta q}{q}}{\frac{\Delta p}{p}} = \frac{p}{q} \times \frac{1}{\frac{\Delta p}{\Delta q}} \qquad (1)$$

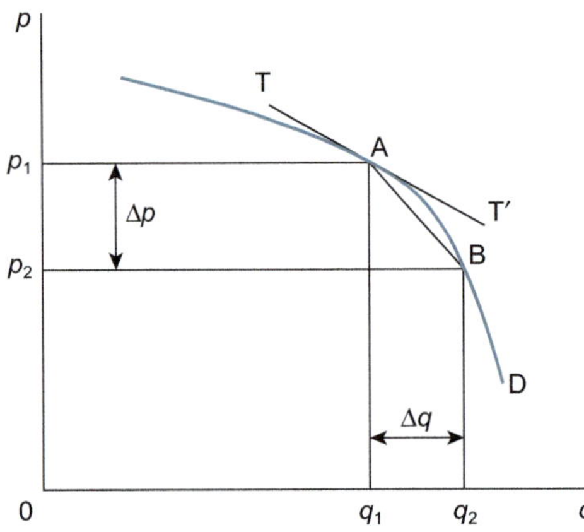

Figure 9.5 Price elasticity of demand at a point

As B gets nearer to A the value of $\Delta p/\Delta q$, which is the slope of the line AB, gets closer to the slope of the tangent TT′ at A. (Note that, as price falls in this example, Δp is negative, giving a negative value for the relevant slopes.) Thus, for an infinitesimally small movement from A

$$\frac{\Delta p}{\Delta q} = \frac{\mathrm{d}p}{\mathrm{d}q} = \text{slope of D at A}$$

Substituting this result into (1), the formula for point elasticity of demand becomes

$$e = \frac{p}{q} \frac{1}{\mathrm{d}p / \mathrm{d}q}$$

While this formula can be readily applied to inverse demand functions, we can rearrange the second element to obtain

$$e = \frac{p}{q} \frac{\mathrm{d}q}{\mathrm{d}p}$$

which is easier to work with when we are given demand functions which have q on the left-hand side.

Example 9.25

What is point elasticity for the inverse demand function $p = 60 - 3q$ when price is 12?

Solution

Differentiating the inverse demand function, we get

$$\frac{\mathrm{d}q}{\mathrm{d}p} = -3$$

Then deriving the underlying demand function

$$3p = 60 - p$$
$$q = \frac{60 - p}{3}$$

When $p = 12$, then

$$q = \frac{60 - 12}{3} = \frac{48}{3} = 16$$

Therefore, point elasticity will be

$$e = \frac{p}{q}\frac{1}{\mathrm{d}p/\mathrm{d}q} = \frac{12}{16} \times \frac{1}{-3} = -0.25$$

This means that if the price was to increase (decrease) by 1%, the quantity demanded would decrease (increase) by 0.25%.

Example 9.26

What is elasticity of demand when price is £676 if a firm's demand function is $q = 60 - 2p^{0.5}$ (where $p \geq 0$, $q \leq 60$)?

Solution

First, we derive $\mathrm{d}q/\mathrm{d}p$ and evaluate when $p = £676$

$$\frac{\mathrm{d}q}{\mathrm{d}p} = -p^{-0.5} = -\left(676^{-0.5}\right) = -0.0385$$

Also, when $p = £676$

$$q = 60 - 2 \times 676^{0.5} = 8$$

Thus,

$$e = \frac{p}{q}\frac{\mathrm{d}q}{\mathrm{d}p} = \frac{676}{8} \times (-0.0385) = -3.25$$

We would therefore say that the demand is elastic when p is £676 because a 1% increase in price leads to a 3.25% decrease in quantity demanded.

QUESTIONS 9.7

1. What is the point elasticity of demand when price is 20 for the demand schedule $p = 45 - 1.5q$? Interpret your result.
2. Explain why the point elasticity of demand decreases in value as one moves down a straight line demand schedule.

3. Given the demand function $q = (1{,}200 - 2p)^{0.5}$, what is elasticity of demand when quantity is 30? Interpret your result.
4. Explain why the demand function $q = 265p^{-1}$ will have the same point elasticity of demand at all prices and say what its value is.

9.8 TAX YIELD

Elementary supply and demand analysis tells us that the effect of a per-unit tax t on a good sold in a competitive market will effectively shift up the supply schedule vertically by the amount of the tax. This will cause the price paid by consumers to rise and the quantity bought to fall. The change in total revenue spent by consumers will depend on the price elasticity of demand.

The Chancellor of the Exchequer may be more interested in the total amount of tax raised for the government, or the tax yield (TY), than total consumer expenditure. If a per-unit tax is increased, the quantity bought will always fall. The question, however, is whether or not this fall in quantity will outweigh the effect on TY of the increase in the amount of tax raised on each unit. To answer this, we need to know the rate of change of the TY with respect to increases in the per-unit tax.

Example 9.27

A competitive market has the demand schedule $p = 92 - 2q$ and the supply schedule $p = 12 + 3q$. What per-unit tax will raise the maximum tax revenue for the government? (All prices are in £.)

Solution

Let the per-unit tax be t. This changes the supply schedule to

$$p = 12 + t + 3q$$

i.e. the intercept on the price axis shifts vertically upwards by the amount t.

We now need to derive a function for q in terms of the tax t. In equilibrium, supply price equals demand price. Therefore,

$$\begin{aligned} 12 + 3q + t &= 92 - 2q \\ 5q &= 80 - t \\ q &= 16 - 0.2t \end{aligned}$$

The tax yield is (amount sold) × (per-unit tax). Therefore,

$$\begin{aligned} \text{TY} = qt &= (16 - 0.2t)t \\ &= 16t - 0.2t^2 \end{aligned}$$

So, the rate of change of TY with respect to t is

$$\frac{\text{dTY}}{\text{d}t} = 16 - 0.4t$$

If dTY/dt > 0, an increase in t will increase TY. However, from the formula for dTY/dt derived earlier, one can see that as the amount of the tax t is increased, the value of dTY/dt falls. Therefore, in order to maximize TY, t should be increased until dTY/dt = 0. Any further increases in t would cause dTY/dt to become negative and cause TY to start to fall. Thus, to maximize tax yield

$$\frac{\text{dTY}}{\text{d}t} = 16 - 0.4t = 0$$
$$16 = 0.4t$$
$$40 = t$$

Therefore, a per-unit tax of £40 will maximize the tax yield.

Rather than working from first principles, as in the previous example, a general formula can be derived for the rate of change of the tax yield with respect to a per-unit tax if both demand and supply schedules are linear. Assume that these schedules are:

$$\text{demand } p = a + bq \qquad \text{supply } p = c + dq$$

where a, b, c and d are parameters and we expect $b < 0$ and $d > 0$.

With a per-unit tax of t, the supply schedule becomes

$$p = c + dq + t$$

Setting supply price equal to demand price we can derive the reduced form equation for TY in terms of the independent variable t and then differentiate it to find the comparative static effect of a change in t. Thus,

$$c + dq + t = a + bq$$
$$q(d - b) = a - c - t$$
$$q = \frac{a - c}{d - b} - \frac{t}{d - b}$$
$$\text{TY} = qt = \left(\frac{a - c}{d - b}\right)t - \frac{t^2}{d - b}$$

$$\frac{\text{dTY}}{\text{d}t} = \frac{a - c}{d - b} - \frac{2t}{d - b} \qquad (1)$$

We can check this formula using the figures from Example 9.27. Given the demand schedule $p = 92 - 2q$ and the supply schedule $p = 12 + 3q$, then

$$a = 92 \quad b = -2 \quad c = 12 \quad d = 3$$

Substituting these values into (1),

$$\frac{\text{dTY}}{\text{d}t} = \frac{92-12}{3-(-2)} - \frac{2t}{5}$$

$$= \frac{80}{5} - \frac{2t}{5} = 16 - 0.4t$$

This is the same as the function derived from first principles in Example 9.27.

QUESTIONS 9.8

1. What level of per-unit tax would maximize the government's tax yield in a competitive market where:

   Demand schedule $p = 180 - 8q$ and
   Supply schedule $p = 25 + 2q$?

2. Assume a market has the demand function $q = 40 - 0.5p$ and the supply function $q = 2p - 4$. The government currently imposes a per-unit tax of £3. If this tax is slightly increased will the tax yield rise or fall?

9.9 THE KEYNESIAN MULTIPLIER

In a simple Keynesian macroeconomic model with no government sector and no foreign trade, it is assumed that

$$Y = C + I \qquad (1)$$
$$C = a + bY \qquad (2)$$

where Y is national income, C is consumption and I is investment, exogenously fixed, and a and b are parameters.

The marginal propensity to consume (MPC) is the rate of change of consumption as national income increases, which is equal to $\text{d}C/\text{d}Y = b$. The multiplier is the rate of change of national income in response to an increase in exogenously determined investment, i.e. $\text{d}Y/\text{d}I$. We can now use differentiation to derive the result that the multiplier is equal to

$$\frac{1}{1-\text{MPC}}$$

Substituting (2) into (1) we get

$$Y = a + bY + I$$
$$Y(1-b) = a + I$$
$$Y = \frac{a+I}{1-b} = \frac{a}{1-b} + \frac{I}{1-b}$$

Therefore, since b = MPC, differentiation gives the multiplier, K, as

$$\frac{dY}{dI} = \frac{1}{1-b} = \frac{1}{1-MPC} = K$$

This multiplier can be used to calculate the increase in investment necessary to achieve any specified increase in national income.

Example 9.28

In a basic Keynesian macroeconomic model it is assumed that $Y = C + I$, where $I = 250$ and $C = 0.75Y$. What is the equilibrium level of Y? What increase in I would be needed to cause Y to increase to 1,200?

Solution

$$\begin{aligned} Y = C + I &= 0.75Y + 250 \\ 0.25Y &= 250 \\ \text{Equilibrium income} \qquad Y &= 1,000. \end{aligned}$$

For any increase (ΔI) in I the resulting increase (ΔY) in Y will be determined by the formula

$$\Delta Y = K\Delta I \qquad (1)$$

where K is the multiplier. We know that

$$K = \frac{1}{1-\text{MPC}}$$

In this example, MPC = dC/dY = 0.75. Therefore,

$$K = \frac{1}{1-0.75} = \frac{1}{0.25} = 4 \qquad (2)$$

The required change in Y is

$$\Delta Y = 1,200 - 1,000 = 200 \qquad (3)$$

Therefore, substituting (2) and (3) into (1) to find ΔI, the required increase in I

$$\begin{aligned} 200 &= 4\Delta I \\ \Delta I &= 50 \end{aligned}$$

Multipliers for other exogenous variables in more complex macroeconomic models can be derived using the same method. However, for differentiation with respect to one exogenous variable the other variables must remain constant and so we shall return to this topic in Chapter 11 when partial differentiation is explained.

QUESTIONS 9.9

1. In a basic Keynesian macroeconomic model it is assumed that $Y = C + I$ where $I = 820$ and $C = 60 + 0.8Y$.
   (a) What is the marginal propensity to consume?
   (b) What is the equilibrium level of Y?
   (c) What is the value of the multiplier?
   (d) What increase in I is required to increase Y to 5,000?
   (e) If this increase takes place will savings $(Y - C)$ still equal I?

10 Unconstrained optimization

Learning objectives

After completing this chapter students should be able to:

- find the maximum or minimum point of a single variable function by differentiation and checking first-order and second-order conditions
- use calculus to help find a firm's profit-maximizing output
- find the optimum order size for a firm wishing to minimize the cost of holding inventories and purchasing costs
- deduce the comparative static effects of different forms of taxes on the output of a profit-maximizing firm.

10.1 FIRST-ORDER CONDITIONS FOR A MAXIMUM

Consider the total revenue function

$$\text{TR} = 60q - 0.2q^2$$

This will take an inverted U-shape similar to that shown in Figure 10.1. If we ask the question 'when is TR at its maximum?' the answer is obviously at M, which is the highest point on the curve. At this maximum position the TR schedule is flat. To the left of M, TR is rising and has a positive slope, and to the right of M, the TR schedule is falling and has a negative slope. At M itself the slope is zero.

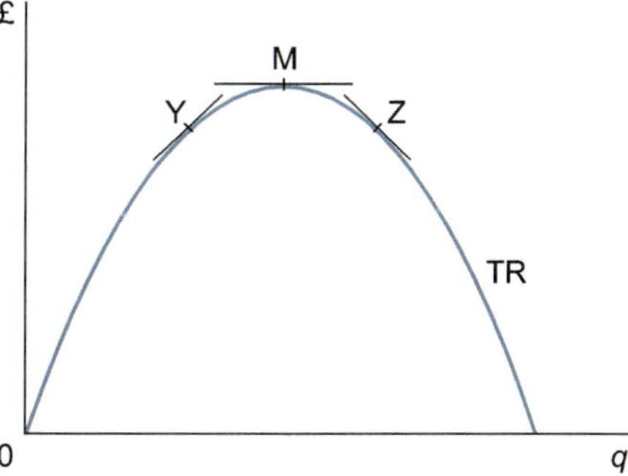

Figure 10.1 Function slope and function maximum

We can therefore say that for a function *of this shape* the maximum point will be where its slope is zero. This 'zero slope' requirement is a necessary **first-order condition** for a maximum.

DOI: 10.4324/9781003360827-10

Zero slope will not guarantee that a function is at a maximum though. A function's slope may be zero for other reasons, as explained in the next section where the necessary additional second-order conditions are explained. However, in this particular example we know for certain that zero slope corresponds to the maximum value of the function.

In Chapter 9 we learned that the slope of a function can be obtained by differentiation. So, for the function

$$\text{TR} = 60q - 0.2q^2$$

$$\text{slope} = \frac{\text{dTR}}{\text{d}q} = 60 - 0.04q$$

This slope is zero when

$$60 - 0.4q = 0$$
$$60 = 0.4q$$
$$150 = q$$

Therefore, TR is maximized when quantity is 150.

QUESTIONS 10.1

1. What output will maximize total revenue if TR = $250q - 2q^2$?
2. If a firm faces the demand schedule $p = 90 - 0.3q$ how much does it have to sell to maximize sales revenue?
3. A firm faces the total revenue schedule TR = $600q - 0.5q^2$
   (a) What is marginal revenue when q is 100?
   (b) When is total revenue at its maximum?
   (c) What price should the firm charge to achieve this maximum TR?
4. For the non-linear demand schedule $p = 750 - 0.1q^2$ what output will maximize the sales revenue?

10.2 SECOND-ORDER CONDITIONS FOR A MAXIMUM

In the example in Section 10.1, it was obvious that the TR function was a maximum when its slope was zero because we knew the function had an inverted U-shape. However, consider the function in Figure 10.2(a). This has a slope of zero at M, but this is its minimum point not its maximum. In the case of the function in Figure 10.2(b) the slope is zero at I, but this is neither a maximum nor a minimum point.

The examples in Figure 10.2 clearly illustrate that although a zero slope is *necessary* for a function to be at its maximum it is not a *sufficient* condition. A zero slope just means that the function is at what is known as a 'stationary point', i.e. its slope is neither increasing nor decreasing. Some stationary points will be turning points, where

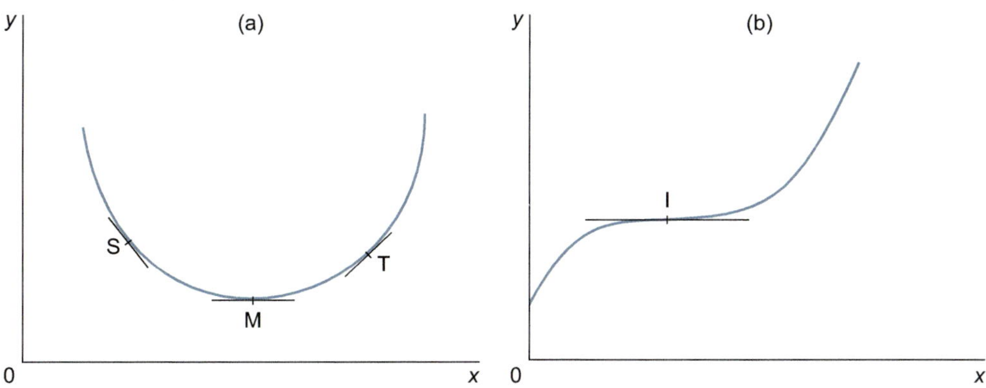

Figure 10.2 Stationary points and function slope

the slope changes from positive to negative (or vice versa), and will be maximum (or minimum) points of the function.

In order to find out whether a function is at a maximum or a minimum when its slope is zero, or a point of inflexion as in Figure 10.2(b), we have to consider what are known as the *second-order* conditions. (The first-order condition for any of the three forms of stationary point is that the slope of the function is zero.) The second-order conditions tell us what is happening to the rate of change of the slope of the function. If the rate of change of the slope is negative it means that the slope decreases as the variable on the horizontal axis is increased. If the slope is decreasing then, for a point where the actual slope is zero, this means that the slope of the function is positive slightly to the left and negative slightly to the right of this point. This is the case in Figure 10.1. The slope is positive at Y, zero at M and negative at Z. Thus, if the rate of change of the slope of a function is negative at the point where the actual slope is zero then that point is a maximum.

This negative rate of change of the slope is the second-order condition for a maximum. Until now, we have just assumed that a function is maximized when its slope is zero if a sketch graph suggests that it takes an inverted U-shape. From now on we shall make this more rigorous check of the second-order conditions to confirm whether a function is maximized at any stationary point.

It is a straightforward exercise to find the rate of change of the slope of a function. We know that the slope of a function $y = f(x)$ can be found by differentiation. Therefore, if we differentiate the function for the slope of the original function, i.e. we differentiate dy/dx, we get the rate of change of the slope. This is known as the **second-order derivative** and is written $\frac{d^2y}{dx^2}$.

Example 10.1

Show that the function $y = 60x - 0.2x^2$ satisfies the second-order condition for a maximum when $x = 150$.

Solution

First, we differentiate this function to find a stationary point where the slope is zero:

$$\frac{dy}{dx} = 60 - 0.4x = 0 \qquad (1)$$

$$x = 150$$

Therefore, the first-order condition for a maximum is met when x is 150.

To get the rate of change of the slope we differentiate (1) with respect to x again, giving

$$\frac{d^2y}{dx^2} = -0.4$$

This second-order derivative will always be negative, whatever the value of x.

Therefore, the second-order condition for a maximum is met and so y must be a maximum when q is 150.

In this example, the second-order derivative did not depend on the value of x at the function's stationary point, but for other functions the value of the second-order derivative may depend on the value of the independent variable.

Example 10.2

Show that TR is a maximum when q is 18 for the non-linear demand schedule.

$$p = 194.4 - 0.2q^2$$

Solution

$$\text{TR} = pq = \left(194.4 - 0.2q^2\right)q = 194.4q - 0.2q^3$$

For a stationary point on this cubic function the slope must be zero and so

$$\frac{d\text{TR}}{dq} = 194.4 - 0.6q^2 = 0$$

$$194.4 = 0.6q^2$$

$$324 = q^2$$

$$18 = q$$

When q is 18 then the second-order derivative is

$$\frac{d^2\text{TR}}{dq^2} = -1.2q = -1.2(18) = -21.6 < 0$$

Therefore, the second-order condition for a maximum is satisfied and TR is a maximum at the stationary point where q is 18.

(Note that in this example the second-order derivative $-1.2q$ will be less than zero for any positive value of q.)

QUESTIONS 10.2

Find stationary points for the following functions and say whether or not they are at their maximum at these points.

(1) $\text{TR} = 720q - 0.3q^2$
(2) $\text{TR} = 225q - 0.12q^3$
(3) $\text{TR} = 96q - q^{1.5}$
(4) $\text{AC} = 51.2q^{-1} + 0.4q^2$

10.3 SECOND-ORDER CONDITIONS FOR A MINIMUM

By similar reasoning to that set out in Section 10.2, if the rate of change of the slope of a function is positive at the point when the slope is zero then the function is at a minimum. This is illustrated in Figure 10.2(a). The slope of the function is negative at S, zero at N and positive at T. As the slope changes from negative to positive, the rate of change of this slope must be positive at the stationary point N.

Example 10.3

Find the minimum point of the average cost function $\text{AC} = 25q^{-1} + 0.1q^2$

Solution

The slope of the AC function will be zero when

$$\frac{\text{dAC}}{\text{d}q} = -25q^{-2} + 0.2q = 0 \qquad (1)$$

$$0.2q = 25q^{-2}$$
$$q^3 = 125$$
$$q = 5$$

The rate of change of the slope at this point is found by differentiating (1), and then evaluating the second-order derivative when $q = 5$, giving

$$\frac{\text{d}^2\text{AC}}{\text{d}q^2} = 50q^{-3} + 0.2 = \frac{50}{125} + 0.2 = 0.6 > 0$$

Therefore, the second-order condition for a minimum value of AC is satisfied when q is 5.

The actual value of AC at its minimum point is found by substituting this value for q back into the original AC function. Thus, when $q = 5$ then

$$\text{AC} = 25q^{-1} + 0.1q^2 = \frac{25}{5} + 0.1 \times 25 = 5 + 2.5 = 7.5$$

QUESTIONS 10.3

Find whether any stationary points exist for the following functions for positive values of q and say whether or not the stationary points are at the minimum values of the function.

1. $AC = 345.6q^{-1} + 0.8q^2$
2. $AC = 600q^{-1} + 0.5q^{1.5}$
3. $MC = 30 + 0.4q^2$
4. $TC = 15 + 27q - 9q^2 + q^3$
5. $MC = 8.25q$

10.4 SUMMARY OF SECOND-ORDER CONDITIONS

If $y = f(x)$ and there is a stationary point where $\frac{dy}{dx} = 0$, then

(i) this point is a **maximum** if $\frac{d^2y}{dx^2} < 0$,

(ii) this point is a **minimum** if $\frac{d^2y}{dx^2} > 0$.

Strictly speaking, (i) and (ii) are conditions for *local* maximums and minimums. It is possible, for example, that a function may take a shape such as that shown in Figure 10.3. This has no true global maximum or minimum, as values of y continue towards plus and minus infinity as shown by the arrows. Points M and N, which satisfy the previous second-order conditions for maximum and minimum, respectively, are therefore just local maximum and minimum points. However, for many examples that you are likely to encounter in economics, any local maximum (or minimum) points will also be global maximum (or minimum) points and so you need not worry about this distinction. If you are uncertain then you can plot a function on a spreadsheet to see the pattern of turning points.

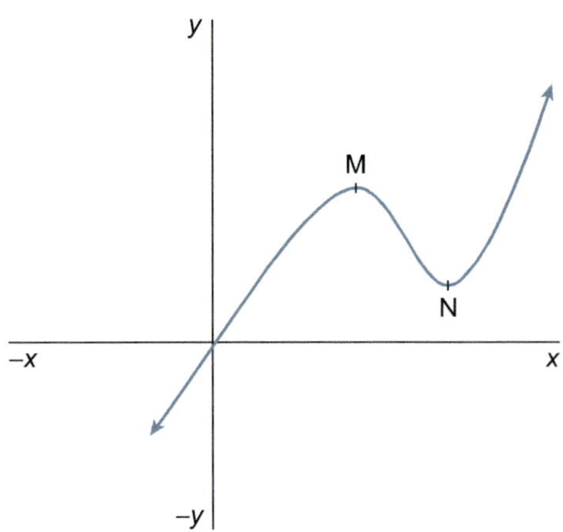

Figure 10.3 Local minimum and local maximum

If $\frac{d^2y}{dx^2} = 0$, there may be an inflexion point that is neither a maximum nor a minimum, such as I in Figure 10.2(b). To check this, one really needs to investigate further, looking at the third, fourth and possibly higher-order derivatives for more complex polynomial functions. However, we will not go into these conditions here. In all the applications given in this text, it will be obvious whether or not functions are at a maximum or minimum at any stationary points.

Some functions do not have maximum or minimum points. Linear functions are an obvious example as they cannot satisfy the first-order conditions for a turning point, i.e. that $dy/dx = 0$, except when they are horizontal lines. Also, the slope of a straight line is always a constant and so the second-order derivative, which represents the rate of change of the slope, will always be zero and so the second-order conditions for a maximum or minimum cannot be satisfied either.

Example 10.4

In Chapter 5 we considered an example of a break-even chart where a firm was assumed to have the total cost function TR = $18q$ and the total revenue function TC = 240 + $14q$. Show that the profit-maximizing output cannot be determined for this firm.

Solution

The profit function will be

$$\pi = \text{TR} - \text{TC} = 18q - (240 + 14q) = 4q - 240$$

Its rate of change with respect to q will be

$$\frac{d\pi}{dq} = 4$$

There is obviously no output level at which the first-order condition that $\frac{d\pi}{dq} = 4$ can be met and so no stationary point exists. Therefore, the profit-maximizing output cannot be determined. In fact, as the value of $\frac{d\pi}{dq}$ is always 4 this means that profit will keep increasing by £4 for every one unit increase in output, ad infinitum.

End-point solutions

There are some situations where there may be exceptions to these first- and second-order conditions for maximum and minimum values of functions. If the domain of a function is restricted, then a maximum or minimum point may be determined by this restriction, giving what is known as an 'end-point' or 'corner' solution. In such cases, the usual rules for optimization set out in this chapter will not apply. For example, suppose a firm faces the total cost function (in £)

$$\text{TC} = 45 + 18q - 5q^2 + q^3$$

For a stationary point its slope will be

$$\frac{d\text{TC}}{dq} = 18 - 10q + 3q^2 = 0$$

However, if we try using the quadratic equation formula to find a value of q for which (1) holds we see that

$$q = \frac{-b \pm \sqrt{b^2 - 4ac}}{2a} = \frac{-(-10) \pm \sqrt{10^2 - 4 \times 18 \times 3}}{2 \times 3}$$

$$= \frac{10 \pm \sqrt{-16}}{6}$$

We cannot find the square root of a negative number and so no solution exists. There is no turning point as no value of q corresponds to a zero slope for this function. However, if the domain of q is restricted to non-negative values, then TC will be at its minimum value of £45 when $q = 0$. Mathematically the conditions for minimization are not met at this point but, from a practical viewpoint, the minimum cost that this firm can ever face is the £45 it must pay even if nothing is produced. This is an example of an end-point solution.

Therefore, when tackling problems concerned with the minimization or maximization of economic variables, you need to ask whether or not there are restrictions on the domain of the variable in question which may give an end-point solution.

QUESTIONS 10.4

1. A firm faces the demand schedule $p = 200 - 2q$ and the total cost function

$$\text{TC} = \frac{2}{3}q^3 - 14q^2 + 222q + 50$$

Derive expressions for the following functions and find out whether they have maximum or minimum points. If they do, find what value of q this occurs at and calculate the actual value of the function at this output.

(a) Marginal cost
(b) Average variable cost
(c) Average fixed cost
(d) Total revenue
(e) Marginal revenue
(f) Profit

2. A firm attempting to expand output in the short-run faces the total product of labour schedule $\text{TP}_\text{L} = 24L^2 - L^3$. At what levels of L will (a) TP_L, (b) MP_L and (c) AP_L be at their maximum levels?
3. Using your knowledge of economics to apply appropriate restrictions on their domain, say whether or not the following functions have maximum or minimum points:
(a) $\text{TC} = 12 + 62q - 10q^2 + 1.2q^3$
(b) $\text{TC} = 6 + 2.5q$
(c) Demand function $q = 712.5 - 2.5p$

10.5 PROFIT MAXIMIZATION

We have already encountered some problems involving the maximization of a profit function. As profit maximization is one of the most common optimization problems that you will encounter in economics, in this section we shall carefully work through the second-order condition for profit maximization and see how it relates to the different intersection points of a firm's MC and MR schedules.

Consider the firm whose marginal cost and marginal revenue schedules are shown by MC and MR in Figure 10.4. At what output will profit be maximized?

Previously we mentioned the rule for profit maximization requires that MC = MR. However, there are two points, X and Y, where MC = MR. Only X satisfies the second rule for profit maximization, which is that MC cuts MR from below at the point of intersection. This corresponds to the second-order condition for a maximum required by the differential calculus, as illustrated in the following example.

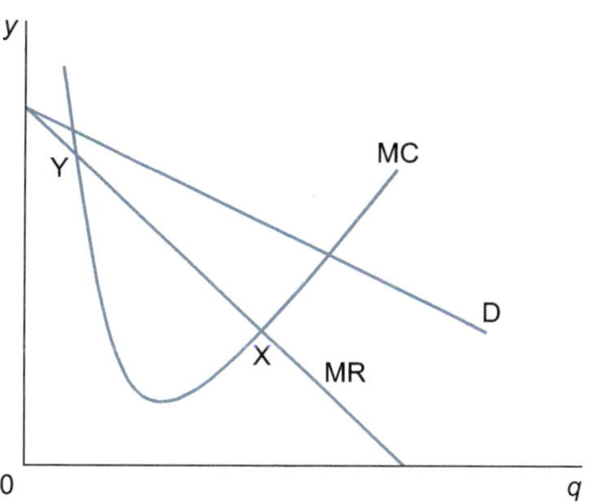

Figure 10.4 Profit maximization: MC = MR

Example 10.5

Find the profit-maximizing output for a firm with the total cost function $TC = 4 + 97q - 8.5q^2 + \frac{1}{3}q^3$ and the total revenue function TR = $58q - 0.5q^2$.

Solution

First, let us derive the MC and MR functions and see where they intersect.

$$MC = \frac{dTC}{dq} = 97 - 17q + q^2 \quad (1)$$

$$MR = \frac{dTR}{dq} = 58 - q \quad (2)$$

Therefore, when MC = MR

$$\begin{aligned} 97 - 17q + q^2 &= 58 - q \\ 39 - 16q + q^2 &= 0 \quad (3) \\ (3 - q)(13 - q) &= 0 \end{aligned}$$

Thus, $q = 3$ or $q = 13$

These are the two outputs at which the MC and MR schedules intersect, but which one satisfies the second rule for profit maximization? To answer this question, the problem can be reformulated by deriving a function for profit and then trying to find its maximum. Thus, profit will be

$$\begin{aligned}\pi = \text{TR} - \text{TC} &= 58q - 0.5q^2 - (4 + 97q - 8.5q^2 + \frac{1}{3}q^3)\\ &= 58q - 0.5q^2 - 4 - 97q + 8.5q^2 - \frac{1}{3}q^3\\ &= -39q + 8q^2 - 4 - \frac{1}{3}q^3\end{aligned}$$

Differentiating and setting equal to zero

$$\frac{d\pi}{dq} = -39 + 16q - q^2 = 0 \tag{4}$$

$$0 = 39 - 16q + q^2 \tag{5}$$

Equation (5) is the same as (3) and therefore has the same two solutions, i.e. $q = 3$ or $q = 13$. To check the second-order conditions, from (4) we can derive the second-order derivative

$$\frac{d^2\pi}{dq^2} = 16 - 2q$$

When $q = 3$ then $d^2\pi/dq^2 = 16 - 6 = 10 > 0$ and so π is a minimum.

When $q = 13$ then $d^2\pi/dq^2 = 16 - 26 = -10 < 0$ and so π is a maximum.

Thus, only one of the intersection points of MR and MC satisfies the second-order condition for a maximum and corresponds to the profit-maximizing output. This will be where MC cuts MR from below. We can prove that this must be so by differentiating (1) and (2) to get:

$$\text{slope of MC} = \frac{d\text{MC}}{dq} = -17 + 2q$$

$$\text{slope of MR} = \frac{d\text{MR}}{dq} = -1$$

When $q = 3$, then the slope of MC is

- $17 + 2(3) = -17 + 6 = -11 < -1$, i.e. steeper negative slope than MR

When $q = 13$, then the slope of MC is

- $17 + 2(13) = 9$, i.e. positive slope

Thus, when $q = 3$, the MC schedule has a steeper negative slope than MR and so must cut it from above. When $q = 13$, MC has a positive slope and so must cut MR from below. Therefore, we have proved that the second-order condition for a maximum is met when MC cuts MR from below.

QUESTIONS 10.5

1. A monopoly faces the total revenue schedule TR = $300q - 2q^2$ and the total cost schedule TC = $12q^3 - 44q^2 + 60q + 30$. Are there two output levels at which MC = MR? If so, which is the profit-maximizing output?
2. If a firm faces the demand schedule p = $120 - 3q$ and the total cost schedule TC = $120 + 36q + 1.2q^2$, what output level will maximize profit?
3. Explain why a firm which is a monopoly seller in a market with the demand function $q = 167 - 2.5p$ and which faces the total cost schedule TC = $220 + 120q - 12q^2 + 0.5q^3$ can never make a positive profit.
4. What is the maximum profit a firm can make if it faces the demand schedule $p = 660 - 3q$ and the total cost schedule TC = $25 + 240q - 72q^2 + 6q^3$?
5. If a firm faces the demand schedule $p = 53.5 - 0.7q$ what *price* will maximize profits if its total cost schedule is TC = $400 + 35q - 6q^2 + 0.1q^3$?

10.6 INVENTORY CONTROL

In Chapter 9 we considered a few applications of differentiation, such as tax yield maximization, without taking second-order conditions into account. We can now look at an application where it is not so obvious whether a function is maximized or minimized when its slope is zero and where second-order conditions must be fully investigated. This application analyzes how the optimum order size can be calculated for a firm wishing to minimize ordering and storage costs.

A manufacturing company has to take into account costs other than the actual purchase price of the components that it uses. These include:

(a) Reorder costs: each order for a consignment of components will involve administration work, delivery, unloading, etc.
(b) Storage costs: the more components a firm has in storage the more storage space will be needed. There is also the opportunity cost of the firm's capital which will be tied up in the components it has paid for.

If a firm only makes a few large orders its storage costs will be high but, on the other hand, if it makes lots of small orders its reorder costs will be high. How then can it decide on the optimum order size? Assume that:

(i) The total annual demand for components (Q) is evenly spread over the year.
(ii) Each order is of equal size q and that inventory levels are run down to zero before the next consignment arrives.
(iii) F is the fixed cost for making each order.
(iv) S is the storage cost per unit per year.

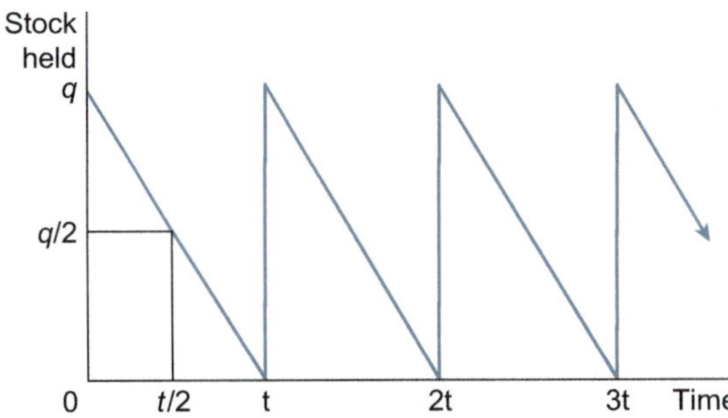

Figure 10.5 Stock levels over time

If each consignment of size q is run down at a constant rate, then the average amount of stock held will be $q/2$. (This is illustrated in Figure 10.5 where t represents the time interval between orders.) Thus, total storage costs for the year will be $(q/2)S$.

The number of orders made in a year will be Q/q. Thus, the total order costs for the year will be $(Q/q)F$.

The firm will wish to choose the order size that minimizes the total of order costs plus storage costs, defined as TC. The mathematical problem is therefore to find the value of q that minimizes

$$\text{TC} = \left(\frac{Q}{q}\right)F + \left(\frac{q}{2}\right)S$$

As Q, F and S are given constants, and remembering that $1/q$ is q^{-1}, differentiating with respect to q gives

$$\frac{\text{dTC}}{\text{d}q} = \frac{-QF}{q^2} + \frac{S}{2} \qquad (1)$$

For a stationary point

$$0 = -\frac{QF}{q^2} + \frac{S}{2}$$

$$\frac{QF}{q^2} = \frac{S}{2}$$

$$\frac{2QF}{S} = q^2$$

Therefore, the optimal order size is

$$q = \sqrt{\frac{2QF}{S}} \qquad (2)$$

Thus, q depends on the square root of the total annual demand Q when F and S are exogenously given constant values.

The second-order conditions now need to be inspected to check that this turning point is a minimum. If (1) is rewritten as

$$\frac{\text{dTC}}{\text{d}q} = -QFq^{-3} + \frac{S}{2}$$

then the second-order derivative is

$$\frac{\text{d}^2\text{TC}}{\text{d}q^2} = 2QFq^{-3} > 0$$

This is positive because Q, F and q all are positive quantities. Thus, any positive value of q that satisfies the first-order condition (2) must also satisfy the second-order condition for a minimum value of TC.

Example 10.6

A firm uses 200,000 units of a component in a year, with demand evenly spread over the year. In addition to the purchase price, each order placed for a batch of components costs £80. Each unit held in stock over a year costs £8. What is the optimum order size?

Solution

The optimum order size is q and so the average stock held is $q/2$. The number of orders is

$$\frac{Q}{q} = \frac{200,000}{q}$$

As each order costs £80 to make and each unit stored for a year costs £8 then

$$\begin{aligned} \text{TC} &= \text{order} + \text{stock holding costs} \\ &= \frac{200,000(80)}{q} + \frac{8q}{2} \\ &= 16,000,000q^{-1} + 4q \end{aligned}$$

For a stationary point

$$\frac{\text{dTC}}{\text{d}q} = -16,000,000q^{-2} + 4 = 0$$

$$4 = \frac{16,000,000}{q^2}$$

$$q^2 = \frac{16,000,000}{4} = 4,000,000$$

$$q = \sqrt{4,000,000} = 2,000$$

The second-order condition for a minimum is met at this stationary point as

$$\frac{\text{d}^2\text{TC}}{\text{d}q^2} = 32,000,000q^{-3} > 0 \quad \text{for any } q > 0$$

Therefore, the optimum order size is 2,000 units.

We could, of course, have solved this problem by just substituting the given values into the formula for optimal order size (2) derived earlier. Thus,

$$q = \sqrt{\frac{2QF}{S}} = \sqrt{\frac{2 \times 200,000 \times 80}{8}} = 2,000$$

QUESTIONS 10.6

In all the following questions, assume that demand is spread evenly over the year and stock is run down to zero before a new order is placed.

1. A firm uses 6,000 tonnes of commodity X every year. The fixed transaction costs involved with each order are £80. Each tonne of X held in stock costs £6 per annum. How many separate orders for X should the firm make during the year?
2. If each order for a batch of components costs £700 to make, storage costs per annum per component are £20 and annual usage is 4,480 components, what is the optimal order size?
3. A firm uses 1,280 units of a component each year. The cost of making an order is £540 and each component held in stock for a year costs the firm £6. What average order size would you advise the firm to make?
4. A firm uses 1,400 units per year of component G. Each order costs £350 to make and average storage costs per unit of G are £20. There is also an extra 'capacity' cost given that the firm has to provide warehousing capable of storing a full order size of q even though this warehousing space will be under-utilized most of the time. This 'capacity' cost will be £15 per unit of G. Adapt the optimal order size formula to include this extra cost and then find the optimal order size for this firm.

10.7 COMPARATIVE STATIC EFFECTS OF TAXES

In Chapter 5 we examined the comparative static effects of taxes on a firm's profit-maximizing output and price when all the relevant functions were linear. Calculus now enables us to extend this analysis to non-linear functions. Having learned how to determine a firm's profit-maximizing output and price by setting up a firm's profit function and then maximizing it, we can now deduce what may happen to these equilibrium values if an exogenous variable changes.

Suppose that a firm operates with the total cost function

$$\text{TC} = 50 + 0.4q^2$$

and is a monopoly facing the demand schedule

$$p = 360 - 2.1q$$

There is no independently determined exogenous variable in this economic model as it currently stands and so, if equilibrium was attained, output and price would remain at their profit-maximizing levels. We shall now examine what would happen to these equilibrium values if the following different forms of tax were imposed on the firm:

(a) a per-unit sales tax
(b) a lump sum tax
(c) a percentage profits tax.

The approach used in each case is to:

- formulate the firm's objective function for the net (after tax) profit that it will be striving to maximize
- find the output when the objective function is maximized, checking both first- and second-order conditions
- specify the profit-maximizing output and price as reduced form functions dependent on the exogenously determined tax
- differentiate to find the impact of a change in the tax on these optimum values.

It is important for you to learn how to set up objective functions from the economic information available and to understand the different impacts that these different types of taxes will have. A common mistake that students sometimes make in this sort of problem is to try to show the effect of a tax by shifting up the supply schedule by the amount of the tax. That method only applies for a sales tax in perfectly competitive markets. This time we have a firm that operates in a monopolistic market (and so there is no supply schedule as such) and some of these taxes are on profit rather than sales.

(a) Per-unit sales tax

If the firm has to pay the government an amount t on each unit of q that it sells, then the total tax it has to pay will be tq. Its total costs, including the tax, will therefore be

$$\text{TC} = 50 + 0.4q^2 + tq$$

Given the demand schedule $p = 360 - 2.1q$ the firm's total revenue function will be

$$\text{TR} = pq = 360 - 2.1q^2$$

The net profit objective function that the firm will wish to maximize will therefore be

$$\begin{aligned}\pi = \text{TR} - \text{TC} &= 360q - 2.1q^2 - \left(50 + 0.4q^2 + tq\right)\\ &= 360q - 2.1q^2 - 50 - 0.4q^2 - tq\\ &= 360q - 2.5q^2 - 50 - tq\end{aligned}$$

Differentiating with respect to q and setting equal to zero to find the first-order condition for a maximum

$$\frac{d\pi}{dq} = 360 - 5q - t = 0 \tag{1}$$

Before proceeding with the comparative static analysis, we can check the second-order conditions to confirm that this stationary point is indeed a maximum. Differentiating (1) again gives

$$\frac{d^2\pi}{dq^2} = -5 < 0$$

and so the second-order condition for a maximum is met.

Returning to the first-order condition (1) in order to find the optimal level of q in terms of t, we write

$$360 - 5q - t = 0 \tag{2}$$

$$360 - t = 5q$$

$$q = 72 - 0.2t \tag{3}$$

This is the reduced form equation for profit-maximizing output in terms of the independent variable t.

Differentiating (3) with respect to t to find the comparative static effect of a change in t on the optimum value of q gives

$$\frac{dq}{dt} = -0.2$$

This means that a one-unit increase in the per-unit sales tax will reduce output by 0.2 units. This comparative static effect is not dependent on any other variable and so at any output level the impact of the tax on q will be the same, as long as it is still profitable for the firm to produce.

The comparative static effect of this tax on price can be found by substituting the function for the optimal level of q in equation (3) into the firm's demand schedule

$$p = 360 - 2.1q$$

Thus,

$$p = 360 - 2.1(72 - 0.2t)$$
$$= 360 - 151.2 + 0.42t$$

Giving the reduced form

$$p = 208.8 + 0.42t$$

Differentiating this with respect to t gives

$$\frac{\mathrm{d}p}{\mathrm{d}t} = 0.42$$

This tells us that the comparative static effect of a £1 increase in the per-unit tax t will be a £0.42 increase in the firm's profit-maximizing price.

(b) A lump sum tax

A lump sum tax is a fixed amount that firms are required to pay to the government. The amount of the tax (T) is not related to sales or profit levels. Before the tax is introduced, the firm in our example faces the total cost and total revenue functions

$$\mathrm{TC} = 50 + 0.4q^2 \quad \text{and} \quad \mathrm{TR} = 360Q - 2.1q^2$$

The imposition of a lump sum tax T will effectively increase fixed costs by the amount of the tax. The firm's total cost function will therefore become

$$\mathrm{TC} = 50 + 0.4q^2 + T$$

and the net profit objective function that the firm attempts to maximize will become

$$\begin{aligned} \pi = \mathrm{TR} - \mathrm{TC} &= 360q - 2.1q^2 - \left(50 + 0.4q^2 + T\right) \\ &= 360q - 2.1q^2 - 50 - 0.4q^2 - T \\ &= 360Q - 2.5q^2 - 50 - T \end{aligned}$$

Differentiating with respect to q and setting equal to zero to find the first-order conditions for a maximum

$$\frac{\mathrm{d}\pi}{\mathrm{d}q} = -5q + 360 = 0 \tag{1}$$

Differentiating (1) again gives

$$\frac{\mathrm{d}^2\pi}{\mathrm{d}q^2} = -5 < 0$$

and so the second-order condition for a maximum is met.

Returning to (1) to find the optimal level of q

$$\begin{aligned} -5q + 360 &= 0 \\ 360 &= 5q \\ q &= 72 \end{aligned} \tag{2}$$

As (2) does not contain any term in T, the firm's profit-maximizing output will always be 72, regardless of the amount of the lump sum tax. Therefore, a change in the lump sum tax T will have no effect on output and consequently it will also have no effect on price.

This is what economic analysis would predict. If a firm has to pay a fixed sum out of its profits, then it would want to be in a position where total gross (before tax) profits are at a maximum in order to maximize net after tax profit. Note, though, that if the lump sum tax was greater than the firm's pre-tax profit then the firm would not be able to pay the tax and might have to close down. It is still possible, though, that the tax might be paid out of accumulated past profits, like the windfall tax that was imposed on some of the UK privatized utility companies in the late 1990s because the government thought that they had earned excessive profits.

(c) Percentage profits tax

If a firm has to pay a proportion of its profits as tax, then it will attempt to maximize net profit which will be

$$\pi = (\text{TR} - \text{TC})(1 - c)$$

where c is the rate of profits tax. (Profits tax is called corporation tax in the UK, so we will use the notation c.)

Thus, for the firm in this example the net profit after tax will be

$$\begin{aligned}\pi &= (\text{TR} - \text{TC})(1 - c)\\ &= (360q - 2.1q^2 - 50 - 0.4q^2)(1 - c)\\ &= (360q - 2.5q^2 - 50)(1 - c)\end{aligned}$$

The term $(1 - c)$ can be treated as a constant that multiplies each of the values in the first set of brackets and so differentiating and setting equal to zero to get first-order condition for profit maximization

$$\frac{\text{d}\pi}{\text{d}q} = (360 - 5q)(1 - c) = 0 \qquad (1)$$

Checking the second-order condition for a maximum

$$\frac{\text{d}^2\pi}{\text{d}q^2} = -5(1 - c) < 0 \quad \text{as long as} \quad 0 < c < 1$$

We would expect a percentage profits tax rate to lie between 0% and 100%. Therefore, c will take a value between 0 and 1 and so the second-order condition for a maximum will be met.

Returning to (1) to find the optimal level of q

$$(360 - 5q)(1 - c) = 0$$

Unless there is a 100% profits tax then

$$(1-c)\neq 0$$

and so it must be the case that

$$\begin{aligned} 360-5q &= 0 \\ q &= 72 \end{aligned} \qquad (2)$$

As (2) does not contain any term in c, we can say that the firm's profit-maximizing output will always be 72, regardless of the amount of the profits tax. Therefore, a change in the rate of profits tax c will have no effect on output. It will therefore also have no effect on price.

This result is what economic analysis would predict and is similar to the case (b) for a lump sum tax. If a firm has to pay a percentage of its profits as tax then it would want to be in a position where total profits before the tax are at a maximum in order to maximize net after tax profit.

QUESTIONS 10.7

1. Derive reduced form equations for equilibrium price and output in terms of

   (a) a per-unit sales tax t
   (b) a lump sum tax T
   (c) a percentage profits tax c

   for each of the following cases, where it is assumed that the demand schedule and total cost function apply to a single firm in an industry.

   (i) $p = 450 - 2q$ and TC $= 20 + 0.5q^2$
   (ii) $p = 200 - 0.3q$ and TC $= 10 + 0.1q^2$
   (iii) $p = 260 - 4q$ and TC $= 8 + 1.2q^2$

2. Derive a reduced form equation that will show thc comparative static effect of a percentage sales tax on a company that faces the demand schedule $p = 680 - 3q$ and the total cost function TC $= 20 + 0.4q^2$.

11 Partial differentiation

Learning objectives

After completing this chapter students should be able to:

- derive the first-order partial derivatives of multi-variable functions
- apply the concept of partial differentiation to production functions, utility functions and the Keynesian macroeconomic model
- derive second-order partial derivatives and interpret their meaning
- check the second-order conditions for maximization and minimization of a function with two independent variables using second-order partial derivatives
- derive the total differential and total derivative of a multi-variable function
- use Euler's theorem to check if the total product is exhausted for a Cobb-Douglas production function.

11.1 PARTIAL DIFFERENTIATION AND THE MARGINAL PRODUCT

For the production function $Q = f(K, L)$ with the two independent variables L and K, the value of the function will change if one independent variable is increased whilst the other is held constant. If K is held constant and L is increased, we will trace out the total product of labour (TP_L) schedule. (TP_L means the output Q that is produced for a given amount of L.) This TP_L schedule will typically take a shape similar to that shown in Figure 11.1.

In your introductory microeconomics course, the marginal product of L (MP_L) was probably defined as the increase in TP_L caused by a one-unit increment in L, assuming K to be fixed at some given level. A more precise definition, however, is that MP_L is the rate of change of TP_L with respect to L. For any given value of L this is the slope of the TP_L function. (Refer back to Section 9.3 if you do not understand why.) Thus, the MP_L schedule in Figure 11.1 is at its maximum when the TP_L schedule is at its steepest, at M, and is zero when TP_L is at its maximum, at N.

Partial differentiation is a technique for deriving the rate of change of a function with respect to increases in one independent variable when all other independent

DOI: 10.4324/9781003360827-11

variables in the function are held constant. Therefore, if the production function $Q = f(K, L)$ is differentiated with respect to L, with K held constant, we get the rate of change of total product with respect to L, in other words MP_L.

The **basic rule for partial differentiation** is that all independent variables, other than the one that the function is being differentiated with respect to, are treated as constants. Apart from this, partial differentiation follows the standard differentiation rules explained in Chapter 9. A curved ∂ is used in a partial derivative to distinguish it from the derivative of a single variable function where a normal letter 'd' is used. For example, the partial derivative of the production function with respect to L is written $\frac{\partial Q}{\partial L}$.

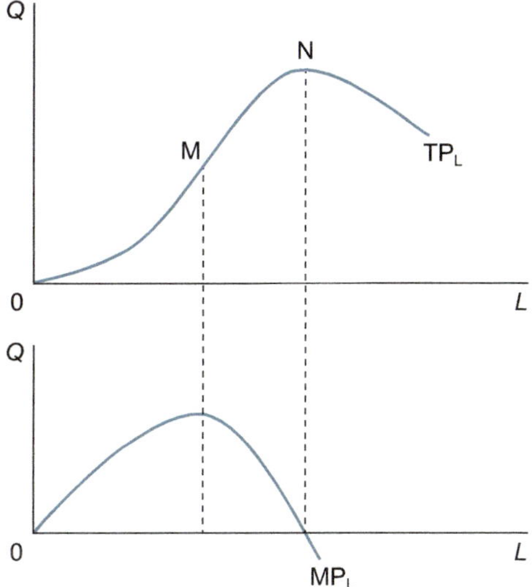

Figure 11.1 Total and marginal product of labour

Example 11.1

If $y = 14x + 3z^2$, find the partial derivatives of this function with respect to x and z.

Solution

The partial derivative of function y with respect to x is

$$\frac{\partial y}{\partial x} = 14$$

(The $3z^2$ disappears as it is treated as a constant. One then just differentiates the term $14x$ with respect to x.)

Similarly, the partial derivative of y with respect to z is

$$\frac{\partial y}{\partial z} = 6z$$

(The $14x$ is treated as a constant and disappears. One then just differentiates the term $3z^2$ with respect to z.)

Example 11.2

Find the partial derivatives of the function $y = 6x^2z$.

Solution

In this function the variable held constant does not disappear as it is multiplied by the other variable. Therefore, treating z as a constant

$$\frac{\partial y}{\partial x} = 12xz$$

And, treating x (and therefore x^2) as a constant

$$\frac{\partial y}{\partial z} = 6x^2$$

Example 11.3

For the production function $Q = 20K^{0.5}L^{0.5}$

(i) derive an expression for MP_L, and
(ii) show that MP_L decreases as one moves along an isoquant by using more L.

Solution

(i) MP_L is found by partially differentiating the production function $Q = 20K^{0.5}L^{0.5}$ with respect to L. Thus,
Note that this MP_L function will continuously slope downward, unlike the MP_L function illustrated in Figure 11.1.
(ii) If the function for MP_L is multiplied top and bottom by $2L^{0.5}$, then we get

$$MP_L = \frac{2L^{0.5}}{2L^{0.5}} \times \frac{10K^{0.5}}{L^{0.5}} = \frac{20K^{0.5}L^{0.5}}{2L} = \frac{Q}{2L} \qquad (1)$$

An isoquant joins combinations of K and L that yield the same output level. Thus, if Q is held constant and L is increased then the function (1) shows us that MP_L will decrease. (Note that moving along an isoquant entails using more L and less K to keep output constant. Although the amount of K used does therefore change, what this result tells us is that with a new lower amount of capital, the MP_L will be lower than it was before.)

We can now see that for any Cobb-Douglas production function taking the format $Q = AK^{\alpha}L^{\beta}$ the law of diminishing marginal productivity holds for each input as long as $0 < \alpha < 1$ and $0 < \beta < 1$. If K is fixed and L is variable, the marginal product of L is found in the usual way by partial differentiation. Thus, when

$$Q = AK^{\alpha}L^{\beta}$$

$$MP_L = \frac{\partial Q}{\partial L} = \beta AK^{\alpha}L^{\beta-1} = \frac{\beta AK^{\alpha}}{L^{1-\beta}}$$

If K is held constant then, given that α, β and A are also constants, the numerator in this expression βAK^α is constant. In the denominator, as L is increased, $L^{1-\beta}$ gets larger (given $0 < \beta < 1$) and so the whole function for MP_L decreases in value, i.e. the marginal product falls as L is increased.

Similarly, if K is increased while L is held constant,

$$MP_K = \frac{\partial Q}{\partial K} = \alpha AK^{\alpha-1}L^\beta = \frac{\alpha AL^\beta}{K^{1-\alpha}}$$

which falls as K increases in value.

When there are more than two inputs in a production function, the same principles still apply. For example, if

$$Q = AX_1^a X_2^b X_3^c X_4^d$$

where X_1, X_2, X_3 and X_4 are inputs, then the marginal product of input X_3 will be

$$\frac{\partial Q}{\partial X_3} = cAX_1^a X_2^b X_3^{c-1} X_4^d = \frac{cAX_1^a X_2^b X_4^d}{X_3^{1-c}}$$

which decreases as X_3 increases in value, *ceteris paribus*.

We can also see that for a production function in the usual Cobb-Douglas format, the marginal product functions will continuously decline towards zero and will never 'bottom out' for finite values of L, i.e. they will never reach a minimum point where the slope is zero. If, for example,

$$Q = 25K^{0.4}L^{0.5}$$

$$MP_L = \frac{\partial Q}{\partial L} = 12.5K^{0.4}L^{-0.5}$$

The first-order condition for a minimum is

$$\frac{\partial Q}{\partial L} = \frac{12.5K^{0.4}}{L^{0.5}} = 0$$

This is satisfied only if $K = 0$ and hence $Q = 0$, or if L becomes infinitely large. Since, for finite values of L, MP_L will still remain positive however large L becomes, this means that on the isoquant map for a two-input Cob-Douglas production function the isoquants will never 'bend back', i.e. there will not be an uneconomic region.

Other formats for production functions are possible though. For example, if

$$Q = 4.6K^2 + 3.5L^2 - 0.012K^3L^3$$

then MP_L will first rise and then fall since

$$MP_L = \frac{\partial Q}{\partial L} = 7L - 0.036K^3L^2$$

The slope of the MP_L function will change from a positive to a negative value as L increases since

$$\text{slope} = \frac{\partial MP_L}{\partial L} = 7 - 0.072K^3L$$

The actual value and position of this MP_L function will depend on the value that the other input K takes.

Example 11.4

If $q = 20x^{0.6}y^{0.2}z^{0.3}$, find the rate of change of q with respect to x, y and z.

Solution

Although there are now three independent variables instead of two, the same rules still apply, this time with two variables treated as constants. Therefore, holding y and z constant

$$\frac{\partial q}{\partial x} = 12x^{-0.4}y^{0.2}z^{0.3}$$

Similarly, holding x and z constant

$$\frac{\partial q}{\partial y} = 4x^{0.6}y^{-0.8}z^{0.3}$$

and holding x and y constant

$$\frac{\partial q}{\partial z} = 6x^{0.6}y^{0.2}z^{-0.7}$$

To avoid making mistakes when partially differentiating a function with several variables it may help if you write in the variables that do not change first and then differentiate. In the previous example, when differentiating with respect to x for instance, this would mean first writing in $y^{0.2}z^{0.3}$ as y and z are held constant.

When a function has a large number of variables, a shorthand notation for the partial derivative is usually used. For example, for the function $f = f(x_1, x_2, \dots, x_n)$, one can write

$$f_1 \text{ instead of } \frac{\partial f}{\partial x_1}, \quad f_2 \text{ instead of } \frac{\partial f}{\partial x_2}, \quad \text{etc.}$$

Example 11.5

Find f_j where j is any input number for the production function

$$f(x_1, x_2, \dots, x_n) = \sum_{i=1}^{n} 6x_i^{0.5}$$

Solution

This function is a summation of several terms. Only one term, the jth, will contain x_j. If one is differentiating with respect to x_j, then all other terms are treated as constants and disappear. Therefore, one only has to differentiate the term $6x_j^{0.5}$ with respect to x_j, giving

$$f_j = 3x_j^{-0.5}$$

This shorthand notation can also be used to express second-order partial derivatives. For example,

$$f_{11} = \frac{\partial^2 f}{\partial x_1^2}$$

Uses of second-order partial derivatives will be explained in Section 11.3.

QUESTIONS 11.1

1. Find $\frac{\partial y}{\partial x}$ and $\frac{\partial y}{\partial z}$ when
   (a) $y = 6 + 3x + 16z + 4x^2 + 2z^2$
   (b) $y = 14x^3z^2$
   (c) $y = 9 + 4xz - 3x^{-2}z^3$
2. Show that the law of diminishing marginal productivity holds for the production function $Q = 12K^{0.4}L^{0.4}$. Will the MP_L schedule take the shape shown in Figure 11.1?
3. Derive formulae for the marginal products of the three inputs in the production function $Q = 40K^{0.3}L^{0.3}R^{0.4}$.
4. Use partial differentiation to explain why the production function $Q = 0.4K + 0.7L$ does not obey the law of diminishing marginal productivity.
5. If $Q = 18K^{0.3}L^{0.2}R^{0.5}$, will the marginal products of any of the three inputs K, L and R become negative?
6. Derive a formula for the partial derivative Q_j , where j is an input number, for the production function

$$Q(x_1, x_2, \ldots, x_n) = \sum_{i=1}^{n} 4x_i^{0.3}.$$

11.2 FURTHER APPLICATIONS OF PARTIAL DIFFERENTIATION

Partial differentiation is basically a mathematical application of the assumption of *ceteris paribus* (i.e. other things being held equal) which is frequently used in economic analysis. Because the economy is a complex system to understand, economists often look at the effect of changes in one variable assuming all other influencing factors remain unchanged.

When the relationship between the different economic variables can be expressed in a mathematical format, then the analysis of the effect of changes in one variable can be discovered via partial differentiation. We have already seen how partial differentiation can be applied to production functions and here we shall examine a few other applications.

Elasticity

In a market the quantity demanded, q, depends on several factors, such as the price of the good (p), average consumer income (m), the price of a complement (p_c), the price of a substitute (p_s) and population (n). This relationship can be expressed as the demand function

$$q = f\left(p, m, p_c, p_s, n\right)$$

In introductory economics courses, price elasticity of demand is usually defined as

$$e = \frac{\text{Percentage change in quantity demanded}}{\text{Percentage change in price}}$$

This definition implicitly assumes *ceteris paribus*, even though there may be no mention of other factors that influence demand. The same implicit assumption is made in the more precise measure of point elasticity of demand with respect to price:

$$e = \frac{p}{q}\frac{1}{\mathrm{d}p/\mathrm{d}q}$$

Recognizing that quantity demanded depends on factors other than price, then point elasticity of demand with respect to price can be more accurately redefined as

$$e = \frac{p}{q}\frac{1}{\partial p/\partial q} = \frac{p}{q}\frac{\partial q}{\partial p}$$

Note that we have employed the **inverse function rule** here. This states that, for any function $y = \mathrm{f}(x)$, as long as $\mathrm{d}y/\mathrm{d}x \neq 0$, then

$$\frac{\mathrm{d}x}{\mathrm{d}y} = \frac{1}{\mathrm{d}y/\mathrm{d}x}$$

This rule can also be used for partial derivatives and so

$$\frac{1}{\partial p/\partial q} = \frac{\partial q}{\partial p}$$

Point elasticity (with respect to own price) can now be determined for specific demand functions that include other explanatory variables.

Example 11.6

For the demand function

$$q = 35 - 0.4p + 0.15m - 0.25p_c + 0.12p_s + 0.003n$$

where the terms are as defined at the start of this section, what is price elasticity of demand when $p = 24$?

Solution

We know the value of p and we can easily derive the partial derivative $\partial q/\partial p = -0.4$. Substituting these values into the elasticity formula

$$e = \frac{p}{q}\frac{\partial q}{\partial p} = \frac{24}{35 - 0.4(24) + 0.15m - 0.25p_c + 0.12p_s + 0.003n}(-0.4)$$
$$= \frac{-9.6}{25.4 + 0.15m - 0.25p_c + 0.12p_s + 0.003n}$$

The actual value of elasticity cannot be calculated until specific values for m, p_c, p_s and n are given. Thus, this example shows that the value of point elasticity of demand with respect to price will depend on the values of other factors that affect demand and which consequently determine the position on the demand schedule.

Other measures of elasticity will also depend on the values of the different variables in the demand function. For example, the basic definition of income elasticity of demand is

$$e_m = \frac{\text{Percentage change in quantity demanded}}{\text{Percentage change in income}}$$

If we assume an infinitesimally small change in income and that all other factors influencing demand are being held constant, then income elasticity of demand can be defined as

$$e_m = \frac{\dfrac{\Delta q}{q}}{\dfrac{\Delta m}{m}} = \frac{m}{q}\frac{\Delta q}{\Delta m} = \frac{m}{q}\frac{\partial q}{\partial m}$$

Thus, for the demand function in Example 11.6, income elasticity of demand will be

$$e_m = \frac{m}{q}\frac{\partial q}{\partial m} = \frac{m}{q}(0.15)$$

If the value of m is given as 30, say, then

$$e_m = \frac{30}{35 - 0.4p + 0.15(30) - 0.25p_c + 0.12p_s + 0.003n}(0.15)$$
$$= \frac{4.5}{39.5 - 0.4p - 0.25p_c + 0.12p_s + 0.003n}$$

Thus, the value of income elasticity of demand will depend on the value of the other factors influencing demand as well as the level of income itself.

Consumer utility functions

The general form of a consumer's utility function is

$$U = U(x_1, x_2, \dots, x_n)$$

where x_1, x_2, …, x_n represent the amounts of the different goods consumed.

Unlike output in a production function, one cannot actually measure utility and this theoretical concept is only of use in making general predictions about the behaviour of 'economic agents', as you should learn in your economics course. Modern economic theory assumes that utility is an 'ordinal concept', meaning that different combinations of goods can be ranked in order of preference but utility itself cannot be quantified in any way. However, economists also work with the concept of 'cardinal' utility where it is assumed that, hypothetically at least, each individual can quantify and compare different levels of their own utility. It is this cardinal utility concept which is used here.

If we assume that only the two goods A and B are consumed, then the utility function will take the form

$$U = \mathrm{U}(A, B)$$

Marginal utility is defined as the rate of change of total utility with respect to the increase in consumption of one good. Therefore, the marginal utility functions for goods A and B, respectively, will be

$$\mathrm{MU}_A = \frac{\partial U}{\partial A} \quad \text{and} \quad \mathrm{MU}_B = \frac{\partial U}{\partial B}$$

Three important principles of utility theory are:

1. The law of diminishing marginal utility says that if, *ceteris paribus*, the quantity consumed of any one good is increased, then eventually its marginal utility will decline.
2. A consumer will consume a good up to the point where its marginal utility is zero if it is a free good, or if a fixed payment is made regardless of the quantity consumed, e.g. water rates.
3. A consumer maximizes satisfaction when each good is consumed up to the point where an extra pound spent on one good will derive the same utility as an extra pound spent on any other good.

Some applications of the first two principles are given in the following examples. We shall return to the third principle in Chapter 12 when we study constrained optimization.

Example 11.7

Find out whether the law of diminishing marginal utility holds for both goods A and B in the following utility functions:

(i) $U = A^{0.6}B^{0.8}$
(ii) $U = 85AB - 1.6A^2B^2$
(iii) $U = 0.2A^{-1}B^{-1} + 5AB$

Solutions

(i) For the utility function $U = A^{0.6}B^{0.8}$, partial differentiation yields the marginal utility functions

$$\text{MU}_A = \frac{\partial U}{\partial A} = 0.6A^{-0.4}B^{0.8} \quad \text{and} \quad \text{MU}_B = \frac{\partial U}{\partial B} = 0.8A^{0.6}B^{-0.2}$$

Thus, MU_A falls as A increases (when B is held constant) and MU_B falls as B increases (when A is held constant). As both marginal utility functions decline, the law of diminishing marginal utility holds.

(ii) For the utility function $U = 85AB - 1.6A^2B^2$, the marginal utility functions will be

$$\text{MU}_A = \frac{\partial U}{\partial A} = 85B - 3.2AB^2$$
$$\text{MU}_B = \frac{\partial U}{\partial B} = 85A - 3.2A^2B$$

Both MU_A and MU_B will be downward-sloping straight lines given that the quantity of the other good is held constant. Therefore, the law of diminishing marginal utility holds.

(iii) When $U = 0.2A^{-1}B^{-1} + 5AB$, then

$$\text{MU}_A = \frac{\partial U}{\partial A} = -0.2A^{-2}B^{-1} + 5B$$
$$\text{MU}_B = \frac{\partial U}{\partial B} = -0.2A^{-1}B^{-2} + 5A$$

As A increases, the term $0.2A^{-2}B^{-1}$ gets smaller. As this term is subtracted from $5B$, which will be constant as B remains unchanged, this means that MU_A rises. Similarly, MU_B will rise as B increases. Therefore, the law of diminishing marginal utility does not hold for this function.

Example 11.8

Given the following utility functions, where the law of diminishing marginal utility holds for each good consumed, how much of A will be consumed if it is a free good? If necessary, give answers in terms of the fixed amount of B.

(i) $U = 96A + 35B - 0.8A^2 - 0.3B^2$
(ii) $U = 72AB - 0.6A^2\,B^2$
(iii) $U = A^{0.3}B^{0.4}$

Solutions

Given that the law of diminishing marginal utility holds for all three functions, we need to try to find the value of A where MU_A is zero. Consumers will not consume extra units of A which have negative marginal utility and hence decrease total utility.

(i) For utility function $U = 96A + 35B - 0.8A^2 - 0.3B^2$ marginal utility of A is zero when

$$MU_A = \frac{\partial U}{\partial A} = 96 - 1.6A = 0$$
$$96 = 1.6A$$
$$60 = A$$

Thus, 60 units of A are consumed if A is free, regardless of the amount of B consumed.

(ii) When $U = 72AB - 0.6A^2B^2$, then MU_A is zero when

$$\frac{\partial U}{\partial A} = 72B - 1.2AB^2 = 0$$
$$72B = 1.2AB^2$$
$$60B^{-1} = A$$

Thus, the amount of A consumed if it is free will depend inversely on the amount of B consumed.

(iii) When $U = A^{0.3}B^{0.4}$ then the marginal utility of A will be

$$MU_A = \frac{\partial U}{\partial A} = 0.3A^{-0.7}B^{0.4}$$

This marginal utility function will decline continuously but, for any non-zero value of B, MU_A will not equal zero unless the amount of A consumed becomes infinitely large. Therefore, no finite solution can be found.

The Keynesian multiplier

If a government sector and foreign trade are introduced, then the basic Keynesian macroeconomic model becomes the accounting identity

$$Y = C + I + G + X - M \qquad (1)$$

and the functional relationships of the consumption function

$$C = cY_d \qquad (2)$$

where c is the marginal propensity to consume, and

$$M = mY_d \qquad (3)$$

where M is imports and m is the marginal propensity to import, and

$$Y_d = (1-t)Y \tag{4}$$

where Y_d is disposable income and t is the tax rate.

Investment I, government expenditure G and exports X are exogenously determined, and c, m and t are given parameters.

By substituting (2), (3) and (4) into (1) to find equilibrium Y we get

$$\begin{aligned} Y &= c(1-t)Y + I + G + X - m(1-t) \\ Y\left[1-c(1-t)+m(1-t)\right] &= I+G+X \\ Y &= \frac{I+G+X}{1-c(1-t)+m(1-m)} = \frac{I+G+X}{1-(c-m)(1-t)} \end{aligned} \tag{5}$$

In the basic Keynesian model without G and X, the investment multiplier is simply dY / dI. However, in this extended model one also has to assume that G and X are constant in order to derive the investment multiplier. Thus, the investment multiplier is found by partially differentiating (5) with respect to I, which gives

$$\frac{\partial Y}{\partial I} = \frac{1}{1-(c-m)(1-t)}$$

You should also be able to see that the government expenditure and export multipliers will also take this format as

$$\frac{\partial Y}{\partial I} = \frac{\partial Y}{\partial G} = \frac{\partial Y}{\partial X} = \frac{1}{1-(c-m)(1-t)}$$

Example 11.9

In a Keynesian macroeconomic system, the following relationships and values hold:

$$\begin{array}{llll} Y = C+I+G+X-M & & & \\ C = 0.8Y_d & M = 0.2Y_d & Y_d = (1-t)Y & \\ t = 0.2 & G = 400 & I = 300 & X = 288 \end{array}$$

What is the equilibrium level of Y? What increase in G would be necessary to increase Y to 2,500? If this increased expenditure takes place, what will happen to

(i) the government's budget surplus/deficit, and
(ii) the balance of payments?

Solution

First, we derive the relationship between C and Y. Thus,

$$C = 0.8Y_d = 0.8(1-t)Y = 0.8(1-0.2)Y \tag{1}$$

Next, we substitute (1) and the other functional relationships and given values into the accounting identity to find the equilibrium Y. Thus,

$$\begin{aligned}
Y &= C + I + G + X - M \\
&= 0.8(1-0.2)Y + 300 + 400 + 288 - 0.21(1-0.2)Y \\
&= 0.64Y + 988 - 0.16Y \\
(1-0.48)Y &= 988 \\
Y &= \frac{988}{0.52} = 1{,}900
\end{aligned}$$

At this equilibrium level of Y the total amount of tax raised will be

$$tY = 0.2(1{,}900) = 380$$

Thus, budget deficit, which is the excess of government expenditure over the amount of tax raised, will be

$$G - tY = 400 - 380 = 20$$

The amount spent on imports will be

$$M = 0.2Y_{\mathrm{d}} = 0.2(0.8Y) = 0.16 \times 1{,}900 = 304$$

and so the balance of payments will be

$$X - M = 288 - 304 = -16$$

i.e. a deficit of 16.

The government expenditure multiplier is

$$\begin{aligned}
\frac{\partial Y}{\partial G} &= \frac{1}{1-(c-m)(1-t)} = \frac{1}{1-(0.8-0.2)(1-0.2)} \\
&= \frac{1}{1-(0.6)(0.8)} = \frac{1}{1-0.48} = \frac{1}{0.52}
\end{aligned} \tag{2}$$

As equilibrium Y is 1,900, the increase in Y required to get to the target level of 2,500 is

$$\Delta Y = 2{,}500 - 1{,}900 = 600 \tag{3}$$

Given that the impact of the multiplier on Y will always be equal to

$$\Delta G \frac{\partial Y}{\partial G} = \Delta Y \tag{4}$$

where ΔG is the change in government expenditure, then substituting (2) and (3) into (4) gives

$$\Delta G \times \frac{1}{0.52} = 600$$
$$\Delta G = 600(0.52) = 312$$

This is the increase in G required to raise Y to 2,500.

At the new level of national income, the amount of tax raised will be

$$tY = 0.2(2{,}500) = 500$$

The new government expenditure level including the 312 increase will be

$$400 + 312 = 712$$

Therefore, the budget deficit will be

$$G - tY = 712 - 500 = 212$$

i.e. there is an increase of 192 in the deficit.

The new level of imports will be

$$M = 0.2(0.8)2{,}500 = 400$$

and so the new balance of payments figure will be

$$X - M = 288 - 400 = -112$$

i.e. the deficit increases by 96.

Cost and revenue functions

Some firms produce several different products. When common production facilities are used, the costs of the individual products will be related and this will be reflected in the total cost schedules. The marginal cost schedules of the individual products can then be derived by partial differentiation.

Example 11.10

A firm produces two goods, with output levels q_1 and q_2, and faces the total cost function

$$\text{TC} = 45 + 125q_1 + 84q_2 - 6q_1^2q_2^2 + 0.8q_1^3 + 1.2q_2^3$$

What are the two relevant marginal cost functions?

Solution

Marginal cost is the rate of change of TC with respect to output. Therefore,

$$MC_1 = \frac{\partial TC}{\partial q_1} = 125 - 12q_1q_2^2 + 2.4q_1^2$$
$$MC_2 = \frac{\partial TC}{\partial q_2} = 84 - 12q_1^2q_2 + 3.6q_2^2$$

These marginal cost schedules show that the level of marginal cost for one good will depend on the amount of the other good that is produced.

Some firms may produce different goods which compete with each other in the market place, or are complements. This means that the price of one good will influence the quantity demanded of the other goods sold by the same firm. Marginal revenue for one good will therefore be the partial derivative of total revenue with respect to the output level of that particular good, assuming that the prices of the other goods are fixed.

Example 11.11

A firm produces goods A and B which are complements. Derive marginal revenue functions for the two goods if the relevant demand schedules are

$$q_A = 850 - 12.5p_A - 3.8p_B \quad \text{and} \quad q_B = 936 - 4.8p_A - 24p_B$$

Solution

Marginal revenue is usually expressed as a function of quantity. Therefore, to derive total and marginal revenue functions, the demand functions are first rearranged to get price as a function of quantity. Thus, for good A

$$q_A = 850 - 12.5p_A - 3.8p_B$$
$$12.5p_A = 850 - 3.8p_B - q_A$$
$$p_A = \frac{850 - 3.8p_B - q_A}{12.5}$$
$$TR_A = p_Aq_A = \left(\frac{850 - 3.8p_B - q_A}{12.5}\right)q_A = \frac{850q_A - 3.8p_Bq_A - q_A^2}{12.5}$$
$$MR_A = \frac{\partial TR}{\partial q_A} = \frac{850 - 3.8p_B - q_A}{12.5} = 68 - 0.304p_B - 0.16q_A$$

Similarly, for good B

$$q_B = 936 - 4.8p_A - 24p_B$$
$$24p_B = 936 - 4.8p_A - q_B$$
$$p_B = 39 - 0.2p_A - \frac{q_B}{24}$$
$$TR_B = p_B q_B = \left(39 - 0.2p_A - \frac{q_B}{24}\right) q_B = 39q_B - 0.2p_A q_B - \frac{q_B^2}{24}$$
$$MR_B = \frac{\partial TR}{\partial q_B} = 39 - 0.2p_A - \frac{q_B}{12}$$

The marginal revenue functions MR_A and MR_B confirm that because the demand functions for the two goods are interrelated, the marginal revenue function for one good will depend on the price level of the other good.

QUESTIONS 11.2

1. The demand function for a good is

$$q = 56.6 - 0.25p - 0.03m + 0.45p_s + 0.6n$$

where q is the quantity demanded per week, p is the price per unit, m is the average weekly income, p_s is the price of a competing good and n is the population in millions. Given values are $p = 65$, $m = 350$, $p_s = 60$ and $n = 24$,
   (a) calculate the price elasticity of demand,
   (b) find out what would happen to (a) if n rose to 26,
   (c) explain why this is an inferior good,
   (d) if producers of the competing product and the manufacturer of this good both increased their prices by the same percentage, what would happen to the quantity demanded (of the original good) assuming that the proportional price change is small and relevant elasticity measures do not alter significantly?
2. Do the following utility functions obey the law of diminishing marginal utility?
   (a) $U = 5A + 8B + 2.2A^2B^2 - 0.3A^3B^3$
   (b) $U = 24A^{0.8}B^{1.2}$
   (c) $U = 6A^{0.7}B^{0.8}$
3. An individual consumes two goods and has the utility function $U = 2A^{0.4}B^{0.4}$, where A and B represent the quantities of the two goods consumed. Will they ever consume either good up to the point where its marginal utility is zero?
4. In a Keynesian macroeconomic model of an economy, using the usual terminology,

$$Y = C + I + G + X - M \qquad Y_d = (1-t)Y \qquad C = 0.75Y_d$$
$$M = 0.25Y_d \qquad I = 820 \qquad G = 960 \qquad t = 0.3 \qquad X = 650$$

what will be the equilibrium value of Y? Use the export multiplier to find out what will happen to the balance of payments if exports exogenously increase by 100.

5. A multiplant firm faces the total cost schedule

$$\text{TC} = 850 + 18q_1 + 25q_2 + 0.6q_1^2q_2 + 1.2q_1q_2^2$$

where q_1 and q_2 are output levels in its two plants. What marginal cost schedule does it face if output in plant 2 is expanded while output in plant 1 is kept unchanged?

6. In a closed economy (i.e. one with no foreign trade) the following relationships hold:

$$C = 0.6Y_d \quad Y_d = (1-t)Y \quad Y = C + I + G$$
$$I = 120 \qquad t = 0.25 \qquad G = 210$$

where C is consumer expenditure, Y_d is disposable income, Y is national income, I is investment, t is the tax rate and G is government expenditure. What is the marginal propensity to consume out of Y? What is the value of the government expenditure multiplier? How much does government expenditure need to be increased to achieve a national income of 700?

11.3 SECOND-ORDER PARTIAL DERIVATIVES

Second-order partial derivatives are found by differentiating the first-order partial derivatives of a function.

When a function has two independent variables, there will be four second-order partial derivatives. Take, for example, the production function

$$Q = 24K^{0.4}L^{0.3}$$

There are two first-order partial derivatives

$$\frac{\partial Q}{\partial K} = 10K^{-0.6}L^{0.3} \quad \text{and} \quad \frac{\partial Q}{\partial L} = 7.5K^{0.4}L^{-0.7}$$

These represent the marginal product functions for K and L. Differentiating these functions for the second time, we get

$$\frac{\partial^2 Q}{\partial K^2} = -6K^{-1.6}L^{0.3} \quad \text{and} \quad \frac{\partial^2 Q}{\partial L^2} = -5.25K^{0.4}L^{-1.7}$$

These second-order partial derivatives represent the rate of change of the marginal product functions. In this example, we can see that the slope of MP_L (i.e. $\partial^2 Q/\partial L^2$) will always be negative (assuming positive values of K and L) and as L increases, with K fixed, the absolute value of this slope diminishes.

We can also find the rate of change of $\partial Q/\partial K$ with respect to changes in L and the rate of change of $\partial Q/\partial L$ with respect to K. These will be

$$\frac{\partial^2 Q}{\partial K \partial L} = 3K^{-0.6}L^{-0.7} \quad \text{and} \quad \frac{\partial^2 Q}{\partial L \partial K} = 3K^{-0.6}L^{-0.7}$$

and are known as 'cross' or 'mixed' partial derivatives. They show how the rate of change of Q with respect to one input alters when the other input changes. In this example, the cross partial derivative $\frac{\partial^2 Q}{\partial L \partial K}$ tells us that the rate of change of MP_L with respect to changes in K will be positive and will fall in value as K increases.

You will also have noted in this example that

$$\frac{\partial^2 Q}{\partial K \partial L} = \frac{\partial^2 Q}{\partial L \partial K}$$

In fact, matched pairs of cross partial derivatives will always be equal to each other.

Thus, for any continuous two-variable function $y = f(x, z)$, there will be four second-order partial derivatives:

(i) $\frac{\partial^2 y}{\partial x^2}$

(ii) $\frac{\partial^2 y}{\partial z^2}$

(iii) $\frac{\partial^2 y}{\partial x \partial z}$

(iv) $\frac{\partial^2 y}{\partial z \partial x}$

with the cross partial derivatives (iii) and (iv) always being equal, i.e.

$$\frac{\partial^2 y}{\partial x \partial z} = \frac{\partial^2 y}{\partial z \partial x}$$

Example 11.12

Derive the four second-order partial derivatives for the production function

$$Q = 6K + 0.3K^2L + 1.2L^2$$

and interpret their meaning.

Solution

The two first-order partial derivatives are

$$\frac{\partial Q}{\partial K} = 6 + 0.6KL \quad \text{and} \quad \frac{\partial Q}{\partial K} = 0.3K^2 + 2.4L$$

and these represent the marginal product functions MP_K and MP_L.

The four second-order partial derivatives are as follows:

(i) $\dfrac{\partial^2 Q}{\partial K^2} = 0.6L$

This represents the slope of the MP_K function. It tells us that the MP_K function will have a constant slope along its length (i.e. it is linear) for any given value of L, but an increase in L will cause an increase in this slope

(ii) $\dfrac{\partial^2 Q}{\partial L^2} = 2.4$

This represents the slope of the MP_L function and tells us that MP_L is a straight line with slope 2.4. This slope does not depend on the value of K.

(iii) $\dfrac{\partial^2 Q}{\partial K \partial L} = 0.6K$

This tells us that MP_K increases if L is increased. The rate at which MP_K rises as L is increased will depend on the value of K.

(iv) $\dfrac{\partial^2 Q}{\partial L \partial K} = 0.6K$

This tells us that MP_L will increase if K is increased and that the rate of this increase will depend on the value of K. Thus, although the slope of the MP_L schedule will always be 2.4, from (ii), its actual position will depend on the amount of K used.

Some other applications of second-order partial derivatives are given next.

Example 11.13

A firm sells two competing products whose demand functions are

$$q_1 = 120 - 0.8p_1 + 0.5p_2 \quad \text{and} \quad q_2 = 160 - 0.4p_1 - 12p_2$$

How will the price of good 2 affect the marginal revenue of good 1?

Solution

To find the total revenue function for good 1 (TR_1) in terms of q_1 we first need to derive the inverse demand function $p_1 = f(q_1, p_2)$. Thus, given

$$\begin{aligned} q_1 &= 120 - 0.8p_1 + 0.5p_2 \\ 0.8p_1 &= 120 + 0.5p_2 - q_1 \\ p_1 &= 150 + 0.625p_2 - 1.25q_1 \\ \text{TR}_1 &= p_1 q_1 \\ &= (150 + 0.625p_2 - 1.25q_1)q_1 \\ &= 150q_1 + 0.625p_2 q_1 - 1.25q_1^2 \end{aligned}$$

Therefore,

$$\text{MR}_1 = \frac{\partial \text{TR}_1}{\partial q_1} = 150 + 0.625p_2 - 2.5q_1$$

By partially differentiating this marginal revenue function with respect to q_1 we can see that it will have a constant slope of −2.5 regardless of the value of p_2 or the amount of q_1 sold.

The effect of a change in p_2 on MR_1 is shown by the cross partial derivative

$$\frac{\partial^2 \text{TR}_1}{\partial q_1 \partial p_2} = 0.625$$

Thus, an increase in p_2 of one unit will cause an increase in the marginal revenue from good 1 of 0.625, i.e. although the slope of the MR_1 schedule remains constant at −2.5, its position shifts upward if p_2 rises.

(Note that in order to answer this question, we have formulated the total revenue for good 1 as a function of one price and one quantity, i.e. $\text{TR}_1 = f(q_1, p_2)$.)

Example 11.14

A firm operates with the production function $Q = 820K^{0.3}L^{0.2}$ and can buy inputs K and L at £65 and £40 per unit, respectively. If it can sell its output at a fixed price of £12 per unit, what is the relationship between increases in L and total profit? Will a change in K affect the extra profit derived from marginal increases in L?

Solution

$$\begin{aligned} \text{TR} &= \text{PQ} = 12\left(820K^{0.3}L^{0.2}\right). \\ \text{TC} &= P_K K + P_L L = 65K + 40L \end{aligned}$$

Therefore, profit will be

$$\begin{aligned} \pi &= \text{TR} - \text{TC} \\ &= 12\left(820K^{0.3}L^{0.2}\right) - (65K + 40L) \\ &= 9{,}840K^{0.3}L^{0.2} - 65K - 40L \end{aligned}$$

The effect of an increase in L on profit is shown by the first-order partial derivative:

$$\frac{\partial \pi}{\partial L} = 1{,}968K^{0.3}L^{-0.8} - 40 \qquad (1)$$

This effect will be positive as long as

$$1{,}968K^{0.3}L^{-0.8} > 40$$

However, if L is continually increased while K is held constant, the value of the term $1{,}968K^{0.3}L^{-0.8}$ will eventually fall below 40 and so $\partial\pi/\partial L$ will become negative.

To determine the effect of a change in K on the marginal profit function with respect to L, we need to differentiate (1) with respect to K, giving

$$\frac{\partial^2 \pi}{\partial L \partial K} = 0.3\left(1{,}968K^{-0.7}L^{-0.8}\right) = 590.4K^{-0.7}L^{-0.8}$$

This cross partial derivative will be positive as long as both K and L take positive values. Thus, an increase in K will have a positive effect on the extra profit generated by marginal increases in L, which is what we would expect since a higher level of K will allow greater output for any given amount of L. The magnitude of this impact will depend on the values of K and L.

Second-order and cross partial derivatives can also be derived for functions with three or more independent variables. For a function with three independent variables, such as $y = f(w, x, z)$ there will be the three second-order partial derivatives

$$\frac{\partial^2 y}{\partial w^2}, \quad \frac{\partial^2 y}{\partial x^2} \quad \text{and} \quad \frac{\partial^2 y}{\partial z^2}$$

plus the six cross partial derivatives

$$\frac{\partial^2 y}{\partial w \partial x} = \frac{\partial^2 y}{\partial x \partial w}, \quad \frac{\partial^2 y}{\partial x \partial z} = \frac{\partial^2 y}{\partial z \partial x} \quad \text{and} \quad \frac{\partial^2 y}{\partial w \partial z} = \frac{\partial^2 y}{\partial z \partial w}$$

These are arranged in pairs because, as with the two-variable case, cross partial derivatives will be equal if the two stages of differentiation involve the same two variables.

Example 11.15

For the production function $Q = 32K^{0.5}L^{0.25}R^{0.4}$ derive all the second-order and cross partial derivatives and show that the cross partial derivatives with respect to each possible pair of independent variables will be equal to each other.

Solution

The three first-order partial derivatives will be

$$\frac{\partial Q}{\partial K} = 16K^{-0.5}L^{0.25}R^{0.4}$$

$$\frac{\partial Q}{\partial L} = 8K^{0.5}L^{-0.75}R^{0.4}$$

$$\frac{\partial Q}{\partial R} = 12.8K^{0.5}L^{0.25}R^{-0.6}$$

The second-order partial derivatives will be

$$\frac{\partial^2 Q}{\partial K^2} = -8K^{-1.5}L^{0.25}R^{0.4}$$

$$\frac{\partial^2 Q}{\partial L^2} = -6K^{0.5}L^{-1.75}R^{0.4}$$

$$\frac{\partial^2 Q}{\partial R^2} = -7.68K^{0.5}L^{0.25}R^{-1.6}$$

plus the six cross partial derivatives:

$$\frac{\partial^2 Q}{\partial K \partial L} = 4K^{-0.5}L^{-0.75}R^{0.4} = \frac{\partial^2 Q}{\partial L \partial K}$$

$$\frac{\partial^2 Q}{\partial L \partial R} = 3.2K^{0.5}L^{-0.75}R^{-0.6} = \frac{\partial^2 Q}{\partial R \partial L}$$

$$\frac{\partial^2 Q}{\partial R \partial K} = 6.4K^{-0.5}L^{0.25}R^{-0.6} = \frac{\partial^2 Q}{\partial K \partial R}$$

Second-order derivatives for multi-variable functions are needed to check second-order conditions for optimization, as explained in the next section.

QUESTIONS 11.3

1. For the production function $Q = 8K^{0.6}L^{0.5}$ derive a function for the slope of the marginal product of L. What effect will a marginal increase in K have upon this MP_L function?
2. Derive all the second-order and cross partial derivatives for the production function $Q = 35KL + 1.4LK^2 + 3.2L^2$ and interpret their meaning.
3. A firm operates three plants with the joint total cost function

$$TC = 58 + 18q_1 + 9q_2q_3 + 0.004q_1^2q_3^2 + 1.2q_1q_2q_3$$

Find all the second-order partial derivatives for TC and demonstrate that the cross partial derivatives can be arranged in three equal pairs.
What, if anything, can be deduced about the effect of an increase in output of q_2 on the cost of producing any given output level for q_1?

11.4 UNCONSTRAINED OPTIMIZATION: FUNCTIONS WITH TWO VARIABLES

For the two variable function $y = f(x, z)$ to be at a maximum or at a minimum, the first-order conditions which must be met are

$$\frac{\partial y}{\partial x}=0 \quad \text{and} \quad \frac{\partial y}{\partial z}=0$$

These are similar to the first-order conditions for optimization of a single variable function that were explained in Chapter 10. To be at a maximum or minimum, the function must be at a stationary point with respect to changes in both variables.

The second-order conditions and the reasons for them were relatively easy to explain in the case of a function of one independent variable. However, when two or more independent variables are involved, the rationale for all the second-order conditions is not quite so straightforward. We shall therefore just state these second-order conditions here and give a brief intuitive explanation for the two variable cases before looking at some applications. The second-order conditions for the optimization of multi-variable functions with more than two variables using matrix algebra are explained in Chapter 16.

For the optimization of two variable functions there are two sets of second-order conditions. For any function $y = f(x, z)$

(1) $$\frac{\partial^2 y}{\partial x^2}<0 \quad \text{and} \quad \frac{\partial^2 y}{\partial z^2}<0 \quad \text{for a maximum}$$

$$\frac{\partial^2 y}{\partial x^2}>0 \quad \text{and} \quad \frac{\partial^2 y}{\partial z^2}>0 \quad \text{for a minimum}$$

These are similar to the second-order conditions for the optimization of a single variable function. The rate of change of a function (i.e. its slope) must be decreasing at a stationary point for that point to be a maximum and it must be increasing for a stationary point to be a minimum. The difference here is that these conditions must hold with respect to changes in both independent variables.

(2) The other second-order condition is

$$\frac{\partial^2 y}{\partial x^2}\frac{\partial^2 y}{\partial z^2}>\left(\frac{\partial^2 y}{\partial x \partial z}\right)^2$$

This must hold at both maximum *and* minimum stationary points.

To get an idea of the reason for this condition, imagine a three-dimensional model with x and z being measured on the two axes of a graph and y being measured by the height above the flat surface on which the x and z axes are drawn. For a point to be the peak of the y 'hill' then, as well as the slope being zero at this point, one needs to ensure that, whichever direction one moves in, the height will fall and the slope will become steeper. Similarly, for a point to be the minimum of a y 'trough' then, as well as the slope being zero, one needs to ensure that the height will rise and the slope will become steeper whichever direction one moves in. As moves can be made in directions other than parallel to the two axes, it can be mathematically proved that the condition

$$\frac{\partial^2 y}{\partial x^2} \times \frac{\partial^2 y}{\partial z^2} > \left(\frac{\partial^2 y}{\partial x \partial z}\right)^2$$

satisfies these requirements as long as the other second-order conditions for a maximum or minimum also hold.

Note also that all these conditions refer to the requirements for *local* maximum or minimum values of a function, which may or may not be *global* maxima or minima. (Refer back to Section 10.4 if you cannot remember the difference between these two concepts.)

Let us now look at some applications of these rules for the unconstrained optimization of a function with two independent variables.

Example 11.16

A firm produces two products which are sold in two separate markets with the demand schedules

$$p_1 = 600 - 0.3q_1 \quad \text{and} \quad p_2 = 500 - 0.2q_2$$

Production costs are related and the firm faces the total cost function

$$\text{TC} = 16 + 1.2q_1 + 1.5q_2 + 0.2q_1q_2$$

If the firm wishes to maximize total profits, how much of each product should it sell? What will the maximum profit level be?

Solution

The total revenue will be

$$\begin{aligned}\text{TR} = \text{TR}_1 + \text{TR}_2 &= p_1q_1 + p_2q_2 \\ &= (600 - 0.3q_1)q_1 + (500 - 0.2q_2)q_2 \\ &= 600q_1 - 0.3q_1^2 + 500q_2 - 0.2q_2^2\end{aligned}$$

Therefore, profit is

$$\begin{aligned}\pi &= \text{TR} - \text{TC} \\ &= 600q_1 - 0.3q_1^2 + 500q_2 - 0.2q_2^2 - (16 + 1.2q_1 + 1.5q_2 + 0.2q_1q_2) \\ &= 600q_1 - 0.3q_1^2 + 500q_2 - 0.2q_2^2 - 16 - 1.2q_1 - 1.5q_2 - 0.2q_1q_2 \\ &= -16 + 598.8q_1 - 0.3q_1^2 + 498.5q_2 - 0.2q_2^2 - 0.2q_1q_2\end{aligned}$$

First-order conditions for maximization of this profit function are

$$\frac{\partial \pi}{\partial q_1} = 598.8 - 0.6q_1 - 0.2q_2 = 0 \tag{1}$$

and

$$\frac{\partial \pi}{\partial q_2} = 498.5 - 0.4q_2 - 0.2q_1 = 0 \tag{2}$$

Simultaneous equations (1) and (2) can now be solved to find the optimal values of q_1 and q_2.

Multiplying (2) by 3	$1{,}495.5 - 1.2q_2 - 0.6q_1 = 0$
Rearranging (1)	$598.8 - 0.2q_2 - 0.6q_1 = 0$
Subtracting gives	$896.7 - q_2 = 0$
Giving the optimal value	$896.7 = q_2$

Substituting this value for q_2 into (1)

$$598.8 - 0.6q_1 - 0.2(896.7) = 0$$
$$598.8 - 179.34 = 0.6q_1$$
$$419.46 = 0.6q_1$$
$$699.1 = q_1$$

Checking second-order conditions by differentiating (1) and (2) again:

$$\frac{\partial^2 \pi}{\partial q_1^2} = -0.6 < 0 \quad \text{and} \quad \frac{\partial^2 \pi}{\partial q_2^2} = -0.4 < 0$$

This satisfies one set of second-order conditions for a maximum.

The cross partial derivative will be

$$\frac{\partial^2 \pi}{\partial q_1 \partial q_2} = -0.2$$

Therefore,

$$\left(\frac{\partial^2 \pi}{\partial q_1^2}\right)\left(\frac{\partial^2 \pi}{\partial q_2^2}\right) = (-0.6)(-0.4) = 0.24$$

This is greater than the value of

$$\left(\frac{\partial^2 \pi}{\partial q_1 \partial q_2}\right)^2 = (-0.2)^2 = 0.04$$

and so the remaining second-order condition for a maximum is satisfied.

The actual profit is found by substituting the optimum values $q_1 = 699.1$ and $q_2 = 896.7$ into the profit function. Thus,

$$\begin{aligned}\pi &= -16 + 598.8q_1 - 0.3q_1^2 + 498.5q_2 - 0.2q_2^2 - 0.2q_1q_2 \\ &= -16 + 598.8(699.1) - 0.3(699.1)^2 + 498.5(896.7) - 0.2(896.7)^2 \\ &\quad - 0.2(699.1)(896.7) \\ &= £432{,}797.02\end{aligned}$$

Example 11.17

A firm sells two products which are partial substitutes for each other. If the price of one product increases, then the demand for the other substitute product rises. The prices of the products (in £) are p_1 and p_2 and their respective demand functions are

$$q_1 = 517 - 3.5p_1 + 0.8p_2 \quad \text{and} \quad q_2 = 770 - 4.4p_2 + 1.4p_1$$

What price should the firm charge for each product to maximize its total sales revenue?

Solution

For this problem it is more convenient to express total revenue as a function of price rather than quantity. Thus,

$$\begin{aligned}\text{TR} &= \text{TR}_1 + \text{TR}_2 = p_1q_1 + p_2q_2 \\ &= p_1(517 - 3.5p_1 + 0.8p_2) + p_2(770 - 4.4p_2 + 1.4p_1) \\ &= 517p_1 - 3.5p_1^2 + 0.8p_1p_2 + 770p_2 - 4.4p_2^2 + 1.4p_1p_2 \\ &= 517p_1 - 3.5p_1^2 + 770p_2 - 4.4p_2^2 + 2.2p_1p_2\end{aligned}$$

First-order conditions for a maximum are

$$\frac{\partial \text{TR}}{\partial p_1} = 517 - 7p_1 + 2.2p_2 = 0 \qquad (1)$$

and

$$\frac{\partial \text{TR}}{\partial p_2} = 770 - 8.8p_2 + 2.2p_1 = 0 \qquad (2)$$

Multiplying (1) by 4	$2{,}068 - 28p_1 + 8.8p_2 = 0$
Rearranging and adding (2)	$770 + 2.2p_2 - 8.8p_2 = 0$
gives	$2{,}838 - 25.8p_1 = 0$
	$2{,}838 = 25.8p_1$
	$110 = p_1$

Substituting this value of p_1 into (1)

$$\begin{aligned}517 - 7(110) + 2.2p_2 &= 0 \\ 2.2p_2 &= 253 \\ p_2 &= 115\end{aligned}$$

Checking second-order conditions:

$$\frac{\partial^2 \text{TR}}{\partial p_1^2} = -7 < 0, \quad \frac{\partial^2 \text{TR}}{\partial p_2^2} = -8.8 < 0 \quad \text{and} \quad \frac{\partial^2 \text{TR}}{\partial p_1 \partial p_2} = 2.2$$

$$\left(\frac{\partial^2 \text{TR}}{\partial q_1^2}\right)\left(\frac{\partial^2 \text{TR}}{\partial q_2^2}\right) = (-7)(-8.8) = 61.6 > 4.84 = (2.2)^2 = \left(\frac{\partial^2 \text{TR}}{\partial q_1 \partial q_2}\right)^2$$

Therefore, all second-order conditions for a maximum value of total revenue are satisfied when p_1 = £110 and p_2 = £115.

Example 11.18

A multiplant monopoly operates two plants whose total cost schedules are

$$\text{TC}_1 = 8.5 + 0.03q_1^2 \quad \text{and} \quad \text{TC}_2 = 5.2 + 0.04q_2^2$$

If it faces the demand schedule

$$p = 60 - 0.04q$$

where $q = q_1 + q_2$. How much should it produce in each plant to maximize profits?

Solution

The total revenue is

$$\text{TR} = pq = (60 - 0.4q)q = 60q - 0.04q^2$$

Substituting $(q_1 + q_2)$ for q gives

$$\begin{aligned}\text{TR} &= 60(q_1 + q_2) - 0.04(q_1 + q_2)^2 \\ &= 60q_1 + 60q_2 - 0.04q_1^2 - 0.08q_1q_2 - 0.04q_2^2\end{aligned}$$

Thus, subtracting the two total cost schedules, profit is

$$\begin{aligned}\pi &= \text{TR} - \text{TC} = \text{TR} - (8.5 + 0.03q_1^2) - (5.2 + 0.04q_2^2) \\ &= 60q_1 + 60q_2 - 0.04q_1^2 - 0.08q_1q_2 - 0.04q_2^2 - 8.5 - 0.03q_1^2 - 5.2 - 0.04q_2^2 \\ &= -13.7 + 60q_1 + 60q_2 - 0.07q_1^2 - 0.08q_2^2 - 0.08q_1q_2\end{aligned}$$

First-order conditions for a maximum value of π require and

$$\frac{\partial \pi}{\partial q_1} = 60 - 0.14q_1 - 0.08q_2 = 0 \tag{1}$$

and

$$\frac{\partial \pi}{\partial q_2} = 60 - 0.1q_2 - 0.16q_2 = 0 \tag{2}$$

Multiplying (1) by 2 $\quad 120 - 0.28q_1 - 0.16q_2 = 0$

Rearranging and subtracting (2) $\quad 60 - 0.08q_1 - 0.16q_2 = 0$

$$\begin{aligned} 60 - 0.2q_1 &= 0 \\ 60 &= 0.2q_1 \\ 300 &= q_1 \end{aligned}$$

Substituting this value of q_1 into (1)

$$\begin{aligned} 60 - 0.14(300) - 0.0q_2 &= 0 \\ 18 &= 0.08q_2 \\ 225 &= q_2 \end{aligned}$$

Checking second-order conditions:

$$\frac{\partial^2 \pi}{\partial q_1^2} = -0.14 < 0, \quad \frac{\partial^2 \pi}{\partial q_2^2} = -0.16 < 0 \quad \text{and} \quad \frac{\partial^2 \pi}{\partial q_2 \, \partial q_2} = -0.08$$

$$\left(\frac{\partial^2 \pi}{\partial q_1^2}\right)\left(\frac{\partial^2 \pi}{\partial q_2^2}\right) = (-0.14)(-0.16) = 0.0224 > 0.0064 = (-0.08)^2 = \left(\frac{\partial^2 \pi}{\partial q_1 \, \partial q_2}\right)^2$$

Therefore, all second-order conditions are satisfied for profit maximization when $q_1 = 300$ and $q_2 = 225$.

We can also check that the total profit is positive for these output levels. Total output is

$$q = q_1 + q_2 = 300 + 225 = 525$$

Substituting this value into the demand schedule

$$p = 60 - 0.04q = 60 - 0.04(525) = 39$$

Therefore,

$$\begin{aligned} \text{TR} &= pq = 39(525) = 20{,}475 \\ \text{TC} &= \text{TC}_1 + \text{TC}_2 = \; 8.5 + 0.03(300)^2 \;] + [\, 5.2 + 0.04(225)^2 \\ &= 2{,}708.5 + 2{,}030.2 = 4{,}738.7 \end{aligned}$$

$$\pi = \text{TR} - \text{TC} = 20{,}475 - 4{,}738.7 = £15{,}736.30$$

Note that this method could also be used to solve the multiplant monopoly problems in Chapter 5 that only involved linear functions. The unconstrained optimization method used here is, however, a more general method that can be used for both linear and non-linear functions.

Example 11.19

A firm sells its output in a perfectly competitive market at a fixed price of £200 per unit. It buys the two inputs K and L at prices of £42 per unit and £5 per unit, respectively, and faces the production function

$$q = 3.1K^{0.3}L^{0.25}$$

What combination of K and L should it use to maximize profit?

Solution

$$\begin{aligned} \text{TR} &= pq = 200\left(3.1K^{0.3}L^{0.25}\right) = 620K^{0.3}L^{0.25} \\ \text{TC} &= 42K + 5L \end{aligned}$$

Therefore, the profit function the firm wishes to maximize is

$$\pi = \text{TR} - \text{TC} = 620K^{0.3}L^{0.25} - 42K - 5L$$

First-order conditions for a maximum require

$$\frac{\partial \pi}{\partial K} = 186K^{-0.7}L^{0.25} - 42 = 0 \quad \text{and} \quad \frac{\partial \pi}{\partial L} = 155K^{0.3}L^{-0.75} - 5 = 0$$

giving

$$\begin{aligned} 186L^{0.25} &= 42K^{0.7} \\ L^{0.25} &= \frac{42}{186}K^{0.7} \end{aligned} \tag{1}$$

and

$$\begin{aligned} 155K^{0.3} &= 5L^{0.75} \\ 31K^{0.3} &= L^{0.75} \end{aligned} \tag{2}$$

Taking (1) to the power of 3

$$L^{0.75} = \left(\frac{42}{186}K^{0.7}\right)^3 = \frac{42^3}{186^3}K^{2.1} \tag{3}$$

Setting (3) equal to (2)

$$\begin{aligned} \frac{42^3}{186^3}K^{2.1} &= 31K^{0.3} \\ K^{1.8} &= \frac{31(186)^3}{42^3} = 2{,}692.481 \\ K &= 80.471179 \end{aligned} \tag{4}$$

Substituting (4), the value for K, into (1)

$$L^{0.25} = \frac{42}{186}(80.471179)^{0.7} = 4.8717455$$
$$L = \left(L^{0.25}\right)^4 = (4.8717455)^4 = 563.29822$$

Therefore, first-order conditions suggest that the optimum values are (to 2 decimal places) $L = 563.3$ and $K = 80.47$.

Checking second-order conditions:

$$\frac{\partial^2 \pi}{\partial K^2} = (-0.7)186K^{-1.7}L^{0.25}$$
$$= -130.2(80.47)^{-1.7}(563.3)^{0.25} = -0.3653576 < 0$$
$$\frac{\partial^2 \pi}{\partial L^2} = (-0.75)155K^{0.3}L^{-1.75}$$
$$= -116.25(80.47)^{0.3}(563.3)^{-1.25} = -0.0066572 < 0$$
$$\frac{\partial^2 \pi}{\partial K \partial L} = (0.25)186K^{-0.7}L^{-0.75}$$
$$\left(\frac{\partial^2 \pi}{\partial K^2}\right)\left(\frac{\partial^2 \pi}{\partial L^2}\right) = (-0.3653576)(-0.0066572) = 0.0024323$$
$$\left(\frac{\partial^2 \pi}{\partial K \partial L}\right)^2 = (0.0186404)^2 = 0.0003475$$
$$= 46.5(80.47)^{-0.7}(563.3)^{-0.75} = 0.0186404$$

Therefore, as

$$\frac{\partial^2 \pi}{\partial K^2}\frac{\partial^2 \pi}{\partial L^2} > \left(\frac{\partial^2 \pi}{\partial K \partial L}\right)^2$$

all second-order conditions for maximum profit are satisfied when $K = 80.47$ and $L = 563.3$.

The actual profit will be

$$\pi = 620K^{0.3}L^{0.25} - 42K - 5L$$
$$= 620(80.47)^{0.3}(563.3)^{0.25} - 42(80.47) - 5(563.3)$$
$$= 11{,}265.924 - 3{,}379.74 - 2{,}816.5 = £5{,}069.68$$

Note that in this problem, and other similar ones in this section, the indices in the Cobb-Douglas production function add up to less than unity, giving decreasing returns to scale and hence rising average and marginal (long-run) cost schedules. If there were increasing returns to scale and the average and marginal cost schedules continued to fall, a firm facing a fixed price would wish to expand output indefinitely and so no profit-maximizing solution would be found by this method.

Example 11.20

A multiplant monopoly operates two plants whose total cost schedules are

$$\mathrm{TC}_1 = 36 + 0.003q_1^3 \quad \text{and} \quad \mathrm{TC}_2 = 45 + 0.005q_2^3$$

If its total output is sold in a market where the demand schedule is $p = 320 - 0.1q$, where $q = q_1 + q_2$, how much should it produce in each plant to maximize total profits?

Solution

The total revenue is

$$\mathrm{TR} = pq = (320 - 0.1q)q = 320q - 0.1q^2$$

Substituting $q_1 + q_2 = q$ gives

$$\begin{aligned}\mathrm{TR} &= 320(q_1 + q_2) - 0.1(q_1 + q_2)^2 \\ &= 320q_1 + 320q_2 - 0.1\left(q_1^2 + 2q_1q_2 + q_2^2\right) \\ &= 320q_1 + 320q_2 - 0.1q_1^2 - 0.2q_1q_2 - 0.1q_2^2\end{aligned}$$

Thus, profit will be

$$\begin{aligned}\pi &= \mathrm{TR} - \mathrm{TC} = \mathrm{TR} - \mathrm{TC}_1 - \mathrm{TC}_2 \\ &= \left(320q_1 + 320q_2 - 0.1q_1^2 - 0.2q_1q_2 - 0.1q_2^2\right) - \left(36 + 0.005q_1^3\right) - \left(45 + 0.005q_2^3\right) \\ &= 320q_1 + 320q_2 - 0.1q_1^2 - 0.2q_1q_2 - 0.1q_2^2 - 36 - 0.003q_1^3 - 45 - 0.005q_2^3\end{aligned}$$

First-order conditions for a maximum require

$$\frac{\partial \pi}{\partial q_1} = 320 - 0.2q_1 - 0.2q_2 - 0.009q_1^2 = 0 \tag{1}$$

and

$$\frac{\partial \pi}{\partial q_2} = 320 - 0.2q_1 - 0.2q_2 - 0.015q_2^2 = 0 \tag{2}$$

Subtracting (2) from (1)

$$-0.009q_1^2 + 0.015q_2^2 = 0$$

$$q_2^2 = \left(\frac{-0.009}{0.015}\right)q_1^2 = 0.6q_1^2 \qquad (3)$$

$$q_2 = \sqrt{0.6q_1^2} = 0.7746q_1$$

Substituting (3) for q_2 in (1)

$$320 - 0.2q_1 - 0.2(0.7746q_1) - 0.009q_1^2 = 0$$

$$320 - 0.2q_1 - 0.15492q_1 - 0.009q_1^2 = 0$$

$$0 = 0.009q_1^2 + 0.35492q_1 - 320$$

Using the quadratic formula to solve (4)

$$q_1 = \frac{-b \pm \sqrt{b^2 - 4ac}}{2a} = \frac{0.35492 \pm \sqrt{(0.35492)^2 \quad 4(0.009)(\quad 320)}}{0.018}$$

$$= \frac{-0.35492 \pm \sqrt{11.64598}}{0.018} = \frac{-0.35492 \pm 3.412619}{0.018}$$

Disregarding the negative solution, this gives plant 1 output

$$q_1 = \frac{3.057699}{0.018}$$

$$= 169.87216 = 169.87 \text{ (to 2 dp)}$$

Substituting this value for q_1 into (3)

$$q_2 = 0.7746(169.87216) = 131.58 \text{ (to 2 dp)}$$

Checking second-order conditions:

$$\frac{\partial^2 \pi}{\partial q_1^2} = -0.2 - 0.018q_1 = -0.2 - 0.018(169.87) = -3.25766 < 0$$

$$\frac{\partial^2 \pi}{\partial q_1^2} = -0.2 - 0.03q_2 = -0.2 - 0.03(131.58) = -4.1474 < 0$$

$$\frac{\partial^2 \pi}{\partial q_1 q_2} = -0.2$$

Thus, using the shorthand notation for the previous second-order derivatives,

$$(\pi_{11})(\pi_{22}) = (-3.25766)(-4.1474) = 13.51 > 0.04 = (-0.2)^2 = (\pi_{12})^2$$

Therefore, all second-order conditions for a maximum value of profit are satisfied when q_1 = 169.87 and q_2 = 131.58.

When a function involves more than two independent variables the second-order conditions for a maximum or minimum become even more complex and matrix algebra is needed to check them. However, until we get to Chapter 16, for economic problems involving three or more independent variables we shall just consider how the first-order conditions can be used to determine optimum values. From the way these problems are constructed it will be obvious whether or not a maximum or a minimum value is being sought, and it will be assumed that second-order conditions are satisfied for the values that meet the first-order conditions.

Example 11.21

A firm operates with the production function

$$Q = 95K^{0.3}L^{0.2}R^{0.25}$$

and buys the three inputs K, L and R at prices of £30, £16 and £12 per unit, respectively. If it can sell its output at a fixed price of £4 a unit, what is the maximum profit it can make? (Assume that second-order conditions for a maximum are met at stationary points.)

Solution

$$\begin{aligned}\pi = \text{TR} - \text{TC} &= PQ - (P_K K + P_L L + P_R R)\\ &= 4\left(95K^{0.3}L^{0.2}R^{0.25}\right) - (30K + 16L + 12R)\\ &= 380K^{0.3}L^{0.2}R^{0.25} - 30K - 16L - 12R\end{aligned}$$

First-order conditions for a maximum are

$$\frac{\partial \pi}{\partial K} = 114K^{-0.7}L^{0.2}R^{0.25} - 30 = 0 \qquad (1)$$

$$\frac{\partial \pi}{\partial L} = 76K^{0.3}L^{-0.8}R^{0.25} - 16 = 0 \qquad (2)$$

$$\frac{\partial \pi}{\partial R} = 95K^{0.3}L^{0.2}R^{-0.75} - 12 = 0 \qquad (3)$$

From (1)

$$114L^{0.2}R^{0.25} = 30K^{0.7}$$

$$R^{0.25} = \frac{30K^{0.7}}{114L^{0.2}} \qquad (4)$$

Substituting (4) into (2)

$$
\begin{aligned}
76K^{0.3}L^{-0.8}\left[\frac{30K^{0.7}}{114L^{0.2}}\right] &= 16 \\
76K^{0.3}\left(30K^{0.7}\right) &= 16L^{0.8}\left(114L^{0.2}\right) \\
2,280K &= 1,824L \\
K &= 0.8L
\end{aligned} \tag{5}
$$

Substituting (5) into (4)

$$
R^{0.25} = \frac{30(0.8L)^{0.7}}{114L^{0.2}} = \frac{30(0.8)^{0.7}\,L^{0.5}}{114} \tag{6}
$$

Taking each side of (6) to the power of 3

$$
R^{0.75} = \frac{27,000(0.8)^{2.1}\,L^{1.5}}{114^3}
$$

Inverting

$$
R^{-0.75} = \frac{114^3}{27,000(0.8)^{2.1}\,L^{1.5}} \tag{7}
$$

Substituting (7) and (5) into (3)

$$
\begin{aligned}
95K^{0.3}L^{0.2}R^{-0.75} - 12 &= 0 \\
\frac{95(0.8L)^{0.3}L^{0.2}(114)^3}{27,000(0.8)^{2.1}L^{1.5}} &= 12 \\
\frac{95(0.8)^{0.3}L^{0.3}L^{0.2}(114)^3}{(0.8)^{2.1}L^{1.5}} &= 324,000 \\
\frac{95(114)^3}{(0.8)^{1.8}L} &= 324,000 \\
\frac{95(114)^3}{324,000(0.8)^{1.8}} &= L \\
649.12924 &= L
\end{aligned} \tag{8}
$$

Substituting (8) into (5)

$$
K = 0.8(649.12924) = 519.3034
$$

Substituting (8) into (6)

$$
\begin{aligned}
R^{0.25} &= \frac{30(0.8)^{0.7}(649.12924)^{0.5}}{114} \\
R &= \frac{30^4(0.8)^{2.8}(649.12924)^2}{114^4} = 1,081.882
\end{aligned}
$$

It is assumed that the second-order conditions for a maximum are met when K, L and R take these values.

The maximum profit level will therefore be (taking quantities to 1dp)

$$\begin{aligned}\pi &= \text{TR} - \text{TC} = PQ - (P_K K + P_L L + P_R R)\\ &= 3800(519.3)^{0.3}(649.1)^{0.2}(1{,}081.9)^{0.25} - 30(519.3) - 16(649.1) - 12(1{,}081.9)\\ &= 51{,}929.98 - 15{,}579 - 10{,}385.6 - 12{,}982.8\\ &= £12{,}982.58\end{aligned}$$

QUESTIONS 11.4

(Ensure that you check that second-order conditions are satisfied for these unconstrained optimization problems.)

1. A firm produces two products which are sold in separate markets with the demand schedules

   $$p_1 = 210 - 0.4q_1^2 \qquad p_2 = 491 - 6q_2$$

   Production costs are related and the firm's total cost schedule is

   $$\text{TC} = 32 + 0.8q_1^2 + 0.7q_2^2 + 0.1q_1q_2$$

   How much should the firm sell in each market to maximize total profits?

2. A company produces two competing products whose demand schedules are

   $$q_1 = 219 - 1.8p_1 + 0.5p_2 \text{ and } \qquad q_2 = 303 - 2.1p_2 + 0.8p_1$$

   What price should it charge in the two markets to maximize total sales revenue?

3. A price-discriminating monopoly sells in two separable markets with demand functions

   $$q_1 = 120 - 6p_1 \qquad \text{and } q_2 = 110 - 8p_2$$

   and faces the total cost schedule TC = 4,200 + $0.3q^2$, where $q = q_1 + q_2$. What should it sell in each market to maximize total profit? (*Note that negative quantities are not allowed.*)

4. A monopoly sells its output in two separable markets with the demand schedules

$$p_1 = 20 - \frac{q_1}{6} \qquad \text{and } p_2 = 13.75 - \frac{q_2}{8}$$

If it faces the total cost schedule TC = 74 + 2.26q + 0.01q^2 where $q = q_1 + q_2$, what is the maximum profit it can make?

5. A multiplant monopoly operates two plants whose cost schedules are

$$\text{TC}_1 = 2.4 + 0.015q_1^2 \qquad \text{and TC}_2 = 3.5 + 0.012q_2^2$$

and sells its total output in a market where $p = 32 - 0.02q$.
How much should it produce in each plant to maximize total profits?

6. A firm operates two plants with the total cost schedules

$$\text{TC}_1 = 62 + 0.00018q_1^3 \qquad \text{and TC}_2 = 48 + 0.00014q_2^3$$

and faces the demand schedule $p = 2{,}360 - 0.15q$. To maximize profits, how much should it produce in each plant?

7. A firm faces the production function $Q = 0.8K^{0.4}L^{0.3}$. It sells its output at a fixed price of £450 a unit and can buy the inputs K and L at £15 per unit and £8 per unit, respectively. What input mix will maximize profit?
8. A firm selling in a perfectly competitive market where the ruling price is £40 can buy inputs K and L at prices per unit of £20 and £6, respectively. If it operates with the production function $Q = 21K^{0.4}L^{0.2}$, what is the maximum profit it can make?
9. A firm faces the production function $Q = 2.4K^{0.6}L^{0.2}$, where K costs £25 per unit and L costs £9 per unit, and sells Q at a fixed price of £82 per unit. Explain why it cannot make a profit of more than £20,000, no matter how efficiently it plans its input mix.
10. A firm can buy inputs K and L at £32 and £20 per unit respectively and sell its output at a fixed price of £5 per unit. How should it organize production to ensure maximum profit if it faces the production function $Q = 82K^{0.5}L^{0.3}$?

11.5 TOTAL DIFFERENTIALS AND TOTAL DERIVATIVES

In Chapter 9, when the concept of differentiation was introduced, you learned that the derivative dy/dx measured the rate of change of y with respect to x for infinitesimally small changes in x and y. For any non-linear function $y = \text{f}(x)$, the value of dy/dx will alter if x and y alter. It is therefore not possible to predict the effect of a given increase in x on y with complete accuracy in such cases. However, for a very small change in x (Δx), we can say that it will be approximately true that Δy, the resulting change in y, will be

$$\Delta y = \frac{dy}{dx}\Delta x$$

The closer the function $y = f(x)$ is to a straight line, the more accurate will be the prediction, as the following example demonstrates.

Example 11.22

For the following functions, assume that the value of x increases from 10 to 11. Predict the effect on y using the derivative dy/dx evaluated at the first value of x and check the answer against the new value of the function.

(i) $y = 2x$
(ii) $y = 2x^2$
(iii) $y = 2x^3$

Solution

In all cases the change in x is $\Delta x = 11 - 10 = 1$.

(i) $y = 2x, \quad \frac{dy}{dx} = 2$

Therefore, predicted change in y is

$$\Delta y = \frac{dy}{dx}\Delta x = 1 \times 2 = 2$$

The actual values are $y = 2(10) = 20$ when $x = 10$
$y = 2(11) = 22$ when $x = 11$
The actual change is $22 - 20 = 2$ (accuracy of prediction 100%)

(ii) $y = 2x^2, \quad \frac{dy}{dx} = 4x = 4(10) = 40$

Therefore, predicted change in y is

$$\Delta y = \frac{dy}{dx}\Delta x = 40 \times 1 = 40$$

The actual values are $y = 2(10)^2 = 200$ when $x = 10$
$y = 2(11)^2 = 242$ when $x = 11$
The actual change is $242 - 200 = 42$ (accuracy of prediction 95%)

(iii) $y = 2x^3, \quad \frac{dy}{dx} = 6x^2 = 6(10)^2 = 600$

Therefore, predicted change in y is

$$\Delta y = \frac{dy}{dx}\Delta x = 600 \times 1 = 600$$

$$\text{The actual values are } y = 2(10)^3 = 2{,}000 \text{ when } x = 10$$
$$y = 2(11)^3 = 2{,}662 \text{ when } x = 11$$
$$\text{The actual change is } 2{,}662 - 2{,}000 = 662 \text{ (accuracy of prediction 91\%)}$$

This method of predicting approximate actual changes in a variable can itself be useful for practical purposes. However, in economic theory this mathematical method is taken a stage further and helps yield some important results.

Total differentials

If the changes in variables x and y become infinitesimally small, then even for non-linear functions

$$\Delta y = \frac{\mathrm{d}y}{\mathrm{d}x}\Delta x$$

These infinitesimally small changes in x and y are known as 'differentials'. When y is a function of more than one independent variable and there are infinitesimally small changes in all variables, then the total change will be the sum of the impacts of each independent variable. For example, for the function $y = \mathrm{f}(x, z)$, the total change in y resulting from infinitesimally small changes in the independent variables x and y will be

$$\Delta y = \frac{\partial y}{\partial x}\Delta x + \frac{\partial y}{\partial z}\Delta z$$

This is known as the 'total differential' as it shows the total effect on y of changes in all independent variables.

It is usual to write dy, dx, dz, etc., to represent infinitesimally small changes instead of Δy, Δx, Δz, which usually represent small, but finite, changes. Thus,

$$\mathrm{d}y = \frac{\partial y}{\partial x}\mathrm{d}x + \frac{\partial y}{\partial z}\mathrm{d}z$$

Example 11.23

What is the total differential of $y = 6x^2 + 8z^2 - 0.3xz$?

Solution

The total differential is

$$\mathrm{d}y = \frac{\partial y}{\partial x}\mathrm{d}x + \frac{\partial y}{\partial z}\mathrm{d}z = (12x - 0.3z)\mathrm{d}x + (16z - 0.3x)\mathrm{d}z$$

We can now demonstrate some examples of how the concept of a total differential can be used in economics.

In production theory, the slope of an isoquant represents the marginal rate of technical substitution (MRTS) between two inputs. The use of the total differential can help demonstrate that the MRTS will equal the ratio of the marginal products of the two inputs.

In introductory economics texts the MRTS of K for L (usually written as MRTS_{KL}) is defined as the amount of K that would be needed to compensate for the loss of one unit of L so that the production level remains unchanged. This is only an approximate measure though and more accuracy can be obtained when the MRTS_{KL} is defined at a point on an isoquant. For infinitesimally small changes in K and L the MRTS_{KL} measures the rate at which K needs to be substituted for L to keep output unchanged, i.e. it is equal to the negative of the slope of the isoquant at the point corresponding to the given values of K and L, when K is measured on the vertical axis and L on the horizontal axis.

For any given output level, K is effectively a function of L (and vice versa) and so, moving along an isoquant,

$$\text{MRTS}_{\text{KL}} = -\frac{\text{d}K}{\text{d}L} \tag{1}$$

For the production function $Q = \text{f}(K, L)$, the total differential is

$$\text{d}Q = \frac{\partial Q}{\partial K}\text{d}K + \frac{\partial Q}{\partial L}\text{d}L$$

If we are looking at a movement along the same isoquant then output is unchanged and so $\text{d}Q$ is zero and thus

$$\frac{\partial Q}{\partial K}\text{d}K + \frac{\partial Q}{\partial L}\text{d}L = 0$$

$$\frac{\partial Q}{\partial K}\text{d}K = -\frac{\partial Q}{\partial L}\text{d}L \tag{2}$$

$$-\frac{\text{d}K}{\text{d}L} = \frac{\dfrac{\partial Q}{\partial L}}{\dfrac{\partial Q}{\partial K}}$$

We already know that $\partial Q/\partial L$ and $\partial Q/\partial K$ represent the marginal products of K and L. Therefore, from (1) and (2),

$$\text{MRTS}_{\text{KL}} = \frac{\text{MP}_{\text{L}}}{\text{MP}_{\text{K}}}$$

Euler's theorem

Another use of the total differential is to prove *Euler's theorem* and demonstrate the conditions for the 'exhaustion of the total product'. This relates to the marginal productivity theory of factor pricing and the normative idea of what might be considered a 'fair wage', which was debated for many years by political economists.

Consider a firm that uses several different inputs. Each will contribute a different amount to total production. One suggestion for what might be considered a 'fair wage' was that each input, including labour, should be paid the 'value of its marginal product' (VMP). This is defined, for any input i, as marginal product (MP_i) multiplied by the price that the finished good is sold at (P_Q), i.e.

$$VMP_i = P_Q MP_i$$

Any such suggestion is, of course, a normative concept and the value judgements on which it is based can be questioned. However, what we are concerned with here is whether it is even *possible* to pay each input the value of its marginal product. If it is not possible, then it would not be a practical idea to set this as an objective even if it seemed a 'fair' principle.

Before looking at Euler's theorem we can illustrate how the conditions for product exhaustion can be derived for a Cobb-Douglas production function with two inputs. This example also shows how the product price is irrelevant to the product exhaustion question and it is the properties of the production function that matter.

Assume that a firm sells its output Q at a given price P_Q and that $Q = AK^{\alpha} L^{\beta}$ where A, α and β are constants satisfying the usual conditions. If each input was paid a price equal to the value of its marginal product then the prices of the two inputs K and L would be

$$P_K = VMP_K = P_Q \times MP_K = P_Q \frac{\partial Q}{\partial K} \quad \text{and} \quad P_L = VMP_L = P_Q \times MP_L = P_Q \frac{\partial Q}{\partial L}$$

The total expenditure on inputs would therefore be

$$\begin{aligned} TC = KP_K + LP_L &= K\left(P_Q \frac{\partial Q}{\partial K}\right) + L\left(P_Q \frac{\partial Q}{\partial L}\right) \\ &= P_Q\left(K\frac{\partial Q}{\partial K} + L\frac{\partial Q}{\partial L}\right) \end{aligned} \quad (1)$$

Total revenue from the sale of the firm's output will be

$$TR = P_Q Q$$

Total expenditure on inputs (which are paid the value of their marginal product) will equal total revenue when TR = TC. Therefore,

$$P_Q Q = P_Q\left(K\frac{\partial Q}{\partial K} + L\frac{\partial Q}{\partial L}\right)$$

Cancelling P_Q, this gives

$$Q = K\frac{\partial Q}{\partial K} + L\frac{\partial Q}{\partial L} \quad (2)$$

Thus, the **conditions of product exhaustion** are based on the physical properties of the production function. If (2) holds then the product is exhausted. If it does not hold then there will be either not enough revenue or surplus revenue.

For the Cobb-Douglas production function $Q = AK^{\alpha} L^{\beta}$ we know that

$$\frac{\partial Q}{\partial K} = \alpha AK^{\alpha-1}L^{\beta} \quad \text{and} \quad \frac{\partial Q}{\partial L} = \beta AK^{\alpha}L^{\beta-1}$$

Substituting these values into (2), this gives

$$\begin{aligned} Q &= K\left(\alpha AK^{\alpha-1}L^{\beta}\right)+L\left(\beta AK^{\alpha}L^{\beta-1}\right) \\ &= \alpha AK^{\alpha}L^{\beta}+\beta AK^{\alpha}L^{\beta} \\ &= \alpha Q+\beta Q \\ Q &= Q(\alpha+\beta) \end{aligned}$$

The condition required for (3) to hold is that $\alpha + \beta = 1$. This means that product exhaustion occurs for a Cobb-Douglas production function when there are constant returns to scale.

We can also see from (3) and (1) that:

(i) when there are decreasing returns to scale and $\alpha + \beta < 1$, then

$TC = P_Q\,(\alpha + \beta)Q < P_Q\,Q = TR$ and so there will be a surplus left over if all inputs are paid their VMP, and

(ii) when there are increasing returns to scale and $\alpha + \beta > 1$, then

$TC = P_Q(\alpha + \beta)Q > P_Q\,Q = TR$ and so there will not be enough revenue to pay each input its VMP.

Euler's theorem also applies to the case of a general production function

$$Q = \mathrm{f}\left(x_1, x_2, \ldots, x_n\right)$$

The previous example showed that the price will always cancel in the TR and TC formulae and what we are interested in is whether

$$Q = x_1\frac{\partial Q}{\partial x_1}+x_2\frac{\partial Q}{\partial x_2}+\ldots+x_n\frac{\partial Q}{\partial x_n} \tag{1}$$

Using the notation

$$\mathrm{f}_1=\frac{\partial Q}{\partial x_1},\ \mathrm{f}_2=\frac{\partial Q}{\partial x_2},\ \ldots,\ \text{etc.}$$

the total differential of this production function will be

$$\mathrm{d}Q=\mathrm{f}_1\mathrm{d}x_1+\mathrm{f}_2\mathrm{d}x_2+\ldots+\mathrm{f}_n\mathrm{d}x_n \tag{2}$$

Assume that all inputs are increased by the same proportion λ. Thus,

$$\frac{\mathrm{d}x_i}{x_i}=\lambda \quad \text{for all } i \tag{3}$$

$$\text{and so} \quad \mathrm{d}x_i=\lambda$$

Substituting (3) into (2) gives

$$\begin{aligned} dQ &= f_1\lambda x_1 + f_2\lambda x_2 + \ldots + f_n\lambda x_n \\ &= \lambda\left(f_1 dx_1 + f_2 dx_2 + \ldots + f_n dx_n\right) \\ \frac{dQ}{\lambda} &= f_1 x_1 + f_2 x_2 + \ldots + f_n x_n \end{aligned} \tag{4}$$

Multiplying top and bottom of the left-hand side of (4) by Q gives

$$\left(\frac{1}{\lambda}\frac{dQ}{Q}\right)Q = f_1 x_1 + f_2 x_2 + \ldots + f_n x_n$$

Thus, product exhaustion will only hold if

$$\frac{1}{\lambda}\frac{dQ}{Q} = 1$$

If this result does hold it means that output increases by the same proportion as the inputs, and

$$\frac{dQ}{Q} = \lambda$$

i.e. there are constant returns to scale.

If there are decreasing returns to scale, output increases by a smaller proportion than the inputs. Therefore,

$$\frac{dQ}{Q} < \lambda$$

and so

$$\frac{1}{\lambda}\frac{dQ}{Q} < 1$$

This means that

$$f_1 x_1 + f_2 x_2 + \ldots + f_n x_n Q$$

so that if each input is paid the value of its marginal product there will be some surplus left over.

Similarly, if there are increasing returns to scale then

$$\frac{dQ}{Q} > \lambda$$

Therefore,

$$\frac{1}{\lambda}\frac{dQ}{Q} > 1$$

and

$$f_1x_1 + f_2x_2 + \ldots + f_nx_n > Q$$

which means that the total cost of paying each input the value of its marginal product will sum to more than the total revenue earned, i.e. it will not be possible.

To sum up, Euler's theorem proves that if each input is paid the value of its marginal product the total cost of the inputs will

(i) equal total revenue if there are constant returns to scale;
(ii) be less than total revenue if there are decreasing returns to scale;
(iii) be greater than total revenue if there are increasing returns to scale.

Example 11.24

Is it possible for a firm to pay each input the value of its marginal product if it operates with the production function $Q = 14K^{0.6}L^{0.8}$?

Solution

If each input is paid its VMP then the price of input K will be

$$P_K = \text{VMP}_K = P\text{MP}_K = P\frac{\partial Q}{\partial K} = P\left(8.4K^{-0.4}L^{0.8}\right)$$

where P is the price of the final product. For input L,

$$P_L = \text{VMP}_L = P\text{MP}_L = P\frac{\partial Q}{\partial L} = P\left(11.2K^{0.6}L^{-0.2}\right)$$

The total cost of inputs will therefore be

$$\begin{aligned}\text{TC} = P_K K + P_L L &= P\left(8.4K^{-0.4}L^{0.8}\right)K + P\left(11.2K^{0.6}L^{-0.2}\right)L \\ &= P\left(8.4K^{0.6}L^{0.8}\right) + P\left(11.2K^{0.6}L^{0.8}\right) \\ &= P\left(8.4K^{0.6}L^{0.8} + 11.2K^{0.6}L^{0.8}\right) \\ &= P\left(19.6K^{0.6}L^{0.8}\right)\end{aligned}$$

The total revenue from selling the product will be

$$\text{TR} = PQ = P\left(14K^{0.6}L^{0.8}\right)$$

Therefore,

$$\frac{\text{TC}}{\text{TR}} = \frac{P\left(19.6K^{0.6}L^{0.8}\right)}{P\left(14K^{0.6}L^{0.8}\right)} = \frac{19.6}{14} = 1.4$$

Thus, the total revenue is not enough to pay each input the value of its marginal product. This checks out with the predictions of Euler's theorem, given that there are increasing returns to scale for this production function.

QUESTIONS 11.5

1. Derive the total differentials of the following production functions.
   (a) $Q = 20K^{0.6}L^{0.4}$
   (b) $Q = 48K^{0.3}L^{0.2}R^{0.4}$
   (c) $Q = 6K^{.8} + 5L^{0.7} + 0.8K^2L^2$
2. If each input is paid the value of its marginal product, will this exhaust a firm's total revenue if the relevant production function is
   (a) $Q = 4K + 1.5L$?
   (b) $Q = 8K^{0.4}L^{0.3}$?
   (c) $Q = 3.5K^{0.25}L^{0.35}R^{0.3}$?
3. If $y = 40x^{0.4}z^{0.3}$ and $x = 5z^{0.25}$, find the total effect of a change in z on y.
4. A consumer spends all her income on the two goods A and B. The quantity of good A bought is determined by the demand function $Q_A = f(P_A, P_B, M)$ where P_A and P_B are the prices of the two goods and M is real income.
   A change in the price of A will also affect real income M via the function $M = g(P_A, P_B, £M)$ where £M is money income.
   Derive an expression for the total effect of a change in P_A on Q_A.

Total derivatives

In partial differentiation it is assumed that one variable changes while all other independent variables are held constant. However, in some instances there may be a connection between the independent variables and so this *ceteris paribus* assumption will not apply. For example, in a production function the amount of one input used may affect the amount of another input that can be used with it. From the total differential of a function, we can derive a total derivative which can cope with this additional effect.

Assume $y = f(x, z)$ and also that $x = g(z)$.

Thus, any change in z will affect y:

(a) directly via the function $f(x, z)$, and
(b) indirectly by changing x via the function $g(z)$, which in turn will affect y via the function $f(x, z)$.

The total differential of $y = f(x, z)$ is

$$dy = \frac{\partial y}{\partial x}dx + \frac{\partial y}{\partial z}dz$$

Dividing through by dz gives

$$\frac{\mathrm{d}y}{\mathrm{d}z} = \frac{\partial y}{\partial x}\frac{\mathrm{d}x}{\mathrm{d}z} + \frac{\partial y}{\partial z}$$

The first term shows the indirect effect of z, via its impact on x, and the second term shows the direct effect.

Example 11.25

If $Q = 25K^{0.4}L^{0.5}$ and $K = 0.8L^2$ what is the total effect of a change in L on Q? Identify the direct and indirect effects.

Solution

The total differential is

$$\mathrm{d}Q = \frac{\partial Q}{\partial K}\mathrm{d}K + \frac{\partial Q}{\partial L}\mathrm{d}L$$

The total derivative with respect to L will be

$$\frac{\mathrm{d}Q}{\mathrm{d}L} = \frac{\partial Q}{\partial K}\frac{\mathrm{d}K}{\mathrm{d}L} + \frac{\partial Q}{\partial L} \tag{1}$$

From the functions given in the question we can derive

$$\frac{\partial Q}{\partial K} = 10K^{-0.6}L^{0.5}, \qquad \frac{\partial Q}{\partial L} = 12.5K^{0.4}L^{-0.5} \text{ and } \qquad \frac{\mathrm{d}K}{\mathrm{d}L} = 1.6L$$

Substituting these derivatives into (1), we get

$$\begin{aligned}\frac{\mathrm{d}Q}{\mathrm{d}L} &= \left(10K^{-0.6}L^{0.5}\right)1.6L + 12.5K^{0.4}L^{-0.5}\\ &= 16K^{-0.6}L^{1.5} + 12.5K^{0.4}L^{-0.5}\end{aligned}$$

The first term shows the indirect effect of changes in L on Q and the second term shows the direct effect.

12 Constrained optimization

Learning objectives

After completing this chapter students should be able to:

- solve constrained optimization problems by the substitution method
- use the Lagrange method to set up and solve constrained maximization and constrained minimization problems
- apply the Lagrange method to resource allocation problems in economics.

12.1 CONSTRAINED OPTIMIZATION AND RESOURCE ALLOCATION

Chapters 10 and 11 dealt with the optimization of functions without any constraints imposed. However, in economics we often come across resource allocation problems that involve the optimization of some variable subject to certain limitations. For example, a firm may try to maximize output subject to a budget constraint for expenditure on inputs, or it may wish to minimize costs subject to a specified output being produced. We have already seen in Chapter 6 how constrained optimization problems with linear constraints and objective functions can be tackled using linear programming. This chapter now explains how problems involving the constrained optimization of non-linear functions can be tackled, using partial differentiation.

We shall consider two methods:

(i) constrained optimization by substitution, and
(ii) the Lagrange multiplier method.

The Lagrange multiplier method can be used for most types of constrained optimization problems. The substitution method is mainly suitable for problems where a function with only two variables is maximized or minimized subject to one constraint. We shall consider this simpler substitution method first.

DOI: 10.4324/9781003360827-12

12.2 CONSTRAINED OPTIMIZATION BY SUBSTITUTION

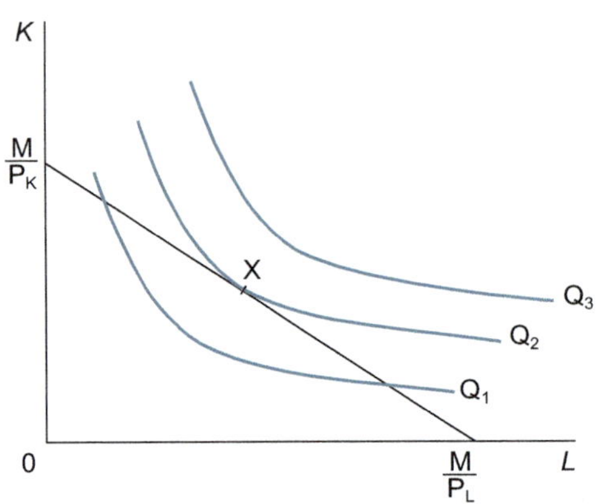

Figure 12.1 Finding the highest feasible output

Consider the example of a firm that wishes to maximize output $Q = f(K, L)$, with a fixed budget M for purchasing inputs K and L at set prices P_K and P_L. This problem is illustrated in Figure 12.1. The firm needs to find the combination of K and L that will allow it to reach the optimum point X which is on the highest possible isoquant within the budget constraint with intercepts $\frac{M}{P_K}$ and $\frac{M}{P_L}$.

To determine a solution for this type of economic resource allocation problem we have to reformulate it as a mathematical constrained optimization problem. The following examples show ways in which this can be done.

Example 12.1

A firm faces the production function $Q = 12K^{0.4}L^{0.4}$ and can buy the inputs K and L at prices per unit of £40 and £5, respectively. If it has a budget of £800 what combination of K and L should it use in order to produce the maximum possible output?

Solution

The problem is to maximize the function $Q = 12K^{0.4}L^{0.4}$ subject to the budget constraint

$$40K + 5L = 800 \qquad (1)$$

(In all problems in this chapter, it is assumed that each constraint 'bites', i.e. the entire budget is used in this example.)

The theory of the firm tells us that a firm is optimally allocating a fixed budget if the last £1 spent on each input adds the same amount to output, i.e. marginal product over price should be equal for all inputs. This optimization condition can be written as

$$\frac{MP_K}{P_K} = \frac{MP_L}{P_L} \qquad (2)$$

The marginal products can be determined by partial differentiation:

$$MP_K = \frac{\partial Q}{\partial K} = 4.8K^{-0.6}L^{0.4} \tag{3}$$

$$MP_L = \frac{\partial Q}{\partial L} = 4.8K^{0.4}L^{-0.6} \tag{4}$$

Substituting (3) and (4), and the given prices for P_K and P_L into (2)

$$\frac{4.8K^{-0.6}L^{0.4}}{40} = \frac{4.8K^{0.4}L^{-0.6}}{5}$$

Dividing both sides by 4.8 and multiplying by 40 gives

$$K^{-0.6}L^{0.4} = 8K^{0.4}L^{-0.6}$$

Multiplying both sides by $K^{0.6}L^{0.6}$ gives

$$L = 8K \tag{5}$$

Substituting (5) for L into the budget constraint (1) gives

$$\begin{aligned} 40K + 5(8K) &= 800 \\ 40K + 40K &= 800 \\ 80K &= 800 \end{aligned}$$

Thus, the optimal value of K is

$$K = 10$$

and, from (5), the optimal value of L is

$$L = 80$$

Note that although this method allows us to derive optimum values of K and L that satisfy condition (2), it does not provide a check on whether this is a unique solution, i.e. there is no second-order condition check. However, it may be assumed that in all the problems in this section the objective function is maximized (or minimized depending on the question) when the basic economic rules for an optimum are satisfied.

This method is not the only way of tackling this problem by substitution. An alternative approach, explained next, is to encapsulate the constraint within the function to be maximized, and then maximize this new objective function.

Example 12.1 (reworked)

Solution

From the budget constraint

$$\begin{aligned} 40K + 5L &= 800 \\ 5L &= 800 - 40K \qquad (1) \\ L &= 160 - 8K \qquad (2) \end{aligned}$$

Substituting (2) into the objective function $Q = 12K^{0.4}L^{0.4}$ gives

$$Q = 12K^{0.4}(160 - 8K)^{0.4} \qquad (3)$$

We are now faced with the unconstrained optimization problem of finding the value of K that maximizes the function (3) which has the budget constraint (1) 'built in' to it by substitution. This requires us to set $dQ/dK = 0$. However, it is not straightforward to differentiate the function in (3), and we must wait until further topics in calculus have been covered before proceeding with this solution (see Chapter 13, Example 13.9).

To make sure that you understand the basic substitution method, we shall use it to tackle another constrained maximization problem.

Example 12.2

A firm faces the production function $Q = 20K^{0.4}L^{0.6}$. It can buy inputs K and L for £400 a unit and £200 a unit, respectively. What combination of L and K should be used to maximize output if its input budget is constrained to £6,000?

Solution

$$\text{MP}_L = \frac{\partial Q}{\partial L} = 12K^{0.4}L^{-0.4} \qquad \text{and } \text{MP}_K = \frac{\partial Q}{\partial K} = 8K^{-0.6}L^{0.6}$$

Optimal input mix requires

$$\frac{\text{MP}_\text{L}}{P_\text{L}} = \frac{\text{MP}_\text{K}}{P_\text{K}}$$

Therefore,

$$\frac{12K^{0.4}L^{-0.4}}{200} = \frac{8K^{-0.6}L^{0.6}}{400}$$

Cross-multiplying gives

$$\begin{aligned} 4{,}800K &= 1{,}600L \\ 3K &= L \end{aligned}$$

Substituting this result into the budget constraint

$$200L + 400K = 6{,}000$$

gives

$$200(3K) + 400K = 6{,}000$$
$$600K + 400K = 6{,}000$$
$$1{,}000K = 6{,}000$$
$$K = 6$$

Therefore

$$L = 3K = 18$$

The examples of constrained optimization considered so far have only involved output maximization when a firm faces a Cobb-Douglas production function, but the same technique can also be applied to other forms of production functions.

Example 12.3

A firm faces the production function

$$Q = 120L + 200K - L^2 - 2K^2$$

for all positive values of Q. It can buy L at £5 a unit and K at £8 a unit and has a budget of £70. What is the maximum output it can produce?

Solution

$$\text{MP}_\text{L} = \frac{\partial Q}{\partial L} = 120 - 2L$$

$$\text{MP}_\text{K} = \frac{\partial Q}{\partial K} = 200 - 4K$$

For optimal input combination

$$\frac{\text{MP}_\text{L}}{\text{P}_\text{L}} = \frac{\text{MP}_\text{K}}{\text{P}_\text{K}}$$

Therefore, substituting MP_K and MP_L and the given input prices

$$\frac{120 - 2L}{5} = \frac{200 - 4K}{8}$$
$$8(120 - 2L) = 5(200 - 4K)$$
$$960 - 16L = 1{,}000 - 20K$$
$$20K = 40 + 16L$$
$$K = 2 + 0.8L \qquad (1)$$

Substituting (1) into the budget constraint

$$5L + 8K = 70$$

gives

$$\begin{aligned} 5L + 8(2 + 0.8L) &= 70 \\ 5L + 16 + 6.4L &= 70 \\ 11.4L &= 54 \\ L &= 4.74 \quad \text{(to 2 dp)} \end{aligned}$$

Substituting this result into (1)

$$K = 2 + 0.8(4.74) = 5.79$$

Therefore, maximum output is

$$\begin{aligned} Q &= 120L + 200K - L^2 - 2K^2 \\ &= 120(4.74) + 200(5.79) - (4.74)^2 - 2(5.79)^2 \\ &= 1{,}637.28 \end{aligned}$$

This technique can also be applied to consumer theory, where utility is maximized subject to a budget constraint.

Example 12.4

The utility a consumer derives from consuming the two goods A and B can be assumed to be determined by the utility function $U = 40A^{0.25}B^{0.5}$. If A costs £4 a unit and B costs £10 a unit and the consumer's income is £600, what combination of A and B will maximize utility?

Solution

The marginal utility of A is

$$\text{MU}_\text{A} = \frac{\partial U}{\partial A} = 10A^{-0.75}B^{0.5}$$

The marginal utility of B is

$$\text{MU}_\text{B} = \frac{\partial U}{\partial B} = 20A^{0.25}B^{-0.5}$$

Consumer theory tells us that total utility will be maximized when the utility derived from the last pound spent on each good is equal to the utility derived from the last pound spent on any other good. This optimization rule can be expressed as

$$\frac{\text{MU}_\text{A}}{P_\text{A}} = \frac{\text{MU}_\text{B}}{P_\text{B}}$$

Therefore, substituting the previous MU functions and the given prices of £4 and £10, this condition becomes

$$\frac{10A^{-0.75}B^{0.5}}{4} = \frac{20A^{0.25}B^{-0.5}}{10}$$

$$100B = 80A$$

$$B = 0.8A \qquad (1)$$

Substituting (1) for B in the budget constraint

$$4A + 10B = 600$$

gives

$$A + 10(0.8A) = 600$$

$$4A + 8A = 600$$

$$12A = 600$$

$$A = 50$$

Thus from (1)

$$B = 0.8(50) = 40$$

The substitution method can also be used for **constrained minimization** problems. If output is given and a firm is required to minimize the cost of this output, then one variable can be eliminated from the production function before it is substituted into the cost function which is to be minimized.

Example 12.5

A firm operates with the production function $Q = 4K^{0.6}L^{0.4}$ and buys inputs K and L at prices per unit of £40 and £15, respectively. What is the cheapest way of producing 600 units of output?

Solution

The output constraint is

$$600 = 4K^{0.6}L^{0.4}$$

Therefore,

$$\frac{150}{K^{0.6}} = L^{0.4}$$

$$\left(\frac{150}{K^{0.6}}\right)^{2.5} = L$$

$$\frac{275{,}567.6}{K^{1.5}} = L \qquad (1)$$

The total cost of inputs, which is to be minimized, is

$$\mathrm{TC} = 40K + 15L \qquad (2)$$

Substituting (1) into (2) gives

$$\text{TC} = 40K + 15(275{,}567.6)K^{-1.5}$$

Differentiating and setting equal to zero to find a stationary point

$$\frac{\text{dTC}}{\text{d}K} = 40K - 22(275{,}567.6)K^{-2.5} = 0 \tag{3}$$

$$40 = \frac{22.5(275{,}567.6)}{K^{2.5}}$$

$$K^{2.5} = \frac{22.5(275{,}567.6)}{40} = 155{,}006.78$$

$$K = \sqrt[2.5]{155{,}006.78} = 119.16268$$

Substituting this value into (1) gives

$$L = \frac{275{,}567.6}{(119.1628)^{1.5}} = 211.84478$$

This time we can check the second-order condition for minimization. Differentiating (3) again gives

$$\frac{\text{d}^2\text{TC}}{\text{d}K^2} = (2.5)22.5(275{,}567.6)K^{-3.5} > 0 \text{for any } K > 0$$

This confirms that these values minimize TC. We can also check that these values give 600 when substituted back into the production function.

$$Q = 4K^{0.6}L^{0.4} = 4(119.16268)^{0.6}(211.84478)^{0.4} = 600$$

Thus, cost minimization is achieved when K = 119.16 and L = 212.84 (to 2 dp) and so total production costs will be

$$\text{TC} = 40(119.16) + 15(211.84) = £7{,}944$$

QUESTIONS 12.1

1. If a firm has a budget of £378 what combination of K and L will maximize output given the production function $Q = 40K^{0.6}L^{0.3}$ and prices for K and L of £20 per unit and £6 per unit, respectively?
2. A firm faces the production function $Q = 6K^{0.4}L^{0.5}$. If it can buy input K at £32 a unit and input L at £8 a unit, what combination of L and K should it use to maximize production if it is constrained by a fixed budget of £36,000?

3. A consumer spends all her income of £120 on the two goods A and B. Good A costs £10 a unit and good B costs £15. What combination of A and B will she purchase if her utility function is $U = 4A^{0.5}B^{0.5}$?
4. If a firm faces the production function $Q = 4K^{0.5}L^{0.5}$ what is the maximum output it can produce for a budget of £200 if the price of K is £4 per unit and L costs £2 per unit?
5. A firm faces the production function $Q = 2K^{0.2}L^{0.6}$ and can buy L at £240 a unit and K at £4 a unit.
   (a) If it has a budget of £16,000 what combination of K and L should it use to maximize output?
   (b) If it is given a target output of 40 units of Q what combination of K and L should it use to minimize the cost of this output?
6. A firm has a budget of £1,140 and can buy inputs K and L at £3 and £8 respectively a unit. Its output is determined by the production function

$$Q = 6K + 200L - 0.025K^2 - 0.05L^2$$

for positive values of Q. What is the maximum output it can produce?
7. A firm operates with the production function $Q = 30K^{0.4}L^{0.2}$ and buys inputs K and L at £12 per unit and £5 per unit respectively. What is the cheapest way of producing 750 units of output? (Work to nearest whole units of K and L.)

12.3 THE LAGRANGE MULTIPLIER: CONSTRAINED MAXIMIZATION WITH TWO VARIABLES

The best way to explain how to use the Lagrange multiplier method is with an example and so we shall work through the problem in Example 12.1 from the previous section using the Lagrange multiplier method.

The firm is trying to maximize output $Q = 12K^{0.4}L^{0.4}$ subject to the budget constraint $40K + 5L = 800$. The first step is to rearrange the budget constraint so that zero appears on one side of the equality sign:

$$0 = 800 - 40K - 5L \qquad (1)$$

We then write the 'Lagrange equation' or 'Lagrangian' in the form

$$G = (\text{function to be optimized}) + \lambda(\text{constraint})$$

where G is just the value of the Lagrangian function and the Greek letter λ (lambda) is known as the 'Lagrange multiplier'.

Do not worry about where these terms come from or what their actual values are. They are just introduced to help the analysis. Note also that in some other texts a 'curly' L is often used to represent the Lagrange function. This can confuse students because economics problems frequently involve labour, represented by L, as one of the variables in the function to be optimized. This text therefore uses the notation 'G' to avoid this confusion.

In this problem the Lagrange function is thus

$$G = 12K^{0.4}L^{0.4} + \lambda(800 - 40K - 5L) \tag{2}$$

Next, we derive the partial derivatives of G with respect to K, L and λ and set them equal to zero, i.e. find the stationary points of G that satisfy the first-order conditions for a maximum.

$$\frac{\partial G}{\partial K} = 4.8K^{-0.6}L^{0.4} - 40\lambda = 0 \tag{3}$$

$$\frac{\partial G}{\partial L} = 4.8K^{0.4}L^{-0.6} - 5\lambda = 0 \tag{4}$$

$$\frac{\partial G}{\partial \lambda} = 800 - 40K - 5L = 0 \tag{5}$$

You will note that (5) is the same as the budget constraint (1). We now have a set of three linear simultaneous equations in three unknowns to solve for K and L. The Lagrange multiplier λ can be eliminated as, from (3),

$$0.12K^{-0.6}L^{0.4} = \lambda$$

and from (4)

$$0.96K^{0.4}L^{-0.6} = \lambda$$

Therefore,

$$0.12K^{-0.6}L^{0.4} = 0.96K^{0.4}L^{-0.6}$$

Multiplying both sides by $K^{0.6}L^{0.6}$

$$\begin{aligned} 0.12L &= 0.96K \\ L &= 8K \end{aligned} \tag{6}$$

Substituting (6) into (5)

$$\begin{aligned} 800 - 40K - 5(8K) &= 0 \\ 800 &= 80K \\ 10 &= K \end{aligned}$$

Substituting back into (5)

$$800 - 40(10) - 5L = 0$$
$$400 = 5L$$
$$80 = L$$

These are the same values of K and L as those obtained by the substitution method. Thus, the values of K and L that satisfy the first-order conditions for a maximum value of the Lagrangian function G are the values that will maximize output subject to the given budget constraint. We shall just accept this result without going into the proof of why this is so.

Strictly speaking we should now check the second-order conditions in the previous problem to be sure that we actually have a maximum rather than a minimum. These, however, are rather complex, involving an examination of the function at and near the stationary points found, and are discussed in the next section. For the time being you can assume that once the stationary points of a Lagrangian function have been found the second-order conditions for a maximum will automatically be met. Some more examples are worked through so that you can become familiar with this method.

Example 12.6

A firm can buy two inputs K and L at £18 per unit and £8 per unit, respectively, and faces the production function $Q = 24K^{0.6}L^{0.3}$. What is the maximum output it can produce for a budget of £50,000? (Work to nearest whole units of K, L and Q.)

Solution

The budget constraint is $\quad 50{,}000 - 18K - 8L = 0$

and the function to be maximized is $\quad Q = 24K^{0.6}L^{0.3}$

The Lagrangian for this problem is therefore

$$G = 24K^{0.6}L^{0.3} + \lambda(50{,}000 - 18K - 8L)$$

Partially differentiating to find the stationary points of G gives

$$\frac{\partial G}{\partial K} = 14.4K^{-0.4}L^{0.3} - 18\lambda = 0$$

$$\frac{14.4L^{0.3}}{18K^{0.4}} = \lambda \qquad (1)$$

$$\frac{\partial G}{\partial L} = 7.2K^{0.6}L^{-0.7} - 8\lambda = 0$$

$$\frac{7.2K^{0.6}}{8L^{0.7}} = \lambda \qquad (2)$$

$$\frac{\partial G}{\partial \lambda} = 50,000 - 18K - 8L = 0 \tag{3}$$

Setting (1) equal to (2) to eliminate λ

$$\frac{14.4L^{0.3}}{18K^{0.4}} = \frac{7.2K^{0.6}}{8L^{0.7}}$$

$$115.2L = 129.6K$$

$$L = 1.125K \tag{4}$$

Substituting (4) into (3)

$$50,000 - 18K - 8(1.125K) = 0$$

$$50,000 - 18K - 9K = 0$$

$$50,000 = 27K$$

$$1,851.8519 = K \tag{5}$$

Substituting (5) into (4)

$$L = 1.125(1,851.8519) = 2,083.3334$$

Thus, to the nearest whole unit, optimum values of K and L are 1,852 and 2,083 respectively.

We can check that when these whole values of K and L are used the total cost will be

$$\text{TC} = 18K + 8L = 18(1,852) + 8(2,083) = 33,336 + 16,664 = £50,000$$

and so the budget constraint is satisfied. The actual maximum output level will be

$$\text{Q} = 24K^{0.6}L^{0.3} = 24(1,852)^{0.6}(2,083)^{0.3} = 21,697 \text{ units}$$

Although the same mathematical method can be used for various economic applications, you still need to use your knowledge of economics to set up the mathematical problem in the first place. The following example demonstrates another application of the Lagrange method.

Example 12.7

A consumer has the utility function $U = 40A^{0.5}B^{0.5}$. The prices of the two goods A and B are initially £20 and £5 per unit, respectively, and the consumer's income is £600. The price of A then falls to £10. Work out the income and substitution effects of this price change on the amount of A consumed using Hicks's method and say whether A and B are normal or inferior goods.

Solution

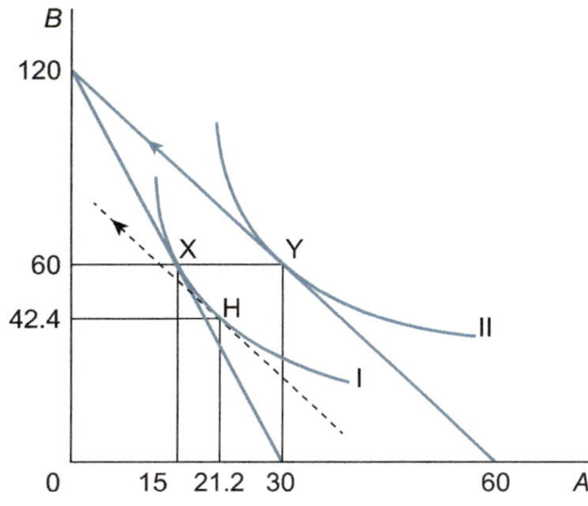

Figure 12.2 Income and substitution effects in consumer choice

To help solve this problem the relevant budget schedules and indifference curves are illustrated in Figure 12.2, although the indifference curves are not accurately drawn to scale. The original optimum is at X. The price fall for A causes the budget line to become flatter and swing round, giving a new equilibrium at Y.

Hicks's method for splitting the total change in A into its income and substitution effects requires one to draw a 'ghost' or 'decomposition' budget line parallel to the new budget line (reflecting the new relative prices) but tangential to the original indifference curve. This is shown by the broken line tangential to indifference curve I at H. From X to H is the substitution effect and from H to Y is the income effect of the price change. This problem requires us to find the corresponding values of A and B for the three tangency points X, Y and H and then to comment on the direction of these changes.

The original equilibrium is the combination of A and B that maximizes the utility function $U = 40A^{0.5}B^{0.5}$ subject to the budget constraint $600 = 20A + 5B$. These values of A and B can be found by deriving the stationary points of the Lagrange function

$$G = 40A^{0.5}B^{0.5} + \lambda(600 - 20A - 5B)$$

Thus,

$$\frac{\partial G}{\partial A} = 20A^{-0.5}B^{0.5} - 20\lambda = 0 \quad \text{giving} \quad A^{-0.5}B^{0.5} = \lambda \tag{1}$$

$$\frac{\partial G}{\partial B} = 20A^{0.5}B^{-0.5} - 5\lambda = 0 \quad \text{giving} \quad 4A^{0.5}B^{-0.5} = \lambda \tag{2}$$

$$\frac{\partial G}{\partial \lambda} = 600 - 20A - 5B = 0 \tag{3}$$

Setting (1) equal to (2)

$$\begin{aligned} A^{-0.5}B^{0.5} &= 4A^{0.5}B^{-0.5} \\ B &= 4A \end{aligned} \tag{4}$$

Substituting (4) into (3)

$$\begin{aligned} 600 - 20A - 5(4A) &= 0 \\ 600 &= 40A \\ 15 &= A \end{aligned}$$

Substituting this value into (4)

$$B = 4(15) = 60$$

Thus, A = 15 and B = 60 at the original equilibrium at X.

When the price of A falls to 10, the budget constraint becomes

$$600 = 10A + 5B$$

and so the new Lagrange function is

$$G = 40A^{0.5}B^{0.5} + \lambda(600 - 10A - 5B)$$

New stationary points will be where

$$\frac{\partial G}{\partial A} = 20A^{-0.5}B^{0.5} - 10\lambda = 0 \quad \text{giving} \quad 2A^{-0.5}B^{0.5} = \lambda \qquad (5)$$

$$\frac{\partial G}{\partial B} = 20A^{0.5}B^{-0.5} - 5\lambda = 0 \quad \text{giving} \quad 4A^{0.5}B^{-0.5} = \lambda \qquad (6)$$

$$\frac{\partial G}{\partial \lambda} = 600 - 10A - 5B = 0 \qquad (7)$$

Setting (5) equal to (6)

$$\begin{aligned} 2A^{-0.5}B^{0.5} &= 4A^{0.5}B^{-0.5} \\ B &= 2A \end{aligned} \qquad (8)$$

Substituting (8) into (7)

$$\begin{aligned} 600 - 10A - 5(2A) &= 0 \\ 600 &= 20A \\ 30 &= A \end{aligned}$$

Substituting this value into (8) gives

$$B = 2(30) = 60$$

Thus, the total effect of the price change is to increase consumption of A from 15 to 30 units and leave consumption of B unchanged at 60.

There are several ways of finding the values of A and B that correspond to point H. We know that H is on the same indifference curve as point X, and therefore the utility function will take the same value at both points. We can find the value of utility at X where A = 15 and B = 60. This will be

$$U = 40A^{0.5}B^{0.5} = 40(15)^{0.5}(60)^{0.5} = 40(900)^{0.5} = 40(30) = 1{,}200$$

Thus, at any point on the indifference curve I

$$40A^{0.5}B^{0.5} = 1{,}200$$
$$B^{0.5} = 30A^{-0.5}$$
$$B = 900A^{-1} \quad (9)$$

The slope of indifference curve I will therefore be

$$\frac{dB}{dA} = -900A^{-2} \quad (10)$$

At point H, the indifference curve I is tangential to the new budget line whose slope will be

$$\frac{-P_A}{P_B} = \frac{-10}{5} = -2 \quad (11)$$

Therefore, from (10) and (11)

$$-900A^{-2} = -2$$
$$450 = A^2$$
$$21.2132 = A$$

Substituting this value into (9)

$$B = 900(21.2132)^{-1} = 42.4264$$

Thus, the substitution effect of A's price fall, from X to H, increases consumption of A from 15 to 21.2 units and decreases consumption of B from 60 to 42.4 units. This effect is negative (i.e. quantity rises when price falls) in line with standard consumer theory.

The income effect, from H to Y, increases consumption of A from 21.2 to 30 units and also increases consumption of B from 42.4 back to its original 60 unit level. As both income effects are positive, both A and B must be normal goods.

QUESTION 12.2

Use the Lagrange method to answer questions 1, 2, 3, 4, 5(a) and 6 from Questions 12.1.

12.4 THE LAGRANGE MULTIPLIER: SECOND-ORDER CONDITIONS

Inasmuch as it involves setting the first derivatives of the objective function equal to zero, the Lagrange method of solving constrained optimization problems is similar to the method of solving unconstrained optimization problems involving functions of several variables that was explained in Chapter 11. However, one cannot simply apply the same set of second-order conditions to check for a maximum or minimum because of the special role that the Lagrange multiplier takes. The mathematics required to prove why this is so, and to explain what additional second-order conditions are necessary for a Lagrangian function to be a maximum or minimum, becomes rather complex. We shall therefore just look at an intuitive explanation of what these conditions involve here. The use of matrix algebra to check second-order conditions in constrained optimization problems will then be explained later, in Chapter 16.

First, consider the conditions for a maximum. If we assume that a function has two independent variables then both the function to be maximized, f(A, B), and the constraint could take on several possible forms, as illustrated in Figure 12.3. These diagrams are all constructed on the same basis as isoquant maps, with I, II and III representing different levels of the objective function, with its value increasing as one moves away from the origin.

In Figure 12.3(a), the objective function is convex to the origin and the constraint CD is linear. Maximization of the objective function occurs at the tangency point T.

In Figure 12.3(b), the objective function is concave to the origin and so it is maximized subject to the linear constraint CD at the corner point C. Thus, the tangency point T does *not* determine the maximum value.

In Figure 12.3(c), the objective function is convex to the origin but the constraint CD is non-linear and more sharply curved than the objective function. Thus, the tangency point T is *not* the maximum value. Higher values of the objective function can be found at points R and S, for example.

From the previous examples we can see that, for the two-variable case, a linear constraint and an objective function that is convex to the origin will ensure maximization at the point of tangency. In other cases tangency may not ensure maximization. If a problem involves the maximization of production subject to a linear budget constraint this means that output is maximized where the slope of the isoquant is equal to the slope of the budget constraint. We have already seen in Chapter 11 how a Cobb-Douglas production function in the form $Q = AK^{\alpha}L^{\beta}$ will correspond to a set of isoquants which continually decline and become flatter as L is increased, i.e. are convex to the origin. Thus, in any constrained optimization problem attempting to maximize a production

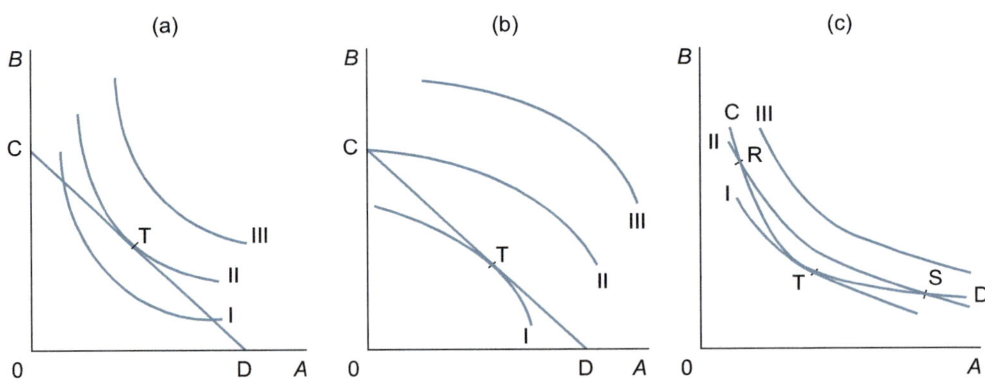

Figure 12.3 Isoquant maps and budget constraints

function in the Cobb-Douglas format subject to a linear budget constraint, the input combination that satisfies the first-order conditions will be a maximum.

If the Lagrangian represents other concepts with similar shaped functions and constraints, such as utility, the same conditions apply. In all such cases one assumes that the independent variables in the objective function must take positive values and so any negative mathematical solutions can be disregarded.

Although we cannot illustrate functions with more than two variables diagrammatically, the same basic principles apply when one is attempting to maximize a function with three or more variables subject to a linear constraint. Thus, any Cobb-Douglas production function with more than two inputs will be at a maximum, subject to a specified linear budget constraint, when the first-order conditions for optimization of the relevant Lagrange equation are met. For the purpose of answering the problems in this chapter, and for most constrained maximization problems that you will encounter in a first-year economics course, it can be assumed that the stationary points of the Lagrange function will satisfy the second-order conditions for a maximum.

The properties of the objective function and constraint that guarantee that a Lagrange function is minimized when first-order conditions are met are the reverse of the properties required for a maximum, i.e. the objective function must be linear and the constraint must be convex to the origin. Thus, to find values of K and L that minimize a budget function in the form

$$\text{TC} = P_K K + P_L L$$

subject to a given output Q^* being produced via the production function $Q = AK^{\alpha}L^{\beta}$, then the corresponding Lagrange function is

$$G = P_K K + P_L L + \lambda\left(Q^* - AK^{\alpha}L^{\beta}\right)$$

and the values of K and L which satisfy the first-order conditions

$$\frac{\partial G}{\partial K} = 0 \quad \text{and} \quad \frac{\partial G}{\partial L} = 0$$

will also satisfy second-order conditions for a minimum.

If you refer back to Figure 12.3(a), you can see the rationale for this rule in the two-input case. If input prices are given and one is trying to minimize the cost of the output represented by isoquant II, then one needs to find the budget constraint with slope equal to the negative of the price ratio which is nearest the origin and still goes through this isoquant. The linear objective function and the constraint convex to the origin guarantee that this will be at the tangency point T. Some examples of minimization problems that use this rule are given in the next section.

To conclude this section, let us reiterate what we have learned about second-order conditions and Lagrangians. When the objective function is in the Cobb-Douglas format and the constraint is linear, then the second-order conditions for a maximum are met at the stationary points of the Lagrange function. This rule is reversed for minimization.

12.5 CONSTRAINED MINIMIZATION USING THE LAGRANGE MULTIPLIER

As was explained in Section 12.4, the same principles used to construct a Lagrange function for a constrained maximization problem are used to construct a Lagrange function for a constrained minimization problem. The difference is that the components of the function are reversed, as is shown in the following examples. In all these cases the constraints and objective function take formats which guarantee that second-order conditions for a minimum are met.

Example 12.8

A firm operates with the production function $Q = 4K^{0.6}L^{0.5}$ and can buy K at £15 a unit and L at £8 a unit. What input combination will minimize the cost of producing 200 units of output?

Solution

The output constraint is $200 = 4K^{0.6}L^{0.5}$ and the objective function to be minimized is the total cost function TC = $15K + 8L$. The corresponding Lagrangian function is therefore

$$G = 15K + 8L + \lambda\left(200 - 4K^{0.6}L^{0.5}\right)$$

Partially differentiating G and setting equal to zero, first-order conditions require

$$\frac{\partial G}{\partial K} = 15 - \lambda 2.4K^{-0.4}L^{-0.5} = 0 \quad \text{giving} \quad \frac{15K^{0.4}}{2.4L^{0.5}} = \lambda \tag{1}$$

$$\frac{\partial G}{\partial L} = 8 - \lambda 2K^{-0.6}L^{-0.5} = 0 \quad \text{giving} \quad \frac{4L^{0.5}}{K^{0.6}} = \lambda \tag{2}$$

$$\frac{\partial G}{\partial \lambda} = 200 - 4K^{0.6}L^{0.5} = 0 \tag{3}$$

Setting (1) equal to (2) to eliminate λ

$$\frac{15K^{0.4}}{2.4L^{0.5}} = \frac{4L^{0.5}}{K^{0.6}}$$

$$15K = 9.6L$$

$$1.5625K = L \tag{4}$$

Substituting (4) into (3)

$$\begin{aligned}
200 - 4K^{0.6}(1.5625K)^{0.5} &= 0 \\
200 &= 4K^{0.6}(1.5625)^{0.5}K^{0.5} \\
\frac{200}{4(1.5625)^{0.5}} &= K^{1.1} \\
40 &= K^{1.1} \\
K &= \sqrt[1.1]{40} = 28.603434
\end{aligned}$$

Substituting this value into (4)

$$\begin{aligned}
1.5625(28.603434) &= L \\
44.692866 &= L
\end{aligned}$$

Thus, the optimal input combination is 28.6 units of K plus 44.7 units of L (to 1 dp). We can check that these input values correspond to the given output level by substituting them back into the production function. Thus,

$$Q = 4K^{0.6}L^{0.5} = 4(28.6)^{0.6}(44.7)^{0.5} = 200$$

which is correct, allowing for rounding error. The actual cost entailed will be

$$\text{TC} = 15K + 8L = 15(28.6) + 8(44.7) = £786.60$$

Example 12.9

The prices of inputs K and L are given as £12 per unit and £3 per unit respectively, and a firm operates with the production function $Q = 25K^{0.5}L^{0.5}$.

(i) What is the minimum cost of producing 1,250 units of output?
(ii) Demonstrate that the maximum output that can be produced for this budget will be the 1,250 units specified in (i).

Solution

This question essentially asks us to demonstrate that the constrained maximization and minimization methods give consistent answers.

(i) The output constraint is that

$$1,250 = 25K^{0.5}L^{0.5}$$

The objective function to be minimized is the cost function

$$\text{TC} = 12K + 3L$$

The corresponding Lagrange function is therefore

$$G = 12K + 3L + \lambda(1{,}250 - 25K^{0.5}L^{0.5})$$

First-order conditions require

$$\frac{\partial G}{\partial K} = 12 - \lambda 12.5K^{-0.5}L^{0.5} = 0 \quad \text{giving} \quad \frac{12K^{0.5}}{12.5L^{0.5}} = \lambda \tag{1}$$

$$\frac{\partial G}{\partial L} = 3 - \lambda 12.5K^{0.5}L^{-0.5} = 0 \quad \text{giving} \quad \frac{3L^{0.5}}{12.5K^{0.5}} = \lambda \tag{2}$$

$$\frac{\partial G}{\partial \lambda} = 1{,}250 - 25K^{0.5}L^{0.5} = 0 \tag{3}$$

Setting (1) equal to (2)

$$\frac{12K^{0.5}}{12.5L^{0.5}} = \frac{3L^{0.5}}{12.5K^{0.5}}$$

$$4K = L \tag{4}$$

Substituting (4) into (3)

$$1{,}250 - 25K^{0.5}(4K)^{0.5} = 0$$

$$1{,}250 = 25K^{0.5}(4)^{0.5}K^{0.5}$$

$$1{,}250 = 50K$$

$$25 = K$$

Substituting this value into (4)

$$4(25) = L$$

$$100 = L$$

When these optimum values of K and L are used the actual minimum cost will be

$$\text{TC} = 12K + 3L = 12(25) + 3(100)$$

$$= 300 + 300 = £600$$

(ii) This part of the question requires us to find the values of K and L that will maximize output subject to a budget of £600, i.e. the answer to (i). The objective function to be maximized is therefore $Q = 25K^{0.5}L^{0.5}$ and the constraint is $12K + 3L = 600$. The corresponding Lagrange equation is thus

$$G = 25K^{0.5}L^{0.5} + \lambda(600 - 12K - 3L)$$

First-order conditions require

$$\frac{\partial G}{\partial K} = 12.5K^{-0.5}L^{0.5} - 12\lambda = 0 \quad \text{giving} \quad \frac{12.5L^{0.5}}{12K^{0.5}} = \lambda \tag{5}$$

$$\frac{\partial G}{\partial L} = 12.5K^{0.5}L^{-0.5} - 3\lambda = 0 \quad \text{giving} \quad \frac{12.5K^{0.5}}{3L^{0.5}} = \lambda \tag{6}$$

$$\frac{\partial G}{\partial \lambda} = 600 - 12K - 3L = 0 \tag{7}$$

Setting (5) equal to (6)

$$\frac{12.5L^{0.5}}{12K^{0.5}} = \frac{12.5K^{0.5}}{3L^{0.5}}$$

$$3L = 12K$$

$$L = 4K \tag{8}$$

Substituting (8) into (7)

$$600 - 12K - 3(4K) = 0$$

$$600 - 12K - 12K = 0$$

$$600 = 24K$$

$$25 = K$$

Substituting this value into (8)

$$L = 4(25) = 100$$

These are the same optimum values of K and L that were found in part (i) earlier. The actual output produced by 25 of K plus 100 of L will be

$$Q = 25K^{0.5}L^{0.5} = 25(25)^{0.5}(100)^{0.5} = 1{,}250$$

which checks out with the amount specified in the question.

Although most of the examples of constrained optimization presented in this chapter are concerned with a firm's output and costs, or a consumer's utility level and income, the Lagrange method can be applied to other areas of economics. For instance, in environmental economics one may wish to find the cheapest way of securing a given level of environmental cleanliness.

Example 12.10

Assume that there are two sources of pollution into a lake. The local water authority can clean up the discharges and reduce pollution levels from these sources but there are, of course, costs involved. The damage effects of each pollution source are measured on a 'pollution scale'. The lower the pollution level the greater the cost of achieving it, as is shown by the cost schedules for cleaning up the two pollution sources:

$$Z_1 = 478 - 2C_1^{0.5} \quad \text{and} \quad Z_2 = 600 - 3C_2^{0.5}$$

where C_1 and C_2 are expenditure levels (in £000s) on reducing pollution and Z_1 and Z_2 are pollution levels. It is assumed that all values of C_1, C_2, Z_1 and Z_2 are non-negative.

To secure an acceptable level of water purity in the lake the water authority's objective is to reduce the total pollution level to 1,000 using the cheapest method. How can this be achieved?

Solution

This can be formulated as a constrained optimization problem where the constraint is the total amount of pollution $Z_1 + Z_2 = 1{,}000$ and the objective function to be minimized is the cost of pollution control TC = $C_1 + C_2$. Thus, the Lagrange function is

$$G = C_1 + C_2 + \lambda(1{,}000 - Z_1 - Z_2)$$

Substituting in the cost functions for Z_1 and Z_2, this becomes

$$G = C_1 + C_2 + \lambda\left[1{,}000 - \left(478 - 2C_1^{0.5}\right) - \left(600 - 3C_2^{0.5}\right)\right]$$
$$G = C_1 + C_2 + \lambda\left(-78 + 2C_1^{0.5} + 3C_2^{0.5}\right)$$

First-order conditions require

$$\frac{\partial G}{\partial C_1} = 1 + \lambda C_1^{-0.5} = 0 \quad \text{giving} \quad \lambda = -C_1^{0.5} \tag{1}$$

$$\frac{\partial G}{\partial C_2} = 1 + \lambda 1.5C_2^{-0.5} = 0 \quad \text{giving} \quad \lambda = \frac{-C_2^{0.5}}{1.5} \tag{2}$$

$$\frac{\partial G}{\partial \lambda} = -78 + 2C_1^{0.5} + 3C_2^{0.5} = 0 \tag{3}$$

Equating (1) and (2)

$$-C_1^{0.5} = \frac{-C_2^{0.5}}{1.5}$$

$$1.5C_1^{0.5} = C_2^{0.5} \tag{4}$$

Substituting (4) into (3)

$$-78 + 2C_1^{0.5} + 3\left(1.5C_1^{0.5}\right) = 0$$
$$2C_1^{0.5} + 4.5C_1^{0.5} = 78$$
$$6.5C_1^{0.5}78 = 78 \tag{5}$$
$$C_1^{0.5} = 12$$
$$C_1 = 144$$

Substituting (5) into (4)

$$C_2^{0.5} = 1.5(12) = 18$$
$$C_2 = 324$$

We can use these optimum pollution control expenditure amounts to check the total pollution level:

$$\begin{aligned} Z_1 + Z_2 &= (478 - 2C_1^{0.5}) + (600 - 3C_2^{0.5}) \\ &= 478 - 2(12) + 600 - 3(18) \\ &= 1{,}000 \end{aligned}$$

which is the required level. Thus, the water authority should spend £144,000 on reducing the first pollution source and £324,000 on reducing the second source.

QUESTIONS 12.3

1. Use the Lagrange multiplier method to answer questions 5(b) and 7 from Questions 12.1.
2. What is the cheapest way of producing 850 units of output if a firm operates with the production function $Q = 30K^{0.5}L^{0.5}$ and can buy input K at £75 a unit and L at £40 a unit?
3. Two pollution sources can be cleaned up if money is spent on them according to the functions $Z_1 = 780 - 12C_1^{0.5}$ and $Z_2 = 600 - 8C_2^{0.5}$ where Z_1 and Z_2 are the pollution levels from the two sources and C_1 and C_2 are expenditure levels (in £000s) on pollution reduction. What is the cheapest way of reducing the total pollution level from 1,380, which is the level without any controls, to 1,000?
4. A firm buys inputs K and L at £70 a unit and £30 a unit respectively and faces the production function $Q = 40K^{0.5}L^{0.5}$. What is the cheapest way it can produce an output of 500 units?

12.6 CONSTRAINED OPTIMIZATION WITH MORE THAN TWO VARIABLES

The same procedures that were used for two-variable problems are also used for applying the Lagrange method to constrained optimization problems with three or more variables. The only difference is that one has a more complex set of simultaneous equations to solve for the optimum values that satisfy the first-order conditions. Although some of these sets of equations may initially look rather awkward to work with, they can usually be greatly simplified and solutions can be found by basic algebra, as the following examples show. As with the two-variable problems, it is assumed that second-order conditions for a maximum (or minimum) are satisfied at stationary points of the Lagrange function in the problems set out here.

Example 12.11

A firm has a budget of £300 to spend on the three inputs X, Y and Z whose prices per unit are £4, £1 and £6, respectively. What combination of their respective quantities, x, y and z, should it employ to maximize output if it faces the production function $Q = 24x^{0.3}y^{0.2}z^{0.3}$?

Solution

The budget constraint is

$$300 - 4x - y - 6z = 0$$

and the objective function to be maximized is

$$Q = 24x^{0.3}y^{0.2}z^{0.3}$$

Thus, the Lagrange function is

$$G = 24x^{0.3}y^{0.2}z^{0.3} + \lambda(300 - 4x - y - 6z)$$

Differentiating with respect to each variable and setting equal to zero gives

$$\frac{\partial G}{\partial x} = 7.2x^{-0.7}y^{0.2}z^{0.3} - 4\lambda = 0 \quad \text{giving } \lambda = 1.8x^{-0.3}y^{0.2}z^{0.3} \qquad (1)$$

$$\frac{\partial G}{\partial y} = 4.8x^{0.3}y^{-0.8}z^{0.3} - \lambda = 0 \quad \text{giving } \lambda = 4.8x^{0.3}y^{-0.8}z^{0.3} \qquad (2)$$

$$\frac{\partial G}{\partial z} = 7.2x^{0.3}y^{0.2}z^{-0.7} - 6\lambda = 0 \quad \text{giving } \lambda = 1.2x^{0.3}y^{0.2}z^{-0.7} \qquad (3)$$

$$\frac{\partial G}{\partial \lambda} = 300 - 4x - y - 6z = 0 \qquad (4)$$

A simultaneous three-linear-equation system in the three unknowns x, y and z can now be set up if λ is eliminated. There are several ways in which this can be done. In the following method used we set (1) and then (3) equal to (2) to eliminate x and z and then substitute into (4) to solve for y. Whichever method is used, the point of the exercise is to arrive at functions for any two of the unknown variables in terms of the remaining third variable.

Thus, setting (1) equal to (2)

$$1.8x^{-0.7}y^{0.2}z^{0.3} = 4.8x^{0.3}y^{-0.8}z^{0.3}$$

Multiplying both sides by $x^{0.7}y^{0.8}$ and dividing by $z^{0.3}$ gives

$$1.8y = 4.8x$$
$$0.375y = x \quad (5)$$

We have now eliminated z and obtained a function for x in terms of y. Next, we need to eliminate x and obtain a function for z in terms of y. To do this we set (2) equal to (3), giving

$$4.8x^{0.3}y^{-0.8}z^{0.3} = 1.2x^{0.3}y^{0.2}z^{-0.7}$$

Multiplying through by $z^{0.7}y^{0.8}$ and dividing by $x^{0.3}$ gives

$$\begin{aligned} 4.8z &= 1.2y \\ z &= 0.25y \end{aligned} \quad (6)$$

Substituting (5) and (6) into the budget constraint (4)

$$\begin{aligned} 300 - 4(0.375y) - y - 6(0.25y) &= 0 \\ 300 - 1.5y - y - 1.5y &= 0 \\ 300 &= 4y \\ 75 &= y \end{aligned}$$

Therefore, from (5) $\quad x = 0.375(75) = 28.125$

and from (6) $\quad z = 0.25(75) = 18.75$

If these optimal values of x, y and z are used then the maximum output will be

$$Q = 24x^{0.3}y^{0.2}z^{0.3} = 24(28.125)^{0.3}(75)^{0.2}(18.75)^{0.3} = 373.1 \text{ units}$$

Example 12.12

A firm uses the three inputs K, L and R to manufacture good Q and faces the production function

$$Q = 50K^{0.4}L^{0.2}R^{0.2}$$

It has a budget of £24,000 and can buy K, L and R at £80, £12 and £10 per unit, respectively. What combination of inputs will maximize its output?

Solution

The objective function to be maximized is $Q = 50K^{0.4}L^{0.2}R^{0.2}$ and the budget constraint is

$$24{,}000 - 80K - 12L - 10R = 0$$

Thus, the Lagrange equation is

$$G = 50K^{0.4}L^{0.2}R^{0.2} + \lambda(24{,}000 - 80K - 12L - 10R)$$

Differentiating this with respect to each variable:

$$\frac{\partial G}{\partial K} = 20K^{-0.6}L^{0.2}R^{0.2} - 80\lambda = 0 \text{ giving } \quad \lambda = 0.25K^{-0.6}L^{0.2}R^{0.2} \qquad (1)$$

$$\frac{\partial G}{\partial L} = 10K^{0.4}L^{-0.8}R^{0.2} - 12\lambda = 0 \text{ giving } \quad \lambda = \frac{10}{12}K^{0.4}L^{-0.8}R^{0.2} \qquad (2)$$

$$\frac{\partial G}{\partial R} = 10K^{0.4}L^{0.2}R^{-0.8} - 10\lambda \text{ giving } \quad \lambda = K^{0.4}L^{0.2}R^{-0.8} \qquad (3)$$

$$\frac{\partial G}{\partial \lambda} = 24{,}000 - 80K - 12L - 10R \qquad (4)$$

Equating (1) and (2) to eliminate R

$$0.25K^{-0.6}L^{0.2}R^{0.2} = \frac{10}{12}K^{0.4}L^{-0.8}R^{0.2}$$

$$3L = 10K$$

$$0.3L = K \qquad (5)$$

Equating (2) and (3) to eliminate K and get R in terms of L

$$\frac{10}{12}K^{0.4}L^{-0.8}R^{0.2} = K^{0.4}L^{0.2}R^{-0.8}$$

$$10R = 12L$$

$$R = 1.2L \qquad (6)$$

Substituting (5) and (6) into (4)

$$24{,}000 - 80(0.3L) - 12L - 10(1.2L) = 0$$

$$24{,}000 = 24L + 12L + 12L$$

$$24{,}000 = 48L$$

$$500 = L$$

Substituting this value for L into (5) and (6)

$$K = 0.3(500) = 150 \quad \text{and} \quad R = 1.2(500) = 600$$

Using these optimal values for K, L and R, the firm's maximum output will be

$$Q = 50K^{0.4}L^{0.2}R^{0.2} = 50(150)^{0.4}(500)^{0.2}(600)^{0.2}$$

$$= 4{,}622 \text{ units}$$

Example 12.13

A firm buys the four inputs K, L, R and M at per-unit prices of £50, £30, £25 and £20, respectively, and operates with the production function

$$Q = 160K^{0.3}L^{0.25}R^{0.2}M^{0.25}$$

What is the maximum output it can make for a total cost of £30,000?

Solution

The relevant Lagrange function is

$$G = 160K^{0.3}L^{0.25}R^{0.2}M^{0.25} + \lambda(30{,}000 - 50K - 30L - 25R - 20M)$$

Differentiating to find stationary points, setting equal to zero and then equating to λ

$$\frac{\partial G}{\partial K} = 48K^{-0.7}L^{0.25}R^{0.2}M^{0.25} - 50\lambda = 0 \text{ giving } \quad \lambda = \frac{48L^{0.25}R^{0.2}M^{0.25}}{50K^{0.7}} \qquad (1)$$

$$\frac{\partial G}{\partial L} = 40K^{0.3}L^{-0.75}R^{0.2}M^{0.25} - 30\lambda = 0 \text{ giving } \quad \lambda = \frac{4K^{0.3}R^{0.2}M^{0.25}}{3L^{0.75}} \qquad (2)$$

$$\frac{\partial G}{\partial R} = 32K^{0.3}L^{0.25}R^{-0.8}M^{0.25} - 25\lambda = 0 \text{ giving } \quad \lambda = \frac{32K^{0.3}L^{0.2}5M^{0.25}}{25R^{0.8}} \qquad (3)$$

$$\frac{\partial G}{\partial M} = 40K^{0.3}L^{0.25}R^{0.2}M^{-0.75} - 20\lambda = 0 \text{ giving } \quad \lambda = \frac{2K^{0.3}L^{0.25}R^{0.2}}{50M^{0.75}} \qquad (4)$$

$$\frac{\partial G}{\partial \lambda} = 30{,}000 - 50K - 30L - 25R - 20M = 0 \qquad (5)$$

Equating (1) and (2)

$$\frac{48L^{0.25}R^{0.2}M^{0.25}}{50K^{0.7}} = \frac{4K^{0.3}R^{0.2}M^{0.25}}{3L^{0.75}}$$

Dividing through by $R^{0.2}M^{0.25}$ and cross-multiplying

$$144L = 200K$$
$$0.72L = K \qquad (6)$$

Note that because it is simpler to divide by 200 than 144, we have expressed K as a fraction of L rather than vice versa. Having done this we must now find R and M in terms of L and so (2) must be equated with (3) and (4) to ensure that L is not cancelled out in each set of equalities. Thus, equating (2) and (3)

$$\frac{4K^{0.3}R^{0.2}M^{0.25}}{3L^{0.75}} = \frac{32K^{0.3}L^{0.25}M^{0.25}}{25L^{0.8}}$$

Cancelling out $K^{0.3}M^{0.25}$ and cross-multiplying

$$100R = 96L$$
$$R = 0.96L \qquad (7)$$

Equating (2) and (4)

$$\frac{4K^{0.3}R^{0.2}M^{0.25}}{3L^{0.75}} = \frac{2K^{0.3}L^{0.25}R^{0.2}}{M^{0.75}}$$

Cancelling $K^{0.3}R^{0.2}$ and cross-multiplying

$$\begin{aligned} 4M &= 6L \\ M &= 1.5L \end{aligned} \qquad (8)$$

Substituting (6), (7) and (8) into (5)

$$\begin{aligned} 30{,}000 - 50(0.72L) - 30L - 25(0.96L) - 20(1.5L) &= 0 \\ 30{,}000 - 36L - 30L - 24L - 30L &= 0 \\ 30{,}000 &= 120L \\ 250 &= L \end{aligned}$$

Substituting this value for L into (6), (7) and (8)

$$K = 0.72(250) = 180, \quad R = 0.96(250) = 240 \quad \text{and} \quad M = 1.5(250) = 375$$

Using these optimal values of L, K, M and R gives the maximum output level

$$\begin{aligned} Q &= 160K^{0.3}L^{0.25}R^{0.2}M^{0.25} \\ &= 160\left(180^{0.3}\right)\left(250^{0.25}\right)\left(240^{0.2}\right)\left(375^{0.25}\right) = 39{,}786.6 \text{ units} \end{aligned}$$

Example 12.14

A firm operates with the production function $Q = 20K^{0.5}L^{0.25}R^{0.4}$. The input prices per unit are £20 for K, £10 for L and £5 for R. What is the cheapest way of producing 1,200 units of output?

Solution

This time output is the constraint such that

$$20K^{0.5}L^{0.25}R^{0.4} = 1{,}200$$

and the objective function to be minimized is the cost function

$$\text{TC} = 20K + 10L + 5R$$

The corresponding Lagrange function is therefore

$$G = 20K + 10L + 5R + \lambda\left(1{,}200 - 20K^{0.5}L^{0.25}R^{0.4}\right)$$

Differentiating to get stationary points

$$\frac{\partial G}{\partial K} = 20 - \lambda 10K^{-0.5}L^{0.25}R^{0.4} = 0 \text{ giving } \quad \lambda = \frac{2K^{0.5}}{L^{0.25}R^{0.4}} \tag{1}$$

$$\frac{\partial G}{\partial L} = 10 - \lambda 5K^{0.5}L^{-0.75}R^{0.4} = 0 \text{ giving } \quad \lambda = \frac{2L^{0.75}}{K^{0.5}R^{0.4}} \tag{2}$$

$$\frac{\partial G}{\partial R} = 5 - \lambda 8K^{0.5}L^{0.25}R^{-0.6} = 0 \text{ giving } \quad \lambda = \frac{5R^{0.6}}{8K^{0.5}L^{0.25}} \tag{3}$$

$$\frac{\partial G}{\partial \lambda} = 1,200 - 20K^{0.5}L^{0.25}R^{0.4} = 0 \tag{4}$$

Equating (1) and (2)

$$\frac{2K^{0.5}}{L^{0.25}R^{0.4}} = \frac{2L^{0.75}}{K^{0.5}R^{0.4}}$$

$$K = L \tag{5}$$

Equating (2) and (3)

$$\frac{2L^{0.75}}{K^{0.5}R^{0.4}} = \frac{5R^{0.6}}{8K^{0.5}L^{0.25}}$$

$$16L = 5R$$

$$3.2L = R \tag{6}$$

Substituting (5) and (6) into (4) to eliminate R and K

$$1,200 - 20(L)^{0.5} L^{0.25} (3.2L)^{0.4} = 0$$

$$1,200 - 20(3.2)^{0.4} L^{1.15} = 0$$

$$60 = 1.5924287L^{1.15}$$

$$37.678296 = L^{1.15}$$

$$23.47 = L$$

Substituting this value for L into (5) and (6) gives

$$K = 23.47 \text{ and } \quad R = 3.2(23.47) = 75.1$$

Checking that these values do give the required 1,200 units of output:

$$Q = 20K^{0.5}L^{0.25}R^{0.4} = 20(23.47)^{0.5}(23.47)^{0.25}(75.1)^{0.4} = 1,200$$

The cheapest cost level for producing this output will therefore be

$$20K + 10L + 5R = 20(23.4) + 10(23.47) + 5(75.1) = £1{,}079.60$$

Example 12.15

A firm that operates with the production function $Q = 45K^{0.4}L^{0.3}R^{0.3}$ has received an order for 75,000 units of its product. It can buy input K at £80 a unit, L at £35 and R at £50. What is the cheapest way it can produce the required amount of output?

Solution

The output constraint is $45K^{0.4}L^{0.3}R^{0.3} = 75{,}000$ and the objective function to be minimized is TC = $80K + 35L + 50R$. The corresponding Lagrange function is thus

$$G = 80K + 35L + 50R + \lambda\left(75{,}000 - 45K^{0.4}L^{0.3}R^{0.3}\right)$$

Differentiating to get first-order conditions for a minimum

$$\frac{\partial G}{\partial K} = 80 - \lambda 18K^{-0.6}L^{0.3}R^{0.3} = 0 \text{ giving } \quad \lambda = \frac{80K^{0.6}}{18L^{0.3}R^{0.3}} \tag{1}$$

$$\frac{\partial G}{\partial L} = 35 - \lambda 13.5K^{0.4}L^{-0.7}R^{0.3} = 0 \text{ giving } \quad \lambda = \frac{35L^{0.7}}{13.5K^{0.4}R^{0.3}} \tag{2}$$

$$\frac{\partial G}{\partial R} = 50 - \lambda 13.5K^{0.4}L^{0.3}R^{-0.7} = 0 \text{ giving } \quad \lambda = \frac{50R^{0.7}}{13.5K^{0.4}L^{0.3}} \tag{3}$$

$$\frac{\partial G}{\partial \lambda} = 75{,}000 - 45K^{0.4}L^{0.3}R^{0.3} = 0 \tag{4}$$

Equating (1) and (2)

$$\frac{80K^{0.6}}{18L^{0.3}R^{0.3}} = \frac{35L^{0.7}}{13.5K^{0.4}R^{0.3}}$$

$$1{,}080K = 630L$$

$$\frac{12K}{7} = L \tag{5}$$

As we have L in terms of K we now need to use (1) and (3) to get R in terms of K. Thus, equating (1) and (3) yields

$$\frac{80K^{0.6}}{18L^{0.3}R^{0.3}} = \frac{50R^{0.7}}{13.5K^{0.4}L^{0.3}}$$

$$1{,}080K = 900R$$

$$1.2K = R \tag{6}$$

Substituting (5) and (6) into (4)

$$75{,}000 - 45K^{0.4}\left(\frac{12K}{7}\right)^{0.3}(1.2K)^{0.3} = 0$$

$$75{,}000 - 45\left(\frac{12}{7}\right)^{0.3}(1.2)^{0.3}K^{0.4}K^{0.3}K^{0.3} = 0$$

$$75{,}000 = 55.871697K$$

$$1{,}342.3612 = K$$

Substituting this value into (5)

$$L = \frac{12}{7}(1{,}342.3612) = 2{,}301.1907$$

Substituting into (6)

$$R - 1.2(1{,}342.3612) = 1{,}610.8334$$

Thus, optimum values are

$$K = 1{,}342.4, \qquad L = 2{,}301.2 \quad \text{and} \quad R = 1{,}610.8 \quad (\text{to 1 dp})$$

Total expenditure on inputs will then be

$$80K + 35L + 50R = 80(1{,}342.4) + 35(2{,}301.2) + 50(1{,}610.8) = £268{,}474$$

QUESTIONS 12.4

1. A firm has a budget of £570 to spend on the three inputs x, y and z whose prices per unit are, respectively, £4, £6 and £3. What combination of x, y and z will maximize output given the production function $Q = 2x^{0.2}y^{0.3}z^{0.45}$?
2. A firm uses inputs K, L and R to manufacture good Q. It has a budget of £828 and its production function (for positive values of Q) is

$$Q = 20K + 16L + 12R - 0.2K^2 - 0.1L^2 - 0.3R^2$$

If P_K = £20, P_L = £10 and P_R = £6 what is the maximum output it can produce? Assume that second-order conditions for a maximum are satisfied for the relevant Lagrangian.
3. What amounts of the inputs x, y and z should a firm use to maximize output if it faces the production function $Q = 2x^{0.4}y^{0.2}z^{0.6}$ and has a budget of £600, given that the prices of x, y and z are, respectively, £4, £1 and £2 per unit?

4. A firm buys the inputs x, y and z for £5, £10 and £2, respectively, per unit. If its production function is $Q = 60x^{0.2}y^{0.4}z^{0.5}$ how much can it produce for an outlay of £8,250?
5. Inputs K, L, R and M cost £10, £6, £15 and £3 per unit, respectively. What is the cheapest way of producing an output of 900 units if a firm operates with the production function $Q = 20K^{0.4}L^{0.3}R^{0.2}M^{0.25}$?
6. A firm faces the production function $Q = 50K^{0.5}L^{0.2}R^{0.25}$ and is required to produce an output level of 1,913 units. What is the cheapest way of doing this if the per-unit costs of inputs K, L and R are £80, £24 and £45, respectively?

13 Further topics in differentiation and integration

Learning objectives

After completing this chapter students should be able to:

- use the chain, product and quotient rules for differentiation
- choose the most appropriate method for differentiating different forms of functions
- check the second-order conditions for optimization of relevant economic functions using the quotient and other rules for differentiation
- integrate basic functions
- use integration to determine total cost and total revenue from marginal cost and marginal revenue functions
- understand how a definite integral relates to the area under a function and apply this concept to calculate consumer surplus
- apply integration to solve problems with continuous discounting
- integrate more complex functions by parts and by substitution.

13.1 OVERVIEW

In this chapter some techniques are introduced that can be used to differentiate functions that are more complex than those encountered in Chapters 9, 10, 11 and 12. These are the chain rule, the product rule and the quotient rule. As you will see in the worked examples, it is often necessary to combine several of these methods to differentiate some functions. The concept of integration is also introduced and developed.

13.2 THE CHAIN RULE

The chain rule is used to differentiate 'functions within functions'. For example, if we have the function

$$y = \mathrm{f}(z)$$

DOI: 10.4324/9781003360827-13

and we also know that there is a second functional relationship

$$z = \mathrm{g}(x)$$

then we can write y as a function of x in the form

$$y = \mathrm{f}[\mathrm{g}(x)]$$

To differentiate y with respect to x in this type of function we use the chain rule which states that

$$\frac{\mathrm{d}y}{\mathrm{d}x} = \frac{\mathrm{d}y}{\mathrm{d}z} \times \frac{\mathrm{d}z}{\mathrm{d}x}$$

One economics example of a function within a function occurs in the marginal revenue productivity theory of the demand for labour, where a firm's total revenue depends on output which, in turn, depends on the amount of labour employed. An applied example is explained later. However, we shall first look at what is perhaps the most frequent use of the chain rule, which is to break down an awkward function artificially into two components in order to allow differentiation via the chain rule. Assume, for example, that you wish to find an expression for the slope of the non-linear demand function

$$p = (150 - 0.2q)^{0.5} \qquad (1)$$

The basic rules for differentiation explained in Chapter 9 cannot cope with this sort of function. However, if we define a new function

$$z = 150 - 0.2q \qquad (2)$$

then (1) can be rewritten as

$$p = z^{0.5} \qquad (3)$$

(Note that in both (1) and (3) the functions are assumed to hold for $p \geq 0$ only, i.e. negative roots are ignored.)

Differentiating (2) and (3) we get

$$\frac{\mathrm{d}z}{\mathrm{d}q} = -0.2 \quad \text{and} \quad \frac{\mathrm{d}p}{\mathrm{d}z} = 0.5z^{-0.5}$$

Thus, using the chain rule and then substituting equation (2) back in for z, we get

$$\frac{\mathrm{d}p}{\mathrm{d}q} = \frac{\mathrm{d}p}{\mathrm{d}z}\frac{\mathrm{d}z}{\mathrm{d}q} = 0.5z^{-0.5}(-0.2) = \frac{-0.1}{z^{0.5}} = \frac{-0.1}{(150 - 0.2q)^{0.5}}$$

Some more examples of the use of the chain rule are set out next.

Example 13.1

The present value of a payment of £1 due in 8 years' time is given by the formula

$$PV = \frac{1}{(1+i)^8}$$

where i is the given interest rate. What is the rate of change of PV with respect to i?

Solution

If we let

$$z = 1 + i \qquad (1)$$

then we can write

$$PV = \frac{1}{z^8} = z^{-8} \qquad (2)$$

Differentiating (1) and (2) gives

$$\frac{dz}{di} = 1 \text{ and } \qquad \frac{dPV}{dz} = -8z^{-9}$$

Therefore, using the chain rule, the rate of change of PV with respect to i will be

$$\frac{dPV}{di} = \frac{dPVdz}{dzdi} = -8z^{-9}$$

$$= \frac{-8}{(1+i)^9}$$

Example 13.2

If $y = (48 + 20x^{-1} + 4x + 0.3x^2)^4$ what is dy/dx?

Solution

Let

$$z = 48 + 20x^{-1} + 4x + 0.3x^2 \qquad (1)$$

and so

$$\frac{dz}{dx} = -20x^2 + 4 + 0.6x \qquad (2)$$

Substituting (1) into the function given in the question

$$y = z^4$$

$$\text{and so } \frac{dy}{dz} = 4z^3 \qquad (3)$$

Therefore, using the chain rule and substituting (2) and (3)

$$\frac{dy}{dx} = \frac{dydz}{dzdx} = 4z^3\left(-20x^2 + 4 + 0.6x\right)$$
$$= 4\left(48 + 20x^{-1} + 4x + 0.3x^2\right)^3\left(-20x^2 + 4 + 0.6x\right)$$

The marginal revenue productivity theory of the demand for labour

In the marginal revenue productivity theory of the demand for labour, the rule for profit maximization is to employ additional units of labour as long as the extra revenue generated by selling the extra output produced by an additional unit of labour exceeds the marginal cost of employing this additional unit of labour. This rule applies in the short-run when inputs other than labour are assumed fixed.

The optimal amount of labour is employed when

$$\text{MRP}_\text{L} = \text{MC}_\text{L}$$

where MRP_L is the marginal revenue product of labour, defined as the additional revenue generated by an additional unit of labour, and MC_L is the marginal cost of an additional unit of labour. The MC_L is normally equal to the wage rate unless the firm is a monopsonist (sole buyer) in the labour market.

If all relevant functions are assumed to be continuous then the previous definitions can be rewritten as

$$\text{MRP}_\text{L} = \frac{\text{dTR}}{\text{d}L} \quad \text{and} \quad \text{MC}_\text{L} = \frac{\text{dTC}_\text{L}}{\text{d}L}$$

where TR is total sales revenue (i.e. pq) and TC_L is the total cost of labour. If a firm is a monopoly seller of a good, then we effectively have to deal with two functions in order to derive its MRP_L function since total revenue will depend on output, i.e. TR = f(q), and output will depend on labour input, i.e. q = f(L). Therefore, using the chain rule

$$\text{MRP}_\text{L} = \frac{\text{dTR}}{\text{d}L} = \frac{\text{dTR}}{\text{d}q} \times \frac{\text{d}q}{\text{d}L} \tag{1}$$

We already know that

$$\frac{\text{dTR}}{\text{d}q} = \text{MR} \quad \text{and} \quad \frac{\text{d}q}{\text{d}L} = \text{MP}_\text{L}$$

Therefore, substituting these into (1),

$$\text{MRP}_\text{L} = \text{MR} \times \text{MP}_\text{L}$$

This is the rule for determining the profit-maximizing amount of labour which you should encounter in your microeconomics course.

Example 13.3

A firm is a monopoly seller of good q and faces the demand schedule $p = 200 - 2q$, where p is the price in pounds, and the short-run production function $q = 4L^{0.5}$. If it can buy labour at a fixed wage of £8, how many units of L should be employed to maximize profit, assuming other inputs are fixed?

Solution

Using the chain rule, we need to derive a formula for MRP_L in terms of L and then set it equal to £8, given that MC_L is fixed at this wage rate. As

$$\text{MRP}_\text{L} = \frac{\text{dTR}}{\text{d}L} = \frac{\text{dTR}}{\text{d}q} \times \frac{\text{d}q}{\text{d}L} \qquad (1)$$

we need to find $\frac{\text{dTR}}{\text{d}q}$ and $\frac{\text{d}q}{\text{d}L}$.

Given $p = 200 - 2q$, then

$$\text{TR} = pq = (200 - 2q)q = 200q - 2q^2$$

Therefore,

$$\frac{\text{dTR}}{\text{d}q} = 200 - 4q \qquad (2)$$

Given $q = 4L^{0.5}$, then the marginal product of labour will be

$$\frac{\text{d}q}{\text{d}L} = 2L^{-0.5} \qquad (3)$$

Thus, substituting (2) and (3) into (1)

$$\text{MRP}_\text{L} = (200 - 4q)2L^{-0.5}(400 - 8q)L^{-0.5}$$

As all units of L cost £8, setting this function for MRP_L equal to the wage rate we get

$$\frac{400 - 8q}{L^{-0.5}} = 8$$

$$400 - 8q = 8L^{0.5} \qquad (4)$$

Substituting the production function $q = 4L^{0.5}$ into (4), as we are trying to derive a formula in terms of L, gives

$$400 - 8(4L^{0.5}) = 8L^{0.5}$$
$$400 - 32L^{0.5} = 8L^{0.5}$$
$$400 = 40L^{0.5}$$
$$10 = L^{0.5}$$
$$100 = L$$

Therefore, the optimal employment level is 100.

In the previous example, the idea of a 'short-run production function' was used to simplify the analysis, where the input of capital (K) was implicitly assumed to be fixed. Now that you understand how an MRP_L function can be derived we can work with full production functions in the format $Q = \text{f}(K, L)$. The effect of one input increasing while the other is held constant can now be shown by the relevant *partial derivative*

$$\text{MP}_\text{L} = \frac{\partial Q}{\partial L}$$

The same chain rule can be used for partial derivatives, and full and partial derivatives can be combined, as in the following examples.

Example 13.4

A firm operates with the production function $q = 45K^{0.7}L^{0.4}$ and faces the demand function $p = 6{,}980 - 6q$. Derive its MRP_L function.

Solution

By definition $\text{MRP}_\text{L} = \dfrac{\partial \text{TR}}{\partial L}$, where K is assumed fixed.

We know that

$$\text{TR} = pq = (6{,}980 - 6q)q = 6{,}980q - 6q^2$$

Therefore,

$$\frac{\text{dTR}}{\text{d}q} = 6{,}980 - 12q \tag{1}$$

From the production function $q = 45K^{0.7}L^{0.4}$ we can derive

$$\text{MP}_\text{L} = \frac{\partial q}{\partial L} = 18K^{0.7}L^{-0.6} \tag{2}$$

Using the chain rule and substituting (1) and (2)

$$\text{MRP}_\text{L} = \frac{\partial \text{TR}}{\partial L} = \frac{\text{dTR}}{\text{d}q}\frac{\partial q}{\partial L} = (6{,}980 - 12q)18K^{0.7}L^{-0.6} \tag{3}$$

As we wish to derive MRP_L as a function of L, we substitute the production function given in the question into (3) for q. Thus,

$$\begin{aligned}\text{MRP}_\text{L} &= \left[6{,}980 - 12\left(45K^{0.7}L^{0.4}\right)\right]18K^{0.7}L^{-0.6} \\ &= 125{,}640K^{0.7}L^{-0.6} - 9{,}720K^{1.4}L^{-0.2}\end{aligned}$$

Note that the value MRP_L will depend on the amount that K is fixed at, as well as the value of L.

Point elasticity of demand

The chain rule can help the calculation of point elasticity of demand for some non-linear demand functions.

Example 13.5

Find point elasticity of demand when $q = 10$ if $p = (120 - 2q)^{0.5}$.

Solution

Point elasticity is defined as

$$e = \frac{p}{q} \times \frac{1}{\frac{dp}{dq}} \tag{1}$$

Create a new variable $z = 120 - 2q$. Thus, $p = z^{0.5}$ and so, by differentiating:

$$\frac{dz}{dq} = -2 \quad \text{and} \quad \frac{dp}{dz} = 0.5z^{-0.5}$$

Therefore,

$$\begin{aligned}\frac{dp}{dq} &= \frac{dp}{dz}\frac{dz}{dp} \\ &= 0.5z^{-0.5}(-2) \\ &= 0.5(120 - 2q)^{-0.5}(-2) \\ &= \frac{-1}{(120 - 2q)^{0.5}}\end{aligned}$$

and so, inverting this result,

$$\frac{1}{dp/dq} = -(120 - 2q)^{0.5}$$

When $q = 10$, then from the original demand function price can be calculated as

$$p = (120 - 20)^{0.5} = 100^{0.5} = 10$$

Thus, substituting these results into formula (1), point elasticity will be

$$e = \frac{10}{10}(-1)(120 - 2q)^{0.5} = -(120 - 20)^{0.5} = -100^{0.5} = -10$$

Sometimes it may be possible to simplify an expression in order to be able to differentiate it, but one may instead use the chain rule if it is more convenient. The same result will be obtained by both methods, of course.

Example 13.6

Differentiate the function $y = (6 + 4x)^2$.

Solution

(i) By multiplying out

$$y = (6+4x)^2 = 36 + 48x + 16x^2$$

Therefore,

$$\frac{dy}{dx} = 48 + 32x$$

(ii) Using the chain rule, let $z = 6 + 4x$ so that $y = z^2$. Thus,

$$\frac{dy}{dx} = \frac{dy}{dz}\frac{dz}{dx} = 2z \times 4 = 2(6+4x)4 = 48 + 32x$$

QUESTIONS 13.1

1. A firm operates in the short-run with the production function $q = 2L^{0.5}$ and faces the demand schedule $p = 60 - 4q$ where p is price per unit in pounds. If it can employ labour at a wage rate of £4 per hour, how much should it employ to maximize profit?
2. If a supply schedule is given by $p = (2 + 0.05q)^2$ show

   (a) by multiplying out, and
   (b) by using the chain rule,

   that its slope is 2.2 when q is 400.

3. The return R on a sum M invested at $i\%$ for 3 years is given by the formula

   $$R = M(1+i)^3$$

   What is the rate of change of R with respect to i?
4. If $y = (3 + 0.6x^2)^{0.5}$ what is dy/dx?
5. If a firm faces the total cost function $TC = (6 + x)^{0.5}$, what is its marginal cost function?
6. A firm operates with the production function $q = 0.4K^{0.5}L^{0.5}$ and sells its output in a market where it is a monopoly with the demand schedule $p = 60 - 2q$. If K is fixed at 25 units and the wage rate is £7 per unit of L, derive the MRP_L function and work out how much L the firm should employ to maximize profit.

7. A firm faces the demand schedule $p = 650 - 3q$ and the production function $q = 4K^{0.5}L^{0.5}$ and has to pay £8 per unit to buy L. If K is fixed at 4 units how much L should the firm use if it wishes to maximize profits?
8. If a firm operates with the total cost function $TC = 4 + 10(9 + q^2)^{0.5}$, what is its marginal cost when q is 4?
9. Given the production function $q = (6K^{0.5} + 0.5L^{0.5})^{0.3}$, find MP_L when K is 16 and L is 576.

13.3 THE PRODUCT RULE

The product rule allows one to differentiate two functions which are multiplied together.

If $y = uv$ where u and v are both functions of x, then according to the product rule

$$\frac{dy}{dx} = u\frac{dv}{dx} + v\frac{du}{dx}$$

As with the chain rule, one may find it convenient to split a single awkward function into two artificial functions even if these functions do not have any particular economic meaning. The following examples show how this rule can be used.

Example 13.7

If $y = (7.5 + 0.2x^2)(4 + 8x^{-1})$ what is dy/dx?

Solution

This function could in fact be multiplied out and differentiated without using the product rule. However, let us first use the product rule and then we can compare the answers obtained by the two methods. They should, of course, be the same.

We are given the function

$$y = \left(7.5 + 0.2x^2\right)\left(4 + 8x^{-1}\right)$$

so let

$$u = 7.5 + 0.2x^2 \quad \text{and} \quad v = 4 + 8x^{-1}$$

Therefore,

$$\frac{du}{dx} = 0.4x \quad \text{and} \quad \frac{dv}{dx} = -8x^{-2}$$

Thus, using the product rule and substituting these results in, we get

$$\frac{dy}{dx} = u\frac{dv}{dx} + v\frac{du}{dx}$$
$$= (7.5 + 0.2x^2)(-8x^{-2}) + (4 + 8x^{-1})0.4x$$
$$= -60x^{-2} - 1.6 + 1.6x + 3.2$$
$$= 1.6 + 1.6x - 60x^{-2}$$

The alternative method of differentiation is to multiply out the original function. Thus,

$$y = (7.5 + 0.2x^2)(4 + 8x^{-1}) = 30 + 60x^{-1} + 0.8x^2 + 1.6x$$

and so

$$\frac{dy}{dx} = -60x^{-2} + 1.6x + 1.6$$

The answers obtained using both approaches are the same, as we expected.

When it is not possible to multiply out the different components of a function then one must use the product rule to differentiate. One may also need to use the chain rule to help differentiate the different sub-functions.

Example 13.8

A firm faces the non-linear demand schedule $p = (650 - 0.25q)^{1.5}$. What output should it sell to maximize total revenue?

Solution

When the demand function in the question is substituted for p then

$$\text{TR} = pq = (650 - 0.25q)^{1.5}q$$

To differentiate TR using the product rule, first let

$$u = (650 - 0.25q)^{1.5} \quad \text{and } v = q$$

Thus, employing the chain rule to differentiate u

$$\frac{du}{dq} = 1.5(650 - 0.25q)^{0.5}(-0.25) = -0.375(650 - 0.25q)^{0.5}$$

and differentiating v we get $\frac{dv}{dq} = 1$.

Then, using the product rule,

$$\frac{d\text{TR}}{dq} = u\frac{dv}{dq} + v\frac{du}{dq}$$
$$= (650 - 0.25q)^{1.5} + q(-0.375)(650 - 0.25q)^{0.5}$$

$$= (650 - 0.25)^{0.5}(650 - 0.25q - 0.375q)$$
$$= (650 - 0.25)^{0.5}(650 - 0.625q)$$

For a stationary point

$$\frac{dTR}{dq} = (650 - 0.25)^{0.5}(650 - 0.625q) = 0$$

Therefore, either

$$650 - 0.25q = 0 \quad \text{or} \quad 650 - 0.625q = 0$$
$$2{,}600 = q \quad \text{or} \quad 1{,}040 = q$$

We now need to check which of these values of q satisfies the second-order condition for a maximum. (You should immediately be able to see why it will not be 2,600 by observing what happens when this quantity is substituted into the demand function.)

To derive d^2TR/dq^2 we need to use the product rule again to differentiate dTR/dq:

$$\frac{dTR}{dq} = (650 - 0.25)^{0.5}(650 - 0.625q)$$

Let

$$u = (650 - 0.25q)^{0.5} \quad \text{and} \quad v = 650 - 0.625q$$

then using the chain rule we get

$$\frac{du}{dq} = 0.5(650 - 0.25q)^{-0.5}(-0.25) = -0.125(650 - 0.25q)^{-0.5}$$

and $\frac{dv}{dq} = -0.625$.

Now we can move to employing the product rule

$$\frac{d^2TR}{dq^2} = u\frac{dv}{dq} + v\frac{du}{dq}$$
$$= (650 - 0.25q)^{0.5}(-0.625) + (650 - 0.25q)(-0.125)(650 - 0.25q)^{-0.5}$$
$$= \frac{(650 - 0.25q)(-0.625) + (650 - 0.625q)(-0.125)}{(650 - 0.25q)^{0.5}}$$

Substituting the value $q = 1{,}040$ into this gives

$$\frac{d^2TR}{dq^2} = \frac{(300)(-0.625) + 0}{390^{0.5}} = -12.34 < 0$$

Therefore, the second-order condition is met and TR is maximized when $q = 1{,}040$.

Example 13.9

At what level of K is the function $Q = 12K^{0.4}(160 - 8K)^{0.4}$ at a maximum? (*This is Example 12.1 reworked, which was not completed in the last chapter.*)

Solution

We need to differentiate the function $Q = 12K^{0.4}(160 - 8K)^{0.4}$ to check the first-order condition for a maximum. To use the product rule, let

$$u = 12K^{0.4} \quad \text{and} \quad v = (160 - 8K)^{0.4}$$

and so

$$\frac{du}{dK} = 4.8K^{-0.6} \quad \text{and} \quad \frac{dv}{dK} = 0.4(160 - 8K)^{-0.6}(-8)$$
$$= -3.2(160 - 8K)^{-0.6}$$

Therefore,

$$\frac{dQ}{dK} = 12K^{0.4}(-3.2)(160 - 8K)^{-0.6} + (160 - 8K)^{0.4}\,4.8K^{-0.6}$$
$$= \frac{-38.4K + (160 - 8K)4.8}{(160 - 8K)^{0.6}K^{0.6}}$$
$$= \frac{768 - 76.8K}{(160 - 8K)^{0.6}K^{0.6}}$$

Setting this equal to zero for a stationary point must mean

$$768 - 76.8K = 0$$
$$K = 10$$

As we have already left this example in mid-solution once already, it will not do any harm to leave it once again. Although the second-order condition could be worked out using the product rule it is more convenient to use the quotient rule in this case and so we shall continue this problem later, in Example 13.13.

Example 13.10

In a perfectly competitive market the demand schedule is $p = 120 - 0.5q^2$ and the supply schedule is $p = 20 + 2q^2$. If the government imposes a per-unit tax t on the good sold in this market, what level of t will maximize the government's tax yield?

Solution

With the tax the supply schedule shifts upwards by the amount of the tax and becomes

$$p = 20 + 2q^2 + t$$

In equilibrium, demand price equals supply price. Therefore,

$$120 - 0.5q^2 = 20 + 2q^2 + t$$
$$100 - t = 2.5q^2$$
$$40 - 0.4t = q^2$$
$$(40 - 0.4t)^{0.5} = q$$

$$(40 - 0.4t)^{0.5} = q \qquad (1)$$

The government's tax yield (TY) is tq. Substituting (1) for q, this gives

$$\text{TY} = t(40 - 0.4t)^{0.5} \qquad (2)$$

We need to set $\text{dTY/d}t = 0$ for the first-order condition for maximization of TY. From (2), let

$$u = t \quad \text{and} \quad v = (40 - 0.4t)^{0.5}$$

giving

$$\frac{\text{d}u}{\text{d}t} = 1 \qquad \frac{\text{d}v}{\text{d}t} = 0.5(40 - 0.4t)^{-0.5}(-0.4) = -0.2(40 - 0.4t)^{-0.5}$$

Therefore, using the product rule

$$\frac{\text{dTY}}{\text{d}t} = t(-0.2)(40 - 0.4t)^{-0.5} + (40 - 0.4t)^{0.5}$$
$$= \frac{-0.2t + 40 - 0.4t}{(40 - 0.4t)^{0.5}}$$

and

$$\frac{40 - 0.6t}{(40 - 0.4t)^{0.5}} = 0 \qquad (3)$$

For finite values of t the first-order condition (3) will only hold when

$$40 - 0.6t = 0$$
$$66.67 = t$$

To check second-order conditions for this stationary point we need to find $\frac{\text{d}^2\text{TY}}{\text{d}t^2}$.
From (3)

$$\frac{\text{dTY}}{\text{d}t} = (40 - 0.6t)(40 - 0.4t)^{-0.5}$$

To differentiate using the product rule, let

$$u = 40 - 0.6t \quad \text{and} \quad v = (40 - 0.4t)^{-0.5}$$

giving

$$\frac{du}{dt} = -0.6 \qquad \frac{dv}{dt} = -0.5(40 - 0.4t)^{-1.5}(-0.4)$$

Therefore,

$$\frac{d^2TY}{dt^2} = (40 - 0.6t)\left[0.2(40 - 0.4t)^{-1.5}\right] + (40 - 0.4t)^{-0.5}(-0.6) \tag{4}$$

When $t = 66.67$, then $40 - 0.6t = 0$ and so the first term in (4) disappears, giving

$$\frac{d^2TY}{dt^2} = [40 - 0.4(66.67)]^{-0.5}(-0.6) = -0.1644 < 0$$

Therefore, the second-order condition for a maximum is satisfied when $t = 66.67$. Maximum tax revenue is raised when the per-unit tax is £66.67.

QUESTIONS 13.2

1. If $y = (6x + 7)^{0.5}(2.6x^2 - 1.9)$, what is dy/dx?
2. What output will maximize total revenue given the non-linear demand schedule $p = (60 - 2q)^{1.5}$?
3. Derive a function for the marginal product of L given the production function $Q = 85(0.5K^{0.8} + 3L^{0.5})^{0.6}$.
4. If $Q = 120K^{0.5}(250 - 0.5K)^{0.3}$ at what value of K will $dQ/dK = 0$? (That is, find the first-order condition for maximization of Q.)
5. In a perfectly competitive market the demand schedule is $p = 600 - 4q^{0.5}$ and the supply schedule is $p = 30 + 6q^{0.5}$. What level of a per-unit tax levied on the good sold in this market will maximize the government's tax yield?
6. For the demand schedule $p = (60 - 0.1q)^{0.5}$:
   (a) derive an expression for the slope of the demand schedule;
   (b) demonstrate that this slope gets flatter as q increases from 0 to 600;
   (c) find the output at which total revenue is a maximum.

13.4 THE QUOTIENT RULE

The quotient rule allows one to differentiate two functions where one function is divided by the other function.

If $y = u/v$ where u and v are functions of x, then according to the quotient rule

$$\frac{dy}{dx} = \frac{v\frac{du}{dx} - u\frac{dv}{dx}}{v^2}$$

Example 13.11

What is $\frac{dy}{dx}$ if $y = \frac{4x^2}{8+0.2x}$?

Solution

Defining relevant sub-functions and differentiating them

$$u = 4x^2 \quad \text{and} \quad v = 8 + 0.2x$$

$$\frac{du}{dx} = 8x \qquad \frac{dv}{dx} = 0.2$$

Therefore, according to the quotient rule,

$$\frac{dy}{dx} = \frac{v\frac{du}{dx} - u\frac{dv}{dx}}{v^2} = \frac{(8+0.2x)8x - 4x^2(0.2)}{(8+0.2x)^2}$$

$$= \frac{64x + 1.6x^2 - 0.8x^2}{(8+0.2x)^2} = \frac{64x + 0.8x^2}{(8+0.2x)^2}$$

This solution could also have been found using the product rule, since any function in the form $y = u/v$ can be written as $y = uv^{-1}$. We can check this by reworking Example 13.11 and differentiating the function $y = 4x^2(8 + 0.2x)^{-1}$.
Defining relevant sub-functions and differentiating them

$$u = 4x^2 \quad \text{and} \quad v = (8 + 0.2x)^{-1}$$

$$\frac{du}{dx} = 8x \qquad \frac{dv}{dx} = -0.2(8+0.2x)^{-2}$$

(Note that we use the chain rule to find dv/dx.)
Thus, using the product rule

$$\frac{dy}{dx} = u\frac{dv}{dx} + v\frac{du}{dx} = 4x^2\left[-0.2(8+0.2x)^{-2}\right] + (8+0.2x)^{-1}8x$$

$$= \frac{-0.8x^2 + (8+0.2x)8x}{(8+0.2x)^2} = \frac{-0.8x^2 + 64x + 1.6x^2}{(8+0.2x)^2}$$

$$= \frac{64x + 0.8x^2}{(8+0.2x)^2}$$

The answers obtained using both rules are identical, as expected.

Whether one chooses to use the quotient rule or the product rule depends on the functions to be differentiated. Only practice will give you an idea of which will be easier to use for specific examples.

Example 13.12

Derive a function for marginal revenue (in terms of q) if a monopoly faces the non-linear demand schedule $p = \frac{252}{(4+q)^{0.5}}$.

Solution

$$\text{TR} = pq = \frac{252q}{(4+q)^{0.5}}$$

Defining

$$u = 252q \quad \text{and} \quad v = (4+q)^{0.5}$$

$$\text{gives} \quad \frac{\text{d}u}{\text{d}q} = 252 \qquad \frac{\text{d}v}{\text{d}q} = 0.5(4+q)^{-0.5}$$

Therefore, using the quotient rule

$$\text{MR} = \frac{\text{dTR}}{\text{d}q} = \frac{v\frac{\text{d}u}{\text{d}q} - u\frac{\text{d}v}{\text{d}q}}{v^2} = \frac{(4+q)^{0.5}\,252q(0.5)(4+q)^{-0.5}}{4+q}$$

$$= \frac{(4+q)252 - 126q}{(4+q)^{1.5}} = \frac{1{,}008 + 126q}{(4+q)^{1.5}}$$

All three rules are useful when solving economic problems. We can now see how the quotient rule can be used to check the second-order condition in the unfinished Example 13.9.

Example 13.13

The objective is to find the value of K which maximizes $Q = 12K^{0.4}(160 - 8K)^{0.4}$. In Example 13.9, first-order conditions were satisfied when

$$\frac{\text{d}Q}{\text{d}K} = \frac{768 - 76.8K}{(160-8K)^{0.6}\,K^{0.6}} = 0$$

which holds when $K = 10$.

To derive $\text{d}^2 Q/\text{d}K^2$ let $u = 768 - 76.8K$ and $v = (160 - 8K)^{0.6}K^{0.6}$. Therefore,

$$\frac{\text{d}u}{\text{d}K} = -76.8 \tag{1}$$

and, using the product rule,

$$\frac{\text{d}v}{\text{d}K} = (160-8K)^{0.6}\,0.6K^{-0.4} + K^{0.6}\,0.6(160-8K)^{-0.4}(-8)$$

$$= \frac{(160-8K)0.6 - 4.8K}{K^{0.4}(160-8K)^{0.4}} = \frac{96 - 9.6K}{K^{0.4}(160-8K)^{0.4}} \tag{2}$$

Therefore, using the quotient rule and substituting (1) and (2)

$$\frac{\text{d}^2Q}{\text{d}K^2} = \frac{(160-8K)^{0.6}\,K^{0.6}(-76.8) - (768 - 768.K)\frac{96-9.6K}{K^{0.4}(160-8K)^{0.4}}}{(160-8K)^{1.2}\,K^{1.2}}$$

$$= \frac{(160-8K)K(-76.8) - 76.8(10-K)9.6(10-K)}{(160-8K)^{1.6}\,K^{1.6}}$$

At the stationary point when $K = 10$ several terms become zero, giving

$$\frac{d^2Q}{dK^2} = \frac{-76.8(800)}{(800)^{1.6}} < 0$$

Therefore, the second-order condition for a maximum is satisfied when $K = 10$.

Minimum average cost

In your introductory economics course, you were probably given an intuitive geometrical explanation of why a marginal cost schedule cuts a U-shaped average cost curve at its minimum point. The quotient rule can now be used to prove this rule.

In the short run, with only one variable input, assume that total cost (TC) is a function of q. Thus, MC = dTC/dq (as explained in Chapter 9) and, by definition, AC = TC/q.

To differentiate AC using the quotient rule let

$$u = \text{TC} \qquad \text{and} \qquad v = q$$

giving

$$\frac{du}{dq} = \frac{d\text{TC}}{dq} = \text{MC} \qquad \frac{dv}{dq} = 1$$

Therefore, using the quotient rule, first-order conditions for a minimum are

$$\frac{d\text{AC}}{dq} = \frac{q\text{MC} - \text{TC}}{q^2} = 0 \qquad (1)$$

$$q\text{MC} - \text{TC} = 0 \qquad (\text{or } q \to \infty, \text{ which we disregard}) \qquad (2)$$

$$\text{MC} = \frac{\text{TC}}{q} = \text{AC}$$

Therefore, MC = AC when AC is at a stationary point.

To check second-order conditions we need to find $\frac{d^2\text{AC}}{dq^2}$. From (1) we know that

$$\frac{d\text{AC}}{dq} = \frac{q\text{MC} - \text{TC}}{q^2}.$$

Again, use the quotient rule and let $u = q\text{MC} - \text{TC}$ and $v = q^2$ so that

$$\frac{du}{dq} = \left(q\frac{d\text{MC}}{dq} + \text{MC}\right) - \text{MC} = q\frac{d\text{MC}}{dq}$$

and $$\frac{dv}{dq} = 2q$$

Therefore,

$$\frac{d^2\text{AC}}{dq^2} = \frac{q^2\left(q\frac{d\text{MC}}{dq}\right) - (q\text{MC} - \text{TC})2q}{q^4} \qquad (3)$$

The first-order condition for a minimum is satisfied when qMC = TC, from (2). Substituting this result into (3) the second term in the numerator disappears and we get

$$\frac{d^2AC}{dq^2} = \frac{q^2\left(q\frac{dMC}{dq}\right)}{q^4} = \frac{1dMC}{qdq}$$

$$\frac{1}{q}\frac{dMC}{dq} > 0 \quad \text{when} \quad \frac{dMC}{dq} > 0$$

Therefore, the second-order condition for a minimum is satisfied when MC = AC and MC is rising. Thus, although MC may cut AC at another point when MC is falling, when MC is rising it cuts AC at its minimum point.

Individual labour supply

Not all of you will have encountered the theory of individual labour supply. Nevertheless, you should now be able to understand the following example which shows how the utility-maximizing combination of work and leisure hours can be found when an individual's utility function, wage rate and maximum working day are specified.

Example 13.14

In the theory of individual labour supply it is assumed that an individual derives utility from both leisure (L) and income (I). Income is determined by hours of work (H) multiplied by the hourly wage rate (w), i.e. $I = wH$.

Assume that each day a total of 12 hours is available for an individual to split between leisure and work, the wage rate is given as £4 an hour and that the individual's utility function is $U = L^{0.5}I^{0.75}$. How will this individual balance leisure and income to maximize utility?

Solution

A maximum working day of 12 hours means that hours of work $H = 12 - L$. Therefore, given an hourly wage of £4, income earned will be

$$I = wH = w(12 - L) = 4(12 - L) = 48 - 4L \tag{1}$$

Substituting (1) into the utility function

$$U = L^{0.5}I^{0.75} = L^{0.5}(48 - 4L)^{0.75} \tag{2}$$

To differentiate U using the product rule let

$$u = L^{0.5} \qquad \text{and} \qquad v = (48 - 4L)^{0.75}$$

giving

$$\frac{du}{dL} = 0.5L^{-0.5} \qquad \frac{dv}{dL} = 0.75(48 - 4L)^{-0.25}(-4)$$
$$= -3(48 - 4L)^{-0.25}$$

Therefore,

$$\frac{dU}{dL} = L^{0.5}\left[-348 - 4L^{-0.25}\right] + 48 - 4L^{0.75}\left(0.5L^{-0.5}\right)$$

$$= \frac{-3L + 48 - 4L0.5}{48 - 4L^{0.25}L^{0.5}} \qquad (3)$$

$$= \frac{24 - 5L}{48 - 4L^{0.25}L^{0.5}} = 0$$

for a stationary point. Therefore,

$$24 - 5L = 0$$
$$24 = 5L$$
$$4.8 = L$$

and so

$$H = 12 - 4.8 = 7.2 \text{ hours}$$

To check the second-order condition we need to differentiate (3). Let

$$u = 24 - 5L \quad \text{and} \quad v = (48 - 4L)^{0.25} L^{0.5}$$

giving

$$\frac{du}{dL} = -5$$

and

$$\frac{dv}{dL} = (48 - 4L)^{0.25}0.5L^{-0.5} + L^{0.5}0.25(48 - 4L)^{-0.75}(-4)$$

$$= \frac{(48 - 4L)0.5 - L}{L^{0.5}(48 - 4L)^{0.75}}$$

$$= \frac{24 - 3L}{L^{0.5}(48 - 4L)^{0.75}}$$

Therefore, using the quotient rule,

$$\frac{d^2U}{dL^2} = \frac{(48 - 4L)^{0.25}L^{0.5}(-5) - (24 - 5L)\left[(24 - 3L)/L^{0.5}(48 - 4L)^{0.75}\right]}{(48 - 4L)^{0.5}L}$$

When $L = 4.8$ then $24 - 5L = 0$ and so the second part of the numerator disappears. Then, dividing through top and bottom by $(48 - 4L)^{0.25}L^{0.5}$ we get

$$\frac{d^2U}{dL^2} = \frac{-5}{(48-4L)^{0.25}L^{0.5}} = -0.985 < 0$$

and so the second-order condition for maximization of utility is satisfied when 7.2 hours are worked and 4.8 hours are taken as leisure.

QUESTIONS 13.3

1. If $y = \dfrac{(3x+0.4x^2)}{(8-6x^{1.5})^{0.5}}$ what is $\dfrac{dy}{dx}$?
2. Derive a function for marginal revenue for the demand schedule

$$p = \frac{720}{(25+q)^{0.5}}$$

3. Using your answer from Question 13.2.4, show that the second-order condition for a maximum value of the function $Q = 120^{0.5}(250 - 0.5K)^{0.3}$ is satisfied when K is 312.5 and evaluate d^2Q/dK^2.
4. For the demand schedule $p = (800 - 0.4q)^{0.5}$, find which value of q will maximize total revenue, using the quotient rule to check the second-order condition.
5. Assume that an individual can choose the number of hours per day that they work up to a maximum of 12 hours. This individual attempts to maximize the utility function $U = L^{0.4}I^{0.6}$ where L is defined as hours not worked out of the 12-hour maximum working day, and I is income, equal to hours worked (H) times the hourly wage rate of £15. What mix of leisure and work will be chosen?
6. Show that when a firm faces a U-shaped short-run average variable cost (AVC) schedule, its marginal cost schedule will always cut the AVC schedule at its minimum point when MC is rising.

13.5 INTEGRATION

Integrating a function means finding another function that gives the first function when it is differentiated. It is basically differentiation in reverse and is sometimes referred to as anti-differentiation. The rules for integration are the reverse of those for differentiation. Economists like to express their ideas using graphs which show relationships between various variables, but sometimes it is important to put more precise values on analyzed functions. This is where the mathematical technique of integration comes in. It is commonly used by economists to calculate areas in graphs and to analyze probability distributions, of which you will learn more in a statistics course. The remainder of this chapter will look at some of the basic rules and methods of integration and show their applications to problems in economics and finance.

Assume that you wish to integrate the function

$$f'(x) = 12x + 24x^2$$

This means that you wish to find a function $y = f(x)$ such that

$$\frac{dy}{dx} = f'(x) = 12x + 24x^2$$

From your knowledge of differentiation, you should be able to work out that if

$$y = 6x^2 + 8x^3$$

then

$$\frac{dy}{dx} = 12x + 24x^2$$

However, although this is one solution, the same derivative can be obtained from other functions. For example, if $y = 35 + 6x^2 + 8x^3$ then we also get

$$\frac{dy}{dx} = 12x + 24x^2$$

In fact, whatever constant term is in the function the same derivative will be obtained. Since constant numbers disappear when a function is differentiated and we cannot know what constant should appear in an integrated function, we add a 'constant of integration' (C) to our solution. Because there is an indefinite number of possible solutions, this technique is often referred to as **indefinite integration**.

The notation used for integration is

$$y = \int f'(x)dx$$

This means that y is the integral of the function $f'(x)$. The sign $\int$ is known as the integration sign. The 'dx' signifies that if y is differentiated with respect to x, the result will be $f'(x)$. We can therefore write the integral of the previous example as

$$y = \int \left(12x + 24x^2\right)dx = 6x^2 + 8x^3 + C$$

The general rule for the integration of individual terms in an expression is

$$\int ax^n dx = \frac{ax^{n+1}}{n+1} + C$$

where a and n are given parameters and $n \neq -1$.

(The special case when $n = -1$, i.e. the integral $\int(1/x)\mathrm{d}x$, will be dealt with in Chapter 15 when we cover exponential functions.)

As you may have already noticed, if the function to be integrated is a sum or difference of two or more elements, then this rule is applied to each element individually. (This is analogous to differentiation as explained in Section 9.2.)

It is always a good idea to check whether your solution differentiates back to the initial function that is being integrated.

Example 13.15

Solve the integral $\int x^4\,\mathrm{d}x$.

Solution

We need to find a function y which differentiates to $f'(x) = x^4$.
When a function is differentiated, we subtract 1 from the exponent. Thus, to integrate, we add 1 to the power of our function and so

$$x^{4+1} = x^5$$

Differentiating this gives

$$(x^5)' = 5x^4$$

This is not the same as $f(x)$. The result is five times too large. Dividing by 5 we get

$$y = \frac{1}{5}x^5$$

Now, if we differentiate y we get

$$\frac{\mathrm{d}y}{\mathrm{d}x} = 5 \times \frac{1}{5}x^{5-1} = x^4$$

although we still need to add the constant of integration to our result. The final solution of the integration is therefore

$$\int x^4 \mathrm{d}x = \frac{1}{5}x^5 + C$$

To check whether our solution is correct we differentiate it, giving

$$\left(\frac{1}{5}x^5 + C\right)' = x^4 + 0 = x^4$$

This is the original function, and so our solution is correct.

Using this technique you should now be able to check the integrals in the following example.

Example 13.16

Solve the following integrals:	Solutions:
(i) $\int 30x^4 dx$	$y = 6x^5 + C$
(ii) $\int (24 + 7.2x) dx$	$y = 24x + 3.6x^2 + C$
(iii) $\int 0.5x^{-0.5} dx$	$y = x^{0.5} + C$
(iv) $\int (48x - 0.4x^{-1.4}) dx$	$y = 24x^2 + x^{-0.4} + C$
(v) $\int (65 + 1.5x^{-2.5} + 1.5x^2) dx$	$y = 65x - x^{-1.5} + 0.5x^3 + C$

In earlier chapters, we have seen how the differentiation of total cost, total revenue and other functions gives the corresponding marginal function. For example,

$$\frac{dTC}{dq} = MC \quad \text{and} \quad \frac{dTR}{dq} = MR$$

Therefore, the integration of the marginal function will give the corresponding total function, apart from the unknown constant. However, we can often work out the constant if given some additional information.

Total cost functions can usually be split into fixed and variable components. The integral of marginal cost will give total variable costs plus a constant of integration which should be equal to total fixed cost (TFC). For example, if we are given the information that total variable cost is

$$TVC = 25q - 6q^2 + 0.8q^3$$

and total fixed cost is $TFC = 10$

then, by definition $TC = TVC + TFC$

$$= 10 + 25q - 6q^2 + 0.8q^3$$

Since constants differentiate to zero, marginal cost can be found by differentiating either total cost or total variable cost. In this case it will be

$$MC = \frac{dTC}{dq} = \frac{dTVC}{dq} = 25 - 12q + 2.4q^2$$

Integrating marginal cost will give total variable cost, but to work out total cost we also need to know total fixed cost. For example, if we were given the information that total fixed cost was 10 and that

$$MC = 25 - 12q + 2.4q^2$$

then we could find total variable cost by integration (*in this case we ignore the constant of integration*) as

$$TVC = \int MC dq = 25q - 6q^2 + 0.8q^3$$

Thus,

$$TC = TFC + TVC = 10 + 25q - 6q^2 + 0.8q^3$$

Example 13.17

If a firm spends £650 on fixed costs, what is its total cost function if its marginal cost function is MC = 82 − 16q + 1.8q^2?

Solution

We know that for any cost function

$$\text{TVC} = \int \text{MC}\,\text{d}q \quad (\text{ignoring } C)$$

Therefore,

$$\text{TVC} = \int\left(82 - 16q + 1.8q^2\right)\text{d}q = 82q - 8q^2 + 0.6q^3$$

We know that TC = TFC + TVC and TFC = 650 and therefore

$$\text{TC} = 650 + 82q - 8q^2 + 0.6q^3$$

If one is given a firm's marginal revenue function then one can integrate this to find the total revenue function. For example, if

$$\text{MR} = 360 - 2.5q$$
$$\text{TR} = \int \text{MR}\,\text{d}q = 360q - 1.25q^2 + C$$

When q is zero, TR must also be zero (because TR = pq). Thus C = 0 and

$$\text{TR} = 360q - 1.25q^2$$

Example 13.18

If MR = 520 − 3$q^{0.5}$ what is the corresponding TR function?

Solution

$$\text{TR} = \int \text{MR}\,\text{d}q = \int\left(520 - 3q^{0.5}\right)\text{d}q = 520q - 2q^{1.5}$$

Recall that TR = $p \times q$ and the relationship between price and quantity is described by the demand function. Thus, once the TR function corresponding to a given MR function has been derived, one has to divide it by q to arrive at the inverse demand function, as the following exercise demonstrates.

Example 13.19

What total revenue will a firm earn if it charges a price of £715 and its marginal revenue function is MR = $960 - 0.15q^2$?

Solution

As we have established that the integral of this form of MR function will not have a constant of integration, then

$$\text{TR} = \int \text{MR}dq = \int \left(960 - 0.15q^2\right)dq = 960q - 0.05q^3$$

We now need to find the quantity sold at the price of £715 by using the TR function to find the inverse demand schedule.

Since TR = pq then p = TR/q and so

$$p = \frac{1}{q}\left(960q - 0.05q^3\right) = 960 - 0.05q^2$$

Solving for q:

$$\begin{aligned} 0.05q^2 &= 960 - p \\ q^2 &= 19{,}200 - 20p \\ q &= (19{,}200 - 20p)^{0.5} \end{aligned}$$

When $p = 715$ then

$$q = (19{,}200 - 14{,}300)^{0.5} = 4{,}900^{0.5} = 70$$

and so total revenue will be

$$\text{TR} = pq = 715(70) = £50{,}050$$

If both MC and MR functions are known, then one can use integration to work out what the actual profit is at any given level of output, provided that TFC is also known.

Example 13.20

If a firm faces the marginal cost schedule MC = $180 + 0.3q^2$ and the marginal revenue schedule MR = $540 - 0.6q^2$, and total fixed costs are £65, what is the maximum profit it can make? (Assume that the second-order condition for a maximum is met.)

Solution

Profit is maximized when MC = MR. Therefore,

$$180 + 0.3q^2 = 540 - 0.6q^2$$
$$0.9q^2 = 360$$
$$q^2 = 400$$
$$q = 20$$

To find the actual profit (π), we now integrate to get TR and TC and then subtract TC from TR:

$$\text{TR} = \int \text{MR}\,\text{d}q = \int\left(540 - 0.6q^2\right)\text{d}q = 540q - 0.2q^3$$
$$\text{TC} = \int \text{MC}\,\text{d}q + \text{TFC} = \int\left(180 + 0.3q^2\right)\text{d}q + 65 = 180q + 0.1q^3 + 65$$
$$\pi = \text{TR} - \text{TC\&} = 540q - 0.2q^3 - \left(180q + 0.1q^3 + 65\right)$$
$$= 540q - 0.2q^3 - 180q - 0.1q^3 - 65 = 360q - 0.3q^3 - 65$$

Thus, when $q = 20$ the maximum profit level is

$$\pi = 360(20) - 0.3(20)^3 - 65 = £4,735$$

QUESTIONS 13.4

1. Integrate the following functions:
   (a) $25x$
   (b) $5 + 1.2x + 0.15x^2$
   (c) $120x^4 - 60x^3$
   (d) $42 - 18x^{-2}$
   (e) $90x^{0.5} - 44x^{-1.2}$

2. Find the total variable cost functions corresponding to the following marginal cost functions:
   (a) $\text{MC} = 4 + 0.1q$
   (b) $\text{MC} = 42 - 18q + 6q^2$
   (c) $\text{MC} = 35 + 0.9q^2$
   (d) $\text{MC} = 62 - 16q + 1.5q^2$
   (e) $\text{MC} = 185 - 24q + 1.2q^3$

3. Find the corresponding total revenue functions for the marginal revenue functions:
   (a) $\text{MR} = 40 - 4q$
   (b) $\text{MR} = 600 - 25q$

13.6 DEFINITE INTEGRALS

The previous section looked at indefinite integrals and now we are going to study **definite integrals**. Definite integration is a technique most commonly used to find sizes of areas of different types of shapes bounded by non-linear functions. For example, economists use this technique to evaluate consumer and producer surpluses, probabilities of occurrences of various events, investment flows and present values.

We use the same symbol of integration to denote definite integrals but add two values called the limits of integration at the top and bottom of the integration sign. For example, the definite integral $\int_3^8 6x\text{d}x$ gives the size of the area between the function f(x), x axis and two vertical lines at x = 3 and x = 8, as shown by the shaded area in Figure 13.1.

It is calculated as the value of this integral when x is 8 minus its value when x is 3. Thus, given that

$$\int 6x^2\text{d}x = 2x^3 + C$$

then

$$\int_3^8 6x^2\text{d}x = \left[2x^3 + C\right]_3^8 = \left[2(8)^3 + C\right] - \left[2(3)^3 + C\right]$$
$$= 1024 + C - 54 - C = 970$$

In any definite integral the two constants of integration will always cancel out and so they can be omitted.

This example illustrates the usual notation used when evaluating a definite integral of writing the relevant values outside a set of square brackets which contains the integral of the given function. So, given that constants of integration always cancel out, this example can be restated more concisely as

$$\int_3^8 6x^2\text{d}x = \left[2x^3\right]_3^8 = 2(8)^3 - 2(3)^3 = 1{,}024 - 54 = 970$$

The same procedure is used for more complex functions.

Example 13.21

Evaluate the definite integral $\int_5^6 \left(6x^{0.5} - 3x^{-2} + 85x^4\right)\text{d}x$.

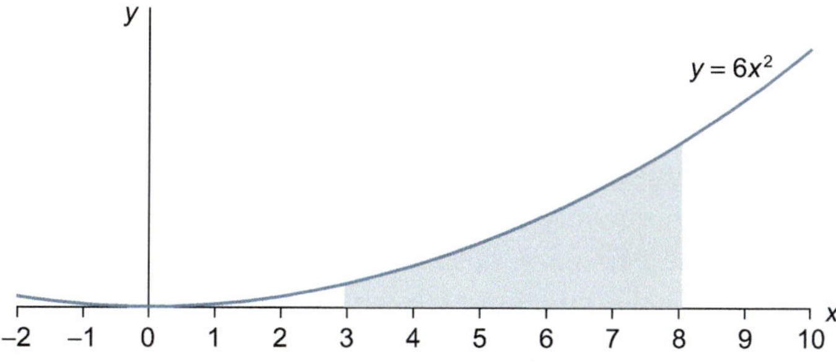

Figure 13.1 Size of an area calculated using definite integrals

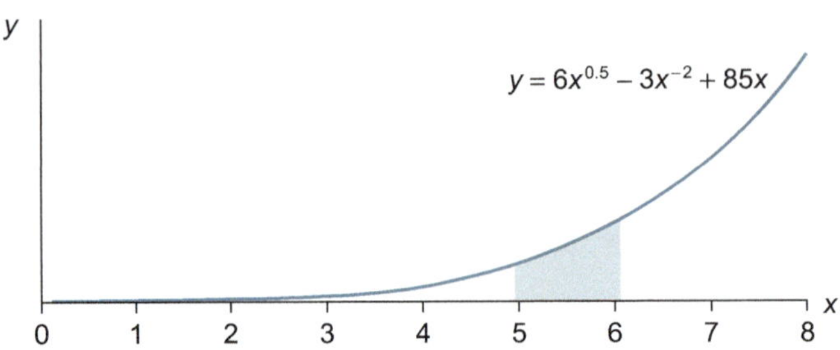

Figure 13.2 Size of an area calculated using definite integrals

Solution

$$\int_5^6 \left(6x^{0.5} - 3x^{-2} + 85x^4\right)\mathrm{d}x = \left[4x^{1.5} + 3x^{-1} + 17x^5\right]_5^6$$
$$= (58.788 + 0.5 + 132{,}192) - (44.721 + 0.6 + 53{,}125)$$
$$= 132{,}251.29 - 53{,}170.32 = 79080.97$$

As before, the solution gives the area between a function and the horizontal axis that is between the limits of integration, as shown in Figure 13.2.

Definite integrals of marginal cost and marginal revenue functions

This concept of the definite integral has several applications in economics. To evaluate TVC from an MC function for a given value of output one simply evaluates the definite integral of MC between zero and the given quantity. For example, assume that you wished to find the value of TVC when $q = 8$ and you are given the marginal cost function MC = 7.5 + $0.3q^2$. This TVC value would be equal to the area under the MC schedule between zero and the given quantity of 8, which is

$$\int_0^8 (7.5 + 0.3q^2)\mathrm{d}q = \left[7.5q + 0.1q^3\right]_0^8 = 60 + 51.2 = 111.2$$

We can also see that the increase in TVC between two quantities will be equal to the area under the corresponding MC schedule between the given quantities. Assume that the marginal cost function is MC = $6x^2$, where x is output and cost is in £ and you wish to determine the increase in TVC when output is increased from 3 to 8 units. This will be the shaded area in Figure 13.1 which we have already found to be 970 'square units', or £970 when the function represents MC. This must be so because this area represents the definite integral $\left[2x^3\right]_3^8$ which is the value of TVC when quantity is 8 minus its value when quantity is 3.

The definite integral of a function between two given quantities has been shown to be equal to the area under the function between the two quantities but above the horizontal axis. If a function takes negative values, i.e. it goes below the horizontal axis, then definite integration evaluates the size of an area below the horizontal axis but

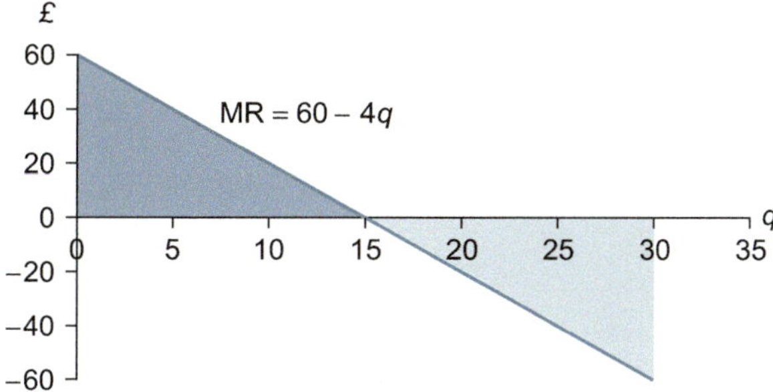

Figure 13.3 Marginal revenue function: integrating to find the value of total revenue

above the function, and the result has a negative sign. For example, Figure 13.3 shows a graph of the marginal revenue function MR = 60 − 4q. If we integrate it over the interval from $q = 0$ to $q = 15$, then we will obtain the value of total revenue obtained from selling 15 units which will be

$$\int_0^{15} (60 - 4q)\,dq = \left[60q - 2q^2\right]_0^{15} = 900 - 450 = 450$$

If we want to know by how much the total revenue will change if the sales are increased from 15 to 30 units, then we evaluate the definite integral of the MR function over the interval from $q = 15$ to $q = 30$, giving

$$\int_{15}^{30} (60 - 4q)\,dq = \left[60q - 2q^2\right]_{15}^{30} = (1800 - 1800) - (900 - 450) = -450$$

Thus, an increase in sales from 15 to 30 units will lead to a decrease in total revenue of £450. This result has a negative sign because over this output range the MR function lies below the horizontal axis, as shown in Figure 13.3.

You should also be able to work out that when output is 30 units the firm's total revenue will be zero because the positive revenue of £450 from selling the first 15 units will be cancelled out by the drop in revenue from selling a further 15 units. Can you think of a reason for this?

Consumer and producer surplus

We can also use definite integration to determine consumer and producer surplus. Consumer surplus is defined as the area below a demand schedule but above the market price. This difference between what consumers are willing to pay for a good and what they actually have to pay is often used as a measure of welfare. Figure 13.4 shows a non-linear inverse demand function $p = 30 - q^2$. The shaded area represents the consumer surplus when price is £21, which corresponds to $q = 3$.

To find the value of this consumer surplus we first need to find the size of the whole area under the demand schedule between zero and an output of 3. Using definite integration this area will give

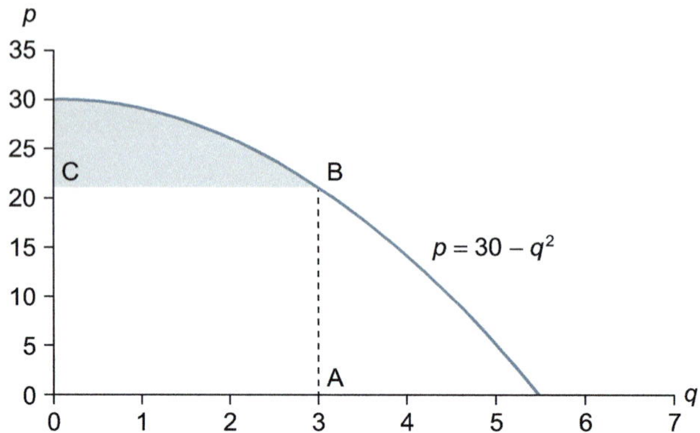

Figure 13.4 Consumer surplus

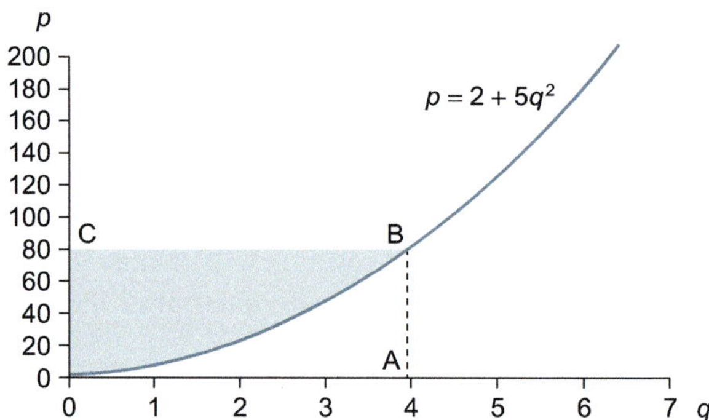

Figure 13.5 Producer surplus

$$\int_0^3 \left(30 - q^2\right) \mathrm{d}q = \left[30q - \tfrac{1}{3}q^3\right]_0^3 = 90 - 9 = 81$$

Since we are only interested in the area between the demand curve and the market price level, we need to deduct the value of the area given by the rectangle 0ABC.

Thus, the consumer surplus in our example is

$$\text{CS} = 81 - (3 \times 21) = 81 - 63 = 18$$

By analogy, producer surplus is given by the area above a supply curve and below the market price. For the non-linear inverse supply function $p = 2 + 5q^2$ the producer surplus when 4 units are supplied and price is £82 will be the shaded area shown in Figure 13.5.

This producer surplus is calculated by subtracting the size of an area between the supply curve and the axes, up to the quantity supplied, from a rectangular area limited by the axes, price level and quantity supplied (area 0ABC minus area 0AB in Figure 13.5). Thus, first we find the size of the rectangular area 0ABC:

$$\text{area 0ABC} = 4 \times 82 = 328$$

and then subtract a definite integral of the inverse supply function over the interval from the origin to $q = 4$. Therefore, producer surplus will be

$$PS = 328 - \int_0^4 (2 + 5q^2)\,dq = \left[2q + \tfrac{5}{3}q^3\right]_0^4 = 328 - (8 + 106.67) = 213.33$$

Example 13.22

A competitive market is described by the non-linear demand schedule $p = 1{,}946 - q^2$ and the supply schedule $p = 2 + 5q^2$. Find the following:

(i) equilibrium quantity and price
(ii) consumer surplus
(iii) producer surplus.

Solution

(i) Equating demand and supply prices to find the equilibrium quantity

$$\begin{aligned} 1{,}946 - q^2 &= 2 + 5q^2 \\ 6q^2 &= 1{,}944 \\ q^2 &= 324 \\ q &= 18 \qquad (\text{reject } q = -18) \end{aligned}$$

Substituting $q = 18$ into the inverse demand function to obtain the equilibrium price,

$$p = 1{,}946 - q^2 = 1{,}946 - 18^2 = 1{,}622$$

(ii) To calculate consumer surplus when the market is in equilibrium, we evaluate the definite integral of the inverse demand function between $q = 0$ and the equilibrium quantity and then subtract consumers' expenditure given by the rectangular area below the market price. Thus, consumer surplus will be

$$\begin{aligned} CS &= \int_0^{18} (1{,}946 - q^2)\,dq - 18 \times 1{,}622 \\ &= \left[1{,}946q - \tfrac{1}{3}q^3\right]_0^{18} - 29{,}196 = (35{,}028 - 1{,}944) - 29{,}196 = 3{,}888 \end{aligned}$$

(iii) To calculate producer surplus we subtract the definite integral of the inverse supply function from consumers' expenditure found in part (ii). Thus,

$$\begin{aligned} PS &= 29{,}196 - \int_0^{18} (2 + 5q^2)\,dq \\ &= 29{,}196 - \left[2q + \frac{5}{3}q^3\right]_0^{18} = 29{,}196 - (36 + 9{,}720) = 19{,}440 \end{aligned}$$

Investment flows and capital formation

The technique of definite integration is also useful in the realm of finance where it is used to appraise investment flows and discounting. When we invest capital in a project, we expect to receive returns (sometimes called 'money flows') during the project's duration. These flows will change the value of capital stock, and the rate of this change is referred to as net investment. Since the value of accumulated capital, K, changes with time, it can be expressed as a function of time, $K(t)$. If we differentiate this function with respect to t we will obtain the amount by which the capital stock is increasing from one year to another, i.e. the net investment

$$I(t) = \frac{dK}{dt}$$

In some situations, we may know the net investment function because, for example, the project promises particular returns in each year of its duration. Instead, we may want to calculate the capital formation, i.e. the amount by which the value of capital stock changes during the period from t_1 to t_2. To do this we need to evaluate the definite integral

$$\int_{t1}^{t2} I(t)dt$$

which tells us the value of all returns from the project within the specified time period.

Example 13.23

The returns from a project are described by the net investment function $I(t) = 15{,}000t^{-0.5}$, where t is a year in the project's duration. Find the following:

(i) the value of the return in year 4
(ii) the capital formation from the end of year 1 to the end of year 5
(iii) the number of years needed to accumulate £60,000.

Solution

(i) To find the value of returns in any time period we need to substitute the period number for t. Here we have

$$I(4) = 15{,}000t^{-0.5} = 15{,}000 \times 4^{-0.5} = \frac{15{,}000}{2} = 7{,}500$$

Thus, the money flow generated by the project in year 4 is £7,500.

(ii) In order to calculate the capital formation we need to evaluate the definite integral

$$\int_1^5 15{,}000t^{-0.5}dt = \left[30{,}000t^{0.5}\right]_1^5 = 67{,}082.04 - 30{,000} = 37{,}082.04$$

Thus, the project will generate returns totalling £37,082.04 between the end of year 1 and the end of year 5.

(iii) We need to calculate the number of years, T, required to accumulate £60,000. The investment project starts to generate returns from the beginning of the first year, so we need to integrate the net investment function over the time interval from zero to T and set it equal to £60,000. Therefore,

$$\int_0^T 15,000t^{-0.5}dt = 60,000$$
$$\left[30,000t^{0.5}\right]_0^T = 60,000$$
$$30,000T^{0.5} = 60,000$$
$$T^{0.5} = 2$$
$$T = 4$$

This solution shows that it will take exactly 4 years from the project's commencement for the capital formation to reach £60,000. In other words, £15,000 invested today will generate net returns of £60,000 over the next 4 years.

QUESTIONS 13.5

1. Given the non-linear demand schedule $p = 600 - 6q^{0.5}$ and the corresponding marginal revenue function MR $= 600 - 9q^{0.5}$, use definite integrals to find:
   (a) total revenue when q is 2,500;
   (b) the change in total revenue when q increases from 2,025 to 2,500;
   (c) consumer surplus when q is 2,500 and price is £300;
   (d) the change in consumer surplus when q increases from 2,025 to 2,500 owing to a price fall from £330 to £300.
2. If a firm faces the marginal cost function MC $= 40 - 18q + 4.5q^2$ what would be the increase in total cost if output were increased from 30 to 40?
3. Given the net investment function $I(t) = 22,000t^{0.5}$
   (a) find the capital formation from the end of the third year to the end of the seventh year;
   (b) how long will it take for the returns to accumulate to £132,000?
4. If investment flows are described by the net investment function $I(t) = 1,200t^{0.2}$
   (a) calculate the capital formation that occurs between the end of year 2 and the end of year 5;
   (b) what is the number of years required for capital formation to reach £24,000?

13.7 INTEGRATION BY SUBSTITUTION AND INTEGRATION BY PARTS

Some functions take more complex forms and cannot be integrated using the simple rules described in the previous sections. In such cases we often resort to either of the two methods: integration by substitution or by parts. By the end of this section you will be able to decide which method is the appropriate one as well as apply them to solve complex integrals.

Method of substitution

The method of **integration by substitution** follows from the chain rule of differentiation and is used when the function to be integrated can be simplified by a substitution of a single variable, here u, for a part of the function. Consider the integral

$$\int (7x+4)^6 \,\mathrm{d}x$$

We can simplify this by replacing the expression inside the brackets by u, so

$$u = 7x + 4$$

We need to make one more adjustment before we can proceed with integrating. After the substitution we get $\int u^6 \mathrm{d}x$, however in order to integrate functions of u we need to have du at the end of our integral. Thus, we will derive an expression for dx in terms of du following these steps:

Differentiate $u = 7x + 4$ with respect to x, giving

$$\frac{\mathrm{d}u}{\mathrm{d}x} = 7$$

Re-arrange this equation to get dx on the left-hand side

$$\mathrm{d}u = 7\mathrm{d}x$$
$$\mathrm{d}x = \frac{1}{7}\mathrm{d}u$$

Now replace dx by $\frac{1}{7}\mathrm{d}u$ in the integral, to give

$$\int u^6 \left(\frac{1}{7}\right) \mathrm{d}u$$

This is now easy to integrate using the method learned in Section 13.5, starting by putting the constant value outside the integral. Thus,

$$\frac{1}{7}\int u^6 \mathrm{d}u = \frac{1}{7}\left(\frac{1}{7}u^7\right) + C = \frac{1}{49}u^7 + C$$

Remembering that the original function was in terms of x not u, to get the final solution we need to replace u by $(7x + 4)$, giving the solution

$$\int (7x+4)^6 \,\mathrm{d}x = \frac{(7x+4)^7}{49} + C$$

Example 13.24

Use the method of substitution to solve the integral $\int \frac{1}{(2x-1)^2}\mathrm{d}x$.

Solution

Substitute the term $u = 2x - 1$ to get the integral

$$\int \frac{1}{u^2}\mathrm{d}x = \int u^{-2}\mathrm{d}x$$

To replace dx with du, differentiate u with respect to x and then re-arrange the resulting equation to isolate dx, giving

$$\frac{\mathrm{d}u}{\mathrm{d}x} = 2 \quad \text{and thus} \quad \mathrm{d}x = \frac{1}{2}\,\mathrm{d}u$$

Substituting for dx gives us an integral which is easy to solve, as

$$\int u^{-2}\left(\frac{1}{2}\right)\mathrm{d}u = \frac{1}{2}\int u^{-2}\mathrm{d}u = -\frac{1}{2}u^{-1} + C$$

Finally, substitute back for u to get the solution:

$$\int \frac{1}{(2x-1)^2}\mathrm{d}x = -\frac{1}{2(2x-1)} + C$$

The method of integration by substitution is particularly useful when the function we want to integrate can be decomposed into two parts which are multiplied by each other and one of these parts is the derivative of the other. For example, for

$$\int \left(3x^2 + 4x^3\right)\left(6x + 12x^2\right)\mathrm{d}x$$

the function to be integrated consists of two multiplicative parts and we can see that the content of the second bracket is the derivative of the first bracket. Thus, we can integrate it by substitution. We start by replacing the first part with

$$u = 3x^2 + 4x^3 \qquad (1)$$

As before, we differentiate u with respect to x in order to replace dx by du, and so

$$\frac{\mathrm{d}u}{\mathrm{d}x} = 6x + 12x^2$$

Multiplying both sides of this equation by dx gives

$$\mathrm{d}u = \left(6x + 12x^2\right)\mathrm{d}x \qquad (2)$$

Note that the right-hand side of this equation is exactly the same as the second part of the original integral together with dx. Thus we can substitute (1) for $(3x^2 + 4x^3)$ and (2) for $(6x + 12x^2)\mathrm{d}x$ to get a new integral which is simple to solve

$$\int u\mathrm{d}u = \frac{1}{2}u^2 + C$$

Now we only need to substitute back for u to get the final solution

$$\int (3x^2 + 4x^3)(6x + 12x^2)\mathrm{d}x = \frac{1}{2}(3x^2 + 4x^3)^2 + C$$

You can differentiate this solution using the chain rule in order to check if this is correct.

Integration by parts

This technique is also used to integrate more complex functions that consist of two multiplicative parts, i.e. elements which are multiplied by each other. It follows directly from the product rule of differentiation covered earlier in this chapter and can be best explained using a simple example.

Assume you want to solve the following integral by parts

$$\int x(x^3 + 1)\mathrm{d}x$$

We will use the **integration by parts formula**:

$$\int u\mathrm{d}v = uv - \int v\mathrm{d}u$$

To begin we need to identify the parts u and dv in our integral and then compute du and v. Thus, let

$$u = x \quad \text{and} \quad \mathrm{d}v = (x^3 + 1)\mathrm{d}x$$

then

$$\frac{\mathrm{d}u}{\mathrm{d}x} = 1 \quad \text{and} \quad v = \int \mathrm{d}v$$

$$\mathrm{d}u = \mathrm{d}x \qquad v = \int (x^3 + 1)\mathrm{d}x = \frac{1}{4}x^4 + x$$

(We can ignore the constant of integration at this stage but we need to remember to add it to the final result.)

We can now substitute into the integration by parts formula and solve, giving

$$\int u\mathrm{d}v = uv - \int v\mathrm{d}u = \int x\left(x^3+1\right)\mathrm{d}x = x\left(\frac{1}{4}x^4+x\right) - \int\left(\frac{1}{4}x^4+x\right)\mathrm{d}x$$

$$= \frac{1}{4}x^5 + x^2 - \left(\frac{1}{20}x^5 + \frac{1}{2}x^2\right) + C$$

$$= \frac{1}{5}x^5 + \frac{1}{2}x^2 + C$$

As usual, it is a good idea to check if our result is correct by differentiating it. Thus,

$$\left(\frac{1}{5}x^5 + \frac{1}{2}x^2 + C\right)' = x^4 + x = x\left(x^3+1\right)$$

and so we know our integral is correct.

The technique of integration by parts is particularly useful for integrating composite exponential functions (covered in Chapter 15) and other 'non-standard' functions. It is frequently used in conjunction with the substitution method to tackle integrals of more complex functions that you should be familiar with by now.

Example 13.25

Solve the integral $\int \frac{z^3}{\left(z^2+5\right)^3}\mathrm{d}z$.

Solution

This integral requires the use of both methods. Note that if we differentiate the expression in the bracket, we get $2z$. In order to be able to use the method of substitution we want to have z instead of z^3 in the numerator, therefore we rewrite our initial functions as

$$\int z^2 \frac{z}{\left(z^2+5\right)^3}\mathrm{d}z$$

Now we have two parts multiplied by each other and we can apply the integration by parts formula, $\int u\mathrm{d}v = uv - \int v\mathrm{d}u$, where

$$u = z^2 \quad \text{and} \quad \mathrm{d}v = \frac{z}{\left(z^2+5\right)^3}\mathrm{d}z$$

then

$$\frac{\mathrm{d}u}{\mathrm{d}z} = 2z \quad \text{and} \quad v = \int \mathrm{d}v$$

$$\mathrm{d}u = 2z\mathrm{d}z \qquad v = \int \frac{z}{\left(z^2+5\right)^3}\mathrm{d}z \tag{1}$$

We use integration by substitution to solve (1).

Let $w = z^2 + 5$, then $\frac{\mathrm{d}w}{\mathrm{d}z} = 2z$ and $z\mathrm{d}z = 0.5\mathrm{d}w$. Substituting these into (1) gives

$$\int \frac{0.5\mathrm{d}w}{w^3} = \frac{-1}{4w^2} \quad \text{and thus} \quad v = \frac{-1}{4\left(z^2+5\right)^2}$$

We now have all the parts needed in the integration by parts formula

$$\int z^2 \frac{z}{\left(z^2+5\right)^3}\,\mathrm{d}z = \frac{-z^2}{4\left(z^2+5\right)^2} - \int \frac{-2z}{4\left(z^2+5\right)^2}\,\mathrm{d}z \tag{2}$$

To solve the last integral in (2) we need to use the substitution method again. As before we use $w = z^2 + 5$ and $z\mathrm{d}z = 0.5\mathrm{d}w$ giving

$$\begin{aligned}\int \frac{-2z}{4\left(z^2+5\right)^2}\,\mathrm{d}z = -\frac{1}{2}\int \frac{z}{\left(z^2+5\right)^2}\,\mathrm{d}z = -\frac{1}{2}\int \frac{0.5dw}{w^2} &= \frac{1}{4w}+C \\ &= \frac{1}{4\left(z^2+5\right)}+C\end{aligned} \tag{3}$$

Substituting (3) into (2) will give us the final solution:

$$\begin{aligned}\int z^2 \frac{z}{\left(z^2+5\right)^3}\,\mathrm{d}z &= \frac{-z^2}{4\left(z^2+5\right)^2} - \frac{1}{4\left(z^2+5\right)} + C \\ &= \frac{-z^2}{4\left(z^2+5\right)^2} - \frac{z^2+5}{4\left(z^2+5\right)^2} + C \\ &= \frac{-2z^2-5}{4\left(z^2+5\right)^2} + C\end{aligned}$$

QUESTIONS 13.6

1. Use the method of substitution to solve the integrals:

   (a) $\int (21+6x)^3\,\mathrm{d}x$

   (b) $\int \frac{3}{(4-3m)^2}\,\mathrm{d}m$

   (c) $\int \left(8x+6x^5\right)^3\left(8+30x^4\right)\mathrm{d}x$

   (d) $\int \frac{3x^2}{\left(x^3-6\right)^4}\,\mathrm{d}x$

   (e) $\int \frac{20y^7+1}{\left(5y^8+2y\right)^3}\,\mathrm{d}y$

2. Use the method of integration by parts to solve the integrals:

   (a) $\int x^5\left(x^4+2x\right)\mathrm{d}x$ (b) $\int x^{-4}\left(x^2+3\right)\mathrm{d}x$ (c) $\int x\left(6+2x^2\right)\mathrm{d}x$

3. Decide which method of integration is more appropriate in each case and solve the following integrals:

   (a) $\int \frac{2}{(3+4x)^2}\,\mathrm{d}x$ (b) $\int x^5\sqrt{x^3+1}\,\mathrm{d}x$ (c) $\int (7+2x)^5\,\mathrm{d}x$

14 Dynamics and difference equations

Learning objectives

After completing this chapter students should be able to:

- demonstrate how a time lag can affect the pattern of adjustment to equilibrium in some basic economic models
- construct spreadsheets to plot the time path of dependent variables in economic models with simple lag structures
- set up and solve linear first-order difference equations
- apply the difference equation solution method to the cobweb, Keynesian and Bertrand models involving a single lag
- identify the stability conditions in these models.

14.1 DYNAMIC ECONOMIC ANALYSIS

In earlier chapters much of the economic analysis used has been comparative statics. This entails the comparison of different (static) equilibrium situations, with no mention of the mechanism by which price, quantity or other variables adjust to their new equilibrium values. The branch of economic analysis that looks at how variables adjust between equilibrium values is known as 'dynamics', and this chapter gives an introduction to some simple dynamic economic models.

The ways in which markets adjust over time vary tremendously. In some financial markets, such as commodity futures exchanges, prices are changed every few seconds and adjustments to new equilibrium prices are almost instantaneous. In other markets the adjustment process may be a slow trial and error process over several years, in some cases so slow that price and quantity hardly ever reach their proper equilibrium values because supply and demand schedules shift before equilibrium has been reached. There is therefore no one economic model that can explain the dynamic adjustment process in all markets.

DOI: 10.4324/9781003360827-14

The simple dynamic adjustment models explained here will give you an idea of how adjustments can take place between equilibria in certain types of markets and how mathematics can be used to calculate the values of variables at different points in time during the adjustment process. They are only very basic models, however, designed to give you an introduction to this branch of economics. The mathematics required to analyze more complex dynamic models goes beyond the scope of this text.

In this chapter, time is considered as a discrete variable and the dynamic adjustment process between equilibria is seen as a step-by-step process. (The distinction between discrete and continuous variables was explained in Section 8.1.) This enables us to calculate different values of the variables that are adjusting to new equilibrium levels:

(i) using a spreadsheet, and
(ii) using the mathematical concept of 'difference equations'.

Models that assume a process of continual adjustment are considered in Chapter 15, using 'differential equations'.

14.2 THE COBWEB: ITERATIVE SOLUTIONS

In some markets, particularly agricultural markets, supply cannot immediately expand to meet increased demand. Crops have to be planted and grown and livestock takes time to raise. Some manufactured products can also take a while to produce when orders suddenly increase. The cobweb model takes into account this delayed response on the supply side of a market by assuming that quantity supplied now (Q_t^s) depends on the price in the previous time period (P_{t-1}), i.e.,

$$Q_t^s = \mathrm{f}(P_{t-1})$$

where the subscripts denote the time period. Consumer demand for the same product (Q_t^d), however, is assumed to depend on the current price, i.e.,

$$Q_t^d = \mathrm{f}(P_t)$$

This is a reasonable picture of many agricultural markets. The quantity offered for sale this year depends on what was planted at the start of the growing season, which in turn depends on last year's price. Consumers look at current prices, though, when deciding what to buy.

The cobweb model also assumes that:

- the market is perfectly competitive
- supply and demand are both linear schedules.

Before we go any further, it must be stressed that this model does *not* explain how price adjusts in all competitive markets, or even in all perfectly competitive agricultural markets. It is a simple model with some highly restrictive assumptions that can only explain how price adjusts in these particular circumstances. Some markets may have a more complex lag structure, e.g. $Q_t^s = f(P_{t-1}, P_{t-2}, P_{t-3})$, or may not have linear demand and supply functions. You should also not forget that intervention in agricultural markets, such as the EU Common Agricultural Policy, usually means that price is not competitively determined and hence the cobweb assumptions do not apply. Having said all this, the cobweb model can still give a fair idea of how price and quantity adjust in many markets with a delayed supply.

The assumptions of the cobweb model mean that the demand and supply functions can be specified in the format

$$Q_t^d = a + bP_t \qquad \text{and} \qquad Q_t^s = c + dP_{t-1}$$

where *a*, *b*, *c* and *d* are parameters specific to individual markets.

Note that, as demand schedules slope down from left to right, the value of *b* is expected to be negative. As supply schedules usually cut the price axis at a positive value (and therefore the quantity axis at a negative value if the line were theoretically allowed to continue into negative quantities) the value of *c* will also usually be negative. Remember that these functions have *Q* as the dependent variable, but in supply and demand analysis *Q* is usually measured along the horizontal axis.

Although desired quantity demanded only equals desired quantity supplied when a market is in equilibrium, it is always true that actual quantity bought equals actual quantity sold. In the cobweb model it is assumed that in any one time period producers supply a given amount Q_t^s. Thus, there is effectively a vertical short run supply schedule at the amount determined by the previous time period's price. Price then adjusts so that all the produce supplied is bought by consumers. This adjustment means that

$$Q_t^d = Q_t^s$$

Therefore,

$$\begin{aligned} a + bP_t &= c + dP_{t-1} \\ bP_t &= c - a + dP_{t-1} \\ P_t &= \frac{c-a}{b} + \frac{d}{b}P_{t-1} \end{aligned}$$

This is what is known as a '**linear first-order difference equation**'. A difference equation expresses the value of a variable in one time period as a function of its value in earlier periods; in this case

$$P_t = f(P_{t-1})$$

It is clearly a linear relationship as the terms $(c - a)/b$ and d/b will each take a single numerical value in an actual example. It is 'first order' because only a single lag on the previous time period is built into the model and the coefficient of P_{t-1} is a simple constant. In the next section we will see how this difference equation can be used to derive an expression for P_t in terms of t.

Before doing this, let us first get a picture of how the cobweb price adjustment mechanism operates using a numerical example.

Example 14.1

In an agricultural market where the assumptions of the cobweb model apply, the demand and supply schedules are

$$Q_t^{\rm d} = 400 - 20P_t \quad \text{and} \quad Q_t^{\rm s} = -50 + 10P_{t-1}$$

A long-run equilibrium has been established for several years but then one year there is an unexpectedly good crop and output rises to 160. Explain how price will behave over the next few years following this one-off 'shock' to the market.

(Note: in this example and in most other examples in this chapter, no specific units of measurement for P or Q are given in order to keep the analysis as simple as possible. In actual applications, of course, price will be measured in currency units, e.g. £, and quantity in physical units, e.g. thousands of tonnes.)

Solution

In long-run equilibrium, price and quantity will remain unchanged each time period. Denoting equilibrium values with the suffix *, this means that:

(i) The long-run equilibrium price $P^* = P_t = P_{t-1}$

and (ii) The long-run equilibrium quantity $Q^* = Q_t^{\rm d} = Q_t^{\rm s}$

Therefore, when the market is in equilibrium

$$Q^* = 400 - 20P^* \quad \text{and} \quad Q^* = -50 + 10P^*$$

Equating to solve for P^* and Q^* gives

$$400 - 20P^* = -50 + 10P^*$$
$$450 = 30P^*$$
$$15 = P^*$$
$$Q^* = 400 - 20P^* = 400 - 300 = 100$$

These values correspond to the point where the supply and demand schedules intersect, as illustrated in Figure 14.1.

If an unexpectedly good crop causes an amount of 160 to be supplied onto the market one year, then this means that the short-run supply schedule effectively becomes the vertical line S_0 in Figure 14.1. To sell this amount, the price has to be reduced to P_0, corresponding to the point A where S_0 cuts the demand schedule.

Producers will then plan production for the next time period on the assumption that P_0 is the ruling price. The amount supplied will therefore be Q_1, corresponding to point B on the supply schedule. However, in the next time period when this reduced supply quantity Q_1 is put onto the market it will sell for price P_1, corresponding to point on the demand schedule C. Further adjustments in quantity and price are shown by points D, E, F, etc. These trace out a cobweb pattern (hence the 'cobweb' name) which converges on the long-run equilibrium point where the supply and demand schedules intersect.

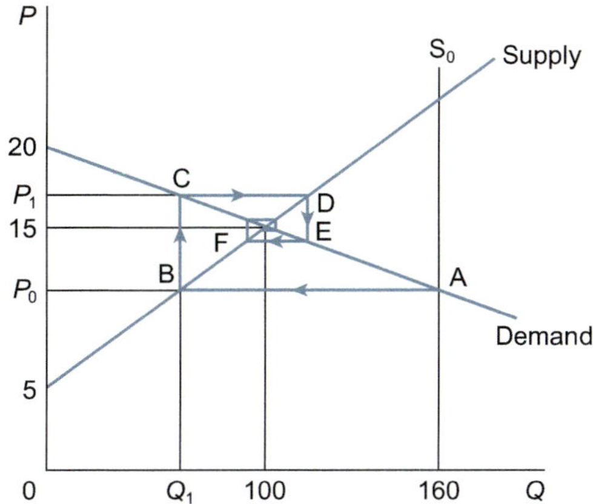

Figure 14.1 Market adjustments in the cobweb model

In some markets, price will not always return towards its long-run equilibrium level, as we shall see later when some other examples are considered. However, first let us concentrate on finding the actual pattern of price adjustment in this particular example.

Approximate values for the first few prices could be read off the graph in Figure 14.1, but as price converges towards the centre of the cobweb it gets difficult to read values accurately. We shall therefore calculate the first few values of P manually, so that you can become familiar with the mechanics of the cobweb model, and then set up a spreadsheet that can rapidly calculate patterns of price adjustment over a much longer period.

Quantity supplied in each time period is calculated by simply entering the previously ruling price into the market's supply function

$$Q_t^{s} = -50 + 10P_{t-1}$$

but how is this price calculated? There are two ways this can be done:

(a) from first principles, using the given supply and demand schedules, and
(b) using a difference equation, in the format $P_t = f(P_{t-1})$ derived earlier,

and both these methods are explained next.

(a) To calculate from first principles, we start with the demand function

$$Q_t^{d} = 400 - 20P_t$$

which can be rearranged to give the inverse demand function

$$P_t = 20 - 0.05Q_t^{\text{d}}$$

The model assumes that a fixed quantity Q_t^{s} arrives on the market each time period and then price adjusts until $Q_t^{\text{d}} = Q_t^{\text{s}}$. Thus, P_t can be found by inserting the current quantity supplied Q_t^{s} into the function for P_t. Assuming that the initial disturbance to the system when Q^{s} rises to 160 occurs in time period 0, the values of P and Q over the next three time periods can be calculated as follows:

If $Q_0^{\text{s}} = 160$ is the initial value inserted into the inverse demand function, then this gives the initial price in period 0 when 'market shock' occurs as $P_0 = 20 - 0.05Q_0^{\text{s}} = 20 - 0.05(160) = 20 - 8 = 12$.

This price in period 0 then determines quantity supplied in period 1, which is $Q_1^{\text{s}} = -50 + 10P_0 = -50 + 10(12) = -50 + 120 = 70$.

This quantity then determines the market clearing price in period 1, which is $P_1 = 20 - 0.05Q_1^{\text{s}} = 20 - 0.05(70) = 20 - 3.5 = 16.5$.

The same adjustment process then continues for future time periods as follows:

$Q_2^{\text{s}} = -50 + 10P_1 = -50 + 10(16.5) = -50 + 165 = 115$

$P_2 = 20 - 0.05Q_2^{\text{s}} = 20 - 0.05(115) = 20 - 5.75 = 14.25$

$Q_3^{\text{s}} = -50 + 10P_2 = -50 + 10(14.25) = -50 + 142.5 = 92.5$

$P_3 = 20 - 0.05Q_3^{\text{s}} = 20 - 0.05(92.5) = 20 - 4.625 = 15.375$

The pattern of price adjustment is therefore 12, 16.5, 14.25, 15.375, etc., corresponding to the cobweb graph in Figure 14.1. Price initially falls below its long-run equilibrium value of 15 and then converges back towards this equilibrium, alternating above and below it but with the magnitude of the difference becoming smaller each period.

(b) The same pattern of price adjustment can be obtained by using the difference equation

$$P_t = \frac{c-a}{b} + \frac{d}{b}P_{t-1} \qquad (1)$$

and substituting in the given values of a, b, c and d to get

$$P_t = \frac{(-50) - 400}{-20} + \frac{10}{-20}P_{t-1}$$

$$P_t = 22.5 - 0.5P_{t-1} \qquad (2)$$

The original price P_0 still has to be derived by inserting the shock quantity 160 into the inverse demand function, as already explained, which gives

$$P_0 = 20 - 0.05(160) = 12$$

The subsequent prices can then be determined using the difference equation (2), giving

$$P_1 = 22.5 - 0.5P_0 = 22.5 - 0.5(12) = 16.5$$
$$P_2 = 22.5 - 0.5P_1 = 22.5 - 0.5(16.5) = 14.25$$
$$P_3 = 22.5 - 0.5P_2 = 22.5 - 0.5(14.25) = 15.375$$
etc.

As expected, these prices are the same as those calculated by method (a).

A spreadsheet can be set up to calculate price over a large number of time periods. The spreadsheet shown in Table 14.1 can be constructed using the instructions given in Table 14.2. This calculates price each period from first principles, but you can also try to construct your own spreadsheet based on the difference equation approach.

This spreadsheet shows a series of prices and quantities converging on the equilibrium values of 15 for price and 100 for quantity. The first few values can be checked against the manually calculated values and are, as expected, the same. To bring home the point that each price adjustment is smaller than the previous one, the change in price from

Table 14.1 Excel output based on Table 14.2

	A	B	C	D	E	F	G	H
1	Ex.	COBWEB	MODEL					
2	14.1			Qd=a+bPt		Qs=c+dPt		
3								
4		Parameter	a =	400	c =	−50		
5		values	b =	−20	d =	10		
6		Initial shock	Quantity =	160				
7						Equilibrium	Price =	15
8	Time	Quantity	Price	Change		Equilibrium	Quantity =	100
9	t	Qt	Pt	in Pt				
10	0	160	12.00			Stability =>	STABLE	
11	1	70	16.50	4.50				
12	2	115	14.25	−2.25				
13	3	92.5	15.38	1.13				
14	4	103.75	14.81	−0.56				
15	5	98.125	15.09	0.28				
16	6	100.9375	14.95	−0.14				
17	7	99.53125	15.02	0.07				
18	8	100.23438	14.99	−0.04				
19	9	99.882813	15.01	0.02				
20	10	100.05859	15.00	−0.01				

Table 14.2 Setting up a spreadsheet for Example 14.1

CELL	Enter	Explanation
As in Table 14.1	Enter all labels and column headings shown in Table 14.1.	Note: do not enter the word 'STABLE' in cell G10. The stability condition will be deduced by the spreadsheet.
D4	400	These are the parameter values for this example.
D5	−20	
F4	−50	
F5	10	
D6	160	This is initial 'shock' quantity in time period 0.
A10 to A20	Enter numbers from 0 to 10.	These are the time periods.
B10	=D6	Quantity in time period 0 is initial 'shock' value.
C10	=(B10-D$4)/D$5	Calculates P_0, the initial market clearing price. Given that $Q_t^d = a + bP_t$ then $P_t = \left(Q_t^d - a\right) / b$. Note the $ on row for cells D4 and D5 to refer to the same parameters. Format to 2 dp.
C11 to C20	Copy formula from C10 down column.	Will calculate price in each time period (when all quantities in column B calculated).
B11	=F$4+F$5*C10	Calculates quantity in year 1 based on price in previous time period according to supply function $Q_t^s = c + dP_{t-1}$. Format to 2 dp.
D11	=C11-C10	Calculates change in price between time periods.
B12 to B20	Copy formula from B11 down column.	Calculates quantity supplied in each time period.
D12 to D20	Copy formula from D11 down column.	Calculates price change since previous time period.
H7	=(F4-D4)/(D5-F5)	Calculates equilibrium price using the formula $P^* = (c - a)/(b - d)$.
H8	=F4+F5*H7	Calculates equilibrium quantity $Q^* = a + bP^*$.
G10	Enter the formula below	This uses the Excel 'IF' logic function to determine whether $d/(-b)$ is less than 1, greater than 1, or equals 1. This stability criterion is explained later.

=IF(-F5/D5<1, "STABLE",IF(-F5/D5>1,"UNSTABLE","OSCILLATING"))

the previous time period is also calculated. (The price columns are formatted to 2 decimal places so price is calculated to the nearest penny.)

Although the stability of this example is obvious from the way that price converges on its equilibrium value of 15, a stability check is entered which may be useful when this spreadsheet is used for other examples. The stability requirement is that the absolute value $|d/b| < 1$, and the rationale for this is explained later in Section 14.3. Assuming that b is always negative and d is positive, the market will be stable if $d/-b < 1$ and unstable, i.e. price will not converge back to its equilibrium, if $d/-b > 1$.

To understand why price may not always return to its long-run equilibrium level in markets where the cobweb model applies, consider Example 14.2.

Example 14.2

In a market where the usual assumptions of the cobweb model apply, the demand and supply functions are

$$Q_t^d = 120 - 4P_t \quad \text{and} \quad Q_t^s = -80 + 16P_{t-1}$$

If in one time period the long-run equilibrium is disturbed by output unexpectedly rising to a level of 90, explain how price will adjust over the next few time periods.

Solution

The long-run equilibrium price can be determined from the formula

$$P^* = \frac{c-a}{b-d} = \frac{(-80)-120}{-4-16} + \frac{-200}{-20} = 10$$

Thus, the long-run equilibrium quantity is

$$Q^* = 120 - 4P^* = 120 - 4(10) = 80$$

You could use the spreadsheet developed for Example 14.1 to trace out the subsequent pattern of price adjustment but if a few values are calculated manually it can be seen that calculations after period 2 are irrelevant.

Using the standard cobweb model difference equation

$$P_t = \frac{c-a}{b} + \frac{d}{b}P_{t-1} \tag{1}$$

and substituting in the known values, we get

$$P_t = \frac{(-80)-120}{-4} + \frac{16}{-4}P_{t-1} = 50 - 4P_{t-1} \tag{2}$$

The initial price P_0 can be found by inserting the shock quantity of 90 into the demand function. Thus,

$$Q_0^s = 90 = 120 - 4P_0$$
$$4P_0 = 30$$
$$P_0 = 7.5$$

Putting this value into the difference equation (2) we get

$$P_1 = 50 - 4P_0 = 50 - 4(7.5) = 20$$
$$P_2 = 50 - 4P_1 = 50 - 4(20) = -30$$

There is not much point in going any further with the calculations. Assuming that producers will not pay consumers to take goods off their hands, negative prices cannot exist. What has happened is that price has followed the path *ABCD* traced out in Figure 14.2.

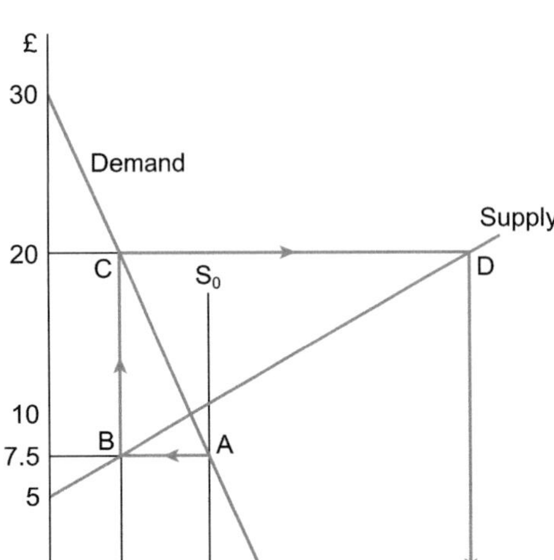

Figure 14.2 Unstable market in the cobweb model

The initial quantity 90 put onto the market causes price to drop to 7.5. Suppliers then reduce supply for the next period to

$$Q_1^s = -80 + 16P_0$$
$$= -80 + 16(7.5) = 40$$

This sells for price $P_1 = 20$ and so supply for the following period is increased to

$$Q_2^s = -80 + 16P_1 = -80 + 16(20) = 240$$

Consumers would only consume 120 even if price were zero (where the demand schedule hits the axis) and so, when this even greater quantity of 240 is put onto the market, price will collapse to zero and there will still be unsold produce. Producers will not wish to supply anything for the next time period if they expect a price of zero and so no further production will take place.

This is clearly an unstable market, but why is there a difference between this market and the stable market considered in Example 14.1? It depends on the slopes of the supply and demand schedules. If the absolute value of the slope of the demand schedule $|b|$ is less than the absolute value of the slope of the supply schedule $|d|$ then the market is stable, and vice versa. These slopes are determined by the parameters b and d. Thus, the stability conditions are

$$\text{Stable: } \left|\frac{d}{b}\right| < 1 \qquad \text{Unstable: } \left|\frac{d}{b}\right| > 1$$

A formal proof of these conditions, based on the difference equation solution method, plus an explanation of what happens when $\left|\frac{d}{b}\right| = 1$, is given in Section 14.3.

Although in theoretical models of unstable markets (such as Example 14.2) price 'explodes' and the market collapses, this may not happen in reality if some of the cobweb model assumptions no longer apply and, for example:

- producers learn from experience and do not simply base production plans for the next period on the current price
- supply and demand schedules are not linear along their entire length
- government intervention takes place to support producers or to help consumers.

Another example of an exploding market is Example 14.3, which is solved using the spreadsheet developed for Example 14.1.

Table 14.3 Using a spreadsheet for Example 14.3

	A	B	C	D	E	F	G	H
1	Ex.	COBWEB	MODEL					
2	14.3			Qd=a+bPt		Qs=c+dPt		
3								
4		Parameter	a =	360	c =	−120		
5		Values	b =	−8	d =	12		
6		Initial shock	quantity =	175				
7						Equilibrium	Price =	24
8	Time	Quantity	Price	Change		Equilibrium	Quantity =	168
9	t	Qt	Pt	in Pt				
10	0	175	23.13			Stability =>	UNSTABLE	
11	1	157.5	25.31	2.19				
12	2	183.75	22.03	−3.28				
13	3	144.375	26.95	4.92				
14	4	203.4375	19.57	−7.38				
15	5	114.84375	30.64	11.07				
16	6	247.73438	14.03	−16.61				
17	7	48.398438	38.95	24.92				
18	8	347.40234	1.57	−37.38				
19	9	−101.10352	57.64	56.06				
20	10	571.65527	−26.46	−84.09				

Example 14.3

In an agricultural market where the cobweb assumptions hold and

$$Q_t^{\mathrm{d}} = 360 - 8P_t \quad \text{and} \quad Q_t^{\mathrm{s}} = -120 + 12P_{t-1}$$

a long-run equilibrium is disturbed by an unexpectedly good crop of 175 units. Use a spreadsheet to trace out the subsequent path of price adjustment.

Solution

When the given parameters and shock quantity are entered, your spreadsheet should look like Table 14.3. This is clearly unstable as both the automatic stability check and the pattern of price adjustments show. According to these figures, the market will continue to operate until the eighth time period following the initial shock. In period 9 nothing will be produced (mathematically the model gives a negative quantity) and the market collapses.

QUESTIONS 14.1

(In all these questions, the assumptions of the cobweb model apply to each market.)

1. The agricultural market whose demand and supply functions are

$$Q_t^{\mathrm{d}} = 240 - 20P_t \quad \text{and} \quad Q_t^{\mathrm{s}} = -33\tfrac{1}{3} + 16\tfrac{2}{3}P_{t-1}$$

is initially in long-run equilibrium. Quantity then falls to 50% of its previous level as a result of an unexpectedly poor harvest. How many time periods will it take for price to return to within 1% of its long-run equilibrium level?

2. In an unstable market, the demand and supply schedules are

$$Q_t^{\mathrm{d}} = 200 - 12.5P_t \quad \text{and} \quad Q_t^{\mathrm{s}} = -60 + 20P_{t-1}$$

A shock reduction of quantity to 80 throws the system out of equilibrium. How long will it take for the market to collapse completely?

3. By tracing out the pattern of price adjustment after an initial shock that disturbs the previously ruling long-run equilibrium, say whether or not the following markets are stable. (Choose your own initial Q^{s} value that is appropriate given these demand and supply functions.)
   (a) $Q_t^{\mathrm{d}} = 150 - 1.5P_t$ and $Q_t^{\mathrm{s}} = -30 + 3P_{t-1}$
   (b) $Q_t^{\mathrm{d}} = 180 - 125P_t$ and $Q_t^{\mathrm{s}} = -20 + P_{t-1}$

14.3 THE COBWEB: DIFFERENCE EQUATION SOLUTIONS

Solving the cobweb difference equation

$$P_t = \frac{c-a}{b} + \frac{d}{b}P_{t-1} \tag{1}$$

means putting it into the format

$$P_t = \mathrm{f}(t)$$

so that the value of P_t at any given time t can be immediately calculated without the need to calculate all the preceding values of P_t.

There are two parts to the solution of this cobweb difference equation:

(i) the new long-run equilibrium price, and
(ii) the *complementary function* that tells us how much price diverges from this equilibrium level at different points in time.

A similar format applies to the solution of any linear first-order difference equation.

The equilibrium solution (i) is also known as the *particular solution* (PS). In general, the particular solution is a constant value about which adjustments in the variable in question take place over time.

The complementary function (CF) tells us how the variable in question, i.e. price in the cobweb model, varies from the equilibrium solution as time changes.

These two elements together give what is called the general solution (GS) to a difference equation, which is the full solution. Thus, we can write

$$\text{GS} = \text{PS} + \text{CF}$$

Finding the particular solution is straightforward. In the long run the equilibrium price P^* holds in each time period and so

$$P^* = P_t = P_{t-1}$$

Substituting P^* into the difference equation

$$P_t = \frac{c-a}{b} + \frac{d}{b}P_{t-1} \tag{1}$$

we get

$$\begin{aligned} P^* &= \frac{c-a}{b} + \frac{d}{b}P^* \\ bP^* &= c - b + dP^* \\ a - c &= (d-b)P^* \\ \frac{a-c}{d-b} &= P^* \end{aligned} \tag{2}$$

This, of course, is the same equilibrium value for price that would be derived in the single-time-period linear supply and demand model where

$$Q^{\mathrm{d}} = a + bP \quad \text{and} \quad Q^{\mathrm{s}} = c + dP$$

To find the complementary function, we return to the difference equation (1) but ignore the first term, which is a constant that does not vary over time, i.e. we just consider the equation

$$P_t = \frac{d}{b}P_{t-1} \tag{3}$$

This may seem rather a strange procedure, but it works, as we shall see later when some numerical examples are tackled.

We then assume that P_t depends on t according to the function

$$P_t = Ak^t \tag{4}$$

where A and k are some (as yet) unknown constants. (Note that in this formula t denotes the power to which k is raised and is not just a time superscript.)

This function applies to all values of t, which means that

$$P_{t-1} = Ak^{t-1} \tag{5}$$

Substituting the formulations (4) and (5) for P_t and P_{t-1} back into equation (3) we get

$$Ak^t = \frac{d}{b}Ak^{t-1}$$

Dividing through by Ak^{t-1} gives

$$k = \frac{d}{b}$$

Putting this result into (4) gives the complementary function as

$$P_t = A\left(\frac{d}{b}\right)^t \tag{6}$$

The value of A cannot be ascertained unless the actual value of P_t is known for a specific value of t. (See following numerical examples.)

The **general solution to the cobweb difference equation** therefore becomes

P_t = particular solution + complementary function = (2) plus (6), giving

$$P_t = \frac{a-c}{d-b} + A\left(\frac{d}{b}\right)^t$$

Stability

From this solution we can see that the stability of the model depends on the value of d/b. If A is a non-zero constant, then there are three possibilities:

(i) If $\left|\frac{d}{b}\right| < 1$ then $\left(\frac{d}{b}\right)^t \to 0$ as $t \to \infty$

This occurs in a stable market. Whatever value the constant A takes the value of the complementary function gets smaller over time. Therefore, the divergence of price from its equilibrium also approaches zero. (Note that it is the absolute value of $|d/b|$ that we consider because b will usually be a negative number.)

(ii) If $\left|\frac{d}{b}\right| > 1$ then $\left(\frac{d}{b}\right)^t \to \infty$ as $t \to \infty$

This occurs in an unstable market. After an initial disturbance, as t increases, price will diverge from its equilibrium level by greater and greater amounts.

(iii) If $\left|\frac{d}{b}\right| = 1$ then $\left|\left(\frac{d}{b}\right)^t\right| = 1$ as $t \to \infty$

Price will neither return to its equilibrium nor 'explode'. Normally, $b < 0$ and $d > 0$, so $d/b < 0$, which means that $d/b = -1$ in this case. Therefore, $(d/b)^t$ will oscillate between +1 and −1 depending on whether or not t is an even or odd number. Price will continually fluctuate between two levels (see Example 14.6).

We can now use this method of obtaining difference equation solutions to answer some specific numerical cobweb model problems.

Example 14.4

Use the cobweb difference equation solution to answer the question in Example 14.1, i.e. what happens in the market where

$$Q_t^{\text{d}} = 400 - 20P_t \quad \text{and} \quad Q_t^{\text{s}} = -50 + 10P_{t-1}$$

if there is a sudden one-off change in Q_t^{s} to 160?

Solution

Substituting the values for this market $a = 400$, $b = -20$, $c = -50$ and $d = 10$ into the general cobweb difference equation solution

$$P_t = \frac{a-c}{d-b} + A\left(\frac{d}{b}\right)^t \tag{1}$$

gives

$$P_t = \frac{400-(-50)}{10-(-20)} + A\left(\frac{10}{-20}\right)^t = \frac{450}{30} + A(-0.5)^t = 15 + A(-0.5)^t \tag{2}$$

To find the value of A, we then substitute in the known value of P_0.

The question tells us that the initial 'shock' output level Q_0 is 160 and so, as price adjusts until all output is sold, P_0 can be calculated by substituting this quantity into the demand schedule. Thus,

$$\begin{aligned} Q_0^{\text{d}} &= 160 = 400 - 20P_0 \\ 20P_0 &= 240 \\ P_0 &= 12 \end{aligned}$$

Substituting this value into the general difference equation solution (2) gives, for time period 0,

$$\begin{aligned} P_0 = 12 &= 15 + A(-0.5)^0 \\ 12 &= 15 + A \qquad \text{since } (-0.5)^0 = 1 \\ A &= -3 \end{aligned}$$

Thus, the complete solution to the difference equation in this example is

$$P_t = 15 - 3(-0.5)^t$$

This is usually called the *definite solution* or the *specific solution* because it relates to a specific initial value.

We can use this solution to calculate the first few values of P_t and compare with those we obtained when answering Example 14.1.

$$P_1 = 15 - 3(-0.5)^1 = 15 + 1.5 = 16.5$$
$$P_2 = 15 - 3(-0.5)^2 = 15 - 3(0.25) = 14.25$$
$$P_3 = 15 - 3(-0.5)^3 = 15 - 3(-0.125) = 15.375$$

As expected, these values are identical to those calculated by the iterative method.

In this particular example, price converges fairly quickly towards its long-run equilibrium level of 15. By time period 9, price will be

$$\begin{aligned} P_9 &= 15 - 3(-0.5)^9 = 15 - 3(-0.0019531) \\ &= 15 + 0.0058594 = 15.01 \text{ (to 2 dp)} \end{aligned}$$

This is clearly a stable solution. In this difference equation solution

$$P_t = 15 - 3(-0.5)^t$$

and so we can see that, as t gets larger, the value of $(-0.5)^t$ approaches zero. This is because

$$\left|\frac{d}{b}\right| = |-0.5| = 0.5 < 1$$

and so the stability condition outlined earlier is satisfied.

Note that because $-0.5 < 0$, the direction of the divergence from the equilibrium value alternates between time periods. This is because for any negative quantity $-x$, it will always be true that

$$x < 0, (-x)^2 > 0, (-x)^3 < 0, (-x)^4 > 0, \text{ etc.}$$

Thus, for even-numbered time periods (in this example) price will be above its equilibrium value, and for odd-numbered time periods price will be below its equilibrium value.

Although in this example price converges towards its long-run equilibrium value, it would never actually reach it if price and quantity were divisible into infinitesimally small units. Theoretically, this is a bit like the case of the 'hopping frog' in Chapter 8 when infinite geometric series were examined. The distance from the equilibrium gets smaller and smaller each time period but it never actually reaches zero. For practical purposes, a reasonable cut-off point can be decided upon to define when a full return to equilibrium has been reached. In this numerical example the difference from the equilibrium is less than 0.01 by time period 9, which is for all intents and purposes a full return to equilibrium if P is measured in £ and pence.

The previous example explained the method of solution of difference equations applied to a simple problem where the answers could be checked against iterative solutions. In other cases, one may need to calculate values for more distant time periods, which are more difficult to calculate manually. This method of solution of difference equations will also be useful for those of you who go on to study intermediate economic theory where some models, particularly in macroeconomics, are based on difference equations in an algebraic format which cannot be solved using a spreadsheet.

We shall now consider another cobweb example which is rather different from Example 14.4 in that

(i) price does not return towards its equilibrium level, and
(ii) the process of adjustment is more gradual over time.

Example 14.5

In a market where the assumptions of the cobweb model hold

$$Q_t^{\text{d}} = 200 - 8P_t \quad \text{and} \quad Q_t^{\text{s}} = -43 + 8.2P_{t-1}$$

The long-run equilibrium is disturbed when quantity suddenly changes to 90. What happens to price in the following time periods?

Solution

In long-run equilibrium

$$Q^* = Q_t^{\text{d}} = Q_t^{\text{s}}$$

and

$$P^* = P_t = P_{t-1}$$

Substituting these equilibrium values and equating demand and supply we can find the new equilibrium price. Thus,

$$200 - 8P^* = Q^* = -43 + 8.2P^*$$
$$243 = 16.2P^*$$
$$15 = P^*$$

This will be an unstable equilibrium as

$$\left|\frac{d}{b}\right| = \left|\frac{8.2}{-8}\right| = 1.025 > 1$$

The difference equation that describes the relationship between price in one period and the next will take the usual cobweb model format

$$P_t = \frac{c-a}{b} + \frac{d}{b}P_{t-1} \qquad (1)$$

where $a = 200$, $b = -8$, $c = -43$ and $d = 8.2$, giving

$$P_t = \frac{-43-200}{-8} + \frac{8.2}{-8}P_{t-1}$$

$$P_t = 30.375 - 1.025P_{t-1}$$

Using the formula derived earlier, the solution to this difference equation will therefore be

$$\begin{aligned}P_t &= \frac{a-c}{d-b} + A\left(\frac{d}{b}\right)^t \\ &= \frac{200-(-43)}{8.2-(-8)} + A\left(\frac{8.2}{-8}\right)^t \\ &= \frac{243}{16.2} + A(-1.025)^t \\ &= 15 + A(-1.025)^t\end{aligned} \qquad (2)$$

The first part of this solution is, of course, the equilibrium value of price, which has already been calculated. To derive the value of *A*, we need to find price in period 0. The quantity supplied is 90 in period 0 and so, to find the price that this quantity will sell for, this value is substituted into the demand function. Thus,

$$\begin{aligned}Q_0^{\mathrm{d}} = 90 &= 200 - 8P_0 \\ 8P_0 &= 110 \\ P_0 &= 13.75\end{aligned}$$

Substituting this value into the general solution (2) we get

$$\begin{aligned}P_0 = 13.75 &= 15 + A(-1.025)^0 \\ 13.75 &= 15 + A \\ -1.25 &= A\end{aligned}$$

Note that, as in Example 14.1, the value of parameter *A* is the difference between the equilibrium value of price and the value it initially takes when quantity is disturbed from its equilibrium level, i.e.

$$A = P_0 - P^* = 13.75 - 15 = -1.25$$

Putting this value of *A* into the general solution (2), the specific solution to the difference equation in this example now becomes

$$P_t = 15 - 1.25(-1.025)^t$$

Using this formula to calculate the first few values of P_t gives

$$\begin{aligned}P_0 &= 15 - 1.25(1.025)^0 = 13.75 \\ P_1 &= 15 + 1.25(1.025)^1 = 16.28 \\ P_2 &= 15 - 1.25(1.025)^2 = 13.69 \\ P_3 &= 15 + 1.25(1.025)^3 = 16.35\end{aligned}$$

We can see that, although price is gradually moving away from its long-run equilibrium value of 15, it is a very slow process. By period 10, price is still above 13.00, as

$$P_{10} = 15 - 1.25(1.025)^{10} = 13.40$$

and it takes until time period 102 before price becomes negative, as the following figures show:

$$P_{100} = 15 - 1.25(1.025)^{100} = 0.23$$
$$P_{101} = 15 + 1.25(1.025)^{101} = 30.14$$
$$P_{102} = 15 - 1.25(1.025)^{102} = -0.51$$

This example is not a particularly realistic picture of an agricultural market as many changes in supply and demand conditions would take place over a 100-year time period. (Also, quantity becomes negative in time period 85 when the market would collapse – check this yourself using a spreadsheet.) However, it illustrates the usefulness of the difference equation solution in immediately computing values for distant time periods without first needing to compute all the preceding values.

The following example illustrates what happens when a market is neither stable nor unstable.

Example 14.6

The cobweb model assumptions hold in a market where

$$Q_t^{\mathrm{d}} = 160 - 2P_t \quad \text{and} \quad Q_t^{\mathrm{s}} = -20 + 2P_{t-1}$$

If the previously ruling long-run equilibrium is disturbed by an unexpectedly low output of 50 in one time period, what will happen to price in the following time periods?

Solution

Substituting the values $a = 160$, $b = -2$, $c = -20$ and $d = 2$ for this market into the cobweb difference equation general solution

$$P_t = \frac{a-c}{d-b} + A\left(\frac{d}{b}\right)^t \tag{1}$$

gives

$$P_t = \frac{160-(-20)}{2-(-2)} + A\left(\frac{2}{-2}\right)^t = \frac{180}{4} + A(-1)^t = 45 + A(-1)^t \tag{2}$$

To determine the value of A, first substitute the given value of 50 for Q_0 into the demand function so that

$$160 - 2P_0 = 50 = Q_0$$
$$110 = 2P_0$$
$$55 = P_0$$

Now substitute this value for P_0 into the general solution (2), so that

$$P_0 = 55 = 45 + A(-1)^0$$
$$55 = 45 + A$$
$$10 = A$$

The specific solution to the difference equation for this example is therefore

$$P_t = 45 + 10(-1)^t$$

Using this formula to calculate the first few values of P_t, we see that

$$P_0 = 45 + 10(-1)^0 = 45 + 10 = 55$$
$$P_1 = 45 + 10(-1)^1 = 45 - 10 = 35$$
$$P_2 = 45 + 10(-1)^2 = 45 + 10 = 55$$
$$P_3 = 45 + 10(-1)^3 = 45 - 10 = 35$$
$$P_2 = 45 + 10(-1)^4 = 45 + 10 = 55$$

etc.

Price therefore continually fluctuates between 35 and 55.

This is the third possibility in the stability conditions examined earlier. In this example

$$\left|\frac{d}{b}\right| = \left|\frac{2}{-2}\right| = |-1| = 1$$

Therefore, as $t \to \infty$, P_t neither converges on its equilibrium level nor explodes until the market collapses. This fluctuation between two price levels from year to year is sometimes observed in certain agricultural markets.

QUESTIONS 14.2

(Assume that the usual cobweb assumptions apply in these questions.)

1. In a market where

$$Q_t^{d} = 160 - 20P_t \quad \text{and} \quad Q_t^{s} = -80 + 40P_{t-1}$$

quantity unexpectedly drops from its equilibrium value to 75. Derive the difference equation which will calculate price in the time periods following this event.

2. If $Q_t^{d} = 180 - 0.9P_t$ and $Q_t^{s} = -24 + 0.8P_{t-1}$, say whether or not the long-run equilibrium price is stable and then use the difference equation method to

calculate price in the 30th time period after a sudden one-off increase in quantity to 117.

3. Given the demand and supply functions

$$Q_t^{\mathrm{d}} = 3450 - 6P_t \quad \text{and} \quad Q_t^{\mathrm{s}} = -729 + 4.5P_{t-1}$$

use difference equations to predict what price will be in the 10th time period after an unexpected drop in quantity to 354, assuming that the market was previously in long-run equilibrium.

14.4 THE LAGGED KEYNESIAN MACROECONOMIC MODEL

In the basic Keynesian model of the determination of national income, if foreign trade and government taxation and expenditure are excluded, the model reduces to the accounting identity

$$Y = C + I \tag{1}$$

and the consumption function is

$$C = a + bY \tag{2}$$

To determine the equilibrium level of national income Y^* we substitute (2) into (1), giving

$$Y^* = a + bY^* + I$$
$$Y^*(1-b) = a + I$$
$$Y^* = \frac{a+I}{1-b}$$

This can be evaluated for given values of parameters a and b and exogenously determined investment I.

If there is a disturbance from this equilibrium, e.g. exogenous investment I alters, then the adjustment to a new equilibrium will not be instantaneous. This is the basis of the well-known multiplier effect. An initial injection of expenditure will become income for another sector of the economy. A proportion of this will be passed on as a further round of expenditure, and so on until the 'ripple effect' fades away.

Because consumer expenditure may not adjust instantaneously to new levels of income, a lagged effect may be introduced. If it is assumed that consumers' expenditure in one time period depends on the income that they received in the previous time period, then the consumption function becomes

$$C_t = a + bY_{t-1} \tag{3}$$

where the subscripts denote the time period.

National income, however, will still be determined by the sum of all expenditure within the current time period. Therefore, the accounting identity (1), when time subscripts are introduced, can be written as

$$Y_t = C_t + I_t \tag{4}$$

From (3) and (4) we can derive a difference equation that explains how Y_t depends on Y_{t-1}. Substituting (3) into (4) we get

$$\begin{aligned} Y_t &= (a + bY_{t-1}) + I_t \\ Y_t &= bY_{t-1} + a + I_t \end{aligned} \tag{5}$$

This difference equation (5) can be solved using the method explained in Section 14.3. However, let us first illustrate how this lagged effect works using a numerical example.

Example 14.7

In a basic Keynesian macroeconomic model it is assumed that initially

$$Y_t = C_t + I_t$$

where $I_t = 134$ is exogenously determined, and

$$C_t = 40 + 0.6Y_{t-1}$$

The level of investment I_t falls to 110 and remains at this level in each time period. Trace out the pattern of adjustment to the new equilibrium value of Y, assuming that the model was initially in equilibrium.

Solution

Although this pattern of adjustment can best be viewed using a spreadsheet, let us first work out the first few steps of the process manually and relate them to the familiar 45° line income-expenditure graph (illustrated in Figure 14.3) often used to show how Y is determined in introductory economics texts.

If the system is initially in equilibrium, then income in one time period is equal to expenditure in the previous time period, and income is the same equilibrium value Y^* in each time period. Thus,

$$Y_t = Y_{t-1} = Y^*$$

Therefore, when the original value of $I_t = 134$ is inserted into the accounting identity the model becomes

$$Y^* = C_t + 134 \tag{1}$$

$$C_t = 40 + 0.6Y^* \qquad (2)$$

By substitution of (2) into (1)

$$Y^* = (40 + 0.6Y^*) + 134$$
$$Y^*(1 - 0.6) = 40 + 134$$
$$0.4Y^* = 174$$
$$Y^* = 435$$

This is the initial equilibrium value of Y before the change in I.

Assume time period 0 is when the drop in I to 110 occurs. Consumption in time period 0 will be based on income earned the previous time period, i.e. when Y was still at the old equilibrium level of 435. Thus,

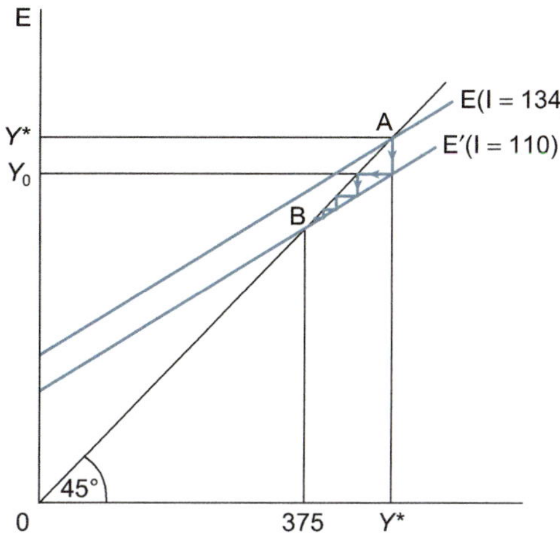

Figure 14.3 Pattern of adjustments in the lagged Keynesian model

$$C_0 = 40 + 0.6(435) = 40 + 261 = 301$$

Therefore,

$$Y_0 = C_0 + I_0 = 301 + 110 = 411$$

In the next time period, the lagged consumption function means that C_1 will be based on Y_0. Thus,

$$Y_1 = C_1 + 110 = (40 + 0.6Y_0) + 110 = 40 + 0.6(411) + 110$$
$$= 40 + 246.6 + 110 = 396.6$$

The value of Y for other time periods can be calculated in a similar fashion:

$$Y_2 = C_2 + I_2 = (40 + 0.6Y_1) + 110$$
$$= 40 + 0.6(396.6) + 110 = 387.96$$

$$Y_3 = C_3 + I_3 = (40 + 0.6Y_2) + 110$$
$$= 40 + 0.6(387.96) + 110 = 382.776$$

and so on. It can be seen that in each time period Y decreases by smaller and smaller amounts as it readjusts towards the new equilibrium value. This new equilibrium value can easily be calculated using the same method as that used earlier to work out the initial equilibrium.

When $I = 110$ and $Y_t = Y_{t-1} = Y^*$ then the model becomes

$$Y^* = C_t + I = C_t + 110$$
$$C_t = 40 + 0.6Y^*$$

By substitution

$$Y^* = \left(40 + 0.6Y^*\right) + 110$$
$$(1 - 0.6)Y^* = 150$$
$$Y^* = \frac{150}{0.4} = 375$$

This path of adjustment is illustrated in Figure 14.3 by the zigzag line with arrows which joins the old equilibrium at A with the new equilibrium at B. (Note that this diagram is not to scale and just shows the direction and relative magnitude of the steps in the adjustment process.)

Unlike the cobweb model described earlier, the adjustment in this Keynesian model is always in the same direction, instead of alternating either side of the final equilibrium. Successive values of Y just approach the equilibrium by smaller and smaller increments because the ratio in the complementary function to the difference equation (explained later) is not negative as it was in the cobweb model. If the initial equilibrium had been below the new equilibrium then, of course, Y would have approached its new equilibrium from below instead of from above.

Further steps in the adjustment of Y in this model are shown in the spreadsheet in Table 14.4, which is constructed as explained in Table 14.5. This clearly shows Y closing in on its new equilibrium as time increases.

Difference equation solution

Let us now return to the problem of how to solve the difference equation

$$Y_t = bY_{t-1} + a + I_t \qquad (1)$$

The general solution can then be applied to numerical problems, such as Example 14.7. By 'solving' this difference equation we mean putting it in the format

$$Y_t = \mathrm{f}(t)$$

so that the value of Y_t can be determined for any given value of t. The basic method is the same as that explained earlier, i.e. the solution is split into two components: the equilibrium or particular solution and the complementary function.

We first need to find the particular solution, which will be the new equilibrium value of Y^*. When this equilibrium is achieved

$$Y_t = Y_{t-1} = Y^*$$

In equilibrium, the single lag Keynesian model

$$C_t = a + bY_{t-1} \qquad (2)$$

Table 14.4 Excel output based on Table 14.5

	A	B	C	D	E
1	Ex.	LAGGED	KEYNESIAN	MODEL	
2	14.7		where	Yt = Ct + It	
3				Ct = a + bYt-1	
4	Parameters				
5	a =	40		Old I value =	134
6	b =	0.6		New I value =	110
7				Old Equil Y =	435
8	Time			New Equil Y =	375
9	T	C	Y		
10	0	301.00	411.00		
11	1	286.60	396.60		
12	2	277.96	387.96		
13	3	272.78	382.78		
14	4	269.67	379.67		
15	5	267.80	377.80		
16	6	266.68	376.68		
17	7	266.01	376.01		
18	8	265.60	375.60		
19	9	265.36	375.36		
20	10	265.22	375.22		
21	11	265.13	375.13		
22	12	265.08	375.08		
23	13	265.05	375.05		
24	14	265.03	375.03		
25	15	265.02	375.02		
26	16	265.01	375.01		
27	17	265.01	375.01		
28	18	265.00	375.00		

and

$$Y_t = C_t + I_t \tag{3}$$

can therefore be written as

$$C_t = a + bY^*$$
$$Y^* = C_t + I_t$$

Table 14.5 Setting up a spreadsheet for Example 14.7

CELL	Enter	Explanation
As in Table 14.4	Enter all labels and column headings.	
B5	40	These are given parameter values for consumption function in this example.
B6	0.6	
E5	134	Original given investment level.
E6	110	New investment level.
D6	160	This is initial 'shock' quantity in time period 0.
A10 to A28	Enter numbers from 0 to 18.	These are the time periods used.
E7	=(B5+E5)/(1-B6)	Calculates initial equilibrium value of Y using formula
		$Y = (a + I)/(1 - b)$.
E8	=(B5+E6)/(1–B6)	Same formula calculates new equilibrium value of Y,
		using new value of I in cell E6.
B10	=B5+B6*E7	Calculates consumption in time period 0 using
		formula $C = a + bY_{t-1}$ where Y_{t-1} is the old
		equilibrium value in cell E7.
C10	=B10+E$6	Calculates Y in time period 0 as sum of current
		consumption value in cell B10 and new investment
		value. Note the \$ on cell E6 to anchor when copied.
B11	=B$5+B$6*C10	Calculates consumption in time period 1 based on
		Y_0 value in cell C10. Note the \$ on cells B5 and B6
		to anchor when copied.
B12 to B28	Copy formula from B11 down column.	Calculates consumption in each time period.
C11 to C28	Copy formula from C10 down column.	Calculates national income Y_t in each time period.

By substitution

$$\begin{aligned} Y^* &= a + bY^* + I_t \\ (1-b)Y^* &= a + I_t \qquad (4) \\ Y^* &= \frac{a + I_t}{1-b} \end{aligned}$$

If the given values of a, b and I_t are put into (4) then the equilibrium value of Y is determined. This is the first part of the difference equation solution.

Returning to the difference equation (1) which we are trying to solve

$$Y_t = bY_{t-1} + a + I_t \qquad (1)$$

If the two constant terms a and I_t are removed then this becomes

$$Y_t = bY_{t-1} \tag{5}$$

To find the complementary function we use the standard method and assume that this solution is in the format

$$Y_t = Ak^t \tag{6}$$

where A and k are unknown parameters. This means that

$$Y_{t-1} = Ak^{t-1} \tag{7}$$

Substituting (6) and (7) into (5) gives

$$\begin{aligned} Ak^t &= bAk^{t-1} \\ k &= b \end{aligned}$$

Thus, the complementary function is

$$Y_t = Ab^t \tag{8}$$

The general solution to the difference equation is the sum of the particular solution (4) and the complementary function (8). Hence,

$$Y_t = \frac{a + I_t}{1 - b} + Ab^t \tag{9}$$

If t is increased, then the value of b^t in the general solution (9) will diminish as long as $|b| < 1$. This condition will be met since b is the marginal propensity to consume which has been estimated to lie between 0 and 1 in empirical studies. Therefore, Y_t will always head towards its new equilibrium value.

The value of the constant A can be determined if an initial value Y_0 is known. Substituting into (9), this gives

$$Y_0 = \frac{a + I_t}{1 - b} + Ab^0$$

Remembering that $b^0 = 1$, this means that

$$A = Y_0 - \frac{a + I_t}{1 - b} \tag{10}$$

Thus, A is the value of the difference between the initial level of income Y_0, immediately after the shock, and its final equilibrium value Y^*.

Putting this result into (9), the general solution to our difference equation becomes

$$Y_t = \frac{a + I_t}{1 - b} + \left(Y_0 - \frac{a + I_t}{1 - b} \right) b^t \tag{11}$$

This may seem to be a rather cumbersome formula, but it is straightforward to use. If you remember that the equilibrium value of Y_t is

$$\frac{a+I_t}{1-b}=Y^*$$

and rewrite (11) as

$$Y_t = Y^* + \left(Y_0 - Y^*\right)b^t \tag{12}$$

you will find it easier to work with.

We can now check that this solution to the lagged Keynesian model difference equation works with the numerical Example 14.7 considered earlier. This model assumed

$$Y_t = C_t + I_t$$

where I_t was initially 134 and

$$C_t = 40 + 0.6Y_{t-1}$$

which corresponded to an initial equilibrium of Y_t of 435.

When I_t was exogenously decreased to 110, the adjustment path towards the new equilibrium value of Y at 375 was worked out by an iterative method. Now let us see what values our difference equation will give.

We have to be careful in determining the initial value Y_0, immediately *after* the increase in investment has taken place. This depends on I_0, which will be the *new* level of investment of 110, and C_0. The level of consumption in period 0 depends on the previously existing equilibrium level of Y_t which was 435 in time period 'minus one'. Therefore,

$$\begin{aligned} C_0 &= a + bY_{t-1} = 40 + 0.6(435) = 301 \\ Y_0 &= C_0 + I_0 = 301 + 110 = 411 \end{aligned} \tag{13}$$

This is the same initial value Y_0 as that calculated in Example 14.7. The new equilibrium value of income is

$$Y^* = \frac{a+I_t}{1-b} = \frac{40+110}{1-0.6} = \frac{150}{0.4} = 375 \tag{14}$$

Substituting (13) and (14) into the formula for the general solution to the difference equation derived earlier

$$Y_t = Y^* + \left(Y_0 - Y^*\right)b^t \tag{12}$$

the general solution for this numerical example becomes

$$\begin{aligned} Y_t &= 375 + (411 - 375)0.6^t \\ &= 375 + 36(0.6)^t \end{aligned}$$

The first few values of Y are thus

$$Y_1 = 375 + 36(0.6) = 375 + 21.6 = 396.6$$
$$Y_2 = 375 + 36(0.6)^2 = 375 + 12.96 = 387.96$$
$$Y_3 = 375 + 36(0.6)^3 = 375 + 7.776 = 382.776$$

These are exactly the same as the answers computed by the iterative method in Example 14.7 and also the same as those produced by the spreadsheet in Table 14.4, which is what one would expect.

This difference equation solution can now be used to calculate Y_t in any given time period. For example, in time period 9 it will be

$$Y_9 = 375 + 36(0.6)^9 = 375 + 0.3628 = 375.3628$$

As t increases in value, eventually the value of $(0.6)^t$ becomes so small that the second term becomes negligible. In the previous example we can say that for all intents and purposes Y_t has effectively reached its equilibrium value of 375 by the ninth time period, although theoretically Y_t would never actually reach 375 if infinitesimally small increments were allowed.

By now, many of you may be thinking that this difference equation method of computing the different values of Y in the adjustment process in a Keynesian macroeconomic model is extremely long-winded and it would be much quicker to compute the values by the iterative method, particularly if a spreadsheet can be used.

In many cases you may be right. However, you must remember that this chapter is only intended to give you an insight into the methods that can be used to trace out the time path of adjustment in dynamic economic models. The mathematical methods of solution explained here can be adapted to tackle more complex problems that cannot be illustrated on a spreadsheet. Also, economists need to set up mathematical formulations for functional relationships in order to estimate the parameters of these functions. Those of you who go on to study econometrics in more depth will discover that the algebraic solutions to difference equations can help in the setting up of models for testing certain dynamic economic relationships.

Now that the general solution to the lagged Keynesian macroeconomic model has been derived, it can be applied to other numerical examples and may even allow you to compute answers more quickly than by setting up a spreadsheet.

Example 14.8

There is initially an equilibrium in the basic Keynesian model

$$Y_t = C_t + I_t$$
$$C_t = 650 + 0.5Y_{t-1}$$

with I_t remaining at 300. Then I_t suddenly increases to 420 and remains there. What will be the actual level of Y six time periods after this change?

Solution

The initial equilibrium in period 'minus 1' before the change is

$$Y_{-1}^* = \frac{a+I_t}{1-b} = \frac{650+300}{1-0.5} = \frac{950}{0.5} = 1,900$$

Therefore, the value of C in time period 0 when the increase in I takes place is

$$C_0 = 650 + 0.5(1,900) = 650 + 950 = 1,600$$

and so the value of Y_t immediately after this shock is

$$Y_0 = C_0 + I_0 = 1,600 + 420 = 2,020$$

The new equilibrium level of Y is

$$Y^* = \frac{a+I_t}{1-b} = \frac{650+420}{1-0.5} = \frac{1,070}{0.5} = 2,140$$

Substituting these values into the general solution for the lagged Keynesian macroeconomic model difference equation, we get the general solution for this example, which is

$$\begin{aligned} Y_t &= Y^* + (Y_0 - Y^*)b^t \\ &= 2,140 + (2,020 - 2,140)0.5^t \\ &= 2,140 - 120(0.5)^t \end{aligned}$$

Therefore, six time periods after the increase in investment

$$\begin{aligned} Y_6 &= 2,140 - 120(0.5)^6 \\ &= 2,140 - 1.875 \\ &= 2,138.125 \end{aligned}$$

Example 14.9

How many time periods will it take Y_t to reach 2,130 in the preceding example?

Solution

We know that $Y_t = 2,130$ and we wish to find t. Thus, substituting this value and the initial value Y_0 and the new equilibrium Y^* calculated in Example 14.8, into the Keynesian model general solution formula

$$Y_t = Y^* + \left(Y_0 - Y^*\right)b^t$$

gives

$$2,130 = 2,140 + (2,020 - 2,140)0.5^t$$
$$-10 = -120(0.5)^t$$
$$0.08333 = (0.5)^t$$

To find t, put this into the log form, which gives

$$\log 0.08333 = t \log 0.5$$
$$\frac{\log 0.083333}{\log 0.5} = t$$
$$3.585 = t$$

Therefore, Y will have exceeded 2,130 by the end of the fourth time period.

Only the most basic lagged Keynesian model has been considered so far in this section. Other possible formulations have been suggested for the ways in which past income levels can determine current expenditure. For example,

$$C_t = a + bY_{t-2}$$

or the 'distributed lag' model

$$C_t = a + b_1 Y_{t-1} + b_2 Y_{t-2}$$

The solutions of these more complex models using difference equations require more advanced mathematical methods than are explained here. You should, however, be able to adapt the spreadsheet set up in Table 14.4 to trace out the adjustment path of income in a distributed lag model with given parameters.

Example 14.10

Use a spreadsheet to estimate Y_t for the 12 time periods after I_t is increased to 140, assuming that Y_t is determined by the distributed lag Keynesian model

$$Y_t = C_t + I_t$$
$$C_t = 320 + 0.5Y_{t-1} + 0.3Y_{t-2}$$

and that the system had previously been in equilibrium with I at 90.

Solution

This is a spreadsheet exercise that you can do yourself by making the necessary adjustments to the formulae that were used to set up the spreadsheet in Table 14.4 when tackling Example 14.7. Be careful in setting up the initial values, however, as the distributed lag means that C will depend on the old equilibrium level of Y up to period 1. Some of the initial values are calculated manually next for you to check against.

The initial equilibrium level Y^* satisfies the equations

$$Y^* = C_t + 90 \quad (1)$$
$$C_t = 320 + 0.5Y^* + 0.3Y^* = 320 + 0.8Y^* \quad (2)$$

By substitution of (2) into (1)

$$Y^* = 320 + 0.8Y^* + 90$$
$$0.2Y^* = 410$$
$$Y^* = 2{,}050 = Y_{t-1} = Y_{t-2}$$

Thus, when I increases to 140

$$C_0 = 320 + 0.5(2{,}050) = 0.3(2{,}050) = 1{,}960$$
$$Y_0 = C_0 + I_0 = 1{,}960 + 140 = 2{,}100$$
$$C_1 = 320 + 0.5(2{,}100) = 0.3(2{,}050) = 1{,}985$$
$$Y_1 = C_1 + I_1 = 1{,}985 + 140 = 2{,}125$$

Your spreadsheet should show Y_t converging on 2,300.

QUESTIONS 14.3

1. A Keynesian macroeconomic model with a single-time-period lag on the consumption function, as described next, is initially in equilibrium and the level of I_t is given at 500.

$$Y_t = C_t + I_t$$
$$C_t = 750 + 0.5Y_{t-1}$$

   I_t is then increased to 650. Use difference equation analysis to find the value of Y_t in the fourth time period after this disturbance to the system. Will it then be within 1% of its new equilibrium level?

2. There is initially an equilibrium in the macroeconomic model

$$Y_t = C_t + I_t$$
$$C_t = 2{,}500 + 0.9Y_{t-1}$$

   with the level of I_t set at 1,100. Investment is then increased to 1,500 where it remains for future time periods. Calculate what the level of Y_t will be in the 40th time period after this investment increase.

3. In a basic Keynesian model with a government sector

$$Y_t = C_t + I_t + G_t C_t = 80 + 0.8Y_{t-1}^{D}$$

   where $I_t = 269$, $G_t = 310$ (exogenously determined government expenditure), and where Y_t^{D} is disposable (after-tax) income. Assume that all income is taxed at a rate of 25%.

Government expenditure is then increased to 450 and kept at this level. What tax revenue can the government expect to raise five time periods after this initial rise in expenditure?

4. Use a spreadsheet to trace out the pattern of adjustment of Y_t towards its new equilibrium value in the model

$$Y_t = C_t + I_t$$
$$C_t = 310 + 0.7Y_{t-1}$$

if I_t is exogenously increased from 240 to 350 and then kept at this new level. Assume the system was initially in equilibrium. What is the value of C in the 14th time period after the increase in I_t?

14.5 DUOPOLY PRICE ADJUSTMENT

An oligopoly is a market with a small number of sellers. It is difficult for economists to predict price and output under oligopoly because firms' reactions to their rivals' actions can vary depending on the strategy they adopt. Firms may naively assume that rivals will not react to whatever pricing policy they themselves operate. They may try to outguess their rivals, or they may collude. You will learn more about these different models in your economics course. Here we will just examine how price may adjust over time in one of the simpler 'naïve' models applied to a duopoly, which is an oligopoly with only two sellers.

Two models, the Cournot model and the Bertrand model, assume that firms do not think ahead. The Cournot model assumes that firms think that their rivals will not change their output in response to their own output decisions and the Bertrand model assumes that firms think that rivals will not change their price.

Without going into the details of the model, the predictions of the Bertrand model can be summarized in terms of the 'reaction functions' shown in Figure 14.4 for two duopolists X and Y. These show the price that will maximize one firm's profits given the value of the other firm's price read off the other axis. For example, X's reaction function R^X slopes up from right to left. If Y's price is higher, then X can get away with a higher price, but if Y lowers its price then X also has to reduce its price otherwise it will lose sales.

This model assumes that both firms have identical cost structures and so the reaction functions are symmetrical, intersecting where they both cross the 45° line representing equal prices. The prediction is that prices will eventually settle at levels P_*^X and P_*^Y, which are equal. The path of adjustment from an initial price P_0^X is shown in Figure 14.4. In time period 1, Y reacts to their rival's P_0^X by setting price P_1^Y; then X sets price P_2^X in time period 2, and so on until P_*^Y and P_*^Y are reached. Note that we assume that the firms adjust their prices in turn and so each firm only sets a new price every other time period. Some applications of this basic model based on Game Theory assume that firms go straight to the intersection point, which is called the 'Nash equilibrium'.

Let us now use our knowledge of difference equations to derive a function that will tell us what the price of one of the firms will be in any given time period, with the aid of a numerical example.

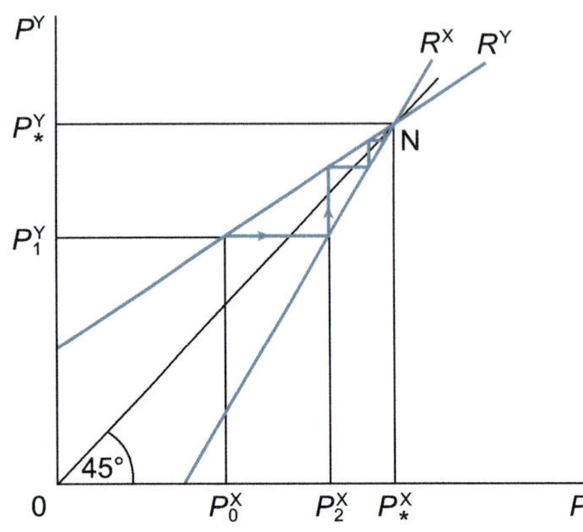

Figure 14.4 Duopoly price adjustment

Example 14.11

Two duopolists, firms X and Y, have the reaction functions

$$P_t^X = 45 + 0.8P_{t-1}^Y \qquad (1)$$

$$P_t^Y = 45 + 0.8P_{t-1}^X \qquad (2)$$

If the assumptions of the Bertrand model hold, derive a difference equation for P_t^X and calculate what P_t^X will be in time period 10 if firm X starts off in time period 0 by setting a price of 300.

Solution

Substituting Y's reaction function (2) into X's reaction function (1), we get

$$P_t^X = 45 + 0.8\left(45 + 0.8P_{t-2}^X\right)$$

$$P_t^X = 45 + 36 + 0.64P_{t-2}^X \qquad (3)$$

$$P_t^X = 81 + 0.64P_{t-2}^X$$

Note that the value P_{t-2}^X appears because we have substituted in the reaction function of Y for period $t - 1$ to correspond to the value of P_{t-1}^Y in X's reaction function, i.e. X is reacting to what Y did in the previous time period which, in turn, depends on what X did in the period before that one and so we have substituted in the equation

$$P_{t-1}^Y = 45 + 0.8P_{t-2}^X$$

The particular solution to the difference equation (3) will be where price no longer changes and

$$P_{t-1}^X = P_t^X = P_*^X$$

which is the equilibrium value of P_t^X. Substituting P_*^X into (3)

$$P_*^X = 81 + 0.64P_*^X$$

$$0.36P_*^X = 81$$

$$P_*^X = 225$$

To find the complementary function we use the standard method of ignoring the constant term and assuming that

$$P_t^X = Ak^t$$

where A and k are constant parameters. Substituting into the difference equation

$$P_t^X = 81 + 0.64P_{t-2}^X \tag{3}$$

where the constant 81 is ignored, gives

$$Ak^t = 0.64Ak^{t-2}$$

Cancelling the common term Ak^{t-2}, we get

$$k^2 = 0.64$$
$$k = 0.8$$

The complementary function is therefore

$$P_t^X = A(0.8)^t$$

Adding the complementary function and particular solution, the general solution to difference equation (1) becomes

$$P_t^X = A(0.8)^t + 225$$

To find the value of A we need to know the specific value of P_t^X at some point in time. The question specifies that initially (i.e. in time period 0) P_t^X is 300 and so

$$P_0^X = 300 = A(0.8)^0 + 225$$
$$300 = A + 225$$
$$A = 75$$

Thus, as in previous difference equation applications, A is the difference between the final equilibrium and the initial value of the variable in question.

The general solution to the difference equation (1) is therefore

$$P_t^X = 75(0.8)^t + 225$$

This can be used to calculate the value of P_t^X every *alternate* time period. (Remember that in the intervening time periods X keeps price constant while Y adjusts its price.) The price adjustment path will therefore be

$$P_0^X = 75(0.8)^0 + 225 = 75 + 225 = 300$$
$$P_2^X = 75(0.8)^2 + 225 = 48 + 225 = 273$$
$$P_4^X = 75(0.8)^4 + 225 = 30.72 + 225 = 255.72$$

etc.

The question asks what price will be in time period 10 and so this can be calculated as

$$\begin{aligned} P_{10}^{X} &= 75(0.8)^{10} + 225 \\ &= 8.05 + 225 \\ &= 233.05 \end{aligned}$$

Example 14.12

Two firms X and Y in an oligopolistic market take a short-sighted view of their situation and set price on the basis of their rivals' price in the previous time period according to the reaction functions

$$\begin{aligned} P_t^{X} &= 300 + 0.75P_{t-1}^{Y} \\ P_t^{Y} &= 300 + 0.75P_{t-1}^{X} \end{aligned}$$

Assume that each adjusts its price every other time period. The market is initially in equilibrium with $P_t^{X} = P_t^{Y} = 1,200$. Firm X then decides to try to improve its profits by raising price to 1,650. Taking into account the reactions to rivals' price changes described in the previous functions, calculate what X's price will be in the eighth time period after its breakaway price rise.

Solution

$$\begin{aligned} P_t^{X} &= 300 + 0.75P_{t-1}^{Y} \\ &= 300 + 0.75\left(300 + 0.75P_{t-2}^{X}\right) \\ &= 300 + 225 + 0.5625P_{t-2}^{X} \\ &= 525 + 0.5625P_{t-2}^{X} \end{aligned} \qquad (1)$$

This difference equation is in the same format as that found in Example 14.11. Its solution will therefore also be in the same format, i.e.

$$P_t^{X} = Ak^{t} + P_*^{X}$$

The equilibrium value P_*^{X} is given in the question as 1,200. This can easily be checked in our difference equation (1) because in equilibrium

$$P_t^{X} = P_{t-1}^{X} = P_*^{X}$$

and so

$$\begin{aligned} P_*^{X} &= 525 + 0.5625P_*^{X} \\ 004375P_*^{X} &= 525 \\ P_*^{X} &= 1,200 \end{aligned}$$

To find the complementary function, let $P_t^{\text{X}} = Ak^t$ and substitute into the difference equation (1) after dropping the constant 525. Thus,

$$Ak^t = 0.5625Ak^{t-2}$$

Cancelling Ak^{t-2} this gives

$$\begin{aligned} k^2 &= 0.5625 \\ k &= 0.75 \end{aligned}$$

Adding the complementary function and particular solution, the general solution to the difference equation becomes

$$P_t^{\text{X}} = A(0.75)^t + 1{,}200$$

The value of A can be found when the initial $P_0^{\text{X}} = 1{,}650$ is substituted. Thus,

$$\begin{aligned} P_0^{\text{X}} = 1{,}650 &= A(0.75)^0 + 1{,}200 \\ 1{,}650 &= A + 1{,}200 \\ 450 &= A \end{aligned}$$

The definite solution to this difference equation is therefore

$$P_t^{\text{X}} = 450(0.75)^t + 1{,}200$$

and so in the eighth period after the initial price rise

$$\begin{aligned} P_8^{\text{X}} &= 450(0.75)^{10} + 1{,}200 \\ &= 45.05 + 1{,}200 \\ &= 1{,}245.05 \end{aligned}$$

QUESTIONS 14.4

1. Two duopolists X and Y react to each other's prices according to the functions

$$\begin{aligned} P_t^{\text{X}} &= 240 + 0.9P_{t-1}^{\text{Y}} \\ P_t^{\text{Y}} &= 240 + 0.9P_{t-1}^{\text{X}} \end{aligned}$$

If firm X sets an initial price of 2,900, what will its price be 20 time periods later? Assume that each firm adjusts price every alternate time period.

2. In an oligopolistic market, the two firms X and Y have the following price reaction functions:

$$P_t^X = 800 + 0.6P_{t-1}^Y$$
$$P_t^Y = 800 + 0.6P_{t-1}^X$$

The usual assumptions of the Bertrand model apply and price is initially in equilibrium at 2,000 for both firms. Firm X then decides to cut its price to 1,500 to try to steal Y's market share. Calculate whether or not P^X will be back within 1% of its equilibrium value within six time periods. (Be careful how you calculate the value of A in the difference equation as this time the initial value is below the equilibrium value of P^X.)

3. In a duopoly where the assumptions of the Bertrand model hold, the two firms' reaction functions are

$$P_t^X = 95.54 + 0.83P_{t-1}^Y$$
$$P_t^Y = 95.54 + 0.83P_{t-1}^X$$

If firm X unexpectedly changes price to 499, derive the solution to the difference equation that determines P_t^X and use it to predict P_t^X in the 12th time period after the initial change.

15 Exponential functions, continuous growth and differential equations

Learning objectives

After completing this chapter students should be able to:

- use the exponential function and natural logarithms to derive the final sum, initial sum and growth rate when continuous growth takes place
- compare and contrast continuous and discrete growth rates
- set up and solve linear first-order differential equations
- use differential equation solutions to predict values in basic market and macroeconomic models
- comment on the stability of economic models where growth is continuous.

15.1 CONTINUOUS GROWTH AND THE EXPONENTIAL FUNCTION

In Chapter 8, growth was treated as a process taking place at discrete time intervals. In this chapter we shall analyze growth as a continuous process, but before we do this it is first necessary to understand the concepts of exponential functions and natural logarithms. The term 'exponential function' is usually used to describe the specific natural exponential function explained next. However, it can also be used to describe any function in the format

$$y = A^x \quad \text{where } A \text{ is a constant and } A > 1$$

This is known as an exponential function to base A. When x increases in value this function obviously increases in value very rapidly if A is a number significantly greater than 1. On the other hand, the value of A^x approaches zero if x takes on larger and

DOI: 10.4324/9781003360827-15

larger negative values. For all values of A it can be deduced from the general rules for exponents (explained in Chapter 2) that $A^0 = 1$ and $A^1 = A$.

Example 15.1

Find the values of $y = A^x$ when A is 2 and x takes the following values:

(a) 0.5 (b) 1 (c) 3 (d) 10 (e) 0 (f) −0.5 (g) −1 (h) −3.

Solutions

(a) $2^{0.5} = 1.41$ (b) $2^1 = 2$ (c) $2^3 = 8$
(d) $2^{10} = 1024$ (e) $2^0 = 1$ (f) $2^{-0.5} = 0.71$
(g) $2^{-1} = 0.50$ (h) $2^{-3} = 0.13$

The natural exponential function

In mathematics there is a special number which yields several useful results when used as a base for an exponential function and is usually represented by the letter 'e'. This number is

$$e = 2.7182818 \quad (\text{to 7 dp})$$

You should have an exponential function key on your calculator, showing as $[e^x]$ or a similar format. Exponential function values can also be calculated on a spreadsheet using the EXP function in Excel. Check that you can use your calculator or computer to obtain the following exponential values:

$$e^1 = 2.7182818 \qquad e^{0.5} = 1.6487213$$
$$e^4 = 54.59815 \qquad e^{-2.624} = 0.0725122$$

In economics, exponential functions to the base e are particularly useful for analyzing growth rates. This number, e, is also used as a base for natural logarithms, explained later in Section 15.4. Although it has already been pointed out that, strictly speaking, the specific function $y = e^x$ should be known as the 'natural exponential function', from now on we shall adopt the usual convention and refer to it simply as the 'exponential function'.

To understand how this rather odd value for e is derived, we return to the method used for calculating the value of an investment developed in Chapter 8. You will recall that the final value (F) of an initial investment (A) deposited for t discrete time periods at an interest rate of i can be calculated from the formula

$$F = A(1+i)^t$$

If the interest rate is 100% then $i = 1$ and the final value becomes

$$F = A(1+1)^t = A(2)^t$$

Assume the initial sum invested $A = 1$. If interest is paid at the end of each year, then after 1 year the final sum will be

$$F_1 = (1+1)^1 = 2$$

In Chapter 8 it was also explained how interest paid monthly at the annual rate divided by 12 will give a larger final return than this nominal annual rate because the interest credited each month will be reinvested. When the nominal annual rate of interest is 100% ($i = 1$) and the initial sum invested is assumed to be 1, the final sum after 12 months invested at a monthly interest rate of $\frac{1}{12}(100\%)$ will be

$$F_{12} = \left(1 + \frac{1}{12}\right)^{12} = 2.6130353$$

If interest was to be credited daily at the rate of $\frac{1}{365}(100\%)$ then the final sum would be

$$F_{365} = \left(1 + \frac{1}{365}\right)^{365} = 2.7145677$$

If interest was to be credited by the hour, and there are 8,760 hours in a 365-day year, the applicable rate would then be $\frac{1}{8760}(100\%)$ and the final sum would be

$$F_{8,760} = \left(1 + \frac{1}{8760}\right)^{8760} = 2.7181209$$

From these calculations we can see that the more frequently that interest is credited the closer the value of the final sum accumulated gets to 2.7182818, the value of e. When interest at a nominal annual rate of 100% is credited at infinitesimally small time intervals then growth is continuous and the final sum will equal e, and thus

$$F_n = \left(1 + \frac{1}{n}\right)^n = 2.7182818 = e \qquad \text{where } n \to \infty$$

This result means that a sum A invested for 1 year at a nominal annual interest rate of 100% credited continuously will accumulate to the final sum of

$$F = \mathrm{e} \times A = 2.7182818A$$

This translates into the annual equivalent rate of

$$\text{AER} = 2.7182818 - 1 = 1.7182818 = 171.83\% \text{ (to 2 dp)}$$

Although bank interest may not actually be credited instantaneously, the crediting of interest on a daily basis is quite common and gives an equivalent annual rate that is practically the same as the continuous rate. (One has to go to the fourth decimal place to find a difference between the two.) Continuous growth, or at least a very close approximation to it, also occurs in other variables relevant to economics, e.g. population and the amount of natural materials mined. Other variables may continuously decline in value over time, e.g. the stock of a non-renewable natural resource.

15.2 ACCUMULATED FINAL VALUES AFTER CONTINUOUS GROWTH

To derive a formula that will give the final sum accumulated after a period of continuous growth, we first assume that growth occurs at several discrete time intervals throughout a year. We also assume that A is the initial sum, r is the nominal annual rate of growth, n is the number of times per year that increments are accumulated and y is the final value. Thus, the nominal growth rate for each fraction of the year will be $\frac{r}{n}$ and the growth over a full year will be $\left(1+\frac{r}{n}\right)^n$.

Using the final sum formula developed in Chapter 8 and taking into account the initial amount A, this means that after t years of growth the final sum will be

$$y = A\left(1+\frac{r}{n}\right)^{nt}$$

To reduce this to a simpler formulation, multiply top and bottom of the exponent by r so that

$$y = A\left(1+\frac{r}{n}\right)^{\left(\frac{n}{r}\right)rt} \qquad (1)$$

If we let $m = \frac{n}{r}$ then $\frac{1}{m} = \frac{r}{n}$ and so (1) can be written as

$$y = A\left(1+\frac{1}{m}\right)^{mrt} = A\left[\left(1+\frac{1}{m}\right)^m\right]^{rt} \qquad (2)$$

Growth becomes continuous as the number of times per year that increments are accumulated increases towards infinity. When $n \to \infty$ then $\frac{n}{r} = m \to \infty$.

Therefore, using the result derived for e in Section 15.1,

$$\left(1+\frac{1}{m}\right)^m \to \text{e as } m \to \infty$$

Substituting this result back into (2) gives

$$y = Ae^{rt}$$

This formula looks much simpler now and it can be used to find the final value of any variable growing continuously at a known annual rate from a given original value.

Example 15.2

Population in a developing country is growing continuously at an annual rate of 3%. If the population is now 4.5 million, what will it be in 15 years' time?

Solution

The final value of the population (in millions) is found by substituting the given numbers into the formula $y = Ae^{rt}$. Thus, with initial value $A = 4.5$, rate of growth $r = 3\% = 0.03$, number of time periods $t = 15$, the final value is

$$y = Ae^{rt} = 4.5e^{0.03(15)} = 4.5e^{0.45} = 4.5 \times 1.5683122 = 7.0574048 \text{ million}$$

Thus, the predicted final population is 7,057,405.

Example 15.3

An economy is forecast to grow continuously at an annual rate of 2.5%. If its GNP is currently €56 billion, what will the forecast for GNP be at the end of the third quarter the year after next?

Solution

In this example: $t = 1.75$ years, $r = 2.5\% = 0.025$, $A = 56$ (€ billion).

Therefore, the final value of GNP will be

$$y = Ae^{rt} = 56e^{0.025(1.75)} = 56e^{0.04375} = 58.504384$$

Thus, the forecast for GNP is €58,504,384,000.

So far, we have only considered positive growth, but the exponential function can also be used to analyze continuous decay if the rate of decline is treated as a negative rate of growth.

Example 15.4

A river flow through a hydroelectric dam is 18 million gallons a day and shrinking continuously at an annual rate of 4%. What will the flow be in 6 years' time?

Solution

The 4% rate of decline becomes the negative growth rate $r = -4\% = -0.04$

We also know the initial value $A = 18$ and the number of time periods $t = 6$. Thus, the final value is

$$y = Ae^{rt} = 18e^{-0.04(6)} = 18e^{-0.24} = 14.16$$

Therefore, the river flow will shrink to 14.16 million gallons per day.

Continuous and discrete growth rates compared

In Section 15.1 it was explained how interest at a rate of 100% credited continuously throughout a year gives an annual equivalent rate of

$$r = e - 1 = 1.7182818 = 171.83\%$$

which is 71.83% above the original 100% rate. However, in practice annual interest rates are usually much lower, so the difference between nominal and annual equivalent rates when interest is credited continuously will be much smaller. This is illustrated in Table 15.1, which shows the maximum difference to be 0.18% when the nominal annual rate of interest is 6%, and that when interest is credited continuously the annual equivalent rate of 6.18% is the same as that when interest is credited on a daily basis if rounded to two decimal places.

Table 15.1 Differences between nominal rates and annual equivalent rates

Interest credited	Frequency rate per annum (n)	Nominal rate $\left(\frac{i}{n}\right)$	Annual equivalent rate $\left(1+\frac{i}{n}\right)^n - 1$
annually	1	6%	6%
6 monthly	2	3%	$(1.03)^2 - 1 = 0.0609 = 6.09\%$
3 monthly	4	1.5%	$(1.015)^4 - 1 = 0.06136 = 6.14\%$
monthly	12	0.5%	$(1.005)^{12} - 1 = 0.06167 = 6.17\%$
daily	365	0.0164%	$(1.00016)^{365} - 1 = 0.061831 = 6.18\%$
continuously	$\to \infty$	$\to 0$	$e^{0.06} - 1 = 0.0618365 = 6.18\%$

QUESTIONS 15.1

1. A country's population is currently 32 million and is growing continuously at an annual rate of 3.5%. What will the population be in 20 years' time if this rate of growth persists?
2. A company launched a successful new product last year. The current weekly sales level is 56,000 units. If sales are expected to grow continually at an annual rate of 12.5%, what will be the expected level of sales 36 weeks from now? (Assume that 1 year is exactly 52 weeks.)
3. Current stocks of mineral M are 250 million tonnes. If these stocks are continually being used up at an annual rate of 9%, what amount of M will remain after 30 years?
4. A renewable natural resource R will allow an estimated maximum consumption rate of 200 million units per annum. Current annual usage is 65 million units. If the annual level of usage grows continually at an annual rate of 7.5% will there be sufficient R to satisfy annual demand after (a) 5 years, (b) 10 years, (c) 15 years, (d) 20 years?
5. Stocks of resource R are shrinking continually at an annual rate of 8.5%. How much will remain in 30 years' time if current stocks are 725,000 units?
6. If €25,000 is deposited in an account where interest is credited on a daily basis that can be approximated to the continuous accumulation of interest at a nominal annual rate of 4.5% what will the final sum be after 5 years?

15.3 CONTINUOUS GROWTH RATES AND INITIAL AMOUNTS

Derivation of continuous rates of growth

The growth rate r can simply be read off from the exponent of a continuous growth function in the format $y = Ae^{rt}$.

To prove that this is the growth rate we can use calculus to derive the rate of change of this exponential growth function.

If variable y changes over time according to the function $y = Ae^{rt}$ then the rate of change of y with respect to t will be the derivative $\frac{dy}{dt}$. However, it is not a straightforward exercise to differentiate this function. For the time being let us accept the result (explained later in Section 15.4) that

$$\text{if} \quad y = e^t \quad \text{then} \quad \frac{dy}{dt} = e^t$$

i.e. the derivative of the exponential function with respect to t is the function itself.

Thus, using the chain rule,

$$\text{when} \quad y = Ae^{rt} \quad \text{then} \quad \frac{dy}{dt} = rAe^{rt}$$

This derivative approximates to the *absolute* amount by which y increases when there is a one-unit increment in time t. However, when analyzing growth rates, we are usually interested in the *proportional* increase in y with respect to its original value. The *rate* of growth is therefore

$$\frac{\frac{dy}{dt}}{y} = \frac{rAe^{rt}}{Ae^{rt}} = r$$

Even though r is the instantaneous rate of growth at any given moment in time, it must be expressed with reference to a time interval, which is usually a year in economic applications, e.g. 4.5% per annum.

Example 15.5

Owing to continuous improvements in technology and efficiency in production, an empirical study found that a factory's output of product Q at any moment in time was determined by the function

$$Q = 40e^{0.03t}$$

where t is the number of years from the base year in the empirical study and Q is the output per year in tonnes. What is the annual growth rate of production?

Solution

When the accumulated amount from continuous growth is expressed by a function in the format $y = Ae^{rt}$ then the growth rate r can simply be read off from the function. Thus, when

$$Q = 40e^{0.03t}$$

the rate of growth is

$$r = 0.03 = 3\%$$

Initial amounts

What if you wished to find the initial amount A that would grow to a given final sum y after t time periods at continuous growth rate r? Given the continuous growth final sum formula

$$y = Ae^{rt}$$

then, by dividing both sides by e^{rt}, we can derive the **exponential growth initial sum formula**

$$A = ye^{-rt}$$

Example 15.6

A parent wants to ensure that their child will have a fund of £35,000 to finance their study at university, which is expected to commence in 12 years' time. They wish to do this by investing a lump sum now. How much will they need to invest if this investment can be expected to grow continuously at an annual rate of 5%?

Solution

Given values are: final amount $y = 35{,}000$, continuous growth rate $r = 5\% = 0.05$ and time period $t = 12$. Thus, the initial sum, using the formula derived earlier, will be

$$\begin{aligned} A = ye^{-rt} &= 35{,}000e^{-0.05(12)} = 35{,}000e^{-0.6} \\ &= 35{,}000 \times 0.5488116 = £19{,}208.41 \end{aligned}$$

Example 15.7

A manager of a wildlife sanctuary wants to ensure that in 10 years' time the number of animals of a particular species in the sanctuary will total 900. How many animals will she need to start with now if this particular animal population grows continuously at an annual rate of 8.5%?

Solution

Given the final amount of $y = 900$, continuous growth rate $r = 8.5\% = 0.085$, and time period $t = 10$, then using the initial sum formula

$$A = ye^{-rt} = 900e^{-0.085(10)} = 900e^{-0.85} = 900 \times 0.4274149 = 384.67$$

Therefore, she will need to start with 385 animals, as you cannot have a fraction of an animal.

Discounting of continuous return flows

The technique shown earlier allows us to calculate the initial invested amount A when there is a single future value and returns are generated only in one time period. However, some investment projects can generate streams of returns over many periods, as we saw in Chapter 8 when discussing net present value (NPV) and annuities. We learned how to find the present value (PV) of an investment which provides a series of regular payments at discrete time intervals. To find the present value of an investment which generates a continuous stream of returns over a period of n years at an annual rate of y pounds per year and with discount rate r we need to evaluate the definite integral

$$PV = \int_0^n ye^{-rt}\,dt$$

In order to be able to apply this formula we need to know how to integrate the function e^t. As noted earlier in this chapter,

$$\text{if} \quad y = e^t \quad \text{then} \quad \frac{dy}{dt} = e^t$$

So it must be true that

$$\int e^t dt = e^t$$

and following from the chain rule

$$\int e^{bt} dt = \frac{1}{b} e^{bt}$$

where b is a constant.

Example 15.8

Find the present value (PV) of an investment project which provides a continuous stream of returns for a period of 8 years at the constant rate of £10,000 a year. Assume the discount rate $r = 5\%$.

Solution

To find the present value we substitute $y = 10{,}000$, $r = 0.05$ and $n = 8$ into the previous formula, giving

$$PV = \int_0^8 10{,}000 e^{-0.06t} dt$$

Next, we solve this definite integral to get the present value of the investment project.

$$PV = 10{,}000 \int_0^8 e^{-0.05t} dt = 10{,}000 \left[\frac{-1}{0.05} e^{-0.05t} \right]_0^8$$

$$= \frac{-10{,}000}{0.05} \left(e^{-0.4} - e^0 \right) = 20{,}000 \left(e^{-0.4} - 1 \right) = 134{,}064$$

Thus, the present value of all the continuous returns generated by this project over the period of 8 years is £134,064.

QUESTIONS 15.2

1. A statistician estimates that a country's population N is growing continuously and can be determined by the function

$$N = 3{,}620{,}000 e^{0.02t}$$

where t is the number of years after 2025. What is the population growth rate? Will population reach 10 million by the year 2075?

2. Assuming that oil stocks will continue to be depleted at the same continuous rate (in proportion to the amount remaining), the amount of oil remaining in an oil field (B), measured in barrels of oil, has been estimated as

$$B = 2{,}430{,}000{,}000\mathrm{e}^{-0.09t}$$

where t is the number of years after 2020. What proportion of the oil stock is extracted each year? How much oil will remain by 2040?

3. An individual wants to ensure that in 15 years' time, when they plan to retire, they will have a pension fund of £240,000. They wish to achieve this by investing a lump sum now, rather than making regular annual contributions. If their investment is expected to grow continuously at an annual rate of 4.5%, how much will they need to invest now?
4. An artificial lake is created with the main aim of making a commercial return from recreational fishing. Allowing for the natural rate of growth of the fish population and the depletion caused by fishing, the number of fish in the lake is expected to shrink continuously by 3.2% a year. How many fish should the owner stock the lake with to ensure that the fish population will still be 500 in 5 years' time, given that it will not be viable to add more fish after the initial stock is introduced?
5. Calculate the present value of an annuity which promises a constant stream of returns over a period of 20 years at a rate of £12,000 a year. Assume the discount rate of 4%.

15.4 NATURAL LOGARITHMS

In Chapter 2 we saw how logarithms to base 10 were defined and utilized in mathematical problems. You will recall that the logarithm of a number to base X is the power to which X must be raised in order to equal that number. Logarithms to the base e have several useful properties and applications in mathematics. These are known as 'natural logarithms', and the usual notation is 'ln' (as opposed to 'log' for logarithms to base 10).

There should be a natural logarithm function key on your calculator, which probably shows as [ln]. Check that you can use this function to derive the following values:

$$\begin{aligned} \ln 1 &= 0 \\ \ln 2.6 &= 0.9555114 \\ \ln 0.45 &= -0.7985 \end{aligned}$$

The rules for using natural logarithms are the same as for logarithms to any other base. For example, to multiply two numbers, their logarithms are added. But how do you then transform the sum of the logarithms back to a number, i.e. what is the 'antilog' of a natural logarithm?

To answer this question, consider the exponential function

$$y = e^x \tag{1}$$

By definition, the natural logarithm of y will be x because that is the power to which e is taken. Thus, we can write

$$\ln y = x \tag{2}$$

If we only know the value of the natural logarithm ln y and wish to find y then, by substituting (2) into (1), we get the result that

$$y = e^x = e^{\ln y}$$

Therefore, y can be found from the natural logarithm ln y by finding the exponential of ln y. For example, if

$$\begin{aligned} &\ln y = 3.214 \\ \text{then} \quad &y = e^{\ln y} = e^{3.214} = 24.8784 \end{aligned}$$

We can check that this is correct by finding the natural logarithm of our answer. Thus,

$$\ln y = \ln 24.8784 = 3.214$$

Although you would not normally need to actually use natural logarithms for basic numeric problems, the next example illustrates how natural logarithms can be used for multiplication.

Example 15.9

Multiply 5,623.76 by 441.873 using natural logarithms.

Solution

Taking natural logarithms and performing multiplication by adding them:

$$\begin{aligned} \ln 5{,}623.760 &= 8.6347558 + \\ \ln 441.873 &= 6.0910225 \\ \hline &\quad 14.725778 \text{ (to 6 dp)} \end{aligned}$$

To transform this logarithm back to its corresponding number we find

$$e^{14.725778} = 2{,}484{,}987.7$$

This answer can be verified by carrying out a straightforward multiplication on your calculator.

Determination of continuous growth rates using natural logarithms

To understand how natural logarithms can help determine rates of continuous growth, consider the following example.

Example 15.10

The consumption of natural mineral resource M has risen from 38 million tonnes (per annum) to 68.4 million tonnes over the last 12 years. If it is assumed that growth in consumption has been continuous, what is the annual rate of growth?

Solution

If growth is continuous then the final consumption level of M will be determined by the exponential function:

$$M = M_0 e^{rt} \tag{1}$$

This time the known values are: the final value $M = 68.4$, the initial consumption value $M_0 = 38$ and $t = 12$, with the rate of growth r being the unknown value that we are trying to determine.

Substituting these known values into (1) gives

$$\begin{aligned} 68.4 &= 38e^{12r} \\ 1.8 &= e^{12r} \end{aligned} \tag{2}$$

In (2), the power to which e must be raised to equal 1.8 is $12r$. Therefore,

$$\ln 1.8 = 12r$$

$$r = \frac{\ln 1.8}{12} = \frac{0.5877867}{12} = 0.049822$$

and so consumption has risen at an annual rate of 4.9%.

A general formula for finding a continuous rate of growth when y, A and t are all known can be derived from the final sum formula. Given

$$y = Ae^{rt}$$

then $$\frac{y}{A} = e^{rt}$$

taking natural logs

$$\ln\left(\frac{y}{A}\right) = rt$$

giving the rate of growth formula

$$r = \frac{1}{t}\ln\left(\frac{y}{A}\right)$$

Example 15.11

Over the last 15 years a country's population has risen continuously at the same annual growth rate from 8.2 million to 11.9 million. What is this rate of growth?

Solution

Entering the known values and using the formula for finding a continuous growth rate r gives

$$r = \frac{1}{t}\ln\left(\frac{y}{A}\right) = \frac{1}{15}\ln\left(\frac{11.9}{8.2}\right)$$

$$= \frac{1}{15}\ln(1.45122) = \frac{1}{15}(0.3724) = 0.02483 = 2.48\%$$

Natural logarithms can also be used to work out rates of decay, which are negative rates of growth.

Example 15.12

The annual catch of fish from a specific sea area is declining continually at a constant rate. Ten years ago, the total annual catch was 940 tonnes and this year the total catch is 784 tonnes. What is the rate of decline?

Solution

If the decline is continuous then the catch C at any point in time will be determined by the function

$$C = C_0 e^{rt}$$

where C_0 is the catch in the initial time period.

Substituting the known values into this function gives

$$\begin{aligned} 784 &= 940e^{10r} \\ 0.8340426 &= e^{10r} \\ \ln 0.8340426 &= 10r \\ -0.1814708 &= 10r \\ -0.0181471 &= r \end{aligned}$$

Therefore, the rate of decline is 1.8%.

Rates of growth and decay can also be determined over time periods of less than a year by employing the same method.

Example 15.13

Consumption of mineral M is known to be increasing continually at a constant rate per annum. The daily rate of consumption was 46.4 tonnes on 1 January and had risen to 47.2 tonnes 3 months later. What is the annual growth rate for consumption of this mineral?

Solution

Three months is a quarter of a year. Thus, using the standard final sum formula for continuous growth

$$\begin{aligned} 47.2 &= 46.4e^{r(0.25)} \\ 1.021645 &= e^{0.25r} \\ \ln 1.021645 &= 0.25r \\ 0.0214141 &= 0.25r \\ 0.0856564 &= r \end{aligned}$$

Therefore, the annual growth rate is 8.57%.

Comparison of discrete and continuous growth

A direct comparison of the continuous growth rate r and the discrete growth rate i that would accumulate the same final sum F over 1 year for a given initial sum A can be found using natural logarithms, as follows:

Continuous growth final sum $\quad F = Ae^r$
Discrete growth final sum $\quad F = A(1+i)$

If we assume both final amounts are the same, then

$$\begin{aligned} Ae^r &= A(1+i) \\ e^r &= (1+i) \end{aligned}$$

Taking logs gives the function for r in terms of i as

$$r = \ln(1+i) \qquad (1)$$

To get i as a function of r, the exponential of each side of (1) is taken, giving

$$\begin{aligned} e^r &= e^{\ln(1+i)} \\ e^r &= 1+i \\ e^r - 1 &= i \end{aligned}$$

Example 15.14

(i) Find the continuous growth rate that, over a year, would correspond to a discrete growth annual rate of:
(a) 0% (b) 10% (c) 50% (d) 100%
(ii) Find the discrete annual growth rates that would correspond to the continuous growth rates (a), (b), (c) and (d) in (i).
Give all answers to 2 decimal places.

Solution

(i) Using the formula $r = \ln(1 + i)$ the answers are:

(a) if $i = 0\% = 0$ then $r = \ln(1 + 0) = \ln 1 = 0\%$

(b) if $i = 10\% = 0.1$ then $r = \ln(1 + 0.1) = \ln 1.1 = 0.09531 = 9.53\%$

(c) if $i = 50\% = 0.5$ then $r = \ln(1 + 0.5) = \ln 1.5 = 0.405465 = 40.55\%$

(d) if $i = 100\% = 1$ then $r = \ln(1 + 1) = \ln 2 = 0.6931472 = 69.31\%$

(ii) Using the formula $i = e^r - 1$ the answers are:

(a) if $r = 0\% = 0$ then $i = e^0 - 1 = 1 - 1 = 0\%$

(b) if $r = 10\% = 0.1$ then $i = e^{0.1} - 1 = 1.10517 - 1 = 0.10517 = 10.52\%$

(c) if $r = 50\% = 0.5$ then $i = e^{0.5} - 1 = 1.64872 - 1 = 0.64872 = 64.87\%$

(d) if $r = 100\% = 1$ then $i = e^1 - 1 = 2.7182818 - 1 = 1.7182818 = 171.83\%$

QUESTIONS 15.3

1. In an advanced industrial economy, population is observed to have grown at a steady rate from 50 to 55 million over the last 20 years. What is the annual rate of growth?
2. If the average quantity of petrol used per week by a typical private motorist has increased from 32.1 litres to 48.4 litres over the last 20 years, what has been the average annual growth rate in petrol consumption assuming that this increase in petrol consumption has been continuous? If, over the same time period, petrol consumption for a typical private car has fallen from 8.75 litres per 100 km to 6.56 litres per 100 km, what has been the average annual growth rate in the distance covered each week by a typical motorist?
3. World reserves of mineral M are observed to have declined from 830 million tonnes to 675 million tonnes over the last 25 years. Assuming this decline to have been continuous, calculate the annual rate of decline and then predict what reserves will be left in 10 years' time.
4. An economy's GNP grows from €5,682 million to €5,727 million during the first quarter of a new government's term of office. If this growth rate persisted through its entire term of office of 4 years, what would GNP be at the time of the next election?
5. If the number of a protected species of animal in a reserve increased continually from 600 in 2014 to 1,450 in 2024, what was the annual growth rate?
6. Over a 12-month period what continuous growth rate is equivalent to a discrete growth rate of 6%?
7. What discrete annual growth rate is equivalent to a continuous growth rate of 6% persisting over 12 months?
8. What interest rate would you prefer to be used to add interest to your savings: 8% applied on a continuous basis or 9% applied once a year?

15.5 DIFFERENTIATION OF LOGARITHMIC FUNCTIONS

We have already used the rule that if $y = e^t$ then $\frac{dy}{dt} = e^t$.

This result can be derived if we accept as given the rule for differentiation of the natural logarithm function, which says that if

$$f(y) = \ln y$$

then $$\frac{df}{dy} = \frac{1}{y}$$

This rule can be proved mathematically but the proof is rather complex. It is not necessary for you to understand it to follow the economic applications in this basic mathematics text and so you are just asked to accept it as given. Note that this rule also helps with the exceptional case in integration not dealt with in Chapter 13, and implies that

$$\int \frac{1}{x} = \ln x$$

Returning to the exponential function, we can write this with the two sides of the equality swapped around, as

$$e^t = y$$

As the natural logarithm of y is the exponent of e, by definition, then

$$t = \ln y$$

Therefore, using the rule for differentiating natural logarithmic functions stated earlier

$$\frac{dt}{dy} = \frac{1}{y} \qquad (1)$$

The inverse function rule in calculus tells us that

$$\frac{dy}{dt} = \frac{1}{\left(\frac{dt}{dy}\right)} \qquad (2)$$

Substituting (1) into (2) gives

$$\frac{dy}{dt} = \frac{1}{\left(\frac{1}{y}\right)} = y$$

Thus, we have shown that when

$$y = e^t$$

then

$$\frac{dy}{dt} = y$$

which is the result we wished to prove.

15.6 CONTINUOUS TIME AND DIFFERENTIAL EQUATIONS

We have already seen how continuous growth rates can be determined and how continuous growth affects the final sum accumulated, but to analyze certain economic models where continuous dynamic adjustment occurs we also need to understand what differential equations are and how they can be solved.

Differential equations contain the derivative of an unknown function. For example,

$$\frac{dy}{dt} = 6y + 27$$

Solving a differential equation in this format entails finding the function y in terms of t. This will enable us to find the value of y for any given value of t.

There are many forms that differential equations can take, but we will confine the analysis here to the case of linear first-order differential equations. First-order means that only first-order derivatives are included. Thus, a first-order differential equation contains terms in $\frac{dy}{dt}$ but not higher-order derivatives such as $\frac{d^2y}{dt^2}$.

Linear means that a differential equation does not contain a product such as $y\left(\frac{dy}{dt}\right)$.

More advanced mathematical economics texts will cover the analysis of higher-order and non-linear differential equations.

As well as containing the first-order derivative, a first-order differential equation will usually also contain the unknown function (y) itself. Thus, a first-order differential equation may contain:

- a constant (although this may be zero)
- the unknown function y
- the first-order derivative $\frac{dy}{dt}$.

At first sight you might think integration would be the way to find the unknown function. However, as a differential equation will include terms in y rather than t the solution is not so straightforward. For example, if we had started with a basic derivative such as

$\frac{dy}{dt} = t$ then we could use integration to find

$y = \int t\,dt = 0.5t^2 + C$ where C is an unknown constant.

But if we start with a differential equation in terms of y instead of t, such as

$$\frac{dy}{dt} = 6y + 27$$

then this method cannot be used to find y.

The next two sections explain how to find solutions to linear first-order differential equations. First, the **homogeneous** case is considered, where there is no constant term and the differential equation to be solved takes the format

$$\frac{dy}{dt} = by \quad \text{where } b \text{ is a constant parameter}$$

Second, the **non-homogeneous** case is considered, where there is a non-zero constant term c and the differential equation to be solved takes the format

$$\frac{dy}{dt} = by + c$$

The information in these forms of differential equations corresponds to some not uncommon situations in economics. We may know the rate at which an economic variable is increasing and its value at a specific time but may not know the direct relationship between its value and the time period.

15.7 SOLUTION OF HOMOGENEOUS DIFFERENTIAL EQUATIONS

The exponential function can help us to derive the solution to a differential equation. In Section 15.4 we learned that the exponential function has the property that if

$$y = e^t \quad \text{then} \quad \frac{dy}{dt} = e^t$$

Thus, using the chain rule for differentiation, for any constant b, if

$$y = e^{bt} \quad \text{then} \quad \frac{dy}{dt} = be^{bt}$$

Therefore, if the differential equation to be solved has no constant term and has the format

$$\frac{dy}{dt} = by$$

then a possible solution is

$$y = e^{bt}$$

because this would give

$$\frac{dy}{dt} = be^{bt} = by$$

For example, if the differential equation to be solved is

$$\frac{dy}{dt} = 5y$$

then one possible solution is

$$y = e^{5t}$$

as this gives

$$\frac{dy}{dt} = 5e^{5t} = 5y$$

However, there are other possible solutions. For example,

$$\text{if} \quad y = 3e^{5t} \quad \text{then} \quad \frac{dy}{dt} = 5\left(3e^{5t}\right) = 5y$$
$$\text{if} \quad y = 3e^{5t} \quad \text{then} \quad \frac{dy}{dt} = 5\left(7e^{5t}\right) = 5y$$

In fact, we can multiply the original solution of e^{5t} by any constant parameter and still get the same solution after differentiation.

Therefore, for any differential equation in the format

$$\frac{dy}{dt} = by$$

the **general solution** can be specified as

$$y = Ae^{bt} \quad \text{where A is arbitrary constant}$$

This must be so since

$$\frac{dy}{dt} = bAe^{bt} = by$$

The actual value of A can be found if the value for y is known for a specific value of t. This will enable us to find the **definite solution**. This is easiest to evaluate when the value of y is known for $t = 0$ as any number taken to the power zero is the number itself.

For example, the general solution to the differential equation

$$\frac{dy}{dt} = 5y$$

will be

$$y_t = Ae^{5t} \tag{1}$$

where y has been given the subscript t to denote the time period that it corresponds to.

If it is known that when $t = 0$ then $y_0 = 12$ then by substituting these values into (1) we get

$$y_0 = 12 = Ae^0$$

As we know that $e^0 = 1$ then

$$12 = A$$

Substituting this value into the general solution (1) we get the definite solution

$$y_t = 12e^{5t}$$

This definite solution can now be used to predict y_t for any value of t. For example, when

$$t = 3$$

then

$$y_3 = 12e^{5(3)} = 12e^{15} = 12(3,269,017.4) = 39,228,208$$

Example 15.15

Solve the differential equation $\frac{dy}{dt} = 1.5y$ if the value of y is 34 when $t = 0$, and then use the solution to predict the value of y when $t = 7$.

Solution

Using the method explained previously, the general solution to this differential equation will be

$$y_t = Ae^{1.5t}$$

When $t = 0$ then $\quad y_0 = 34 = Ae^0$

Therefore, $\quad 34 = A$

The definite solution is thus $\quad y_t = 34e^{1.5t}$

Using this definite solution, we can now predict that when $t = 7$ then

$$y_7 = 34e^{1.5(7)} = 34e^{10.5} = 34(36,315.5) = 1,234,727$$

Differential equation solutions and growth rates

You may have noticed that the solutions to these differential equations have the same format as the functions encountered in Section 15.2 which gave final values after continuous growth for a given time period. This is because what we have done this time is derive the relationship between y and t, starting from the knowledge that $\frac{dy}{dt} = ry$, i.e. that the rate of increase of y (over time) depends on the growth rate r and the specific value of y. This can be a difficult point to grasp, because there are actually two rates involved and it is easy to confuse them.

(i) $\frac{dy}{dt}$ is the rate of increase of y with respect to time t (but over a specified time period it will be a quantity of y rather than a ratio),
(ii) r is the rate of increase of y with respect to its own current value.

When y increases in magnitude over time, larger and larger increases in the value of y each time period will be necessary to maintain the same proportional rate of growth r. In other words, the value of $\frac{dy}{dt}$ must get bigger as t increases.

Table 15.2 illustrates how this happens for the function $y_t = 8e^{0.2t}$, assuming an initial value of 8. To keep the ratio of the increase in y to its current value constant at the 20% rate of growth implicit in this function, the value of $\frac{dy}{dt}$ has to keep increasing. You can check that this must be so by differentiating.

Since $\frac{dy}{dt} = 0.2y_t$ it is obvious that $\frac{dy}{dt}$ must increase if yt does.

Table 15.2 Changes in y over time

t	$y_t = 8e^{0.2t}$	Change in y per time period $\frac{dy}{dt} = 0.2y = 1.6e^{0.2t}$	$\frac{\left(\frac{dy}{dt}\right)}{y_t} = r$
0	8.00	1.6	0.2
1	9.77	1.954	0.2
2	11.93	2.387	0.2
3	14.58	2.915	0.2
4	17.80	3.561	0.2
5	21.75	4.349	0.2

QUESTIONS 15.4

1. For each of the following differential equations, (i) derive the definite solution and (ii) use this solution to predict the value of y when $t = 10$:

   (a) $\frac{dy}{dt} = 0.2y$ with initial value $y_0 = 200$

   (b) $\frac{dy}{dt} = 1.2y$ with initial value $y_0 = 45$

   (c) $\frac{dy}{dt} = -0.4y$ with initial value $y_0 = 14$

   (d) $\frac{dy}{dt} = 1.32y$ with initial value $y_0 = 40$

   (e) $\frac{dy}{dt} = -0.025y$ with initial value $y_0 = 128$

2. The function $y_t = 3e^{0.1t}$ gives the value of y_t at any given time t. When $t = 8$ what is the rate of growth of y
   (a) with respect to itself?
   (b) with respect to time?

15.8 SOLUTION OF NON-HOMOGENEOUS DIFFERENTIAL EQUATIONS

When the constant is not zero and a differential equation takes the format

$$\frac{dy}{dt} = by + c$$

the solution is derived in two parts:

(i) the complementary function, and
(ii) the particular solution.

The **complementary function** (CF) is the same as the solution derived earlier for the case with no constant, i.e. $y_t = Ae^{bt}$

The **particular solution** (PS) is any one particular solution to the complete differential equation. It is also sometimes called the particular integral. For most economic applications you can normally use the final equilibrium value of the unknown function as the particular solution.

Thus, the full solution, which is called the **general solution** (GS), is the sum of these two components, i.e.

$$\text{GS} = \text{CF} + \text{PS}$$

This will be in the format

$$y_t = Ae^{bt} + \text{PS}$$

The value of the arbitrary constant A can be calculated if a value for y is known for a given value of t, and a specific value for A will turn the general solution into a **definite solution** (DS).

In an economic model this definite solution can usually be interpreted as

$$y = \{\textit{Function that show divergence from equilibrium}\} + \{\textit{Equilibrium value}\}$$

The following example explains how this method works.

Example 15.16

Solve the differential equation $\frac{dy}{dt} = 6t + 27$ if the value of y is 18 when $t = 0$.

Solution

To derive the complementary function from the differential equation in the question, we first consider the 'reduced equation' (RE) without the constant term. Thus, in this example the elimination of the constant gives the reduced equation (RE):

$$\frac{dy}{dt} = 6t \qquad \text{(RE)}$$

Using the result derived in the previous section that for any differential equation in the format

$$\frac{dy}{dt} = by \quad \text{then} \quad y_t = Ae^{bt}$$

the solution to the (RE) in this example will therefore be the complementary function

$$y_t = Ae^{6t} \qquad \text{(CF)}$$

To derive the particular solution we consider the situation where the function y reaches its equilibrium value and will not change any more if t increases and so

$$\frac{dy}{dt} = 0$$

The value of y for which this result holds will be a constant, which we can denote by the letter K. This will be the particular solution to the differential equation. In this example, given that

$$\frac{dy}{dt} = 6y + 27$$

if y is constant at value K then

$$\frac{dy}{dt} = 6K + 27 = 0$$
$$K = -4.5 \qquad \text{(PS)}$$

Putting this particular solution (PS) together with the complementary function (CF) gives the general solution

$$y_t = Ae^{6t} - 4.5 \qquad \text{(GS)}$$

As the initial value of y is 18 when $t = 0$ then (remembering that $e^0 = 1$)

$$y_0 = 18 = Ae^0 - 4.5$$
$$18 = A - 4.5$$
$$22.5 = A$$

Putting this value for A into the general solution (GS) gives the definite solution

$$y_t = 22.5e^{6t} - 4.5 \qquad \text{(DS)}$$

If you enter a few values for t you will see that the value of y in this function rapidly becomes extremely large. For example, when $t = 3$ then

$$\begin{aligned} y_3 &= 22.5e^{6(3)} - 4.5 = 22.5e^{18} - 4.5 \\ &= 22.5(65{,}659{,}969) - 4.5 = 1{,}477{,}349{,}303 \end{aligned}$$

Before we investigate the usefulness of this method for the analysis of dynamic economic models, we will work through another example just to make sure that you understand how to arrive at a solution.

Example 15.17

Derive a function for y in terms of t, given the initial value $y_0 = 10$, for the differential equation $\frac{dy}{dt} = -1.5y + 12$.

Solution

The reduced equation without the constant is

$$\frac{dy}{dt} = -1.5y \qquad \text{(RE)}$$

This means that the complementary function will be

$$y_t = Ae^{-1.5t} \qquad \text{(CF)}$$

If y is assumed to equal a constant K, then

$$\frac{dy}{dt} = -1.5K + 12 = 0$$

giving the particular solution

$$K = 8 \qquad \text{(PS)}$$

Putting (CF) and (PS) together, the general solution is therefore

$$y_t = Ae^{-1.5t} + 8 \qquad \text{(GS)}$$

Given the initial value for y_0 we can find A as

$$\begin{aligned} y_0 = 10 &= Ae^0 + 8 \\ 2 &= A \end{aligned}$$

Putting this value for A into the general solution (GS) gives the definite solution

$$y_t = 2e^{-1.5t} + 8 \qquad \text{(DS)}$$

Convergence and stability

Table 15.3 Values of y converge on the equilibrium

t	$y_t = 2e^{-1.5t} + 8$
0	10
1	8.44626
2	8.099574
3	8.022218
4	8.004958
5	8.001106

If the solution to Example 15.17 is used to calculate a few values of y, it can be seen that these converge on the equilibrium value of 8 as t gets larger, as shown in Table 15.3.

Why does this set of values differ from the pattern in Example 15.16 where y_t increases exponentially? The answer is that in any differential equation with a solution in the format

$$y_t = Ae^{bt} + \text{PS}$$

it is the value of the exponent b that determines convergence or divergence.

Convergence towards the particular solution occurs if $\quad b < 0$
Divergence away from the particular solution occurs if $\quad b > 0$

Table 15.4 Values of y diverge from the equilibrium

T	$y = e^t$	$y = e^{-t}$
0	1	1
1	2.718	0.367879
2	7.389	0.135335
3	20.086	0.049787
4	54.598	0.018316
5	148.413	0.006738
6	403.429	0.002479

The reason for this becomes obvious when we compare what happens to the basic exponential functions $y = e^t$ and $y = e^{-t}$ when t increases. As Table 15.4 illustrates, the function e^t expands at an increasing rate whilst the function e^{-t} rapidly diminishes. If any positive value of b multiplies t then the function will be a multiple of the expanding values in the $y = e^t$ column in Table 15.4. On the other hand, a negative value for b will mean that the function will be a multiple of the diminishing values in the e^{-t} column. As the complementary function (CF) normally shows the divergence of an economic variable from equilibrium, if the CF diminishes towards zero then the function as a whole approaches its equilibrium value.

Checking differential equation solutions using a spreadsheet

If you wish to check that you have derived the correct solution to a differential equation, you can use a spreadsheet. Just enter a series of values for t in one column and then enter the formula for the solution in the first cell in the next column, using the Excel EXP formula, and copy it down the column. For example, if the first value for $t = 0$ is in cell A5 then the formula to enter for the first value of the function $y = 2e^{-1.5t} + 8$ from Example 15.16, to go in cell B5, will be:

$$= 2^{*}\text{EXP}\left(-1.5^{*}\text{A5}\right) + 8$$

When $t = 0$ the formula should give the given initial value of y_0 and if the exponent of e is negative then the values of y should converge on the particular solution. If they do not, then you may have made some mistake in your derivation of the solution and it is worth checking through your calculations again.

QUESTIONS 15.5

For each of the following differential equations:
(a) derive the definite solution,
(b) use this solution to predict the value of y when t is 5, and
(c) say whether values of y converge or diverge as t increases.

1. $\dfrac{dy}{dt} = 0.4y - 80$ with initial value $y_0 = 180$

2. $\dfrac{dy}{dt} = -1.5y + 48$ with initial value $y_0 = 12.8$

3. $\dfrac{dy}{dt} = -0.75y + 90$ with initial value $y_0 = 100$

4. $\dfrac{dy}{dt} = 0.08y + 24$ with initial value $y_0 = -225$

15.9 CONTINUOUS ADJUSTMENT OF MARKET PRICE

Assume that in a perfectly competitive market the speed with which price P adjusts towards its equilibrium value depends on how much excess demand there is. This is quite a reasonable proposition. If consumers wish to purchase a lot more produce than suppliers are willing to sell at the current price then there will be great pressure for price to rise, but if there is only a slight shortfall then price adjustment may be sluggish. If excess demand is negative this means that quantity supplied exceeds quantity demanded, in which case price would tend to fall.

To derive the differential equation that describes this process, assume that the demand and supply functions are

$$Q_d = a + bP \quad \text{and} \quad Q_s = c + dP$$

with the parameters a, $d > 0$ and b, $c < 0$.

If r represents the rate of adjustment of P in proportion to excess demand then we can write

$$\frac{dP}{dt} = r(Q_d - Q_s)$$

Substituting the demand and supply functions for Q_d and Q_s gives

$$\begin{aligned}\frac{dP}{dt} &= r[(a+bP)-(c+dP)] \\ &= r(a+bP-c-dP) \\ &= r(b-d)P + r(a-c)\end{aligned}$$

As r, a, b, c and d are all constant parameters, this is effectively a first-order linear differential equation with one term in P plus a constant term. This format is similar to the one in the previous examples, except that it is P that changes over time rather than y, and so the same method of solution can be employed, as the following examples illustrate.

Example 15.18

A perfectly competitive market has the demand and supply functions

$$Q_d = 170 - 8P \quad \text{and} \quad Q_s = -10 + 4P$$

When the market is out of equilibrium the rate of adjustment of price is a function of excess demand such that

$$\frac{dP}{dt} = 0.5(Q_d - Q_s)$$

In the initial time period price P_0 is 10, which is not its equilibrium value. Derive a function for P in terms of t, and comment on the stability of this market.

Solution

Substituting the functions for Q_d and Q_s into the rate of price change function gives

$$\frac{dP}{dt} = 0.5[(170-8P)-(10+4P)] = 0.5(-8-4)P + 0.5(170+10)$$

which simplifies to

$$\frac{dP}{dt} = -6P + 90$$

To solve this linear first-order differential equation we first consider the reduced equation without the constant term

$$\frac{dP}{dt} = -6P \qquad \text{(RE)}$$

The complementary function that is the solution to this RE will be

$$P_t = Ae^{-6t} \qquad \text{(CF)}$$

The particular solution is found by assuming P is equal to a constant K so that

$$\begin{aligned} \frac{dP}{dt} &= -6K + 90 = 0 \\ K &= 15 \end{aligned} \qquad \text{(PS)}$$

This is the market equilibrium price. (Check this yourself using the supply and demand functions and basic linear algebra.)

Putting (CF) and (PS) together gives the general solution

$$P_t = Ae^{-6t} + 15 \qquad \text{(GS)}$$

The value of A can be determined by putting the initial value of 10 for P_0 into the (GS). Thus,

$$\begin{aligned} P_0 = 10 &= Ae^0 + 15 \\ -5 &= A \end{aligned}$$

Using this value for A in (GS) gives the definite solution to this differential equation, which is

$$P_t = -5e^{-6t} + 15 \qquad \text{(DS)}$$

The coefficient of t in this exponential function is the negative number −6. This means that the first term in (DS), i.e. the complementary function, will get closer to zero as t gets larger and so P_t will converge to its equilibrium value of 15. This market is therefore stable.

We can check this by using the previous solution to calculate P_t. For example, when

$$t = 2 \quad \text{then} \quad P_2 = -5e^{-6(2)} + 15 = -5e^{-12} + 15 = 14.99997$$

This is extremely close to the equilibrium price of 15 and so we can say that price returns to its equilibrium value within the first few time periods in this particular market.

In other markets the rate of adjustment may not be so rapid, as the following example demonstrates.

Example 15.19

If the demand and supply functions in a competitive market are

$$Q_d = 50 - 0.2P \quad \text{and} \quad Q_s = -10 + 0.3P$$

and the rate of adjustment of price when the market is out of equilibrium is

$$\frac{dP}{dt} = 0.4(Q_d - Q_s)$$

derive and solve the relevant difference equation to get a function for P in terms of t given that price is 100 in time period 0. Comment on this market's stability.

Solution

Substituting the demand and supply functions into the rate of change function gives

$$\frac{dP}{dt} = 0.4[(50 - 0.2P) - (-10 + 0.3P)] = 0.4(-0.2 - 0.3)P + 0.4(50 + 10)$$

$$\frac{dP}{dt} = -0.2P + 24$$

The reduced equation without the constant term is

$$\frac{dP}{dt} = -0.2P \qquad \text{(RE)}$$

The complementary function will therefore be

$$P_t = Ae^{-0.2t} \qquad \text{(CF)}$$

To find the particular solution we assume that P is equal to a constant K at the equilibrium price and so

$$\frac{dP}{dt} = -0.2K + 24 = 0 \qquad \text{(PS)}$$
$$K = 120$$

The (CF) and (PS) put together give the general solution

$$P_t = Ae^{-0.2t} + 120 \qquad \text{(GS)}$$

As price is 100 in time period 0 then

$$P_0 = 100 = Ae^0 + 120$$
$$-20 = A$$

and so the definite solution to this differential equation is

$$P_t = -20e^{-0.2t} + 120 \qquad \text{(DS)}$$

We can tell that this market is stable as the coefficient of t in the exponential function is the negative number −0.2. However, the sample values calculated next show that the convergence of P_t on its equilibrium value of 120 is relatively slow.

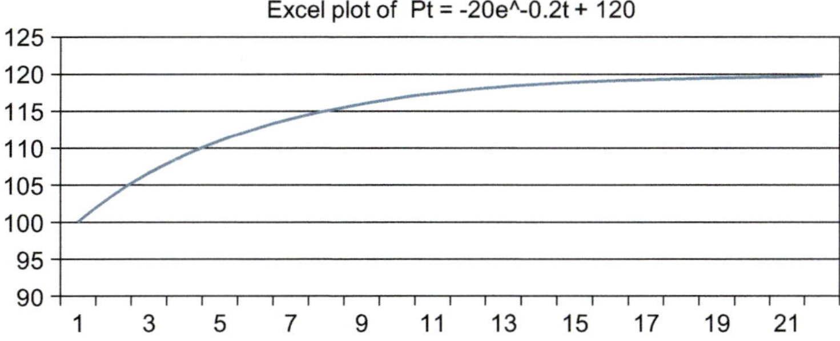

Figure 15.1 Continuous adjustment of market price over time

t	$P_t = -20e^{-0.2t} + 120$
0	100
5	112.642
15	119.004

If a spreadsheet is used to calculate P_t for values of t from 0 to 21 and these are plotted on a graph using the Chart function, it should look like Figure 15.1. Note that this differs from the pattern in the cobweb model considered in Chapter 14, where price alternated above and below its final equilibrium value by smaller and smaller amounts if the market was stable. This time, price gradually approaches its equilibrium value **from one direction only**. A similar time path will occur in other similar market models with continuous price adjustment, although if the initial value is above the equilibrium then price will, obviously, approach this equilibrium from above rather than from below.

QUESTIONS 15.6

1. If the demand and supply functions in a competitive market are

$$Q_d = 35 - 0.5P \quad \text{and} \quad Q_s = -4 + 0.8P$$

and the rate of adjustment of price when the market is out of equilibrium is

$$\frac{dP}{dt} = 0.25(Q_d - Q_s)$$

derive and solve the relevant differential equation to get a function for P in terms of t given that price is 37 in time period 0. Comment on the stability of this market.

2. The demand and supply functions in a competitive market, and the out of equilibrium rate of adjustment of price, are

$$Q_d = 280 - 4P, \quad Q_s = -35 + 8P \quad \text{and} \quad \frac{dP}{dt} = 0.08(Q_d - Q_s)$$

If price is currently 19, derive and solve the relevant differential equation to get a function for P in terms of t and use the solution to explain how close price will be to its equilibrium value after seven time periods.

3. A raw commodity is traded in a market where price adjusts in proportion to excess demand at the rate of 0.28. It has been reliably estimated that

$$Q_d = 95 - 1.8P, \quad Q_s = -12.4 + 2.1P \quad \text{and} \quad \frac{dP}{dt} = 0.28(Q_d - Q_s)$$

where t is measured in months. The current spot price is \$29.35 a tonne. In 4 months' time your company will need to buy a large amount of this commodity. If someone offers you a forward contract and guarantees to supply the amount you will need at a price of \$24.75 would it be worth signing this contract?

4. In a competitive market where the demand and supply functions are

$$Q_d = 560 - 6P \quad \text{and} \quad Q_s = -46 + 28.7P$$

the initial price P_0 is £50. Derive a function for the time path of P and use it to predict price in time period 5 given that price adjusts in proportion to excess demand at the rate

$$\frac{dP}{dt} = 0.01(Q_d - Q_s)$$

How many time periods would you have to wait for the price to drop by £20?

5. A price of \$65 per tonne is currently being quoted for a mineral traded in a competitive commodity market where

$$Q_d = 234 - 3.5P \quad \text{and} \quad Q_s = -7.8 + 2.2P$$

Forecast price when t is 8 if price adjusts at the rate

$$\frac{dP}{dt} = 0.16(Q_d - Q_s)$$

15.10 CONTINUOUS ADJUSTMENT IN A KEYNESIAN MACROECONOMIC MODEL

In a basic Keynesian macroeconomic model, with no foreign trade and no government sector, total expenditure (E) will be the sum of consumer expenditure (C) and exogenously determined investment (I). This model can therefore be specified as

$$E = C + I$$

where consumption $C = a + bY$

and where a and b are given parameters.

$$\text{In equilibrium} \quad E = Y$$
$$\text{and so} \quad Y = C + I$$

However, this macroeconomic system may not always be in equilibrium. For example, if there is an exogenous increase in investment I it may take a while before all the knock-on multiplier effects work through. Let us assume that the speed with which Y adjusts is directly proportional to the difference between total expenditure E and current income Y at ratio r. This relationship can be written as

$$\begin{aligned}\frac{dY}{dt} &= r(E-Y) = r(C+I-Y) \\ &= r(a+bY+I-Y) \\ &= r(b-1)Y + r(a+I)\end{aligned}$$

As r, a, b and I are all constant parameters, this is effectively a differential equation with one term in Y with the constant coefficient $r(b - 1)$ plus another constant term $r(a + I)$. The standard method of solution for first-order linear differential equations can therefore be employed, as shown in the following examples.

Example 15.20

In a basic Keynesian macroeconomic model

$$C = 360 + 0.8Y$$
$$\text{and} \quad I = 120$$

When the system is out of equilibrium the rate of adjustment of Y is

$$\frac{dY}{dt} = 0.25(E-Y) = 0.25(C+I-Y)$$

If national income is initially 2,000, derive a function for Y in terms of t and comment on the stability of this system.

Solution

Substituting the consumption function C and the given level of investment I into the rate of adjustment function gives the differential equation to be solved

$$\begin{aligned}\frac{dY}{dt} &= 0.25(360+0.8Y+120-Y) \\ &= 0.25(480-0.2Y) \\ &= 120-0.05Y\end{aligned}$$

The relevant reduced equation without the constant term is

$$\frac{dY}{dt} = -0.05Y \qquad \text{(RE)}$$

The corresponding complementary function will therefore be

$$Y_t = Ae^{-0.05t} \qquad \text{(CF)}$$

Assuming Y equals a constant value K in equilibrium to determine the particular solution

$$\frac{dY}{dt} = 120 - 0.05K = 0 \qquad \text{(PS)}$$
$$K = 2{,}400$$

The (CF) and (PS) together give the general solution

$$Y_t = Ae^{-0.05t} + 2{,}400 \qquad \text{(GS)}$$

As Y is 2,000 in the initial time period 0, then substituting this known value into the (GS) gives

$$Y_0 = 2{,}000 = Ae^0 + 2{,}400$$
$$-400 = A$$

The definite solution given this initial value is therefore

$$Y_t = -400e^{-0.05t} + 2{,}400 \qquad \text{(DS)}$$

This market is stable because the coefficient of t in the exponential function is negative. The convergence on the equilibrium value of 2,400 is very slow though, as the following values illustrate.

t	$Y_t = -400e^{-0.05t} + 2{,}400$
10	2157.388
20	2252.848
50	2367.166

You will also note that as t increases the values of Y_t approach equilibrium from one direction only.

Now that you are familiar with the different steps of the solution method, we will work through another similar example, but with the initial value above the final equilibrium value.

Example 15.21

In a macroeconomic model where the usual notation applies

$$C = 200 + 0.75Y, \quad E = C + I, \quad I = 180 \quad \text{and} \quad \frac{dY}{dt} = 0.8(E - Y)$$

If $Y_0 = 1{,}200$ derive a function for Y in terms of t and comment on the stability of this model.

Solution

Substituting C and I for E in the rate of adjustment function gives

$$\frac{dY}{dt} = 0.8(C + I - Y) = 0.8(200 + 0.75Y + 80 - Y)$$
$$= 0.8(280 - 0.25Y)$$
$$= 224 - 0.2Y$$

To solve this differential equation, we first set up the reduced equation

$$\frac{dY}{dt} = -0.2Y \qquad \text{(RE)}$$

The complementary function is therefore

$$Y_t = Ae^{-0.2t} \qquad \text{(CF)}$$

Assuming Y equals a constant value K in equilibrium, the particular solution will be

$$\frac{dY}{dt} = 224 - 0.2K = 0$$
$$K = 1{,}120$$

The (CF) and (PS) together give the general solution

$$Y_t = Ae^{-0.2t} + 1{,}120 \qquad \text{(GS)}$$

As Y is 1,200 in the initial time period 0, then

$$Y_0 = 1{,}200 = Ae^0 + 1{,}120$$
$$80 = A$$

The definite solution given this initial value is therefore

$$Y_t = 80e^{-0.2t} + 1{,}120 \qquad \text{(DS)}$$

This market is stable because the coefficient of t in the exponential function is negative. The initial value was higher than the final equilibrium of 1,120 and so values of Y approach equilibrium from above, as the following values illustrate.

t	$Y_t = 80e^{-0.2t} + 1{,}120$
5	1149.43
10	1130.83
20	1121.47

QUESTIONS 15.7

Given that $E = C + I$, and assuming the usual Keynesian macroeconomic model notation applies, derive a function for Y in terms of t for each of the following macroeconomic models and then use it to predict Y when t is 10.

1. $C = 50 + 0.6Y$ and $\frac{dY}{dt} = 0.5(E - Y)$ given $I = 22$ and $Y_0 = 205$
2. $C = 360 + 0.7Y$ and $\frac{dY}{dt} = 0.65(E - Y)$ given $I = 115$ and $Y_0 = 1{,}520$
3. $C = 275 + 0.82Y$ and $\frac{dY}{dt} = 0.2(E - Y)$ given $I = 90$ and $Y_0 = 2{,}040$
4. $C = 48 + 0.53Y$ and $\frac{dY}{dt} = 0.48(E - Y)$ given $I = 18.5$ and $Y_0 = 132$
5. $C = 90 + 0.61Y$ and $\frac{dY}{dt} = 0.45(E - Y)$ given $I = 45$ and $Y_0 = 328$

16 Matrix algebra

Learning objectives

After completing this chapter students should be able to:

- formulate multi-variable economic models in matrix format
- add and subtract matrices
- multiply matrices by a scalar value and by another matrix
- calculate determinants and cofactors
- derive the inverse of a matrix
- use the matrix inverse to solve a system of simultaneous equations
- derive the Hessian matrix of second-order derivatives and use it to check the second-order conditions in an unconstrained optimization problem
- derive the bordered Hessian matrix and use it to check the second-order conditions in a constrained optimization problem
- use input-output analysis to calculate the output and resource requirements for a given level of final demand.

16.1 INTRODUCTION TO MATRICES AND VECTORS

Suppose that you are responsible for hiring cars for your company's staff to use. The weekly hire rates for the five different sizes of cars that are available are:

> Compact – £139, Intermediate – £160, Large – £205, 7-Seater People carrier – £340 and Luxury limousine – £430.

You know that your car hire requirements in the next week will be:

> 4 Compact, 3 Intermediate, 12 Large, 2 People carrier and 1 Luxury limousine.

How would you work out the total car hire bill?

If you worked out total expenditure as

$$4 \times £139 + 3 \times £160 + 12 \times £205 + 2 \times £340 + 1 \times £430 = £4,606$$

DOI: 10.4324/9781003360827-16

then you would be correct. You would have already done a matrix multiplication problem, although you may not have realized it! Before we look at the formal theory of matrices, let us continue with this example.

If you know that your car hire requirements will change from week to week, it can help make calculations clearer if the numbers of cars required in each category are set out in a tabular form, as in Table 16.1. The total car hire bill for each week can then be calculated by multiplying the number of cars to be hired in each category by the corresponding price.

Table 16.1 Car hire requirements

Cars required	Week 1	Week 2	Week 3
Compact	4	7	2
Intermediate	3	5	5
Large	12	9	5
People carrier	2	1	3
Luxury limousine	1	1	2

A **matrix** is defined as an array of numbers (or algebraic symbols) set out in rows and columns. Therefore, the car hire requirements for the 3-week period in this example can be set out as the matrix

$$\mathbf{A} = \begin{bmatrix} 4 & 7 & 2 \\ 3 & 5 & 5 \\ 12 & 9 & 5 \\ 2 & 1 & 3 \\ 1 & 1 & 2 \end{bmatrix}$$

where each row corresponds to a size of car and each column corresponds to a week. Matrices are usually denoted by a capital letter in bold, as for matrix **A**, and the elements of a matrix are enclosed in a set of square or round brackets.

Matrices may also be specified with algebraic terms instead of numbers. Each entry is usually known as an 'element'. The elements in each matrix must form a complete rectangle, without any blank spaces. For example, if there are 5 rows and 3 columns, there must be 3 elements in each row and 5 elements in each column. An element may be zero, though.

The size of a matrix is called its **order**, which is specified as:

$$(\text{number of rows}) \times (\text{number of rows})$$

For example, the matrix **A** has 5 rows and 3 columns and so its order is 5×3.

Matrices with only one column or row are known as **vectors**. These are usually represented by lower-case letters in bold. For example, the set of car rental prices set out earlier can be specified (in £) as the 1×5 **row vector**

$$\mathbf{p} = [139 \quad 160 \quad 205 \quad 340 \quad 430]$$

and the car hire requirements in week 1 can be specified as the 5×1 **column vector**

$$\mathbf{q} = \begin{bmatrix} 4 \\ 3 \\ 12 \\ 2 \\ 1 \end{bmatrix}$$

Matrix addition and subtraction

Matrices that have the same order can be added or subtracted. The addition, or subtraction, is performed on each of the corresponding elements.

Example 16.1

A retailer sells two products, Q and R, in two shops A and B. The number of items sold for the last 4 weeks in each shop are shown in the following two matrices **A** and **B**, where the columns represent weeks and the rows correspond to products Q and R, respectively.

$$\mathbf{A} = \begin{bmatrix} 5 & 4 & 12 & 7 \\ 10 & 12 & 9 & 14 \end{bmatrix} \text{ and } \mathbf{B} = \begin{bmatrix} 8 & 9 & 3 & 4 \\ 8 & 18 & 21 & 5 \end{bmatrix}$$

Derive a matrix for total sales for this retailer for these two products over the last 4 weeks.

Solution

Total sales for each week will simply be the sum of the corresponding elements in matrices **A** and **B**. For example, in week 1 the total sales of product Q will be 5 plus 8. Total combined sales for Q and R can therefore be represented by the matrix

$$\mathbf{T} = \mathbf{A} + \mathbf{B} = \begin{bmatrix} 5 & 4 & 12 & 7 \\ 10 & 12 & 9 & 14 \end{bmatrix} + \begin{bmatrix} 8 & 9 & 3 & 4 \\ 8 & 18 & 21 & 5 \end{bmatrix}$$

$$= \begin{bmatrix} 5+8 & 4+9 & 12+3 & 7+4 \\ 10+8 & 12+18 & 9+21 & 14+5 \end{bmatrix} = \begin{bmatrix} 13 & 13 & 15 & 11 \\ 18 & 30 & 30 & 19 \end{bmatrix}$$

An element of a matrix can be a negative number, as in the next example solution.

Example 16.2

If $\mathbf{A} = \begin{bmatrix} 12 & 30 \\ 8 & 15 \end{bmatrix}$ and $\mathbf{B} = \begin{bmatrix} 7 & 35 \\ 4 & 8 \end{bmatrix}$, what is $\mathbf{A} - \mathbf{B}$?

Solution

$$\mathbf{A} - \mathbf{B} = \begin{bmatrix} 12 & 30 \\ 8 & 15 \end{bmatrix} - \begin{bmatrix} 7 & 35 \\ 4 & 8 \end{bmatrix} = \begin{bmatrix} 12-7 & 30-35 \\ 8-4 & 15-8 \end{bmatrix} = \begin{bmatrix} 5 & -5 \\ 4 & 7 \end{bmatrix}$$

Scalar multiplication

There are two forms of multiplication that can be performed on matrices. A matrix can be multiplied by a specific number (scalar multiplication) or by another matrix (matrix multiplication). Scalar multiplication simply involves the multiplication of each element in a matrix by the scalar value, as in Example 16.3. Matrix multiplication is rather more complex and is explained in Section 16.2.

Example 16.3

The numbers of units of a product sold by a retailer for the last 2 weeks are shown in matrix **A**, where the columns represent weeks and the rows correspond to the two different shop units that sold them.

$$\mathbf{A} = \begin{bmatrix} 12 & 30 \\ 8 & 15 \end{bmatrix}$$

If each item sells for £4, derive a matrix for the total sales revenue for this retailer for these two shop units over this 2-week period.

Solution

Total revenue is calculated by multiplying each element in the matrix of sales quantities **A** by the scalar value 4, the price that each unit is sold at. This can be represented (in £) by the matrix

$$\mathbf{R} = 4\mathbf{A} = \begin{bmatrix} 4\times 12 & 4\times 30 \\ 4\times 8 & 4\times 15 \end{bmatrix} = \begin{bmatrix} 48 & 120 \\ 32 & 60 \end{bmatrix}$$

The scalar value that a matrix is multiplied by may be an algebraic term rather than a specific number value. For example, if the product price in Example 16.3 was specified as p instead of £4 then the total revenue matrix would become

$$\mathbf{R} = \begin{bmatrix} 12p & 30p \\ 8p & 15p \end{bmatrix}$$

Scalar division works in the same way as scalar multiplication, but with each element divided by the relevant scalar value.

Example 16.4

If the set of car rental prices in the vector $\mathbf{p}$ = [139 160 205 340 430] includes VAT (value added tax) at 20% and your company can claim this tax back, what is the vector $\mathbf{v}$ of prices without this tax?

Solution

First of all we need to find the scalar value used to scale down the original vector element values. As the tax rate is 20%, the quoted prices will be 120% times the net price. Therefore, a quoted price divided by 1.2 will be the net price and so the vector of prices (in £) without the tax will be

$$\begin{aligned}\mathbf{v} &= \left(\frac{1}{1.2}\right)\mathbf{p} = \left(\frac{1}{1.2}\right)\begin{bmatrix}139 & 160 & 250 & 340 & 430\end{bmatrix} \\ &= \begin{bmatrix}\left(\frac{1}{1.2}\right)139 & \left(\frac{1}{1.2}\right)160 & \left(\frac{1}{1.2}\right)205 & \left(\frac{1}{1.2}\right)340 & \left(\frac{1}{1.2}\right)430\end{bmatrix} \\ &= \begin{bmatrix}115.83 & 133.33 & 170.83 & 283.33 & 358.33\end{bmatrix}\end{aligned}$$

QUESTIONS 16.1

1. A firm uses three different inputs K, L and R to make two final products X and Y.
   Each unit of X requires 2 units of K, 8 units of L and 23 units of R.
   Each unit of Y requires 3 units of K, 5 units of L and 26 units of R.
   Set up these input requirements in matrix format.
2. 'A vector is a special form of matrix, but a matrix is not a special form of vector.' Is this statement true?
3. For the pairs of matrices that follow, say whether it is possible to add them together and then, where it is possible, derive the matrix **C** = **A** + **B**.

   (a) $\mathbf{A} = \begin{bmatrix} 2 & 35 \\ 18 & 15 \end{bmatrix}$ and $\mathbf{B} = \begin{bmatrix} 4 & 35 \\ 9 & 8 \end{bmatrix}$

   (b) $\mathbf{A} = \begin{bmatrix} 5 & 3 \\ 8 & 1 \end{bmatrix}$ and $\mathbf{B} = \begin{bmatrix} 7 & 0 & 2 \\ 8 & 8 & 1 \end{bmatrix}$

   (c) $\mathbf{A} = \begin{bmatrix} 10 \\ 3 \\ 12 \\ 6 \\ 1 \end{bmatrix}$ and $\mathbf{B} = \begin{bmatrix} 4 \\ 2 \\ 2 \\ -9 \\ 1 \end{bmatrix}$

4. A company sells four products and the sales revenue (in £m) from each product sold through the company's three retail outlets in a year are given in the matrix

   $$\mathbf{R} = \begin{bmatrix} 7 & 3 & 1 & 4 \\ 6 & 3 & 8 & 2.5 \\ 4 & 1.2 & 2 & 0 \end{bmatrix}$$

   If profit earned is always 20% of sales revenue, use scalar multiplication to derive a matrix showing profit on each product for each retail outlet.

16.2 BASIC PRINCIPLES OF MATRIX MULTIPLICATION

If one matrix is multiplied by another matrix, the basic rule is to multiply elements along the rows of the first matrix by the corresponding elements down the columns of the second matrix. The easiest way to understand how this operation works is to first work through some examples that only involve matrices with one row or column, i.e. vectors.

Returning to our car hire example, consider the two vectors

$$\mathbf{p} = \begin{bmatrix} 139 & 160 & 205 & 340 & 430 \end{bmatrix} \text{ and } \mathbf{q} = \begin{bmatrix} 4 \\ 3 \\ 12 \\ 2 \\ 1 \end{bmatrix}$$

The row vector **p** contains the prices of hire cars in each category and the column vector **q** contains the quantities of cars in each category that your company wishes to hire for the week. At the start of this chapter we worked out the total car hire bill as

$$139 \times 4 + 160 \times 3 + 205 \times 12 + 340 \times 2 + 430 \times 1 = £4{,}606$$

In terms of these two vectors, what we have done is multiply the first element in the row vector **p** by the first element in the column vector **q**. Then, going across the row, the second element of **p** is multiplied by the second element down the column of **q**. The same procedure is followed for the other elements until we get to the end of the row and the bottom of the column.

Consider the situation where the car hire prices are still shown by the vector

$$\mathbf{p} = \begin{bmatrix} 139 & 160 & 205 & 340 & 430 \end{bmatrix}$$

but there are now 3 weeks of different car hire requirements, shown by the columns of the matrix

$$\mathbf{A} = \begin{bmatrix} 4 & 7 & 2 \\ 3 & 5 & 5 \\ 12 & 9 & 5 \\ 2 & 1 & 3 \\ 1 & 1 & 2 \end{bmatrix}$$

To calculate the total car hire bill for each of the 3 weeks, we need to find the vector

$$\mathbf{t} = \mathbf{pA}$$

This should have the order 1×3 as there will be one element (i.e. the total car hire bill) for each of the 3 weeks. The first element of **t** is the bill for the first week, which we have already found in the previous example. The car hire bill for the second week is worked out using the same method, but this time the elements across the row vector **p** multiply the elements down the second column of matrix **A**, giving

$$139\times7+160\times5+205\times9+340\times1+430\times1=£4,388$$

The third element is calculated in the same manner, but working down the third column of **A**. The result of this matrix multiplication exercise is therefore

$$\mathbf{t}=\mathbf{pA}=\begin{bmatrix}139 & 160 & 205 & 340 & 430\end{bmatrix}\begin{bmatrix}4 & 7 & 2\\3 & 5 & 5\\12 & 9 & 5\\2 & 1 & 3\\1 & 1 & 2\end{bmatrix}$$

$$=\begin{bmatrix}4606 & 4388 & 3983\end{bmatrix}$$

These examples have shown how the basic principle of matrix multiplication involves the elements across a row vector multiplying the elements down the columns of the matrix being multiplied, and then summing all the products obtained. If the first matrix has more than one row (i.e. it is not a vector) then the same procedure is followed across each row. This means that the number of rows in the final product matrix will correspond to the number of rows in the first matrix.

Example 16.5

Multiply the two matrices $\mathbf{A}=\begin{bmatrix}2 & 3\\8 & 1\end{bmatrix}$ and $\mathbf{B}=\begin{bmatrix}7 & 5 & 2\\4 & 8 & 1\end{bmatrix}$

Solution

Using the method explained earlier, the product matrix will be

$$\mathbf{AB}=\begin{bmatrix}2 & 3\\8 & 1\end{bmatrix}\begin{bmatrix}7 & 5 & 2\\4 & 8 & 1\end{bmatrix}$$

$$=\begin{bmatrix}2\times7+3\times4 & 2\times5+3\times8 & 2\times2+3\times1\\8\times7+1\times4 & 8\times5+1\times8 & 8\times2+1\times1\end{bmatrix}=\begin{bmatrix}26 & 37 & 7\\60 & 48 & 17\end{bmatrix}$$

You now may be wondering what happens if the number of elements along the rows of the first matrix (or vector) does not equal the number of elements in the columns of the matrix that it is multiplying. The answer to this question is that it is **not** possible to multiply two matrices if the number of columns in the first matrix does not equal the number of rows in the second matrix. Therefore, if a matrix **A** has order $m \times n$ and another matrix **B** has order $r \times s$, then the multiplication **AB** can only be performed if $n = r$, in which case the resulting matrix $\mathbf{C} = \mathbf{AB}$ will have order $m \times s$.

This principle is illustrated in Example 16.5. Matrix **A** has order 2 × 2 and matrix **B** has order 2 × 3 and so the product matrix **AB** has order 2 × 3. Some other examples of how the order of different matrices affects the order of the product matrix when they are multiplied are given in Table 16.2.

Table 16.2 Matrix multiplication and order of matrices

A	B	Order of product matrix AB
5 × 3	3 × 2	5 × 2
1 × 8	8 × 1	1 × 1
3 × 5	2 × 4	Matrix multiplication not possible
3 × 4	4 × 3	3 × 3
4 × 3	4 × 3	Matrix multiplication not possible

QUESTIONS 16.2

1. Given the vector $\mathbf{v} = \begin{bmatrix} 2 & 5 \end{bmatrix}$ and matrix $\mathbf{A} = \begin{bmatrix} 6 & 2 \\ 3 & 7 \end{bmatrix}$, find the product matrix **vA**.
2. For the following pairs of matrices, say if it is possible to derive the product matrix **C** = **AB** and, when this is possible, calculate the elements of this product matrix.

   (a) $\mathbf{A} = \begin{bmatrix} 2 & 10 \\ 7 & 15 \end{bmatrix}$ and $\mathbf{B} = \begin{bmatrix} 4 & 2 \\ 9 & 8 \end{bmatrix}$

   (b) $\mathbf{A} = \begin{bmatrix} 5 & 3 \\ 8 & 1 \end{bmatrix}$ and $\mathbf{B} = \begin{bmatrix} 7 & 0 & 2 \\ 12 & 8 & 1 \end{bmatrix}$

   (c) $\mathbf{A} = \begin{bmatrix} 9 \\ 3 \\ 12 \\ 6 \\ 1 \end{bmatrix}$ and $\mathbf{B} = \begin{bmatrix} 4 \\ 0 \\ 2 \\ -9 \\ 1 \end{bmatrix}$

3. A company's input requirements over the next 4 weeks for the three inputs X, Y and Z are given (in numbers of units of each input) by the matrix

   $$\mathbf{R} = \begin{bmatrix} 2 & 0.5 & 1 & 7 \\ 6 & 3 & 8 & 2.5 \\ 4 & 5 & 2 & 0 \end{bmatrix}$$

   The company can buy these inputs from two suppliers, whose prices for the three inputs X, Y and Z are given (in £) by the matrix

   $$\mathbf{P} = \begin{bmatrix} 4 & 6 & 2 \\ 5 & 8 & 1 \end{bmatrix}$$

   where the two rows represent the suppliers and the three columns represent the input prices. Use matrix multiplication to derive a matrix that will give the total input bill for the next 4 weeks for both suppliers.

16.3 MATRIX MULTIPLICATION – THE GENERAL CASE

Now that the basic principles have been explained with some straightforward examples, we can set out a general formula for matrix multiplication that can be applied to more complex matrix multiplication exercises. The general $m \times n$ matrix with any number of rows m and columns n can be written as

$$\mathbf{A} = \begin{bmatrix} a_{11} & a_{12} & . & . & a_{1n} \\ a_{21} & a_{22} & . & . & a_{2n} \\ . & . & . & . & . \\ . & . & . & . & . \\ a_{m1} & a_{m2} & . & . & a_{mn} \end{bmatrix}$$

For each element a_{ij} the subscript i denotes the row number and the subscript j denotes the column number. For example,

a_{11} = element in row 1, column 1
a_{12} = element in row 1, column 2
a_{1n} = element in row 1, column n
a_{mn} = element in row m, column n

If this general $m \times n$ matrix **A** multiplies the general $n \times r$ matrix **B**, then the product will be the $m \times r$ matrix **C**. Thus, we can write

$$\mathbf{AB} = \begin{bmatrix} a_{11} & a_{12} & . & . & a_{1n} \\ a_{21} & a_{22} & . & . & a_{2n} \\ . & . & . & . & . \\ . & . & . & . & . \\ a_{m1} & a_{m2} & . & . & a_{mn} \end{bmatrix} \begin{bmatrix} b_{11} & b_{12} & . & . & b_{1r} \\ b_{21} & b_{22} & . & . & b_{2r} \\ . & . & . & . & . \\ . & . & . & . & . \\ b_{n1} & b_{n2} & . & . & b_{nr} \end{bmatrix} = \begin{bmatrix} c_{11} & c_{12} & . & . & c_{1r} \\ c_{21} & c_{22} & . & . & c_{2r} \\ . & . & . & . & . \\ . & . & . & . & . \\ c_{m1} & c_{m2} & . & . & c_{mr} \end{bmatrix} = \mathbf{C}$$

where

$$\begin{aligned} c_{11} &= a_{11}b_{11} + a_{12}b_{21} + \ldots + a_{1n}b_{n1} \\ c_{12} &= a_{11}b_{12} + a_{12}b_{22} + \ldots + a_{1n}b_{n2} \\ &\quad . \qquad\quad . \qquad\qquad\quad . \\ &\quad . \qquad\quad . \qquad\qquad\quad . \\ c_{mr} &= a_{m1}b_{1r} + a_{m2}b_{2r} + \ldots + a_{mn}b_{nr} \end{aligned}$$

Example 16.6

Find the product matrix **C** = **AB** when

$$\mathbf{A} = \begin{bmatrix} 4 & 2 & 12 \\ 6 & 0 & 20 \\ 1 & 8 & 5 \end{bmatrix} \quad \text{and} \quad \mathbf{B} = \begin{bmatrix} 10 & 0.5 & 1 & 7 \\ 6 & 3 & 8 & 2.5 \\ 4 & 4 & 2 & 0 \end{bmatrix}$$

Solution

Using the general matrix multiplication formula, the elements of the first two rows of the product matrix **C** can be calculated as:

$$
\begin{aligned}
c_{11} &= 4\times 10+2\times 6+12\times 4=40+12+48=100\\
c_{12} &= 4\times 0.5+2\times 3+12\times 4=2+6+48=56\\
c_{13} &= 4\times 1+2\times 8+12\times 2=4+16+24=44\\
c_{14} &= 4\times 7+2\times 2.5+12\times 0=28+5+0=33\\
c_{21} &= 6\times 10+0\times 6+20\times 4=60+0+80=140\\
c_{22} &= 6\times 0.5+0\times 3+20\times 4=3+0+80=83\\
c_{23} &= 6\times 1+0\times 8+20\times 2=6+0+40=46\\
c_{24} &= 6\times 7+0\times 2.5+20\times 0=42+0+0=42
\end{aligned}
$$

Now, try and calculate the elements of the final row yourself. You should get

$$c_{31}=78,\quad c_{32}=44.5,\quad c_{33}=75,\quad c_{34}=27$$

The complete product matrix will therefore be

$$\mathbf{C}=\mathbf{AB}=\begin{bmatrix} 100 & 56 & 44 & 33\\ 140 & 83 & 46 & 42\\ 78 & 44.5 & 75 & 27 \end{bmatrix}$$

Although the calculations for multiplication of small matrices can be done manually fairly quickly, it is now becoming obvious that for large matrices the calculations will be very tedious and time consuming, and so a spreadsheet can be used.

Using a spreadsheet for matrix multiplication

The best way to explain how to use the Excel MMULT formula to multiply two matrices **A** and **B** is to work through an example.

Example 16.7

Given the two matrices $\mathbf{A}=\begin{bmatrix} 8 & 4 & 3\\ 4 & 5 & 6 \end{bmatrix}$ and $\mathbf{B}=\begin{bmatrix} 0.8 & 0.3 & 0.1\\ 0.5 & 0.2 & 0.4\\ 0.3 & 0.2 & 0.1 \end{bmatrix}$, find the product matrix **AB** using an Excel spreadsheet.

Solution

(a) Enter the values of matrices **A** and **B** on a spreadsheet. For example, put the elements of **A** in cells (A2:C3) and the elements of **B** in cells (E2:G4). You can also enter labels for the matrix names in the rows of cells above.

Table 16.3 Matrix multiplication using a spreadsheet

	A	B	C	D	E	F	G	H
1	**Matrix**	A			**Matrix**	**B**		
2	8	4	3		0.8	0.3	0.1	
3	4	5	6		0.5	0.2	0.4	
4					0.3	0.2	0.1	
5	**Matrix**	AB						
6	9	3.8	2.7					
7	8	3.4	3					

(b) Highlight the cells where you want the calculated **AB** product matrix to go. Since the order of **A** is 2 × 3 and the order of **B** is 3 × 3, the product matrix **AB** must have order 2 × 3. You therefore need to highlight a block of cells with 2 rows and 3 columns, such as (A6:C7).

(c) With this cell range still highlighted, enter the formula =MMULT(A2:C3,E2:G4) or use whatever cell ranges apply for your matrices to be multiplied.

(d) Hold down the Ctrl and Shift keys together and press Enter (if you do not do this then the formula will not treat highlighted cells as part of an array, i.e. a matrix).

Your spreadsheet and the computed product matrix **AB** should now be as in Table 16.3.

In the simple example you can check the answers manually. However, once you are satisfied that you can use the MMULT formula properly then you can use it for more complex examples where manual computation would be too time consuming.

Example 16.8

In the spreadsheet in Table 16.4, the MMULT formula has been used to multiply the 5 × 6 matrix **A** by the 6 × 8 matrix **B** to get the 5 × 8 product matrix **AB**. Try entering the matrices **A** and **B** yourself and see if you can use the Excel MMULT formula to get the same product matrix **AB**.

Vectors of coefficients

In economic models, it is common to specify one dependent variable as a function of a vector of explanatory variables, especially when employing econometric analysis to estimate coefficients of these explanatory variables. A typical vector format for a function is $\mathbf{q} = \beta\mathbf{x}$ where β is the vector of coefficients for the exogenous explanatory variables in vector **x**.

For example, assume that the demand for oil in time t is described by the linear function

Table 16.4 Matrix multiplication using a spreadsheet

	A	B	C	D	E	F	G	H	I	J	K	L	M	N	O
1	Matrix A							B							
2	120	160	195	220	285	350		8	9	10	11	12	14	3	2
3	125	165	200	225	290	355		12	13	14	15	16	9	12	5
4	130	170	205	230	150	360		4	5	6	5	6	9	3	7
5	135	175	210	235	200	380		8	9	10	11	12	3	3	4
6	140	180	215	240	110	500		5	6	7	0	3	0	2	3
7								2	3	4	1	4	1	2	0
8								Product							
9								Matrix	AB						
10								7545	8875	10205	7465	10065	5885	4795	4140
11								7740	9100	10460	7680	10330	6065	4920	4245
12								7210	8455	9700	7895	10160	6245	4755	3915
13								7660	8995	10330	8125	10620	6440	5000	4155
14								7610	8995	10380	8455	11060	6735	5165	3975

$$q^t = \beta_0 + \beta_1 x_1^t + \beta_2 x_2^t + \beta_3 x_3^t + \beta_4 x_4^t + \beta_5 x_5^t$$

where the superscript t denotes the time period (rather than an exponent) for all variables and

$$x_1 = \text{price of oil}, \quad x_2 = \text{average income}, \qquad x_3 = \text{price of substitute fuel},$$
$$x_4 = \text{price of a complement(e.g.cars)} \quad \text{and } x_5 = \text{population}$$

This linear demand function for oil in time period t may be specified in a vector format as

$$\mathbf{q}^t = \boldsymbol{\beta}\mathbf{x}^t = \begin{bmatrix} \beta_0 & \beta_1 & \beta_2 & \beta_3 & \beta_4 & \beta_5 \end{bmatrix} \begin{bmatrix} 1 \\ x_1^t \\ x_2^t \\ x_3^t \\ x_4^t \\ x_5^t \end{bmatrix}$$

Note that, although there are five independent explanatory variables in this economic model, the vector of coefficients β has the order 1×6 because there is also a constant term, β_0. The vector of values of the explanatory variables also has 6 elements and thus takes the order 6×1. However, because it multiplies the constant, the first element in the column remains as 1 even though the values of other elements (i.e. the explanatory variables) may change for different time periods. The actual values of the coefficients β_0, β_1, β_2, etc., will be estimated by a method such as Ordinary Least Squares (OLS), which you should come across in your statistics or econometrics modules.

As vector $\boldsymbol{\beta}$ has the order 1×6 and the vector of values of the explanatory variables $\mathbf{x}$ has the order 6×1 then the product matrix $\boldsymbol{\beta}\mathbf{x}$ will have the order 1×1. This means that it will contain the single element q^t which is the predicted output.

Example 16.9

Assume that the demand for oil (in millions of barrels) can be explained by the model $\mathbf{q} = \boldsymbol{\beta}\mathbf{x}$ and the vector of coefficients of the explanatory variables has been reliably estimated as

$$\boldsymbol{\beta} = \begin{bmatrix} \beta_0 & \beta_1 & \beta_2 & \beta_3 & \beta_4 & \beta_5 \end{bmatrix} = \begin{bmatrix} 4.2 & -0.1 & 0.4 & 0.2 & -0.1 & 0.2 \end{bmatrix}$$

Calculate the demand for oil when the vector of explanatory variables is

$$\mathbf{x} = \begin{bmatrix} 1 \\ x_1^t \\ x_2^t \\ x_3^t \\ x_4^t \\ x_5^t \end{bmatrix} = \begin{bmatrix} \text{Constant} \\ \text{Price} \\ \text{Income} \\ \text{Price of substitute} \\ \text{Price of complement} \\ \text{Population (in mln)} \end{bmatrix} = \begin{bmatrix} 1 \\ 30 \\ 18.5 \\ 52 \\ 12.8 \\ 61 \end{bmatrix}$$

Solution

The demand for oil is calculated as

$$\mathbf{q} = \boldsymbol{\beta}\mathbf{x} = \begin{bmatrix} 4.2 & -0.1 & 0.4 & 0.2 & -0.1 & 0.2 \end{bmatrix} \begin{bmatrix} 1 \\ 30 \\ 18.5 \\ 52 \\ 12.8 \\ 61 \end{bmatrix} = [29.92]$$

Thus, the answer is 29.92 million barrels.

You can check the calculations for arriving at this answer manually or using Excel.

QUESTIONS 16.3

1. For each of the pairs of matrices **A** and **B**, use an Excel spreadsheet to find the product matrix **AB**.

(a) $\mathbf{A}=\begin{bmatrix} 4 & 1 & 3 \\ 9 & 8 & 2 \end{bmatrix}$ and $\mathbf{B}=\begin{bmatrix} 2 & 10 & 2 \\ 5 & 5 & 8 \\ 1.5 & 0 & 1 \end{bmatrix}$

(b) $\mathbf{A}=\begin{bmatrix} 7 & 10 & 3 \\ 9 & 5 & 2 \\ 4 & 0 & 5 \end{bmatrix}$ and $\mathbf{B}=\begin{bmatrix} 11 & 2.5 & 1 & 4 \\ 5 & 5 & 8 & 0 \\ 3 & 0 & 1 & 4 \end{bmatrix}$

(c) $\mathbf{A}=\begin{bmatrix} 45 & 34 & 4 & 8 \\ 6 & 7 & 22 & 10 \\ 70 & 3 & 90 & 5 \\ 2 & 2 & 0 & 23 \\ -6 & 5 & 3 & 9 \end{bmatrix}$ and

$$\mathbf{B}=\begin{bmatrix} 2 & 5 & 3 & 4 & 32 & 65 \\ 9 & 5 & 0 & 0 & 9 & 2 \\ 8 & 46 & 1 & 7 & 85 & 31 \\ 4 & 0 & 20 & 24 & 3 & 8 \end{bmatrix}$$

2. The demand for good G depends on a vector of four explanatory variables **x**. There is a linear relationship, including a constant term, between these explanatory variables and g, the amount of good G demanded such that $\mathbf{g} = \boldsymbol{\beta}\mathbf{x}$ where $\boldsymbol{\beta}$ is the vector of coefficients

$$\begin{aligned}\boldsymbol{\beta} &= \begin{bmatrix} \beta_0 & \beta_1 & \beta_2 & \beta_3 & \beta_4 \end{bmatrix} \\ &= \begin{bmatrix} 36 & -0.4 & 0.02 & 1.2 & 0.3 \end{bmatrix}\end{aligned}$$

Calculate the quantity demanded of good G when the vector of values of the explanatory variables is

$$\mathbf{x}=\begin{bmatrix} 1 \\ 14 \\ 8 \\ 82.5 \\ 3.2 \end{bmatrix}$$ where the element x_1 refers to the constant.

16.4 THE MATRIX INVERSE AND THE SOLUTION OF SIMULTANEOUS EQUATIONS

The concept of 'matrix division' is approached in matrix algebra by deriving the inverse of a matrix. One reason to find a matrix inverse is to help solve a set of simultaneous equations specified in matrix format. For example, consider the set of four simultaneous equations:

$$\begin{aligned} 3x_1+8x_2+x_3+2x_4 &= 96 \\ 20x_1-2x_2+4x_3+0.5x_4 &= 69 \\ 11x_1+3x_2+3x_3-5x_4 &= 75 \\ x_1+12x_2+x_3+8x_4 &= 134 \end{aligned}$$

These equations can be represented in matrix format by putting:

- coefficients of the four unknown variables x_1, x_2, x_3 and x_4 into a 4 × 4 matrix **A**
- the four unknown variables themselves into a 4 × 1 vector **x**
- the constant terms from the right-hand side of the equations into the 4 × 1 vector **b**.

They can then be written as

$$\mathbf{Ax} = \begin{bmatrix} 3 & 8 & 1 & 2 \\ 20 & -2 & 4 & 0.5 \\ 11 & 3 & 3 & -5 \\ 1 & 12 & 1 & 8 \end{bmatrix} \begin{bmatrix} x_1 \\ x_2 \\ x_3 \\ x_4 \end{bmatrix} = \begin{bmatrix} 96 \\ 69 \\ 75 \\ 134 \end{bmatrix} = \mathbf{b}$$

If this is not immediately obvious, try working through the matrix multiplication process to get the product matrix **Ax**. Working across the rows of **A**, each element multiplies the elements down the vector of unknown variables x_1, x_2, x_3 and x_4. If you write out the calculations in full for the four elements of the product matrix **Ax** and equate to the corresponding element in vector **b** then you should get the same set of simultaneous equations. For example, multiplying the elements across the first row of **A** by the elements down the column vector **x** gives the first element of **Ax** as

$$3x_1+8x_2+x_3+2x_4$$

so setting this equal to the first element of the product vector **b**, which is 96, gives us the first of our set of simultaneous equations.

You could, of course, solve this set of simultaneous equations by the standard row operations method, but this would take a long time to do manually and there are also certain other advantages from using the matrix method, as you will find out later.

The same matrix format as that derived earlier can be used for the general case. Assume there are n unknown variables $x_1, x_2, \ldots, x_n$, a set of $n \times n$ coefficients a_{11} to a_{nn}, and n constant values $b_1, b_2, b_3, \ldots, b_n$, such that

$$\begin{array}{ccccccccc}
a_{11}x_1 & + & a_{12}x_2 & + & \ldots & + & a_{1n}x_n & = & b_1 \\
a_{21}x_1 & + & a_{22}x_2 & + & \ldots & + & a_{2n}x_n & = & b_2 \\
. & & . & & & & . & & . \\
. & & . & & & & . & & . \\
. & & . & & & & . & & . \\
a_{n1}x_1 & + & a_{n2}x_2 & + & \ldots & + & a_{nn}x_n & = & b_n
\end{array}$$

This system of n simultaneous equations with n unknowns can be written in matrix format as $\mathbf{Ax} = \mathbf{b}$, where $\mathbf{A}$ is the $n \times n$ matrix of coefficients

$$\mathbf{A} = \begin{bmatrix} a_{11} & a_{12} & . & . & a_{1n} \\ a_{21} & a_{22} & . & . & a_{2n} \\ . & . & . & . & . \\ . & . & . & . & . \\ a_{n1} & a_{n2} & . & . & a_{nn} \end{bmatrix}$$

and $\mathbf{x}$ is the vector of unknown variables $\mathbf{x} = \begin{bmatrix} x_1 \\ x_2 \\ . \\ . \\ x_n \end{bmatrix}$

and $\mathbf{b}$ is the vector of constant parameters $\mathbf{b} = \begin{bmatrix} b_1 \\ b_2 \\ . \\ . \\ b_n \end{bmatrix}$

How does this specification of the set of simultaneous equations in the matrix format $\mathbf{Ax} = \mathbf{b}$ help us to solve for the unknown variables in $\mathbf{x}$? If x and A were single terms, instead of vectors and matrices, and $Ax = b$ then basic algebra would suggest that x could be found by simply respecifying the equation as $x = A^{-1}b$. The same logic is used when $\mathbf{x}$, $\mathbf{A}$ and $\mathbf{b}$ are matrices and we try to find $\mathbf{x} = \mathbf{A}^{-1}\mathbf{b}$.

The derivation of the matrix inverse $\mathbf{A}^{-1}$ is, however, a rather involved procedure and it is explained over the next few sections in this chapter. There is no denying that some students will find it hard work ploughing through the analysis. It is worth it, though, because you will learn:

- how to solve large sets of simultaneous equations in a few seconds by using matrix inversion on a spreadsheet
- how to use a set of tools that will be invaluable in the analysis of economic models with more than two variables, particularly when checking the second-order conditions in optimization problems.

Conditions for the existence of the matrix inverse

In Chapter 5 it was explained that in a system of linear simultaneous equations the basic rule for a unique solution to exist is that the number of unknowns must equal the number of equations, and linear dependence between equations must not be present. As long as these conditions hold then matrix analysis can be used to solve for any number of unknown variables. Since the number of unknown variables must equal the number of equations, the matrix of coefficients **A** must be **square**, i.e. the number of rows must equal the number of columns. Also, if we know the values for **A** and **b** and wish to find **x** using the formula $\mathbf{x} = \mathbf{A}^{-1}\mathbf{b}$ then we first have to establish whether the inverse matrix $\mathbf{A}^{-1}$ can actually be determined because in some circumstances it may not exist.

Before we can define what we mean by the inverse of a matrix, we need to introduce the concept of the **identity matrix**. This is any square matrix with each element along the diagonal (from top left to bottom right) being equal to 1 and with all other elements being zero. For example, the 3 × 3 identity matrix is

$$\mathbf{I} = \begin{bmatrix} 1 & 0 & 0 \\ 0 & 1 & 0 \\ 0 & 0 & 1 \end{bmatrix}$$

This identity matrix is the matrix equivalent to the number '1' in standard mathematics. Any matrix multiplied by the identity matrix will give the original matrix. For example,

$$\begin{bmatrix} 7 & 2 & 3 \\ 4 & 8 & 1 \\ 5 & 12 & 4 \end{bmatrix}\begin{bmatrix} 1 & 0 & 0 \\ 0 & 1 & 0 \\ 0 & 0 & 1 \end{bmatrix} = \begin{bmatrix} 7 & 2 & 3 \\ 4 & 8 & 1 \\ 5 & 12 & 4 \end{bmatrix}$$

A matrix **A** can be inverted if there exists an inverse $\mathbf{A}^{-1}$ such that $\mathbf{A}^{-1}\mathbf{A} = \mathbf{I}$, the identity matrix.

Using this definition, we can now see that if

$$\mathbf{Ax} = \mathbf{b}$$

multiplying both sides by $\mathbf{A}^{-1}$ gives

$$\mathbf{A}^{-1}\mathbf{Ax} = \mathbf{A}^{-1}\mathbf{b}$$

Since $\mathbf{A}^{-1}\mathbf{A} = \mathbf{I}$ this means that

$$\mathbf{Ix} = \mathbf{A}^{-1}\mathbf{b}$$

As any matrix or vector multiplied by the identity matrix gives the same matrix or vector then

$$\mathbf{x} = \mathbf{A}^{-1}\mathbf{b}$$

There are several instances **when the inverse of a matrix may not exist**:

First, the **zero**, or **null matrix**, which has all its elements equal to zero. Just as it is not possible to determine the inverse of zero in basic arithmetic, the inverse of the zero matrix **0** cannot be calculated. There are zero matrices corresponding to each possible order. For example, the 2 × 2 zero matrix will be

$$0 = \begin{bmatrix} 0 & 0 \\ 0 & 0 \end{bmatrix}$$

However, if we were trying to solve a set of simultaneous equations, we would be unlikely to start off with a matrix of coefficients that were all zero as this would not tell us very much!

Second, **linear dependence** of two or more rows (or columns) of a matrix will prevent its inverse being calculated. Linear dependence means that all the terms in one row (or column) are the same scalar multiple of the corresponding elements in another row (or column). The reason for this will become obvious when we have worked through the method for finding the inverse, but we can illustrate the problem with a simple example.

Consider the two simultaneous equations

$$8x + 10y = 120 \qquad (1)$$

$$4x + 5y = 60 \qquad (2)$$

All the values of (2) are 0.5 of the values in (1). Clearly, this pair of simultaneous equations cannot be solved by row operations to find the unknowns x and y. If (2) was multiplied by 2 and subtracted from (1), then we would end up with zero on both sides of the equation, which does not tell us anything. This linear dependence would also lead us down the same dead end if we tried to solve using the matrix inverse.

To actually find the inverse of a matrix, we first need to consider some special concepts associated with **square matrices**, namely:

- the determinant
- minors
- cofactors
- the adjoint matrix.

These are explained in the following sections.

QUESTIONS 16.4

1. Identify which of the following sets of simultaneous equations may be suitable for solving by matrix algebra and then put them in appropriate matrix format:

(a) $\begin{aligned} 5x+4y+9z &= 95 \\ 2x+\ \ y+4z &= 32 \\ 2x+5y+4z &= 61 \end{aligned}$

(b) $\begin{aligned} 6x+4y+8z &= 56 \\ 3x+2y+4z &= 28 \\ x-8y+2z &= 34 \end{aligned}$

(c) $\begin{aligned} 5x+4y+2z &= 95 \\ 9x+4y \qquad &= 32 \\ 2x+4y+4z &= 61 \end{aligned}$

(d) $\begin{aligned} 12x+2y+3z &= 124 \\ 6x+7y+\ \ z &= 42 \end{aligned}$

2. Which of the following are identity matrices?

(a) $\begin{bmatrix} 1 & 1 \\ 1 & 1 \end{bmatrix}$

(b) $[1]$

(c) $\begin{bmatrix} 1 & 0 \\ 0 & 1 \end{bmatrix}$

(d) $\begin{bmatrix} 0 & 1 \\ 1 & 0 \end{bmatrix}$

(e) $\begin{bmatrix} 1 & 0 \\ 0 & 1 \\ 1 & 0 \end{bmatrix}$

3. Are there obvious reasons why it may not be possible to derive an inverse for any of the following matrices?

(a) $\begin{bmatrix} 8 & 6 \\ 3 & 1 \end{bmatrix}$

(b) $\begin{bmatrix} 8 & 1 \\ 4 & 5 \\ 7 & 3 \end{bmatrix}$

(c) $\begin{bmatrix} 4 & 2 \\ 2 & 1 \end{bmatrix}$

(d) $\begin{bmatrix} 9 & 9 \\ 1 & 0 \end{bmatrix}$

(e) $\begin{bmatrix} 5 & 11 & 0 \\ -2 & 4 & 0.2 \\ 0 & -5 & 1 \end{bmatrix}$

16.5 DETERMINANTS

For a second-order matrix (i.e. order 2 × 2) the determinant is a number calculated by multiplying the elements in opposite corners and subtracting. The usual notation for a determinant is a set of vertical parallel lines either side of the array of elements, instead of the squared brackets used for a matrix. The determinant of the general 2 × 2 matrix **A**, written as |**A**|, will therefore be:

$$|\mathbf{A}| = \begin{vmatrix} a_{11} & a_{12} \\ a_{21} & a_{22} \end{vmatrix} = a_{11}a_{22} - a_{21}a_{12}$$

Example 16.10

Find the determinant of the matrix $\mathbf{A} = \begin{bmatrix} 5 & 7 \\ 4 & 9 \end{bmatrix}$

Solution

Using the formula defined earlier, the determinant of matrix **A** will be

$$|\mathbf{A}| = \begin{vmatrix} 5 & 7 \\ 4 & 9 \end{vmatrix} = 5 \times 9 - 7 \times 4 = 45 - 28 = 17$$

If any sets of rows or columns of a matrix are **linearly dependent** then the determinant will be zero and we have what is known as a **singular** matrix. For example, if the second row is twice the value of the corresponding elements in the first row and

$$\mathbf{A} = \begin{bmatrix} 5 & 8 \\ 10 & 16 \end{bmatrix}$$

then the determinant $|\mathbf{A}| = \begin{vmatrix} 5 & 8 \\ 10 & 16 \end{vmatrix} = 5 \times 16 - 8 \times 10 = 80 - 80 = 0$

The formula for the matrix inverse (which we will derive later) involves division by the determinant. Therefore, a condition for the inverse of a matrix to exist is that the matrix must be **non-singular**, i.e. the determinant must be different from zero. This condition applies to determinants of any order.

The determinant of a third-order matrix

For the general third-order matrix $\mathbf{A} = \begin{bmatrix} a_{11} & a_{12} & a_{13} \\ a_{21} & a_{22} & a_{23} \\ a_{31} & a_{32} & a_{33} \end{bmatrix}$ the determinant $|\mathbf{A}|$ can be calculated as

$$|\mathbf{A}| = a_{11}\begin{vmatrix} a_{22} & a_{23} \\ a_{32} & a_{33} \end{vmatrix} - a_{12}\begin{vmatrix} a_{21} & a_{23} \\ a_{31} & a_{33} \end{vmatrix} + a_{13}\begin{vmatrix} a_{21} & a_{22} \\ a_{31} & a_{32} \end{vmatrix}$$

This entails multiplying each of the elements in the first row by the determinant of the matrix remaining when the corresponding row and column are deleted. For example, the element a_{11} is multiplied by the determinant of the matrix remaining when row 1 and column 1 are deleted from the original 3 × 3 matrix. If we start from a_{11} then, as we use this method for each element across the row, the sign of each term will be positive and negative alternately. Thus, the second term has a negative sign.

Example 16.11

Derive the determinant of matrix $\mathbf{A} = \begin{bmatrix} 4 & 6 & 1 \\ 2 & 5 & 2 \\ 9 & 0 & 4 \end{bmatrix}$.

Solution

Expanding across the first row using the previous formula, the determinant will be

$$\begin{aligned} |\mathbf{A}| &= 4\begin{vmatrix} 5 & 2 \\ 0 & 4 \end{vmatrix} - 6\begin{vmatrix} 2 & 2 \\ 9 & 4 \end{vmatrix} + 1\begin{vmatrix} 2 & 5 \\ 9 & 0 \end{vmatrix} \\ &= 4(20-0) - 6(8-18) + 1(0-45) \\ &= 80 + 60 - 45 \\ &= 95 \end{aligned}$$

Although the determinants of the third-order matrices were found by expanding along the first row, they could also have been found by expanding along any other row or column. The same principle of multiplying each element along the expansion row (or down the expansion column) by the determinant of the matrix remaining when the corresponding row and column are deleted from the original matrix **A** is employed. This can help make the calculations easier if it is possible to expand along a row or column with one or more elements equal to zero, as in the next example.

However, there are rules regarding **the sign of each term**, which must be followed. These are explained for the general case in the next section. For a third-order determinant it is sufficient to remember that the first term will be positive if you expand along the first or third row or column and the first term will be negative if you expand along the second row or column. The signs of the subsequent terms in the expansion will then alternate.

For example, another way of finding the determinant of the matrix in Example 16.11 is to expand along the third row, which includes a zero and will therefore require less calculation.

Example 16.11 (reworked)

Derive the determinant of matrix $\mathbf{A} = \begin{bmatrix} 4 & 6 & 1 \\ 2 & 5 & 2 \\ 9 & 0 & 4 \end{bmatrix}$ by expanding along the third row.

Solution

Expanding across the third row, the first term will have a positive sign and so

$$\begin{aligned} |\mathbf{A}| &= 9\begin{vmatrix} 6 & 1 \\ 5 & 2 \end{vmatrix} - 0\begin{vmatrix} 4 & 1 \\ 2 & 2 \end{vmatrix} + 4\begin{vmatrix} 4 & 6 \\ 5 & 2 \end{vmatrix} \\ &= 9(12-5) - 0 + 4(20-12) \\ &= 63 + 32 \\ &= 95 \end{aligned}$$

QUESTIONS 16.5

1. Evaluate the following determinants:

(a) $|\mathbf{A}| = \begin{vmatrix} 8 & 2 \\ 3 & 1 \end{vmatrix}$ (b) $|\mathbf{B}| = \begin{vmatrix} 30 & 12 \\ 10 & 4 \end{vmatrix}$

(c) $|\mathbf{C}| = \begin{vmatrix} 5 & 8 \\ -7 & 0 \end{vmatrix}$ (d) $|\mathbf{D}| = \begin{vmatrix} 2 & 5 & 9 \\ 4 & 8 & 3 \\ 1 & 7 & 4 \end{vmatrix}$

(e) $|\mathbf{E}| = \begin{vmatrix} 4 & 3 & 10 \\ 7 & 0 & 3 \\ 12 & 2 & 5 \end{vmatrix}$

16.6 MINORS, COFACTORS AND THE LAPLACE EXPANSION

The Laplace expansion is a method that can be used to evaluate determinants of any order. Before explaining this method, we need to define a few more concepts (some of which we have actually already started using).

Minors

The minor $|\mathbf{M}_{ij}|$ of matrix $\mathbf{A}$ is the determinant of the matrix left when row i and column j are deleted. For example, if the first row and first column are deleted from matrix

$$\mathbf{A} = \begin{bmatrix} a_{11} & a_{12} & a_{13} \\ a_{21} & a_{22} & a_{23} \\ a_{31} & a_{32} & a_{33} \end{bmatrix}$$

the determinant of the remaining matrix will be the minor

$$|\mathbf{M}_{11}| = \begin{vmatrix} a_{22} & a_{23} \\ a_{32} & a_{33} \end{vmatrix}$$

Example 16.12

Find the minor $|\mathbf{M}_{31}|$ of the matrix $\mathbf{A} = \begin{bmatrix} 8 & 2 & 3 \\ 1 & 9 & 4 \\ 4 & 3 & 6 \end{bmatrix}$.

Solution

The minor $|\mathbf{M}_{31}|$ is the determinant of the matrix remaining when the third row and first column have been eliminated from matrix $\mathbf{A}$. Therefore,

$$|\mathbf{M}_{31}| = \begin{vmatrix} 2 & 3 \\ 9 & 4 \end{vmatrix} = 8 - 27 = -19$$

Using this definition of a minor, the formula for the determinant of a third-order matrix expanded across the first row could be specified as

$$|\mathbf{A}| = a_{11}|\mathbf{M}_{11}| - a_{12}|\mathbf{M}_{12}| + a_{13}|\mathbf{M}_{13}|$$

Cofactors

A cofactor is the same as a minor, except that its sign is determined by the row and column that it corresponds to. The sign of cofactor $|\mathbf{C}_{ij}|$ is equal to $(-1)^{i+j}$. Thus, if the row number and column number sum to an even number the cofactor sign will be positive and if they sum to an odd number it will be negative. For example, to derive the cofactor $|\mathbf{C}_{12}|$ for the general third-order matrix $\mathbf{A}$ we eliminate the first row and the second column and then, since $i + j = 3$, we multiply the determinant of the elements that remain by $(-1)^3$. Therefore,

$$|\mathbf{C}_{12}| = (-1)^3 \begin{vmatrix} a_{21} & a_{23} \\ a_{31} & a_{33} \end{vmatrix} = (-1) \begin{vmatrix} a_{21} & a_{23} \\ a_{31} & a_{33} \end{vmatrix}$$

Example 16.13

Find the cofactor $|\mathbf{C}_{22}|$ of the matrix $\mathbf{A} = \begin{bmatrix} 8 & 2 & 3 \\ 1 & 9 & 4 \\ 4 & 3 & 6 \end{bmatrix}$.

Solution

The cofactor $|\mathbf{C}_{22}|$ is the determinant of the matrix remaining when the second row and second column have been eliminated. It will have the sign $(-1)^4$ since $i + j = 4$. The solution is, therefore,

$$|\mathbf{C}_{12}| = (-1)^4 \begin{vmatrix} 8 & 3 \\ 4 & 6 \end{vmatrix} = (+1)(48 - 12) = 36$$

The determinant of a third-order matrix in terms of its cofactors, expanded across the first row, can now be specified as

$$|\mathbf{A}| = a_{11}|\mathbf{C}_{11}| - a_{12}|\mathbf{C}_{12}| + a_{13}|\mathbf{C}_{13}| \qquad (1)$$

Although this looks very similar to the formula for $|\mathbf{A}|$ in terms of its minors, set out earlier, you should note that the sign of the second term is positive. This is because the cofactor itself will have a negative sign.

The Laplace expansion

For matrices of any order n, using the Laplace expansion, the determinant is specified as

$$|\mathbf{A}| = \sum_{i,j=1}^{i,j=n} a_{ij} |\mathbf{C}_{ij}|$$

where the summation from 1 to n takes place across the rows (i) or down the columns (j). If you check the formula (1) for the determinant of a third-order matrix in terms of its cofactors, you will see that this employs the Laplace expansion.

If the original matrix is fourth order or greater, then the first set of cofactors derived by using the Laplace expansion will themselves be third order or greater. Therefore, the Laplace expansion has to be used again to break these cofactors down. This process needs to continue until the determinant is specified in terms of second-order cofactors which can then be evaluated.

With larger determinants this method can involve quite a lot of calculations and so it is usually quicker to use a spreadsheet. But first, let us work through an example by doing the calculations manually to make sure that you understand how this method works.

Example 16.14

Use the Laplace expansion to find the determinant of the 4 × 4 matrix

$$\mathbf{A} = \begin{bmatrix} 8 & 10 & 2 & 3 \\ 0 & 5 & 7 & 10 \\ 2 & 2 & 1 & 4 \\ 3 & 4 & 4 & 0 \end{bmatrix}.$$

Solution

Expanding down the first column (because there is a zero which means one less set of calculations), the first round of the Laplace expansion gives

$$|\mathbf{A}| = 8\begin{vmatrix} 5 & 7 & 10 \\ 2 & 1 & 4 \\ 4 & 4 & 0 \end{vmatrix} - 0\begin{vmatrix} 10 & 2 & 3 \\ 2 & 1 & 4 \\ 4 & 4 & 0 \end{vmatrix} + 2\begin{vmatrix} 10 & 2 & 3 \\ 5 & 7 & 10 \\ 4 & 4 & 0 \end{vmatrix} - 3\begin{vmatrix} 10 & 2 & 3 \\ 5 & 7 & 10 \\ 2 & 1 & 4 \end{vmatrix}$$

A second round of the Laplace expansion is then used to break these third-order cofactors down into second-order cofactors that can be evaluated. The second term is zero and disappears and so

$$\begin{aligned}|\mathbf{A}| &= 8\left(5\begin{vmatrix} 1 & 4 \\ 4 & 0 \end{vmatrix} - 2\begin{vmatrix} 7 & 10 \\ 4 & 0 \end{vmatrix} + 4\begin{vmatrix} 7 & 10 \\ 1 & 4 \end{vmatrix}\right) + 2\left(10\begin{vmatrix} 7 & 10 \\ 4 & 0 \end{vmatrix} - 5\begin{vmatrix} 2 & 3 \\ 4 & 0 \end{vmatrix} + 4\begin{vmatrix} 2 & 3 \\ 7 & 10 \end{vmatrix}\right) \\ &\quad - 3\left(10\begin{vmatrix} 7 & 10 \\ 1 & 4 \end{vmatrix} - 5\begin{vmatrix} 2 & 3 \\ 1 & 4 \end{vmatrix} + 2\begin{vmatrix} 2 & 3 \\ 7 & 10 \end{vmatrix}\right) \\ &= 8[5(-16) - 2(-40) + 4(18)] + 2[10(-40) - 5(-12) + 4(-1)] - 3[10(18) - 5(5) + 2(-1)] \\ &= 8[-80 + 80 + 72] + 2[-400 + 60 - 4] - 3[180 - 25 - 2] \\ &= 8(72) + 2(-344) - 3(153) = 576 - 688 - 459 = -571\end{aligned}$$

Using a spreadsheet to evaluate determinants

It is very straightforward to use the Excel function MDETERM to evaluate determinants. Just type in the matrix and then, in the cell where you want the value of the determinant to appear, enter

$$= \text{MDETERM}(\textit{cell range for matrix})$$

For example, if you had entered the 4 × 4 matrix from Example 16.14 in cells B2 to E5 and you wanted the determinant to appear in cell G2 you would type = MDETERM(B2:E5) in cell G2.

QUESTIONS 16.6

1. For the matrix $\mathbf{A} = \begin{bmatrix} 5 & 0 & 4 \\ 8 & 3 & 6 \\ 2 & 7 & 1 \end{bmatrix}$, evaluate the following minors and cofactors:
   (a) $|\mathbf{M}_{11}|$ (b) $|\mathbf{M}_{33}|$ (c) $|\mathbf{M}_{12}|$ (d) $|\mathbf{C}_{21}|$ (e) $|\mathbf{C}_{13}|$ (f) $|\mathbf{C}_{12}|$

2. Manually calculate the values of the determinants of the matrices **A, B** and **C** and then check your answers using Excel:

$$\mathbf{A} = \begin{bmatrix} 2 & 6 & 2 & 3 \\ 10 & 5 & 7 & 25 \\ 0 & 2 & 1 & 5 \\ 4 & -3 & 4 & 9 \end{bmatrix}, \mathbf{B} = \begin{bmatrix} 8 & 6 & 2 & 1 \\ 3 & 8 & 7 & -4 \\ 0 & -2 & 1 & 5 \\ 4 & 3 & 3 & 2 \end{bmatrix}, \mathbf{C} = \begin{bmatrix} 1 & 5 & 2 & 1 & 1 \\ 6 & 1 & 0 & -4 & 3 \\ 0 & 4 & 7 & 2 & 1 \\ 9 & 2 & 3 & 2 & 2 \\ 0 & 4 & 8 & 0 & 6 \end{bmatrix}$$

16.7 THE TRANSPOSE MATRIX, THE COFACTOR MATRIX, THE ADJOINT AND THE MATRIX INVERSE FORMULA

There are still a few more concepts that are needed before we can determine the inverse of a matrix.

The transpose of a matrix

To get the transpose of a matrix, usually written as $\mathbf{A}^{\mathrm{T}}$, the rows and columns are swapped around, i.e. row 1 becomes column 1 and column 1 becomes row 1, etc. If a matrix is not square then the numbers of rows and columns will alter when it is transposed.

For example, if $\mathbf{A} = \begin{bmatrix} 5 & 20 \\ 16 & 9 \\ 12 & 6 \end{bmatrix}$ then $\mathbf{A}^{\mathrm{T}} = \begin{bmatrix} 5 & 16 & 12 \\ 20 & 9 & 6 \end{bmatrix}$

The matrix of cofactors

If we replace every element in a matrix by its corresponding cofactor then we get the **matrix of cofactors**, usually denoted by **C**.

For example, if $\mathbf{A}=\begin{bmatrix} 2 & 4 & 3 \\ 3 & 5 & 0 \\ 4 & 2 & 5 \end{bmatrix}$ then $\mathbf{C}=\begin{bmatrix} 25 & -15 & -12 \\ -14 & -2 & 12 \\ -15 & 9 & -2 \end{bmatrix}$

To make sure you understand how these numbers were calculated, let us work through some of them. The cofactor $|\mathbf{C}_{ij}|$ of matrix **A** is the determinant of the matrix remaining when row i and column j have been eliminated, with the sign $(-1)^{i+j}$. Thus, some selected elements of the cofactor matrix are

$$c_{11}=|\mathbf{C}_{11}|=(-1)^{(1+1)}\begin{vmatrix} a_{22} & a_{23} \\ a_{32} & a_{33} \end{vmatrix}=(-1)^2\begin{vmatrix} 5 & 0 \\ 2 & 5 \end{vmatrix}=(25-0)=25$$

$$c_{21}=|\mathbf{C}_{21}|=(-1)^{(2+1)}\begin{vmatrix} a_{12} & a_{13} \\ a_{32} & a_{33} \end{vmatrix}=(-1)^3\begin{vmatrix} 4 & 3 \\ 2 & 5 \end{vmatrix}=(-1)(20-6)=-14$$

Check for yourself the calculation of the other elements of **C**.

The adjoint matrix

The adjoint matrix, usually denoted by **AdjA**, is the transpose of the cofactor matrix.

Thus, if $\mathbf{A}=\begin{bmatrix} a_{11} & a_{12} & a_{13} \\ a_{21} & a_{22} & a_{23} \\ a_{31} & a_{32} & a_{33} \end{bmatrix}$ then $\mathbf{AdjA}=\begin{bmatrix} |C_{11}| & |C_{21}| & |C_{31}| \\ |C_{12}| & |C_{22}| & |C_{32}| \\ |C_{13}| & |C_{23}| & |C_{33}| \end{bmatrix}$

Using the cofactor example, we have already shown that for

matrix $\mathbf{A}=\begin{bmatrix} 2 & 4 & 3 \\ 3 & 5 & 0 \\ 4 & 2 & 5 \end{bmatrix}$ the cofactor matrix is $\mathbf{C}=\begin{bmatrix} 25 & -14 & -12 \\ -14 & -2 & 12 \\ -15 & 9 & -2 \end{bmatrix}$

Therefore, the adjoint matrix will be

$$\mathbf{AdjA}=\mathbf{C}^{\mathrm{T}}=\begin{bmatrix} 25 & -14 & -15 \\ -15 & -2 & 9 \\ -12 & 12 & -2 \end{bmatrix}$$

The inverse matrix

The formula for $\mathbf{A}^{-1}$, the inverse of matrix **A**, can now be stated as

$$\mathbf{A}^{-1}=\frac{\mathbf{AdjA}}{|\mathbf{A}|}$$

where matrix **A** is a non-singular matrix, i.e. its determinant $|\mathbf{A}|$ must not be zero.

Example 16.15

Find the inverse matrix $\mathbf{A}^{-1}$ for matrix $\mathbf{A}=\begin{bmatrix} 2 & 4 & 3 \\ 3 & 5 & 0 \\ 4 & 2 & 5 \end{bmatrix}$.

Solution

We have already determined the adjoint for this particular matrix in the previous example. Its determinant $|\mathbf{A}|$ can be evaluated by expanding down the third column as

$$|\mathbf{A}| = 3\begin{vmatrix} 3 & 5 \\ 4 & 2 \end{vmatrix} - 0 + 5\begin{vmatrix} 2 & 4 \\ 3 & 5 \end{vmatrix} = 3(6-20)+5(10-12)$$

$$= 3(-14)+5(-2) = -42-10 = -52$$

Therefore, given that we already know that the $\mathbf{AdjA}=\begin{bmatrix} 25 & -14 & -15 \\ -15 & -2 & 9 \\ -12 & 12 & -2 \end{bmatrix}$, the inverse matrix will be

$$\mathbf{A}^{-1} = \frac{\mathbf{AdjA}}{|\mathbf{A}|} = \frac{1}{-52}\begin{bmatrix} 25 & -14 & -15 \\ -15 & -2 & 9 \\ -12 & 12 & -2 \end{bmatrix} = \begin{bmatrix} -0.48 & 0.27 & 0.29 \\ 0.29 & 0.04 & -0.17 \\ 0.27 & -0.23 & 0.04 \end{bmatrix}$$

The derivation of this matrix inverse has been quite time consuming, but you need to understand this underlying method before learning how to do the calculations on a spreadsheet. However, first let us work through another example from first principles to make sure that you understand each stage of the analysis. This time we will start with a 2 × 2 matrix.

Example 16.16

Find the inverse matrix $\mathbf{A}^{-1}$ for matrix $\mathbf{A}=\begin{bmatrix} 20 & 5 \\ 6 & 2 \end{bmatrix}$.

Solution

Because there are only four elements, the cofactor corresponding to each element of **A** will just be the element in the opposite corner, with the sign $(-1)^{i+j}$. Therefore, the corresponding cofactor matrix will be

$$\mathbf{C}=\begin{bmatrix} 2 & -6 \\ -5 & 20 \end{bmatrix}$$

The adjoint is the transpose of the cofactor matrix and so

$$\mathbf{AdjA}=\begin{bmatrix} 2 & -5 \\ -6 & 20 \end{bmatrix}$$

The determinant of the original matrix **A** is easily calculated as

$$|\mathbf{A}| = 20 \times 2 - 5 \times 6 = 40 - 30 = 10$$

The inverse matrix is thus

$$\mathbf{A}^{-1} = \frac{1}{|\mathbf{A}|}\mathbf{AdjA} = \frac{1}{10}\begin{bmatrix} 2 & -5 \\ -6 & 20 \end{bmatrix} = \begin{bmatrix} 0.2 & -0.5 \\ -0.6 & 2 \end{bmatrix}$$

Derivation of the matrix inverse formula

You can just take the previous formula for the matrix inverse as given and there is no need to work through the proof of this result for the general case. However, we can show how the inverse formula can be derived for the case of a 2 × 2 matrix.

Assume that we wish to invert the matrix $\mathbf{A} = \begin{bmatrix} a & b \\ c & d \end{bmatrix}$

This inverse can be specified as $\mathbf{A}^{-1} = \begin{bmatrix} e & f \\ g & h \end{bmatrix}$

where e, f, g and h are numbers that the inverse formula will calculate.

Multiplying a square non-singular matrix by its inverse will give the identity matrix.

Thus,

$$\mathbf{AA}^{-1} = \begin{bmatrix} a & b \\ c & d \end{bmatrix}\begin{bmatrix} e & f \\ g & h \end{bmatrix} = \begin{bmatrix} ae+bg & af+bh \\ ce+dg & cf+dh \end{bmatrix} = \begin{bmatrix} 1 & 0 \\ 0 & 1 \end{bmatrix} = \mathbf{I}$$

From the calculations for each of the elements of **I** we get the four simultaneous equations

$$ae + bg = 1 \quad (1) \qquad af + bh = 0 \quad (2)$$
$$ce + dg = 0 \quad (3) \qquad cf + dh = 1 \quad (4)$$

The values of the elements of the inverse matrix e, f, g and h in terms of the values of the elements of the original matrix can now be solved by the substitution method.

From (1)

$$ae = 1 - bg$$

and so

$$e = \frac{(1-bg)}{a} \qquad (5)$$

Substituting the result (5) into (3) gives

$$\frac{c(1-bg)}{a}+dg = 0$$
$$c-cbg+dga = 0$$
$$g(ad-bc) = -c$$
$$g = \frac{-c}{ad-bc} \qquad (6)$$

Substituting the result from (6) into (5) gives

$$e=\left(1-\frac{b(-c)}{ad-bc}\right)\frac{1}{a}=\left(\frac{ad-bc+bc}{ad-bc}\right)\frac{1}{a}=\left(\frac{ad}{ad-bc}\right)\frac{1}{a}=\frac{d}{ad-bc} \qquad (7)$$

Using the same substitution method, you can check for yourself that the other two elements of the inverse matrix will be

$$f=\frac{-b}{ad-bc} \qquad (8)$$

and $h=\dfrac{a}{ad-bc}$ (9)

Since the values for e, f, g and h that are derived in (6), (7), (8) and (9) all contain the same term $\dfrac{1}{ad-bc}$ this can be written as a scalar multiplier so that

$$\mathbf{A}^{-1}=\begin{bmatrix} e & f \\ g & h \end{bmatrix}=\frac{1}{ad-bc}\begin{bmatrix} d & -b \\ -c & a \end{bmatrix} \qquad (10)$$

This checks out with the general inverse formula since for matrix $\mathbf{A}=\begin{bmatrix} a & b \\ c & d \end{bmatrix}$, the determinant is $|\mathbf{A}| = ad - bc$

The cofactor matrix is $\mathbf{C}=\begin{bmatrix} d & -b \\ -c & a \end{bmatrix}$ and so the adjoint is $\mathbf{AdjA}=\begin{bmatrix} d & -b \\ -c & a \end{bmatrix}$.

Substituting these results into (10) gives the inverse formula

$$\mathbf{A}^{-1}=\frac{\mathbf{AdjA}}{|\mathbf{A}|}$$

Using a spreadsheet for matrix inversion

Although you need to understand the rationale behind the matrix inversion process, for any actual computations involving a third order or larger matrix, it is quicker to use a spreadsheet rather than do the calculations manually.

To invert a matrix using the Excel MINVERSE formula:

- Enter the matrix that you wish to invert.

- Highlight cells where inverted matrix will go (same dimension as original matrix), and enter in formula bar =MINVERSE(*cell range of matrix to be inverted*).
- Hold down the Ctrl *and* Shift keys together and press Enter. Curly brackets { } will then appear around the formula and the inverted matrix should be calculated in the cells that you have chosen.

QUESTIONS 16.7

1. Derive the inverse matrix $\mathbf{A}^{-1}$ when $\mathbf{A} = \begin{bmatrix} 25 & 15 \\ 10 & 8 \end{bmatrix}$.
2. For the matrix $\mathbf{A} = \begin{bmatrix} 5 & 0 & 2 \\ 3 & 4 & 5 \\ 2 & 1 & 2 \end{bmatrix}$, derive the cofactor matrix **C**, the adjoint matrix **AdjA** and the inverse matrix $\mathbf{A}^{-1}$ by manual calculation.
3. Use a spreadsheet to derive the matrix inverse $\mathbf{A}^{-1}$ for

$$\mathbf{A} = \begin{bmatrix} 4 & 6 & 2 & 3 \\ 10 & 5 & 7 & 20 \\ 0 & 2 & 1 & 5 \\ 4 & -3 & 4 & 12 \end{bmatrix}$$

16.8 APPLICATION OF THE MATRIX INVERSE TO THE SOLUTION OF LINEAR SIMULTANEOUS EQUATIONS

Although small sets of linear equations can be solved by other algebraic techniques, e.g. row operations, we will work through a simple example here to illustrate how the matrix method works before explaining how larger sets of linear equations can be solved using a spreadsheet.

Example 16.17

Use matrix algebra to solve for the unknown variables x_1, x_2 and x_3 given that

$$\begin{aligned} 10x_1 + 3x_2 + 6x_3 &= 76 \\ 4x_1 + 5x_3 &= 41 \\ 5x_1 + 2x_2 + 2x_3 &= 34 \end{aligned}$$

Solution

This set of simultaneous equations can be set up in matrix format as $\mathbf{Ax} = \mathbf{b}$ where

$$\mathbf{Ax} = \begin{bmatrix} 10 & 3 & 6 \\ 4 & 0 & 5 \\ 5 & 2 & 2 \end{bmatrix} \begin{bmatrix} x_1 \\ x_2 \\ x_3 \end{bmatrix} = \begin{bmatrix} 76 \\ 41 \\ 34 \end{bmatrix} = \mathbf{b}$$

To derive the vector of unknowns **x** using the matrix formulation $\mathbf{x} = \mathbf{A}^{-1}\mathbf{b}$, we first have to derive the matrix inverse $\mathbf{A}^{-1}$. The first step is to derive the cofactor matrix, which will be

$$\mathbf{C} = \begin{bmatrix} (0-10) & -(8-25) & (8-0) \\ -(6-12) & (20-30) & -(20-15) \\ (15-0) & -(50-24) & (0-12) \end{bmatrix} = \begin{bmatrix} -10 & 17 & 8 \\ 6 & -10 & -5 \\ 15 & -26 & -12 \end{bmatrix}$$

The adjoint matrix will be the transpose of the cofactor matrix and so

$$\mathbf{AdjA} = \mathbf{C}^{\mathrm{T}} = \begin{bmatrix} -10 & 6 & 15 \\ 17 & -10 & -26 \\ 8 & -5 & -12 \end{bmatrix}$$

The determinant of **A**, expanding along the second row, will be

$$|\mathbf{A}| = \begin{vmatrix} 10 & 3 & 6 \\ 4 & 0 & 5 \\ 5 & 2 & 2 \end{vmatrix} = -4(6-12) + 0 - 5(20-15) = 24 - 25 = -1$$

The matrix inverse will therefore be

$$\mathbf{A}^{-1} = \frac{\mathbf{AdjA}}{|\mathbf{A}|} = \frac{1}{-1}\begin{bmatrix} -10 & 6 & 15 \\ 17 & -10 & -26 \\ 8 & -5 & -12 \end{bmatrix} = \begin{bmatrix} 10 & -6 & -15 \\ -17 & 10 & 26 \\ -8 & 5 & 12 \end{bmatrix}$$

To solve for the vector of unknowns **x** we calculate

$$\mathbf{x} = \mathbf{A}^{-1}\mathbf{b} = \begin{bmatrix} 10 & -6 & -15 \\ -17 & 10 & 26 \\ -8 & 5 & 12 \end{bmatrix}\begin{bmatrix} 76 \\ 41 \\ 34 \end{bmatrix} = \begin{bmatrix} (10\times 76)-(6\times 41)-(15\times 34) \\ (-17\times 76)+(10\times 41)+(26\times 34) \\ (-8\times 76)+(5\times 41)+(12\times 34) \end{bmatrix}$$

$$= \begin{bmatrix} 760-246-510 \\ -1292+410-884 \\ -608+205-408 \end{bmatrix} = \begin{bmatrix} 4 \\ 2 \\ 5 \end{bmatrix} = \begin{bmatrix} x_1 \\ x_2 \\ x_3 \end{bmatrix}$$

You can check that these are the correct values by substituting them for the unknown variables x_1, x_2 and x_3 in the equations given in this problem. For example, substituting into the first equation gives

$$10x_1 + 3x_2 + 6x_3 = 10(4) + 3(2) + 6(5) = 40 + 6 + 30 = 76$$

Using a spreadsheet to solve simultaneous equations

The next example shows how to solve a set of six simultaneous equations with six unknown variables using a spreadsheet. Once you have worked through this example and understood what is involved, it should take you less than a minute to solve similar examples using Excel to do the necessary matrix inversion and multiplication.

Example 16.18

Solve for the unknown variables x_1, x_2, x_3, x_4, x_5 and x_6 given that

$$\begin{aligned}
4x_1 + x_2 + 2x_3 - 17x_4 - 5x_5 + 8x_6 &= 21\\
8x_1 + 9x_2 + 23x_3 + 15x_4 + 11x_5 + 39x_6 &= 593\\
24x_1 + 41x_2 + 9x_3 + 3x_4 + x_6 &= 317\\
6x_1 + 5x_2 - x_4 + 5x_5 - 7x_6 &= 35\\
9x_1 + 11x_2 + 39x_3 + 23x_4 + 15x_5 &= 678\\
28x_1 + 49x_2 + 4x_3 + 5x_4 + 9x_5 + 7x_6 &= 391
\end{aligned}$$

Table 16.5 Using a spreadsheet for Example 16.18

	A	B	C	D	E	F	G	H
1	Example 16.18							
2	**A MATRIX**							**b**
3	4	1	2	−17	−5	8		21
4	8	9	23	15	11	39		593
5	24	41	9	3	0	1		317
6	6	5	0	−1	3	−7		35
7	9	11	39	23	15	0		678
8	28	49	4	5	9	7		391
9	**Inverse A^−1**						**A^−1*b =**	**x**
10	−0.0453	0.08783	0.11969	0.32077	−0.0634	−0.1339	**solution**	5
11	0.02431	−0.0509	−0.0504	−0.1805	0.03194	0.08268	**values**	2
12	0.03398	−0.0162	−1E-17	−0.0512	0.03343	−1E-17		12
13	−0.0723	0.03416	0.06457	0.05788	−0.0253	−0.0591		1
14	0.03184	−0.0257	−0.1339	−0.0156	0.0331	0.11024		8
15	0.00247	0.02302	−5E-18	−0.0118	−0.0137	4.5E-18		4

Solution

Enter the matrix of coefficients **A** and the vector of constant values **b** into a spreadsheet, as shown in Table 16.5. In this table the cells (A3:F8) are used for the **A** matrix and the **b** column vector is in cells (H3:H8) and so the rest of the instructions that follow use these cell references.

Create the inverse matrix $\mathbf{A}^{-1}$ by highlighting a 6 × 6 block of cells (A10:F15) and then typing in the formula =MINVERSE(A3:F8) and making sure both the Ctrl and Shift keys are held down when the Return key is pressed.

To derive the vector of unknowns **x** by finding the product matrix $\mathbf{A}^{-1}\mathbf{b}$, highlight a 6 × 1 column of cells (H10:H15) and then enter =MMULT(A10:F15, H3:H8) and hold down the Ctrl and Shift keys when you hit return.

The vector of unknown variables should be calculated in the six cells of this column. You can now just read off the solution values $x_1 = 5$, $x_2 = 2$, $x_3 = 12$, $x_4 = 1$, $x_5 = 8$ and $x_6 = 4$.

Note that although most numbers in this table have been rounded to 5 decimal places, this would have rounded some very small numbers down to zero, so they have been left in exponent format. For example, the number −1E−17 is −1 divided by 10^{17}.

Estimating the parameters of an economic model

One important application of matrix algebra is to find solutions for unknown variables in econometrics, where estimates of the parameters of an economic model are derived using observations of different values of the variables in the model. Normally, relatively large data sets are used to estimate parameters, and a stochastic (random) error term has to be allowed for. However, to explain the basic principles involved we will work with only three observations and assume no error term. This should help you to understand the more sophisticated models you will encounter if you go on to study intermediate econometric analysis of multi-variable models.

Assume that y is a linear function of three exogenous variables x_1, x_2 and x_3 so that

$$y_i = \beta_1 x_{1i} + \beta_2 x_{2i} + \beta_3 x_{3i}$$

where the subscript i denotes the observation number and β_1, β_2 and β_3 are the parameters whose values we wish to find. There are three observations, which give the values shown next:

Observation number	y	x_1	x_2	x_3
1	240	10	12	20
2	150	5	8	15
3	300	12	18	20

How can these observations be used to estimate the parameters β_1, β_2 and β_3? If the function $y_i = \beta_1 x_{1i} + \beta_2 x_{2i} + \beta_3 x_{3i}$ holds for all three observations (i.e. all three values of i) then there will be three simultaneous equations:

$$240 = \beta_1 10 + \beta_2 12 + \beta_3 20 \quad (1)$$

$$150 = \beta_1 5 + \beta_2 8 + \beta_3 15 \quad (2)$$

$$300 = \beta_1 12 + \beta_2 18 + \beta_3 20 \quad (3)$$

These can be written in matrix format as $\mathbf{y} = \mathbf{X\beta}$

$$\text{where } \mathbf{y} = \begin{bmatrix} 240 \\ 150 \\ 300 \end{bmatrix}, \quad \mathbf{X} = \begin{bmatrix} 10 & 12 & 20 \\ 5 & 8 & 15 \\ 12 & 18 & 20 \end{bmatrix} \quad \text{and} \quad \boldsymbol{\beta} = \begin{bmatrix} \beta_1 \\ \beta_2 \\ \beta_3 \end{bmatrix}$$

Since $\mathbf{X\beta} = \mathbf{y}$

multiplying both sides by inverse $\mathbf{X}^{-1}$ gives $\mathbf{X}^{-1}\mathbf{X\beta} = \mathbf{X}^{-1}\mathbf{y}$

A matrix times its inverse gives the identity matrix. Thus,s $\mathbf{I\beta} = \mathbf{X}^{-1}\mathbf{y}$

and so the vector of parameters $\boldsymbol{\beta}$ will bes $\boldsymbol{\beta} = \mathbf{X}^{-1}\mathbf{y}$

Although we could now finish the calculations using the Excel matrix multiplication process explained earlier, we will continue working through this problem by hand. Note that the notation is different from that used in the previous section because we are trying to find the values of the coefficients in vector $\boldsymbol{\beta}$ rather than the values of the variables y_1, y_2 and y_3, which are already given in vector $\mathbf{y}$.

To find the matrix inverse $\mathbf{X}^{-1}$, we first find the cofactor matrix

$$\mathbf{C} = \begin{bmatrix} 160-270 & -(100-180) & 90-96 \\ -(240-360) & 200-240 & -(180-144) \\ 180-160 & -(150-100) & 80-60 \end{bmatrix}$$
$$= \begin{bmatrix} -110 & 80 & -6 \\ 120 & -40 & 36 \\ 20 & -50 & 20 \end{bmatrix}$$

The adjoint matrix will then be the transpose of this cofactor matrix

$$\mathbf{AdjX} = \mathbf{C}^{\mathrm{T}} = \begin{bmatrix} -110 & 120 & 20 \\ 80 & -40 & -50 \\ -6 & 36 & 20 \end{bmatrix}$$

The determinant of matrix $\mathbf{X}$ can be calculated as

$$|\mathbf{X}| = 10(160-270) - 12(100-180) + 20(90-96) = -260$$

Inserting these values into the formula for the inverse matrix gives

$$\mathbf{X}^{-1} = \frac{\mathbf{AdjX}}{|\mathbf{X}|} = \frac{1}{-260}\begin{bmatrix} -110 & 120 & 20 \\ 80 & -40 & -50 \\ -6 & 36 & 20 \end{bmatrix}$$
$$= \begin{bmatrix} 0.42 & 0.46 & -0.08 \\ -0.3 & 0.15 & 0.19 \\ 0.02 & -0.14 & -0.08 \end{bmatrix}$$

Therefore, the vector of coefficients is

$$\boldsymbol{\beta} = \mathbf{X}^{-1}\mathbf{y} = \begin{bmatrix} 0.42 & 0.46 & -0.08 \\ -0.3 & 0.15 & 0.19 \\ 0.02 & -0.14 & -0.08 \end{bmatrix}\begin{bmatrix} 240 \\ 150 \\ 300 \end{bmatrix} = \begin{bmatrix} 9.23 \\ 6.92 \\ 3.23 \end{bmatrix} = \begin{bmatrix} \beta_1 \\ \beta_2 \\ \beta_3 \end{bmatrix}$$

To check the parameters in vector $\boldsymbol{\beta}$ have been calculated correctly, we can insert the values computed earlier into the first of the set of three simultaneous equations in this example. Thus, from equation (1)

$$y_1 = \beta_1 10 + \beta_2 12 + \beta_3 20 = 9.23(10) + 6.92(12) + 3.23(20) = 240$$

and so the calculated value of 240 for y_1 is correct (allowing for rounding error).

QUESTIONS 16.8

(You can solve questions 1 and 2 manually but Excel should be used for the others.)

1. Use the matrix inverse method to find the unknowns x and y when

$$\begin{aligned} 4x + 6y &= 68 \\ 5x + 20y &= 185 \end{aligned}$$

2. Use matrix algebra to solve for x_1, x_2 and x_3 given that

$$\begin{aligned} 3x_1 + 4x_2 + 3x_3 &= 60 \\ 4x_1 + 10x_2 + 2x_3 &= 104 \\ 4x_1 + 2x_2 + 4x_3 &= 60 \end{aligned}$$

3. Assume that demand for a good (Q) depends on its own price (P), income (M) and the price of a substitute good (S) according to the demand function

$$Q_i = \beta_1 P_i + \beta_2 M_i + \beta_3 S_i$$

where β_1, β_2 and β_3 are parameters whose values are not yet known and the subscript i denotes the observation number. Three observations of the amount Q demanded when P, M and S take on different values are shown next:

Observation number	P	M	S	Q
1	6	5	5	4
2	8	8	6	6.4
3	5	6	4	5.1

Find the values of β_1, β_2 and β_3 by setting up the relevant system of simultaneous equations in matrix format and solving using the inverse matrix.

Use the vector of parameters $\boldsymbol{\beta}$ that you have found to predict the value of Q when P, M and S take the values 7, 9 and 10, respectively, by vector multiplication.

4. Assume that the quantity demanded of oil (Q) depends on its own price (P), income (M), the price of the substitute fuel gas (G), the price of the

complement good cars (C), population (N) and average temperature (T) according to the demand function

$$Q_i = \beta_1 P_i + \beta_2 M_i + \beta_3 G_i + \beta_4 C_i + \beta_5 N_i + \beta_6 T_i$$

where β_1, β_2, β_3, β_4, β_5 and β_6 are parameters, whose values are not yet known, and the subscript i denotes the observation number for Q and the explanatory variables. Six observations of Q when P, M, G, C, N and T take on different values are:

Observation number	P	M	G	C	N	T	Q
1	15	80	12.5	5	4000	18	6.980
2	20	95	14	8	4100	17.4	6.919
3	28	108	11	6	4150	19.2	4.522
4	35	112	16.2	7.5	4230	18.3	4.659
5	36	110	16	8	4215	18.9	4.082
6	30	103	14.5	5.8	4220	19.2	4.981

Find the values of parameters β_1, β_2, β_3, β_4, β_5 and β_6 by setting up the relevant system of simultaneous equations in matrix format and solving using the inverse matrix.

Employing the vector of parameters $\boldsymbol{\beta}$ that you have found, use vector multiplication to predict the value of Q if the explanatory variables take the values

P	M	G	C	N	T
41	148	23	8.2	4890	21.2

16.9 CRAMER'S RULE

Cramer's rule is another method of using matrices for solving sets of simultaneous equations but it finds the values of unknown variables one at a time. This means that it can be easier to use than the matrix inversion method if you only wish to find the value of one unknown variable, although this speed advantage might be not that important if you can use a spreadsheet for matrix inversion and multiplication. However, Cramer's rule is still useful in economics. Those of you who go on to study more advanced mathematical economics will use Cramer's rule to derive predictions from some multi-variable economic models specified in algebraic format.

We already know that a set of n simultaneous equations involving n unknown variables $x_1, x_2, \ldots, x_n$ and n constant values can be specified in matrix format as

$\mathbf{Ax} = \mathbf{b}$ where $\mathbf{A}$ is an $n \times n$ matrix of parameters,
$\mathbf{x}$ is an $n \times 1$ vector of unknown variables, and
$\mathbf{b}$ is an $n \times 1$ vector of constant values

Cramer's rule says that the value of any one of the unknown variables x_i can be found by substituting the vector of constant values $\mathbf{b}$ for the ith column of matrix $\mathbf{A}$ and then dividing the determinant of this new matrix by the determinant of the original $\mathbf{A}$ matrix.

Thus, if the term $\mathbf{A}_i$ is used to denote matrix $\mathbf{A}$ with column i replaced by the vector $\mathbf{b}$ then Cramer's rule gives

$$x_i = \frac{|\mathbf{A}_i|}{|\mathbf{A}|}$$

Example 16.19

Find x_1 and x_2 using Cramer's rule from the following set of simultaneous equations

$$\begin{aligned} 5x_1 + 0.4x_2 &= 12 \\ 3x_1 + \ \ 3x_2 &= 21 \end{aligned}$$

Solution

These simultaneous equations can be represented in matrix format as

$$\mathbf{Ax} = \begin{bmatrix} 5 & 0.4 \\ 3 & 3 \end{bmatrix} \begin{bmatrix} x_1 \\ x_2 \end{bmatrix} = \begin{bmatrix} 12 \\ 21 \end{bmatrix} = \mathbf{b}$$

Using Cramer's rule to find x_1 by substituting the vector $\mathbf{b}$ of constants for column 1 in matrix $\mathbf{A}$ gives

$$x_1 = \frac{|\mathbf{A}_1|}{|\mathbf{A}|} = \frac{\begin{vmatrix} 12 & 0.4 \\ 21 & 3 \end{vmatrix}}{\begin{vmatrix} 5 & 0.4 \\ 3 & 3 \end{vmatrix}}$$
$$= \frac{36 - 8.4}{15 - 1.2} = \frac{27.6}{13.8} = 2$$

In a similar fashion, by substituting vector $\mathbf{b}$ for column 2 in matrix $\mathbf{A}$ we get

$$x_2 = \frac{|\mathbf{A}_2|}{|\mathbf{A}|} = \frac{\begin{vmatrix} 5 & 12 \\ 3 & 21 \end{vmatrix}}{\begin{vmatrix} 5 & 0.4 \\ 3 & 3 \end{vmatrix}}$$
$$= \frac{105 - 36}{15 - 1.2} = \frac{69}{13.8} = 5$$

QUESTIONS 16.9

1. Use Cramer's rule to find the unknowns x and y when

$$24x+2y = 86$$
$$15x+y = 52$$

2. Given the set of simultaneous equations

$$\begin{aligned} 3x_1+4x_2+9x_3 &= 45 \\ 5x_2+2x_3 &= 32 \\ 4x_1+2x_2+4x_3 &= 32 \end{aligned}$$

use Cramer's rule to find the value of x_2 only.

3. In Questions 16.8, use Cramer's rule to find the values of the unknown variables in questions 1 and 2. (Then check that these are the same as those found by the matrix inverse method.)

16.10 SECOND-ORDER CONDITIONS AND THE HESSIAN MATRIX

Matrix algebra can help derive the second-order conditions for optimization exercises involving any number of variables. To explain how, first consider the second-order conditions for unconstrained optimization with only two variables, as encountered in Chapter 11.

If one tries to find a maximum or minimum for the two variable function f(x,y), then the FOC (first-order conditions) for both a maximum and a minimum require that

$$\frac{\partial f}{\partial x}=0 \quad \text{and} \quad \frac{\partial f}{\partial y}=0$$

SOC (second-order conditions) require that

$$\frac{\partial^2 f}{\partial x^2}<0 \quad \text{and} \quad \frac{\partial^2 f}{\partial y^2}<0 \text{ for a maximun}$$

$$\frac{\partial^2 f}{\partial x^2}>0 \quad \text{and} \quad \frac{\partial^2 f}{\partial y^2}>0 \text{ for a minimun}$$

and, for both a maximum and a minimum

$$\left(\frac{\partial^2 f}{\partial x^2}\right)\left(\frac{\partial^2 f}{\partial y^2}\right)>\left(\frac{\partial^2 f}{\partial x \partial y}\right)^2$$

These second-order conditions can be expressed more succinctly in matrix format. For clarity the abbreviated format for specifying second-order partial derivatives is also used, e.g. f_{xx} represents $\frac{\partial^2 f}{\partial x^2}$, f_{xy} represents $\frac{\partial^2 f}{\partial x \partial y}$, etc.

The Hessian matrix

The Hessian matrix contains all the second-order partial derivatives of a function, set out in the format shown in the following examples.

For the two variable function f(x,y) the Hessian matrix will be

$$\mathbf{H} = \begin{bmatrix} f_{xx} & f_{xy} \\ f_{yx} & f_{yy} \end{bmatrix}$$

The Hessian will always be a square matrix with equal numbers of rows and columns.

The **principal minors** of the Hessian matrix are the determinants of the different matrices found by starting with the first element in the first row and then expanding the matrix by adding the next row and column each time. Therefore, for any 2 × 2 Hessian there will be only two principal minors

$$|\mathbf{H}_1| = |f_{xx}| \quad \text{and} \quad |\mathbf{H}_2| = \begin{vmatrix} f_{xx} & f_{xy} \\ f_{yx} & f_{yy} \end{vmatrix}$$

Note that the second-order principal minor is the determinant of the Hessian matrix itself.

The **second-order conditions for a maximum and minimum** can now be specified in terms of the values of the determinants of these principal minors.

SOC for a *maximum* require $|\mathbf{H}_1| < 0$ and $|\mathbf{H}_2| > 0$ (Hessian is *negative definite*)
SOC for a *minimum* require $|\mathbf{H}_1| < 0$ and $|\mathbf{H}_2| > 0$ (Hessian is *positive definite*)

(The terms 'negative definite' and 'positive definite' are used to describe Hessians that meet the requirements specified.)

We can show that these requirements correspond to the second-order conditions for optimization of a two variable function that were set out earlier. For a **maximum** these SOC require

$$f_{xx} < 0, \ f_{yy} < 0 \tag{1}$$

$$\text{and } f_{xx}f_{yy} > (f_{xy})^2 \tag{2}$$

From the Hessian matrix and its principal minors we can deduce that

$$|\mathbf{H}_1| < 0 \text{ means that } f_{xx} < 0 \tag{3}$$

$$|\mathbf{H}_2| > 0 \text{ means that } f_{xx}f_{yy} - f_{xy}f_{yx} > 0 \tag{4}$$

Given that for any pair of cross-partial derivatives $f_{xy} = f_{yx}$ then (4) becomes

$$f_{xx}f_{yy} > \left(f_{xy}\right)^2$$

and so condition (2) is met.

In (2) the term $(f_{xy})^2 > 0$ since any number squared will be greater than zero. Therefore, as $f_{xx}f_{yy}$ is greater than this value, it must be true that

$$f_{xx}f_{yy} > 0 \qquad (5)$$

As we have already shown in (3) that $f_{xx} < 0$ then it must follow from (5) that

$$f_{yy} < 0 \qquad (6)$$

(a negative value must be multiplied by another negative value if the product is positive). Therefore, from (3) and (6), SOC (1) also holds.

Thus, we have shown that the matrix formulation of second-order conditions corresponds to the second-order conditions for optimization of a two variable function that we are already familiar with.

Returning to the price discrimination analysis considered in Chapter 11, we can now solve some applied problems using standard optimization techniques and check second-order conditions using the Hessian matrix.

Example 16.20

A firm has the production function $TC = 120 + 0.1q^2$ and sells its output in two separate markets with demand functions

$$q_1 = 800 - 2p_1 \text{ and } q_2 = 750 - 2.5p_2$$

Find the profit-maximizing output and sales in each market, using the Hessian to check second-order conditions for a maximum.

Solution

From the two demand schedules we can derive

$$p_1 = 400 - 0.5q_1, \quad TR_1 = 400q_1 - 0.5q_1^2, \quad MR_1 = 400 - q_1$$
$$p_2 = 300 - 0.4q_2, \quad TR_2 = 300q_2 - 0.4q_2^2, \quad MR_2 = 300 - 0.8q_2$$

Given that total output $q = q_1 + q_2$, then

$$\begin{aligned} TC &= 120 + 0.1q^2 = 120 + 0.1(q_1 + q_2)^2 \\ &= 120 + 0.1q_1^2 + 0.2q_1q_2 + 0.1q_2^2 \end{aligned}$$

Therefore,

$$\begin{aligned}\pi &= \mathrm{TR}_1 + \mathrm{TR}_2 - \mathrm{TC}\\ &= 400\mathrm{q}_1 - 0.5q_1^2 + 300q_2 - 0.4q_2^2 - 120 - 0.1q_1^2 - 0.2q_1q_2 - 0.1q_2^2\\ &= 400q_1 - 0.6q_1^2 + 300q_2 - 0.5q_2^2 - 120 - 0.2q_1q_2\end{aligned}$$

FOC for a maximum require

$$\frac{\partial\pi}{\partial q_1} = 400 - 1.2q_1 - 0.2q_2 = 0 \quad \text{threrefore} \quad 400 = 1.2q_1 + 0.2q_2 \tag{1}$$

$$\frac{\partial\pi}{\partial q_2} = 300 - q_2 - 0.2q_1 = 0 \quad \text{threrefore} \quad 300 = 0.2q_1 + q_2 \tag{2}$$

To find the optimum values that satisfy the FOC, the simultaneous equations (1) and (2) can be set up in matrix format as

$$\mathbf{Aq} = \begin{bmatrix} 1.2 & 0.2 \\ 0.2 & 1 \end{bmatrix}\begin{bmatrix} q_1 \\ q_2 \end{bmatrix} = \begin{bmatrix} 400 \\ 300 \end{bmatrix} = \mathbf{b}$$

Using Cramer's rule to solve for the sales in each market gives

$$q_1 = \frac{\begin{vmatrix} 400 & 0.2 \\ 300 & 1 \end{vmatrix}}{\begin{vmatrix} 1.2 & 0.2 \\ 0.2 & 1 \end{vmatrix}} = \frac{400 - 60}{1.2 - 0.04} = \frac{340}{1.16} = 293.1$$

$$q_2 = \frac{\begin{vmatrix} 1.2 & 400 \\ 0.2 & 300 \end{vmatrix}}{\begin{vmatrix} 1.2 & 0.2 \\ 0.2 & 1 \end{vmatrix}} = \frac{360 - 80}{1.2 - 0.04} = \frac{280}{1.16} = 241.4$$

To check the second-order conditions we return to the first-order partial derivatives and then find the second-order partial derivatives and the cross-partial derivatives. Thus, from

$$\frac{\partial\pi}{\partial q_1} = 400 - 1.2q_1 - 0.2q_2 \quad \text{and} \quad \frac{\partial\pi}{\partial q_2} = 300 - q_2 - 0.2q_1$$

we get

$$\frac{\partial^2\pi}{\partial q_1^2} = -1.2 \quad \frac{\partial^2\pi}{\partial q_1\,\partial q_2} = -0.2 \quad \text{and} \quad \frac{\partial^2\pi}{\partial q_2^2} = -1 \quad \frac{\partial^2\pi}{\partial q_2\,\partial q_1} = -0.2$$

The Hessian matrix is therefore

$$\mathbf{H} = \begin{bmatrix} \pi_{11} & \pi_{12} \\ \pi_{21} & \pi_{22} \end{bmatrix} = \begin{bmatrix} -1.2 & -0.2 \\ -0.2 & -1 \end{bmatrix}$$

and the determinants of the principal minors are

$$|\mathbf{H}_1| = -1.2 < 0$$

and

$$|\mathbf{H}_2| = \begin{vmatrix} -1.2 & -0.2 \\ -0.2 & -1 \end{vmatrix} = 1.2 - 0.04 = 1.16 > 0$$

As $|\mathbf{H}_1| < 0$ and $|\mathbf{H}_2| > 0$, the Hessian is negative definite. Therefore, SOC for a maximum are met.

Third-order Hessians

For a three variable function $y = \mathrm{f}(x_1, x_2, x_3)$ the Hessian will be the 3×3 matrix of second-order partial derivatives

$$\mathbf{H} = \begin{bmatrix} f_{11} & f_{12} & f_{13} \\ f_{21} & f_{22} & f_{23} \\ f_{31} & f_{32} & f_{33} \end{bmatrix}$$

and the determinants of the three principal minors will be

$$|\mathbf{H}_1| = |\mathrm{f}_{11}|, \quad |\mathbf{H}_2| = \begin{vmatrix} f_{11} & f_{12} \\ f_{21} & f_{22} \end{vmatrix}, \quad |\mathbf{H}_3| = \begin{vmatrix} f_{11} & f_{12} & f_{13} \\ f_{21} & f_{22} & f_{23} \\ f_{31} & f_{32} & f_{33} \end{vmatrix}$$

The SOC conditions for unconstrained optimization of a three variable function are:

(a) For a maximum $|\mathbf{H}_1| < 0, |\mathbf{H}_2| > 0$ and $|\mathbf{H}_3| < 0$ (Hessian is negative definite)

(b) For a minimum $|\mathbf{H}_1|, |\mathbf{H}_2|$ and $|\mathbf{H}_3|$ are all > 0 (Hessian is positive definite)

Example 16.21

A multiplant monopoly produces the quantities q_1, q_2 and q_3 in the three plants that it operates and faces the profit function

$$\pi = -24 + 839q_1 + 837q_2 + 835q_3 - 5.05q_1^2 - 5.03q_2^2$$
$$-5.02q_3^2 - 10q_1q_2 - 10q_1q_3 - 10q_2q_3$$

Find the output levels in each of its three plants q_1, q_2 and q_3 that will maximize profit and use the Hessian to check that second-order conditions are met.

Solution

Differentiating this function π with respect to q_1, q_2 and q_3, and setting equal to zero to find the optimum values where the first-order conditions are met, we get:

$$\pi_1 = 839 - 10.1q_1 - 10q_2 - 10q_3 = 0 \quad (1)$$
$$\pi_2 = 837 - 10q_1 - 10{,}06q_2 - 10q_3 = 0 \quad (2)$$
$$\pi_2 = 835 - 10q_1 - 10q_2 - 10.04q_3 = 0 \quad (3)$$

These conditions can be rearranged to get

$$\begin{aligned} 839 &= 10.1q_1 + 10q_2 + 10q_3 \\ 837 &= 10q_1 + 10.06q_2 + 10q_3 \\ 835 &= 10q_1 + 10q_2 + 10.04q_3 \end{aligned}$$

These simultaneous equations can be specified in matrix format and solved by the matrix inversion method to get the optimum values of q_1, q_2 and q_3 as 42, 36.6 and 4.9, respectively.

Differentiating (1), (2) and (3) again we can derive the Hessian matrix of second-order partial derivatives

$$\mathbf{H} = \begin{bmatrix} \pi_{11} & \pi_{12} & \pi_{13} \\ \pi_{21} & \pi_{22} & \pi_{23} \\ \pi_{31} & \pi_{32} & \pi_{33} \end{bmatrix} = \begin{bmatrix} -10.1 & -10 & -10 \\ -10 & -10.06 & -10 \\ -10 & -10 & -10.04 \end{bmatrix}$$

The determinants of the three principal minors will therefore be

$$|\mathbf{H}_1| = |\pi_{11}| = -10.1$$

$$|\mathbf{H}_2| = \begin{vmatrix} \pi_{11} & \pi_{12} \\ \pi_{21} & \pi_{22} \end{vmatrix} = \begin{vmatrix} -10.1 & -10 \\ -10 & -10.06 \end{vmatrix} = 101.606 - 100 = 1.606$$

$$|\mathbf{H}_3| = \begin{vmatrix} \pi_{11} & \pi_{12} & \pi_{13} \\ \pi_{21} & \pi_{22} & \pi_{23} \\ \pi_{31} & \pi_{32} & \pi_{33} \end{vmatrix} = \begin{vmatrix} -10.1 & -10 & -10 \\ -10 & -10.06 & -10 \\ -10 & -10 & -10.04 \end{vmatrix} = -0.1242$$

(You can check the $\mathbf{H}_3$ determinant calculations using the Excel MDETERM function.) This Hessian is therefore negative definite as

$$|\mathbf{H}_1| = -10.1 < 0, \ |\mathbf{H}_2| = 1.606 > 0, \ |\mathbf{H}_3| = -0.1242 < 0$$

and so the second-order conditions for a maximum are met.

Higher order Hessians

Although you will not be asked to use the Hessian to tackle any problems in this text that involve more than three variables, for your future reference the general SOC conditions that apply to a Hessian of any order are:

(a) Maximum
Principal minors alternate in sign, starting with $|\mathbf{H}_1| < 0$ (negative definite). Thus, a principal minor $|\mathbf{H}_i|$ of order i should have the sign $(-1)^i$

(b) Minimum
All principal minors $|\mathbf{H}_i| > 0$ (positive definite).

QUESTIONS 16.10

1. A firm that sells in two separate markets has the profit function

$$\pi = -120 + 245q_1 - 0.3q_1^2 + 120q_2 - 0.4q_2^2 - 0.18q_1q_2$$

where q_1 and q_2 are sales in the two markets. Find the profit-maximizing sales in each market, using the Hessian to check second-order conditions for a maximum.

2. Find the values of q_1 and q_2 that will maximize the profit function

$$\pi = -12 + 152q_1 - 0.25q_1^2 + 196q_2 - 0.2q_2^2 - 0.1q_1q_2$$

and check that second-order conditions are met using the Hessian matrix.

3. A firm producing three products faces the following profit function. Find the amounts of the three products q_1, q_2 and q_3 that will maximize profit and use the Hessian to check that second-order conditions are met.

$$\begin{aligned}\pi &= -73 + 242q_1 + 238q_2 + 238q_3 - 8.4q_1^2 - 8.25q_2^2 \\ &\quad - 8.1q_3^2 - 16q_1q_2 - 16q_1q_3 - 16q_2q_3\end{aligned}$$

4. A monopoly operates three plants with total cost schedules

$$\begin{aligned}TC_1 &= 40 + 0.1q_1 + 0.4q_1^2, \quad TC_2 = 18 + 3q_2 + 0.02q_2^2, \\ TC_3 &= 30 + 4q_3 + 0.01q_3^2\end{aligned}$$

and faces the market demand schedule

$$p = 250 - 2q \quad \text{where} \quad q = q_1 + q_2 + q_3$$

Set up the profit function and then use it to determine how much the firm should make in each plant to maximize profit, using the Hessian to check that second-order conditions are met.

16.11 CONSTRAINED OPTIMIZATION AND THE BORDERED HESSIAN

In Chapter 12 the solution of constrained optimization problems using the Lagrange multiplier method was explained, but the explanation of how to check if second-order conditions for constrained optimization are met was put on hold. Now that the concept of the Hessian has been covered, we are ready to investigate how the related concept of the **bordered Hessian** can help determine if the second-order conditions are met when the Lagrange method is used.

If second-order partial derivatives are taken for a Lagrange constrained optimization objective function and put into a matrix format this will give what is known as the bordered Hessian. For example, to maximize a utility function $U(X_1, X_2)$ subject to the budget constraint

$$M - P_1X_1 - P_2X_2 = 0$$

the Lagrange equation will be

$$G = U(X_1, X_2) + \lambda(M - P_1X_1 - P_2X_2)$$

Taking first-order derivatives and setting equal to zero we get the first-order conditions:

$$G_1 = U_1 - \lambda P_1 = 0 \quad (1)$$
$$G_2 = U_2 - \lambda P_2 = 0 \quad (2)$$
$$G_\lambda = M - P_1X_1 - P_2X_2 = 0 \quad (3)$$

These are used to solve for the optimum values of X_1 and X_2 when actual values are specified for the parameters.

Differentiating (1), (2) and (3) again with respect to X_1, X_2 and λ gives the bordered Hessian matrix of second-order partial derivatives

$$\mathbf{H}_B = \begin{bmatrix} U_{11} & U_{12} & -P_1 \\ U_{21} & U_{22} & -P_2 \\ -P_1 & -P_2 & 0 \end{bmatrix}$$

You can see that the bordered Hessian $\mathbf{H_B}$ has one more row and one more column than the ordinary Hessian. In this, and most other constrained maximization examples that you will encounter, the extra row and column each contain the negative of the prices of the variables in the constraint.

Although it is possible to use the Lagrange method to tackle constrained optimization problems with several constraints, we will only consider problems with one constraint here.

The second-order conditions for optimization of a Lagrangian with one constraint require that for:

Maximization

- If there are two variables in the objective function (i.e. $\mathbf{H_B}$ is 3×3) then the determinant $|\mathbf{H_B}| > 0$.
- If there are three variables in the objective function (i.e. $\mathbf{H_B}$ is 4×4) then the determinant $|\mathbf{H_B}| < 0$ **and** the determinant of the naturally ordered principal minor of $|\mathbf{H_B}| > 0$. (The naturally ordered principal minor is the matrix remaining when the first row and column have been eliminated from $\mathbf{H_B}$.)

Minimization

- If there are two variables in the objective function the determinant $|\mathbf{H_B}| < 0$.
- If there are three variables in the objective function then the determinant $|\mathbf{H_B}| < 0$ and the determinant of the naturally ordered principal minor of $|\mathbf{H_B}| < 0$.

Example 16.22

An individual has the utility function $U = 4X^{0.5}Y^{0.5}$ and can buy good X at £2 a unit and good Y at £8 a unit. If their budget is £100, find the combination of X and Y that they should purchase to maximize utility and check that second-order conditions are met using the bordered Hessian matrix.

Solution

The Lagrange function is

$$G = 4X^{0.5}Y^{0.5} + \lambda(100 - 2X - 8Y)$$

Differentiating and setting equal to zero to get the FOC for a maximum

$$G_X = 2X^{-0.5}Y^{0.5} - 2\lambda = 0 \quad (1)$$
$$G_Y = 2X^{0.5}Y^{-0.5} - 8\lambda = 0 \quad (2)$$
$$G_\lambda = 100 - 2X - 8Y = 0 \quad (3)$$

From (1)

$$X^{-0.5}Y^{0.5} = \lambda$$

From (2)

$$0.25X^{0.5}Y^{-0.5} = \lambda$$

Therefore,

$$X^{-0.5}Y^{0.5} = 0.25X^{0.5}Y^{-0.5}$$

Multiplying both sides by $4X^{0.5}Y^{0.5}$

$$4Y = X \quad (4)$$

Substituting (4) into (3)

$$100 - 2(4Y) - 8Y = 0$$
$$Y = 6.25$$

and thus from (4)

$$X = 25$$

Differentiating (1), (2) and (3) again gives the bordered Hessian of second-order partial derivatives

$$\mathbf{H}_{\mathrm{B}} = \begin{bmatrix} U_{XX} & U_{XY} & -P_X \\ U_{YX} & U_{YY} & -P_Y \\ -P_X & -P_Y & 0 \end{bmatrix} = \begin{bmatrix} -X^{-1.5}Y^{0.5} & X^{-0.5}Y^{-0.5} & -2 \\ X^{-0.5}Y^{-0.5} & -X^{0.5}Y^{-1.5} & -8 \\ -2 & -8 & 0 \end{bmatrix}$$
$$= \begin{bmatrix} -0.02 & 0.08 & -2 \\ 0.08 & -0.32 & -8 \\ -2 & -8 & 0 \end{bmatrix}$$

The determinant of this bordered Hessian, expanding along the third row, is

$$\begin{aligned} |\mathbf{H}_{\mathrm{B}}| &= -2(-0.64-0.64)+8(0.16+0.16) \\ &= 2.56+2.56=5.12>0 \end{aligned}$$

and so the second-order conditions for a maximum are satisfied.

To illustrate the use of the bordered Hessian to check the second-order conditions required for constrained optimization involving three variables, we shall just consider an example without any specific format for the objective function.

Example 16.23

If a firm is attempting to maximize output $Q = Q(x, y, z)$ subject to a budget of £5,000 where the prices of the inputs x, y and z are £8, £12 and £6, respectively, what requirements are there for the relevant bordered Hessians to ensure that second-order conditions for optimization are met?

Solution

The Lagrange objective function will be

$$G=Q(x,y,z)+\lambda(5000-8x-12y-6z)$$

As there are three variables in the objective function and $\mathbf{H}_\mathbf{B}$ is 4×4 then the second-order conditions for a maximum require that the determinant of the bordered Hessian of second-order partial derivatives $|\mathbf{H}_\mathbf{B}| < 0$. Therefore,

$$|\mathbf{H}_{\mathrm{B}}| = \begin{vmatrix} Q_{xx} & Q_{xy} & Q_{xz} & -8 \\ Q_{yx} & Q_{yy} & Q_{yz} & -12 \\ Q_{zx} & Q_{zy} & Q_{zz} & -6 \\ -8 & -12 & -6 & 0 \end{vmatrix} < 0$$

As $\mathbf{H}_B$ is 4 × 4, the second-order conditions for a maximum also require that the determinant of the naturally ordered principal minor of $\mathbf{H}_B > 0$. Thus, when the first row and column have been eliminated from $\mathbf{H}_B$, this problem also requires that

$$\begin{vmatrix} Q_{yy} & Q_{yz} & -12 \\ Q_{zy} & Q_{zz} & -6 \\ -12 & -6 & 0 \end{vmatrix} > 0$$

Constrained optimization with any number of variables and constraints

All the constrained optimization problems that you will encounter in this text have only one constraint and usually do not have more than three variables in the objective function. However, it is possible to set up more complex Lagrange functions with many variables and more than one constraint.

Second-order conditions requirements for optimization for the general case with m variables in the objective function and r constraints are that the naturally ordered **border preserving** principal minors of dimension m of $\mathbf{H}_B$ must have the following sign

$$\begin{array}{ll} (-1)^{m-r} & \text{for a maximum} \\ (-1)^{r} & \text{for a minimum} \end{array}$$

'Border preserving' means not eliminating the borders added to the basic Hessian, i.e. the last column and the bottom row, which do not contain second-order partial derivatives and typically show the prices of the variables. These requirements only apply to the principal minors of order ≥ $(1 + 2r)$

For example, assume that the problem is to maximize a utility function $U = U(X_1, X_2, X_3)$ subject to the budget constraint $M = P_1X_1 + P_2X_2 + P_3X_3$. Since there is only one constraint ($r = 1$), we just need to consider the principal minors of order greater than 3 since

$$(1+2r) = (1+2) = 3$$

As the full bordered Hessian in this example with three variables is fourth order then only $\mathbf{H}_B$ itself plus the first principal minor need be considered, as this is the only principal minor with order equal to or greater than 3.

The second-order conditions will therefore require that for a maximum:

For the bordered Hessian $m = 4$ and so $|\mathbf{H}_B|$ must have the sign

$$(-1)^{m-r} = (-1)^{4-1} = (-1)^3 = -1 < 0$$

and the determinant of the third-order naturally ordered principal minor of $|\mathbf{H}_B|$ must have the sign

$$(-1)^{m-r} = (-1)^{3-1} = (-1)^2 = +1 > 0$$

These are the same as the basic rules for the three variable case stated earlier.

The last set of problems, see Questions 16.11, just require you to use the bordered Hessian to check that second-order conditions for optimization are met in some numerical examples with a small number of variables to familiarize you with the method. Those students who go on to study further mathematical economics will find that this method will be extremely useful in more complex constrained optimization problems.

QUESTIONS 16.11

1. A firm has the production function $Q = K^{0.5}L^{0.5}$ and buys input K at £12 a unit and input L at £3 a unit and has a budget of £600. Use the Lagrange method to find the input combination that will maximize output, checking that second-order conditions are satisfied by using the bordered Hessian.
2. A firm operates with the production function $Q = 25K^{0.5}L^{0.4}$ and buys input K at £20 a unit and input L at £8 a unit. Use the Lagrange method to find the input combination that will minimize the cost of producing 400 units of Q, using the bordered Hessian to check that second-order conditions are satisfied.
3. A consumer has the utility function $U = 20X^{0.5}Y^{0.4}$ and buys good X at £10 a unit and good Y at £2 a unit. If their budget constraint is £450, what combination of X and Y will maximize utility? Check that second-order conditions are satisfied by using the bordered Hessian.
4. A consumer has the utility function $U = 4X^{0.75}Y^{0.25}$ and can buy good X at £12 a unit and good Y at £2 a unit. Find the combination of X and Y that they should purchase to minimize the cost of achieving a utility level of 580 and check that second-order conditions are met using the bordered Hessian matrix.

16.12 INPUT-OUTPUT ANALYSIS

A country's economy consists of many different sectors that are interrelated and depend on each other. For example, the metal manufacturing industry may require oil as an input, and the oil extraction industry may require metal products to develop oil wells. Input-output analysis is a mathematical method used to model these interrelationships between industrial sectors and to calculate what impact any changes in final demand for different industries' products will have on the total required output from each sector and on the demand for primary resources. Although in the past this method has been seen as more applicable to state planned economies it is still a useful tool for the governments of market economies to help examine the possible impacts of economic policies on different industrial sectors. For example, an input-output model may highlight any potential bottlenecks to economic growth. The main developments can be attributed to the Economics Nobel Prize winner Wassily Leontief, whose name is normally associated with input-output analysis.

As with other economic models, some simplifying assumptions are made, and the basic input-output analysis model assumes that:

(i) Final demand for each industry is given, i.e. it is exogenously determined.
(ii) Each industry produces just one product.
(iii) Inputs are used in fixed proportions in all production processes.
(iv) Constant returns to scale apply in all industries at all output levels.
(v) There is no foreign trade.
(vi) There is at least one primary resource, which may have limited availability.
(vii) An industry may use its own output as an input.
(viii) There is no separate profit or surplus and total output equals total input.

These assumptions, of course, are not all realistic and they also contradict some of the assumptions made in other methods of economic analysis. However, given that Leontief's input-output analysis essentially attempts to provide a general equilibrium model of a whole economic system, some simplification is obviously required. Some of these assumptions can be relaxed in more advanced adaptations of input-output analysis, but here we will just work through the basic model.

Economies have many different industrial sectors and so matrix algebra is invaluable in helping to calculate the different output levels and resource requirements in a multisector input-output model. However, to explain the basic principles involved we will first use linear algebra to work through a simple example with just two industrial sectors and one primary resource.

Input-output tables and coefficients

The first step in constructing any real input-output model would be to collect data on the flows between industry sectors. This would require the measurement of all outputs and inputs in the same units, such as £m. Also, all values are flows per time period.

Let us assume values have been observed as in the input-output table in Table 16.6. This shows, for example, that:

Table 16.6 Input-output flow table

Input-output table flow		**Destination of output**			**£m**
		Y	**Z**	**D**	**Total**
Inputs used	Y	420	150	270	840
	Z	84	200	216	500
	R	336	150	0	486
	Total	840	500	486	

- Reading across the first row, we can see that total output of industry Y is 840, with 420 units being used up within industry Y itself, 150 being used by industry Z and 270 being consumed as final demand (D).
- Reading down the first column, we can see that industry Y uses 420 units of its own output, 84 of industry Z output and 336 of primary resource R, which total to 840 units of input usage.

From this input-output flow table we can construct the input-output coefficient table shown in Table 16.7. As one of the assumptions of input-output analysis is that there are always constant returns to scale, these coefficients can be found by dividing each input flow by the total output. For example:

- If an output of 840 of Y requires an input of 84 units from industry Z, then one unit of Y will require 84/840 = 0.1 units of Z.
- If an output of 500 of Z requires an input of 150 units of primary resource R then one unit of Z will require 150/500 = 0.3 units of R.

Table 16.7 Input-output coefficient matrix

Input-output coefficient matrix		**Destination of output**	
		Y	**Z**
Inputs used	Y	0.5	0.3
	Z	0.1	0.4
	R	0.4	0.3

Impact of changes in final demand on resource requirements

Now that we have the input-output coefficient table, we can work out what happens to the input flows if final demand changes. If the economy has only a fixed amount of the primary resource R that it can use each time period, then we can calculate whether this would be sufficient to meet any increased final demand.

Example 16.24

Given the input-output coefficient table as in Table 16.7, if this economy can use a maximum of 600 units of primary resource each time period is it feasible for final demand to grow so that final consumption is 324 of Y and 261 of Z?

Solution

Given the specified final demand levels for industries Y and Z and the coefficients which show input requirements per unit then:

Industry Y total output is $Y = 0.5Y + 0.3Z + 324$ (1)

Industry Z total output is $Z = 0.1Y + 0.4Z + 261$ (2)

This set of linear simultaneous equations with two unknowns can be easily solved.

$$\begin{aligned} &\text{From (1)} \quad 0.5Y = 0.3Z + 324 \\ &\text{so} \qquad\qquad Y = 0.6Z + 648 \end{aligned} \tag{3}$$

Substituting equation (3) into (2) for Y gives

$$\begin{aligned} Z &= 0.1(0.6Z + 648) + 0.4Z + 261 \\ Z &= 0.06Z + 64.8 + 0.4Z + 261 \\ Z &= 0.46Z + 325.8 \\ 0.54Z &= 325.8 \\ Z &= 603.33 \end{aligned}$$

Substituting this value of Z into (3) gives the total amount of Y produced as

$$Y = 0.6(603.33) + 648 = 1{,}010$$

Therefore, the total amount of primary resource R required will be

$$R = 0.4Y + 0.3Z = 0.4(1010) + 0.3(603.33) = 404 + 181 = 585$$

This is below the maximum limit of 600 units of R available and so this new final demand level is feasible.

Input-output solution using matrix algebra

Before considering the more general case and more complex examples, let us first rework Example 16.24 using matrix algebra. We have already seen how to derive the simultaneous equations

$$\begin{aligned} Y &= 0.5Y + 0.3Z + 324 \\ Z &= 0.1Y + 0.4Z + 261 \end{aligned}$$

By bringing all terms apart from the given final demand levels to the left-hand side, these can be rewritten as

$$\begin{aligned} (1-0.5)Y - 0.3Z &= 324 \\ -0.1Y + (1-0.4)Z &= 261 \end{aligned}$$

Ignoring the coefficients for the primary input requirements, the set of input-output coefficients can be specified as the matrix

$$\mathbf{A} = \begin{bmatrix} 0.5 & 0.3 \\ 0.1 & 0.4 \end{bmatrix}$$

and final demand can be specified as the vector

$$\mathbf{d} = \begin{bmatrix} 324 \\ 261 \end{bmatrix}$$

Therefore, the previous simultaneous equations can be rewritten in matrix format as

$$\left(\begin{bmatrix} 1 & 0 \\ 0 & 1 \end{bmatrix} - \begin{bmatrix} 0.5 & 0.3 \\ 0.1 & 0.4 \end{bmatrix} \right) \begin{bmatrix} Y \\ Z \end{bmatrix} = \begin{bmatrix} 324 \\ 261 \end{bmatrix}$$

or as

$$(\mathbf{I} - \mathbf{A})\mathbf{x} = \mathbf{d}$$

where **I** is the identity matrix and **x** is the vector of unknown values for outputs Y and Z.

If we now multiply both sides by the inverse matrix $(\mathbf{I} - \mathbf{A})^{-1}$ this gives the solution as the vector of outputs

$$\mathbf{x} = (\mathbf{I} - \mathbf{A})^{-1}\mathbf{d}$$

The matrix inversion and other calculations needed to arrive at the values of **x** can be done on a spreadsheet, in a similar fashion to the method used for solving multiple simultaneous equations with matrix algebra explained in Section 16.8. Spreadsheets will be used for more complex examples in the rest of this section and would also make this two-industry example easier to solve. However, just to show how fundamental methods of matrix algebra can be applied to this type of problem, we will continue to solve this particular problem manually.

First, we calculate the values in matrix $(\mathbf{I} - \mathbf{A})$ giving

$$(\mathbf{I} - \mathbf{A}) = \left(\begin{bmatrix} 1 & 0 \\ 0 & 1 \end{bmatrix} - \begin{bmatrix} 0.5 & 0.3 \\ 0.1 & 0.4 \end{bmatrix} \right) = \begin{bmatrix} 0.5 & -0.3 \\ -0.1 & 0.6 \end{bmatrix}$$

then we derive the adjoint matrix, which for this 2 × 2 matrix is

$$\boldsymbol{Adj}(\boldsymbol{I} - \boldsymbol{A}) = \begin{bmatrix} 0.6 & -(-0.3) \\ -(-0.1) & 0.5 \end{bmatrix} = \begin{bmatrix} 0.6 & 0.3 \\ 0.1 & 0.5 \end{bmatrix}$$

Thus, using the inverse matrix formula derived in Section 16.6, we can find the inverse as

$$(\mathbf{I} - \mathbf{A})^{-1} = \frac{\mathbf{Adj}(\mathbf{I} - \mathbf{A})}{|\mathbf{I} - \mathbf{A}|} = \frac{\begin{vmatrix} 0.6 & 0.3 \\ 0.1 & 0.5 \end{vmatrix}}{0.6 \times 0.5 - (-0.1)(-0.3)}$$

$$= \frac{\begin{vmatrix} 0.6 & 0.3 \\ 0.1 & 0.5 \end{vmatrix}}{0.3 - 0.03} = \frac{\begin{vmatrix} 0.6 & 0.3 \\ 0.1 & 0.5 \end{vmatrix}}{0.27}$$

Finally, the solutions for the values of the output vector **x** can be calculated as

$$\mathbf{x} = (\mathbf{I} - \mathbf{A})^{-1}\mathbf{d} = \frac{1}{0.27}\begin{bmatrix} 0.6 & 0.3 \\ 0.1 & 0.5 \end{bmatrix}\begin{bmatrix} 324 \\ 261 \end{bmatrix}$$

$$= \frac{1}{0.27}\begin{bmatrix} 0.6 \times 324 + 0.3 \times 261 \\ 0.1 \times 324 + 0.5 \times 261 \end{bmatrix}$$

$$= \frac{1}{0.27}\begin{bmatrix} 272.7 \\ 162.9 \end{bmatrix} = \begin{bmatrix} 1010 \\ 603.3 \end{bmatrix}$$

These values for the total outputs of industries Y and Z are, of course, the same as those arrived at by linear algebra in Example 16.24.

QUESTIONS 16.12

1. An economist collects input and output data on a simple economy with two industries Y and Z, plus a natural resource R, and uses this data to construct Table 16.8, showing how much input is used by each industry and where output goes, including final demand D.

Table 16.8 Input-output flow table for Question 1

Input-output flow table		Destination of output		
		Y	Z	D
Inputs used	Y	1095	495	600
	Z	438	660	552
	R	657	495	0

   (a) Demonstrate that total output equals total input for this economy, assuming all flows are measured in £m.
   (b) Construct an input-output coefficient matrix for this economy.
   (c) What will be the impact on the total output of both industries if the final demand for industry Y exogenously increases to 690?

2. Assume that a simple economy with two industries Y and Z and one primary resource R has the Leontief input-output coefficient matrix shown in Table 16.9.

Table 16.9 Input-output coefficient matrix for Question 2

Input-output coefficient matrix		Destination of output	
		Y	Z
Inputs used	Y	0.6	0.2
	Z	0.1	0.5
	R	0.3	0.3

   (a) What total production levels of Y and Z would be necessary to sustain final demand levels of 150 of Y and 120 of Z, and how much of the primary resource R would be needed?
   (b) What changes in production levels and primary resource usage would be necessary if final demand for Z is increased to 138?

3. For an economy with two industries A and B, and one primary resource R, which has the Leontief input-output coefficient matrix shown in Table 16.10:

Table 16.10 Input-output coefficient matrix for Question 3

Input-output coefficient matrix		Destination of output	
		A	B
Inputs used	A	0.60	0.80
	B	0.25	0.10
	R	0.15	0.10

   (a) Find how much of the primary resource R would be needed to produce final demand levels of 500 of A and 800 of B, and what the total production levels of A and B would then be.
   (b) If final demand for A is increased to 620 what would the new total production levels of A and B be and how much R would be needed?

16.13 MULTIPLE INDUSTRY INPUT-OUTPUT MODELS

Spreadsheet packages like Excel can be used to find matrix inverses and to multiply matrices and vectors and so can be used for any practical applications of input-output analysis. Before considering the general case for an *n* industry economy, let us first work through a numerical three-industry economy example.

Example 16.25

Assume that a simple economic system has only the three industrial sectors Q, S and T, plus a primary labour resource L. The Leontief input-output coefficient matrix is as in Table 16.11 and the usual assumptions of input-output analysis apply.

(i) What total output will be required from each industrial sector to satisfy a final demand of 240 for industry Q's output, 300 for industry S's output and 125 for industry T's output?
(ii) How much of the primary labour resource L will be needed in total to produce these output levels?

Table 16.11 Leontief input-output coefficient matrix

IP-OP coefficients		Destination of output		
		Q	S	T
Inputs used	Q	0.4	0.3	0.2
	S	0.1	0.2	0.4
	T	0.2	0.1	0.2
	L	0.3	0.4	0.2
	Total	1	1	1

Solution

Total output levels will be determined by the three simultaneous equations

$$Q = 0.4Q + 0.3S + 0.2T + 240$$
$$S = 0.1Q + 0.2S + 0.4T + 300$$
$$T = 0.2Q + 0.1S + 0.2T + 125$$

Bringing like-terms together gives

$$(1-0.4)Q - 0.3S - 0.2T = 240$$
$$-0.1Q + (1-0.2)S - 0.4T = 300$$
$$-0.2Q - 0.1S + (1-0.2)T = 125$$

These simultaneous equations can be written in matrix format as $(\mathbf{I} - \mathbf{A})\mathbf{x} = \mathbf{d}$ where

A represents the input-output coefficient matrix $\mathbf{A} = \begin{bmatrix} 0.4 & 0.3 & 0.2 \\ 0.1 & 0.2 & 0.4 \\ 0.2 & 0.1 & 0.2 \end{bmatrix}$

x represents the unknown total output vector $\mathbf{x} = \begin{bmatrix} Q \\ S \\ T \end{bmatrix}$

d represents the final demand vector $\mathbf{d} = \begin{bmatrix} 240 \\ 300 \\ 125 \end{bmatrix}$

and **I** is the 3 × 3 identity matrix.

If we now multiply both sides by the inverse matrix $(\mathbf{I} - \mathbf{A})^{-1}$ this gives the solution as the vector of outputs $\mathbf{x} = (\mathbf{I} - \mathbf{A})^{-1}\mathbf{d}$

The way that a spreadsheet can be used to calculate the total output values of **x** is illustrated in Table 16.12. First, the known values are entered, so:

- The input-output coefficients are entered in cells C4 to E7, including coefficients for L, and totals are calculated in row 8, to check that each column adds to 1.
- The given final demand values are entered in cells G4 to G6.
- A 3 × 3 identity matrix is entered in cells I4 to K6.

Next, the **I** – **A** matrix is calculated by entering the formula I4 – C4 in cell B11 and then copying through the range B11 to cell D13. To invert this **I** – **A** matrix to get the inverse matrix $(\mathbf{I} - \mathbf{A})^{-1}$, highlight the 3 × 3 range of cells F11 to H13, enter the formula =MINVERSE and then the cell range B11 to D13 within the brackets of the formula, and then remember to hold down both the Ctrl and Shift keys when you hit Enter.

To then multiply the $(\mathbf{I} - \mathbf{A})^{-1}$ matrix by the final demand vector, highlight the three cells J11 to J13 and enter the formula =MMULT. Within this formula's brackets, first put the $(\mathbf{I} - \mathbf{A})^{-1}$ cell range F11 to H13 and then the vector **d** cell range G4 to G6 and then hold down both the Ctrl and Shift keys when you hit Enter. The values calculated in the three cells J11 to J13 will be the total output levels for Q, R and T, which answers part (i) of this question.

To calculate the total amount of L required in cell J15, use the MMULT formula to multiply the row vector of L coefficients in cells C7 to E7 by the final demand column vector in cells J11 to J13, giving the answer 665.

To check that all outputs add up to the value of all inputs we can also construct the input-output flow table in cells C17 to G21. If the total output levels calculated earlier are put into cells C21 to E21 (e.g. by copying and transposing and pasting the values from cells J11 to J13), then the actual inputs from each industrial sector and L can be calculated by multiplying by the relevant coefficients. The simplest way

Table 16.12 Using a spreadsheet for Example 16.25

	A	B	C	D	E	F	G	H	I	J	K
1		IP – OP Coefficient Table									
2			Destination of Output								
3			Q	S	T		D		Identity Matrix		
4	Inputs used	Q	0.4	0.3	0.2		240		1	0	0
5		S	0.1	0.2	0.4		300		0	1	0
6		T	0.2	0.1	0.2		125		0	0	1
7		L	0.3	0.4	0.2						
8		Total	1	1	1						
9										×	
10		I – A MATRIX				(I –A) MINVERSE				MI × d	Output
11		0.6	−0.3	−0.2		2.16	0.935	1.007		924.5	Q
12		−0.1	0.8	−0.4		0.58	1.583	0.935		729.9	S
13		−0.2	−0.1	0.8		0.61	0.432	1.619		478.6	T
14											
15			Destination of Output							665	L
16			Q	S	T	d	Total				
17	Inputs used	Q	369.8	219.0	95.7	240	924.5				
18		S	92.4	146.0	191.4	300	729.9				
19		T	184.9	73.0	95.7	125	478.6				
20		L	277.3	291.9	95.7	0	665				
21		Total	924.5	729.9	478.6	665					

to do this is to enter the formula =C4*C$21 in cell C17 and then copy through the range from C17 to E20, with the $ sign anchoring the calculations to the values in row 21.

If the given final demand is entered again in cells F17 to F21 and the total input usage is calculated in cells G17 to G20 by summing across each row, then the totals calculated should equal the total output levels already found earlier.

The previous examples only contained a small number of industries, but any realistic input-output model will contain a much larger number. However, the matrix method of solution can be employed to cope with any number of industries. Before considering some more complex examples we shall first set out how input-output analysis can be applied to the general case of an economy with n industries.

Ignoring any primary input requirements and assuming n industries, we can construct the coefficient matrix

$$\mathbf{A} = \begin{bmatrix} a_{11} & a_{12} & \cdot & \cdot & \cdot & a_{1n} \\ a_{21} & a_{22} & \cdot & \cdot & \cdot & a_{2n} \\ \cdot & \cdot & \cdot & \cdot & \cdot & \cdot \\ \cdot & \cdot & \cdot & \cdot & \cdot & \cdot \\ \cdot & \cdot & \cdot & \cdot & \cdot & \cdot \\ a_{n1} & a_{n2} & \cdot & \cdot & \cdot & a_{nn} \end{bmatrix}$$

where a_{ij} = is the amount of industry i output required to produce one unit of industry j output.

If x_i = total production of industry i,
d_i = given final demand for industry i,

then using the coefficients in matrix **A** to compute total output for each industry

$$\begin{aligned}
x_1 &= a_{11}x_1 + a_{12}x_2 + \ldots + a_{1n}x_n + d_1 \\
x_2 &= a_{21}x_1 + a_{22}x_2 + \ldots + a_{2n}x_n + d_2 \\
&\vdots \\
x_n &= a_{n1}x_1 + a_{n2}x_2 + \ldots + a_{nn}x_n + d_n
\end{aligned}$$

Rearranging these equations gives

$$\begin{aligned}
(1-a_{11})x_1 - a_{12}x_2 \ldots - a_{1n}x_n &= d_1 \\
-a_{21}x_1 + (1-a_{22})x_2 \ldots - a_{2n}x_n &= d_2 \\
&\vdots \\
-a_{n1}x_1 - a_{n2}x_2 \ldots + (1-a_{nn})x_n &= d_n
\end{aligned}$$

This last set of equations can be represented in matrix form by

$$(\mathbf{I}-\mathbf{A})\mathbf{x}=\mathbf{d}$$

where **I** is the $n \times n$ identity matrix,

$$\mathbf{x} = \begin{bmatrix} x_1 \\ x_2 \\ \cdot \\ \cdot \\ x_n \end{bmatrix}$$ is the vector of industry total outputs,

and $$\mathbf{d} = \begin{bmatrix} d_1 \\ d_2 \\ \cdot \\ \cdot \\ d_n \end{bmatrix}$$ is the vector of industry final demand.

For any given set of final demands **d** we can calculate the vector of final outputs **x** from the matrix equation

$$(\mathbf{I}-\mathbf{A})\mathbf{x} = \mathbf{d}$$

as

$$\mathbf{x} = (\mathbf{I}-\mathbf{A})^{-1}\mathbf{d}$$

For a solution to exist the matrix (**I** − **A**) must be non-singular, i.e. its determinant must be non-zero so that it can be inverted.

The Hawkins-Simon conditions

It is possible that when trying to compute industry outputs in an input-output model:

(i) no solution is possible because the **I** − **A** matrix is singular and cannot be inverted
(ii) industry output quantities are computed but have negative values.

Either of these situations would mean that the input-output model could not be used. Situation (i) would be akin to trying to divide by zero and situation (ii) would mean that the input-output model has no practical use because real industries obviously cannot produce negative outputs.

The Hawkins-Simons conditions state the constraints which must be put on the parameters of an input-output model to prevent such situations arising and for the model to compute non-negative outputs for all industries. The Hawkins-Simons conditions basically require that all the leading principal minors of the (**I** − **A**) matrix are strictly positive. This will ensure that the inverse matrix $(\mathbf{I}-\mathbf{A})^{-1}$ exists and all elements of $(\mathbf{I}-\mathbf{A})^{-1}$ are non-negative.

The leading principal minors are the determinants of certain sections of the (**I** − **A**) matrix, starting with the top left element by itself, then expanding by one row and one column each time. To illustrate how this works, take the (**I** − **A**) matrix from Example 16.25, where we know there are positive solutions for all industries.

Given $\mathbf{I}-\mathbf{A}=\begin{bmatrix} 0.6 & -0.3 & -0.2 \\ -0.1 & 0.8 & -0.4 \\ -0.2 & -0.1 & 0.8 \end{bmatrix}$, then the leading principal minors' determinants are

$$|\mathbf{M}_1| = |0.6| = 0.6, \quad |\mathbf{M}_2| = \begin{vmatrix} 0.6 & -0.3 \\ -0.1 & 0.8 \end{vmatrix} = 0.48 - 0.03 = 0.45$$

$$\begin{aligned} |\mathbf{M}_3| &= \begin{vmatrix} 0.6 & -0.3 & -0.2 \\ -0.1 & 0.8 & -0.4 \\ -0.2 & -0.1 & 0.8 \end{vmatrix} \\ &= 0.6\begin{vmatrix} 0.8 & -0.4 \\ -0.1 & 0.8 \end{vmatrix} - (-0.3)\begin{vmatrix} -0.1 & -0.4 \\ -0.2 & -0.8 \end{vmatrix} + (-0.2)\begin{vmatrix} -0.1 & 0.8 \\ -0.2 & -0.1 \end{vmatrix} \\ &= 0.6(0.64-0.04) + 0.3(-0.08-0.08) - 0.2(0.01+0.16) \\ &= 0.36 - 0.048 - 0.034 = 0.278 \end{aligned}$$

These principal minor determinants are all positive and so, if these checks had been made in advance of computing the industry outputs via matrix algebra, we would know that we would get positive outputs for all industries. These pre-checks were useful in the days when manual computation was required and they are still relevant to some more advanced aspects of input-output modelling. However, since computer software can now be used to invert matrices and compute answers to quite large input-output models within a matter of seconds, there is usually no real need to manually check in advance as to whether or not the properties of the **I** − **A** coefficient matrix satisfy the Hawkins-Simons conditions. If the spreadsheet matrix inversion and multiplication computations yield no or negative industry output values, then clearly the model is not feasible.

Professional economists who specialize in this field may encounter problems such as those outlined earlier, particularly if the data used to estimate inter-industry coefficients is incomplete or inaccurate, and they may have to investigate whether or not the Hawkins-Simons conditions are met. However, a full explanation and proof of why these Hawkins-Simons conditions guarantee feasible solutions would be too complex to present here and would go far beyond the scope of this introductory mathematics text.

QUESTIONS 16.13

1. The Leontief input-output coefficient matrix in Table 16.13 applies to a basic economic system with only the three industrial sectors F, G and H, plus a primary labour resource L. If the usual assumptions of input-output analysis apply:

Table 16.13 Leontief input-output coefficient matrix for Question 1

IP-OP coefficients		Destination of output		
		F	G	H
Inputs used	F	0.2	0.5	0.1
	G	0.3	0.1	0.4
	H	0.1	0.2	0.2
	L	0.4	0.2	0.3

   (a) What total output will be required from each industrial sector to satisfy a final demand of 480 for industry F output, 270 for industry G output and 540 for industry H output? (Assume all quantities are in £m.)
   (b) How much of the primary labour resource L will be needed in total to produce these output levels?
   (c) If final demand in each sector changes to 500 for industry F output, 300 for industry G output and 600 for industry H output, what will the total output levels in each of these sectors change to?

2. An economy has three industrial sectors Q, R and T, plus a primary labour resource L. The usual assumptions of input-output analysis apply and this economy has the Leontief input-output coefficient matrix in Table 16.14. Calculate the total output that will be required in each

Table 16.14 Leontief input-output coefficient matrix for Question 2

IP-OP coefficients		Destination of output		
		Q	R	T
Inputs used	Q	0.25	0.40	0.35
	R	0.10	0.10	0.10
	T	0.15	0.30	0.35
	L	0.50	0.20	0.20

sector and the labour resource requirements to produce enough to satisfy final demands of 170 for Q, 425 for R and 340 for T.

3. Observations of the outputs in a simple economy with the three industrial sectors Q, R and T, and labour resource L reveal the flows between each sector and final demand as in Table 16.5. Derive the Leontief input-output coefficient matrix for this economy and then use this to help find the total output that will be required in each sector and the labour resource requirements to produce enough to satisfy final demands of 90 for Q, 45 for R and 50 for T.

Table 16.15 Input-output flow table for Question 3

IP-OP flows		Destination of output				Total
		Q	R	T	D	
Inputs used	Q	50	60	70	70	250
	R	40	30	50	30	150
	T	75	45	40	40	200
	L	85	15	40	0	140
	Total	250	150	200	140	

4. A small economy has only the two industrial sectors, Y and Z, and uses no primary resource. Explain why it is not possible to estimate total outputs for these industries for any given levels of final demand if the input-output coefficient matrix takes the format

$$\begin{array}{ccc} IP-OP & Y & Z \\ Y & 0.5 & 0.5 \\ Z & 0.5 & 0.5 \end{array}$$

5. An economy has five industrial sectors V, W, X, Y and Z plus a primary resource R. If the usual assumptions of input-output analysis apply, and the input-output coefficient matrix is as shown in Table 16.16, find the total output (to 1 decimal place) that will be required in each sector and the labour resource requirements to produce enough to satisfy final demands for each industry as shown in Table 16.16.

Table 16.16 Input-output coefficient matrix for Question 5

IP-OP coefficients		Destination of output					Final demand
		V	W	X	Y	Z	
Inputs used	V	0.30	0.25	0.40	0.30	0.20	420
	W	0.20	0.10	0.05	0.10	0.10	360
	X	0.05	0.20	0.10	0.10	0.10	285
	Y	0.10	0.10	0.05	0.30	0.25	240
	Z	0.25	0.10	0.10	0.05	0.15	190
	R	0.10	0.25	0.30	0.15	0.20	

6. Another five-sector economy has the input-output coefficient matrix and final demand levels as shown in Table 16.17. Find the total output that will be required in each sector and the labour resource (L) requirements to produce these final demand levels.

Table 16.17 Leontief input-output coefficient matrix for Question 6

IP-OP coefficients		Destination of output					Final demand
		V	W	X	Y	Z	
Inputs used	V	0.10	0.10	0.25	0.60	0.10	180
	W	0.40	0.30	0.25	0.10	0.10	250
	X	0.05	0.10	0.10	0.05	0.20	400
	Y	0.10	0.20	0.10	0.05	0.00	510
	Z	0.25	0.20	0.20	0.05	0.50	345
	L	0.10	0.20	0.10	0.15	0.10	

7. A self-sufficient country has the eight industrial sectors of Agriculture, Mining, Iron and Steel, Engineering, Vehicle Production, Armaments, Construction, Transport and Services plus a primary resource R. If the assumptions of input-output analysis apply, and the input-output coefficient matrix is as shown in Table 16.18, find the total output that will be required in each sector and the labour resource requirements to produce enough to satisfy final demands for each industry as shown in Table 16.18.

Table 16.18 Leontief input-output coefficient matrix for Question 7

IP-OP coefficients		Destination of output									Final demand
		AGR	MIN	I&S	ENG	VEH	ARM	CON	TRAN	SER	
Inputs used	AGR	0.1	0	0	0	0	0	0.2	0.1	0.05	120
	MIN	0.05	0.1	0.05	0.1	0.05	0.05	0	0.1	0	0
	I&S	0	0.1	0.1	0.3	0.2	0.1	0.1	0.1	0	0
	ENG	0	0.1	0.2	0.2	0.3	0.2	0.1	0.1	0.05	0
	VEH	0.1	0.1	0.05	0.1	0.05	0.2	0.1	0.1	0.1	75
	ARM	0	0	0	0	0	0	0	0	0	60
	CON	0	0	0	0	0	0	0	0	0.05	155
	TRAN	0.05	0.1	0.1	0.05	0.05	0.05	0.1	0.1	0.1	84
	SER	0.05	0.1	0.1	0.05	0.05	0.1	0.1	0.05	0.05	92
	R	0.65	0.4	0.4	0.2	0.3	0.3	0.3	0.35	0.6	

Answers

CHAPTER 2

2.1 1. 3,555 2. 865 3. 92,920 4. 23

2.2 1. 919 2. 225 3. 164 4. 627 5. 440 6. 101

2.3 1. 840 2. 17 3. 172 4. 122 5. £13,800 6. 598
7. £176

2.4 1. $\frac{73}{168}$ 2. $\frac{101}{252}$ 3. $4\frac{4}{5}$ 4. $\frac{19}{30}$ 5. $2\frac{13}{21}$ 6. $4\frac{37}{60}$
7. $14\frac{1}{12}$ 8. $3\frac{12}{13}$ 9. $18\frac{39}{40}$ 10. $1\frac{1}{2}$

2.5 1. (a) $-\frac{1}{3}$ (b) $-\frac{5}{7}$ (c) $-1\frac{2}{5}$ (d) -3 (e) -11 2. -1 3. (a) -1
(b) -5 4. $-15, -4\frac{1}{3}, -2\frac{1}{5}, -1\frac{2}{7}, -\frac{7}{9}, -\frac{5}{11}, -\frac{3}{13}, -\frac{1}{15}$

2.6 1. 36.914 2. 751.4 3. 435.1096 4. 36,082 5. 0.09675
6. 610 7. 140 8. (a) 0.1 (b) 0.001 (c) 0.000001 9. (a) 0.452
(b) 2.431 (c) 0.075 (d) 0.002 10. 0.625, 62.5%

2.7 1. −2 2. −24 3. −33 4. 0.45 5. 0.35 6. −117
7. −330 8. 3,600 9. $\frac{-157}{140}$ 10. $\frac{17}{16}$

2.8 1. 0.25 2. 123 3. 6 4. 64 5. 11.641754 6. 531,441
7. 0.0015328 8. 36 9. −618.47021 10. 25.000655

2.9 1. ±25 2. 2 3. 0.2 4. 7 5. 2.4494897 6. 96
7. 10 8. 5.2780316 9. 0.03423 10. 87.977857

2.10 1. 3 2. 12 3. 4 4. $8\times\log 100 = 8\times 2 = 16$
5. $\ln 7 + \ln 8 = 1.9459 + 2.0794 = 4.0254$
6. $\log_4 64 - \log_4 256 = 3 - 4 = -1$ 7. 93.696376 8. 4.38228
9. 5.1331868

CHAPTER 3

3.1 1. (a) $0.01x$ (b) $0.5x$ (c) $0.5x$ 2. $0.01rx + 0.5wx + 0.5mx$ 3. (a) $\frac{x}{12}$ (b) $\frac{xp}{12}$ 4. (a) $0.1x$ kg (b) $0.3x$ kg (c) $x(0.1m + 0.3\text{p})$
5. $0.5w + 0.25$ 6. $20.5x + 10.5y$ 7. $3q–6000$

3.2 1. 456 2. 77.312 3. $r+z$, 5.6% 4. 1.094 5. £465.58
6. £2,100 7. (a) $1{,}000 + 0.8(M - 1{,}000) = 200 + 0.8M$ (b) £2,600

3.3 1. $30x + 4$ 2. $24x - 18y + 7xy - 12$ 3. $6x + 5y - 650$
4. $27H - 360$

3.4 1. $6x^2 - 24x$ 2. $x^2+4x + 9$ 3. $6x + 5y - 650$
4. $42x^2 - 16y^2 - 34xy + 6y$ 5. $33x + 2y - 20y^2 + 62xy - 21$
6. $120 + 2x + 54y + 40z - x^2 + 6y^2 + xy + 4xz + 8yz$ 7. $200q - 2q^2$
8. $53x + 46y$ 9. $8x^2 + 60x + 76$ 10. $4{,}000 + 150x$

3.5 1. $(x + 4)^2$ 2. $(x - 3y)^2$ 3. Does not factorize
4. Does not factorize

3.6 1. $3x + 7 - 20x^{-1}$ 2. $x + 9$ 3. $4y + x + 12$ 4. $200x^{-1} + 21$
5. $179x$ 6. $2(x + 3) + 4 - x - 3 - x + 2 = 9$

3.7 1. 4 2. $\frac{1}{11}$ 3. 7 4. 14 5. 82 6. 60p
7. 33% 8. 40p 9. £3,200 10. 4 m 11. 26

3.8 1. $\frac{1}{n}\sum_{i=1}^{n} H_i$, 173.7cm 2. 35 3. 60

4. $\sum_{i=1}^{n} 6{,}000(0.9)^{i-1}$, 16,260 tonnes 5. (a) $\frac{1}{n}\sum_{i=1}^{n} R_i$, £4,425

(b) $\frac{1}{3}\sum_{n-3}^{n-1} R_i$, £4,933 6. 13.25%, 8.2% 7. 3.14%

3.9 1. (a) $\le$ (b) $<$ (c) $\ge$ (d) $>$ 2. (a) $>$ (b) $\ge$ (c) $>$ (d) $>$
3. (a) $Q_1 < Q_2$ (b) $Q_1 = Q_2$ (c) $Q_1 > Q_2$ 4. $P_2 > P_1$

CHAPTER 4

4.1 1. (a) Quantity demanded depends on price of tea, average expenditure, etc.
(b) Q_t dependent, all others independent. Assuming tea is a normal good,

then the coefficient on P likely to be negative, coefficients on Y, A, N and C likely to be positive. 2. (a) 202 (b) 7 (c) 6, $x \geq 0$ 3. Yes; no

4.2 1. °F = 32 + 1.8°C 2. $P = 2{,}400 - 2Q$ 3. It is not monotonic, e.g. TR = 200 when $q = 5$ or 10 4. T = $(0.0625X - 25)2$; no

4.3 (Answers to 1 to 5 give intercepts on axes.) 1. $x = -12, y = 6$

2. $x = 3\frac{1}{3}, y = -40$ 3. $P = 60, Q = 300$ 4. $P = 150, Q = 750$

5. $K = 24, L = 40$ 6. Goes through origin only.

7. Goes through ($Q = 0$, TC = 200) and ($Q = 10$, TC = 250)

8. Horizontal line at TFC = 75 9. (a) and (d); both slope upwards and have positive intercepts on P axis

4.4 1. $Q = 90 - 5P$; 50; $Q \geq 0, P \geq 0$ 2. $C = 30 + 0.75Y$

3. By £20 to £100 4. $P = 12 - 0.015Q$ 5. £6,440

4.5 1. 3.75, 0.75, 0.375, −0.75 2. $P = 12, Q = 40$; £4.50; 10

3. (a) 2/3 (b) 3 4. (c) (i) (a) (ii) (b) 5. (a), (d)

6. APC = $400Y^{-1} + 0.5 > 0.5$ = MPC 7. (a) 0.263 (b) 0.714 (c) 1.667

4.6 1. −1.5 (a) becomes −1 (b) becomes −1 (c) no change (d) no change

2. $K = 100, L = 160, P_K = £8, P_L = £5$ 3. Cost £520 > budget; P_L reduced by £10 to £30 4. (a) −10 (b) −1 (c) −0.1 (d) −0.025 (e) 0

5. No change 6. Height £120, base 12, slope − 10 = −(wage)

4.7 1. Graph 2. Graph 3. Steeper 4. Like $y = x^{-1}$; £260

4.8 1. Graphs 2. $\pi = 50q - 100 - 0.4q^3$; inverted U

3. (a) AC = $40 + 3250q^{-1}$ (b) Original firms' π per unit = £27.50, but new firms' AC = £170 > price

4.9 1–4. Plot graphs

4.10 1. (1)(a) $16L^{-1}$ (b) 0.16 (c) constant (2)(a) $57{,}243.34L^{-15}$ (b) 57.243 (c) constant (3)(a) $322.54L^{-1}$ (b) 3.2254 (c) increasing (4)(a) $3{,}125L^{-1.25}$ (b) 9.882 (c) increasing (5)(a) $23{,}415{,}916L^{-1.6667}$ (b) 10,868.71 (c) decreasing (6)(a) $4{,}093.062L^{-1.7714}$ (b) 1.173 (c) decreasing

2. $\log Q = \log A + \alpha\log K + \beta\log L + \gamma\log R$

4.11 1. MR = $33.33 - 0.00667Q$ for $Q \geq 500$ 2. MR = $76 - 0.222Q$ for $Q \geq 22.5$ 3. MR = $80 - 0.555Q$ for $Q \geq 562.5$ 4. MC = $30 + 0.0714Q$ for $Q \geq 56$ 5. MC = $56 + 0.1333Q$ for $Q \geq 30$

6. MC = $3 + 0.0714Q$ for $Q \geq 59$

CHAPTER 5

5.1 1. $q = 40, p = 6$ 2. $x = 67, y = 17$ (approximately) 3. No solution exists.

5.2 1. $q = 118, p = 256$ 2. (a) $q = 80, p = 370$ (b) q falls to 78, p rises to 376 3. (a) 40 (b) rises to 50 4. $x = 2.102, y = 62.25$

5.3 1. $A = 24, B = 12$ 2. 200 3. $x = 190, y = 60$

5.4 1. $x = 30, y = 60$ 2. $A = 6, B = 36$ 3. $x = 25, y = 20$

5.5 1. $x = 24, y = 14.4, z = 19.2$ 2. $x = 4, y = 6, z = 4$
3. $A = 6, B = 22, C = 2$ 4. $x = 17, y = 4, z = 8$
5. $A = 82.5, B = 35, C = 6, D = 9$

5.6 1. $q = 500, p = 275$ 2. $K = 17.5, L = 16, R = 10$
3. (a) p rises from £8 to £10 (b) p rises to £9 4. Y = £3,750 m; government deficit £150 m 5. Y = £1,625 m; balance of payments deficit £15 m 6. $L = 80, w = 52$

5.7 1. $p = 184 + 0.2a, q = 43.2 - 0.06a, p = 216, q = 52.8$
2. $p = 84 + 0.2t, q = 32 - 0.4t, p = 85, q = 30$ 3. $p = 122.4 + 0.2t, q = 13.8 - 0.1t, p = 123.4, q = 13.3$ 4. (a) $Y = 100/(0.25 + 0.75t), Y = 250$ (b) $Y = 110/(0.25 + 0.75t), Y = 275$ 5. $p = (4{,}200 + 3{,}800v)/(9 + 5v)$, $q = (750 - 50v)/(9 + 5v)$, 489.23, 76.15

5.8 1. $q_1 = 60, q_2 = 80, p_1 = £10, p_2 = £8$ 2. $q_1 = 40, q_2 = 50, p_1 = £6, p_2 = £4$ 3. $p_1 = £8.75, q_1 = 60, p_2 = £6.10, q_2 = 550$ 4. £81 for extra 65 units 5. £7.50 for extra 25 units 6. $q_1 = 48, q_2 = 39, p_1 = £12, p_2 = £8.87$ 7. (a) 190 units (b) £175 for extra 75 units

5.9 1. $q_1 = 180, q_2 = 200, p = £39$ 2. $q_1 = 1{,}728, q_2 = 780, p = £190.70$
3. $q_1 = 1{,}510, q_2 = 1{,}540, q_A = 800, q_B = 2{,}250, P_A = £500, P_B = £625$
4. $q_1 = 160, q_2 = 600, q_A = 293\frac{1}{3}, q_B = 266\frac{2}{3}, q_C = 200,$
$P_A = £95, P_B = £80, P_C = £60$
5. $q_1 = 15.47, q_2 = 27.34, q_3 = 26.17, p = £14.20$

CHAPTER 6

6.1 1. 8.4 of A, 4.64 of B (tonnes); (a) no change (b) no B, 12.16 of A
2. $A = 13, B = 27$ 3. 12 of A, 5 of B 4. 22.5 of A, 7.5 of B
5. 6 of A, 32 of B 6. 13.64 of A, 21.82 of B; £7092; surplus 2.72 of R, 22.72 of mix additive 7. Produce 15 of A, 21 of B 8. 30 of A, B = 0 9. Objective function parallel to first constraint 10. 24,000 shares in X, 18,000 shares in Y, return £8,640

6.2 1. $C = 70$ when $A = 1, B = 1.5$, slack in $x = 30$ 2. $A = 3, B = 0$
3. $Q = 2.5, R = 1.5$; excesses 62.5 mg of B, 27.5 mg of C 4. 10 of A, 5 of B; space for 50 extra loads of X 5. Zero R, 15 tonnes of T; G exceeds by 45kg 6. 100 of A, 40 of B

6.3 1. 2 of A, 1 of B 2. 7.5 of X and 15 of Y 3. 8

CHAPTER 7

7.1 1. 2 or 3 2. 10 or 60 3. When $q = 2$ 4. 0.5 5. 9

7.2 1. 10 or −12.5 2. £16.353 3. (a) 1.01 or 98.99 (b) 11.27 or 88.73 (c) no solution exists 4. 2.5 or 10

7.3 1. $x = 15, y = 15$ or $x = -3, y = 249$ 2. $x = 1.75, y = 3.15$ or $x = -1.53$, $y = 20.97$ 3. 16.4 4. $q_1 = 3.2, q_2 = 4.8, p_1 = £136, p_2 = £96$ 5. $p_1 = £15, q_1 = 80, p_2 = £8.50, q_2 = 70$

7.4 1. 52 2. 109 3. 10

CHAPTER 8

8.1 1. £4,630.50 2. £314.70 3. £17,623.16 4. £744.71 5. £40,441.40 6. £5,030.03

8.2 1. £43,747.41; 12.68% 2. £501,159.74; 7.44% 3. (a) as APR 11.35% 4. £2,083.61; 19.25% 5. £625; 6.09% 6. 19.28% 7. 0.01467% 8. £494,531.25; 4.5%

8.3 1. £6,301.69 2. £355.89 3. No, A = £9,106.27 4. £6,851.65 5. (a) £9,638.58 (b) £11,579.83 (c) £13,318.15 6. 5 7. 5.27 years 8. 12.1 years 9. 5.45 years 10. 3.42 years 11. (a) 10.7% (b) 9.5% (c) 13.7.5% (d) 10.3% (e) 8.4% 12. 0.8% 13. (b) as PV = £5,269.85

8.4 1. (a) £90.75 (b) −£100.07 (c) −£474.01 (d) £622.86 (e) £1,936.87 (f) £877.33 (g) £791.25 (h) £992.16 2. B, PV = £6,569.10 3. (a) All viable (b) A best, NPV = £6,824.68 4. No, NPV = £1,277.40 5. Yes, NPV = £4,363.45 6. (a) Yes, NPV = £610.02 (b) no, NPV = −£522.30 7. B, NPV = £856.48

8.5 1. $r_A = 20\%$, $r_B = 41.6\%$, $r_C = 20\%$, $r_D = 20\%$; B consistently best 2. A $r_A = 21.25\%$, $r_B = 20.42\%$ B NPV_B = £2,698.94, NPV_A = £2,291.34 3. IRR = 16.93%

8.6 1. (a) 2.5, 781.25, 50,857.3 (b) 3, 121.5, 14,762 (c) 1.4, 10.756, 139.6 (d) 0.8, 19.66, 267.8 (e) 0.75, 0.57, 9.06 2. 5,741 3. A, £1,149.32; B, £2,980.91; C, £45,216.47 4. Yes, NPV = £3,774.71 5. £4,149.20

8.7 1. (a) not convergent (b) converges on 600 (c) not convergent (d) converges on 54 (e) converges on 961.54 (f) not convergent 2. £3,076.92 3. Yes, NPV = £50,000 4. (a) £240,000 (b) £120,000 (c) £80,000 (d) £60,000 5. £3,500

8.8 1. (a) £5,886.54 (b) £14,716.35 (c) £33,111.79 2. Income rises to (a) £7,551.43 (b) £18,878.59 (c) £42,476.82 3. Falls from £25,677.63 to £14,677.61 4. Yes, surplus left = £30,532.3 5. No, shortfall = £16,040.27 6. Increases income; changes shortfall £12,733.44 to surplus £33,456.44 7. Yes, surplus £73,423.28 8. No, shortfall £23,193.72 9. (a) Not feasible, shortfall £47,287.67 (b) Feasible, surplus £15,387.46 10. 13 years 11. 22 years

8.9 1. £11,746.43 2. £15,284.30 3. (a) £26,032.02 (b) £5,791.77 4. £1,627.96 5. £2,215.26 6. £261.09

8.10 1. £152.59 2. £197.38 3. £191.46 4. £1,426 5. £224,486 6. £134,315 7. (a) 12.2% (b) 9.7% 8. Cash price better (PV of repayments = £9,103 + £2,995 deposit = £12,098) 9. Yes as APR = 71.3% 10. A best as B's true APR = 5.0138%

8.11 1. £33,539.78 2. £17,711.69 3. £4,504.38 4. £17,022.50 5. £312.27 6. £21,428.20 (pension fund £291,216.20)

8.12 1. £44,461.68 2. £4,221.84 3. £17,877.62 4. £32,900.77 5. £778.60 6. £154.04 extra

8.13 1. 6.82 years 2. After 15.21 years 3. 4% 4. Yes, sum of infinite GP = 1,300 million tonnes 5. 4.85%

8A 1. £100.70 2. £89.03 3. £97.40 4. (i) Yc = 1.94% (ii) YS = 3.0556% (iii) YTM = 2.9174% 5. £53.98 6. PE £38.50 < £40 market price so not a good investment 7. £80

CHAPTER 9

9.1 1. $36x^2$ 2. 192 3. 21.6 4. $260x^4$

9.2 1. $3x^2 + 60$ 2. 250 3. $-4x^{-2} - 4$ 4. 1 5. $0.2x^{-3} + 0.6x^{-0.7}$

9.3 1. $120 - 6q$, 20 2. 25 3. 14,400 4. £200

9.4 1. 7.5 2. $12q^2 - 40q + 60$ 3. (a) $1.5q^2 - 6q + 25$ (b) $0.5q^2 - 3q + 25 + 20q^{-1}$ (c) $q - 3 - 20q^{-2}$ 4. MC constant at 0.8

9.5 1. 4 2. (a) 80 (b) 158.33 (c) 40 or 120 3. 6

9.6 1. $50-\frac{2}{3}q$ 2. 900 3. $24 - 1.2q^2$

9.7 1. 0.8 2. Proof 3. 0.16667 4. 1

9.8 1. £77.50 2. Rise, maximum TY when t = £39

9.9 1. (a) 0.8 (b) 4,400 (c) 5 (d) 120 (e) Yes, both 940

CHAPTER 10

10.1 1. 62.5 2. 150 3. (a) 500 (b) 600 (c) 300 4. 50

10.2 1. 1,200, max. 2. 25, max. 3. 4,096, max. 4. 4, not max.

10.3 1. 6, min. 2. 14.4956, min. 3. 0, min. 4. 3, not min. 5. No stationary point exists

10.4 1. (a) MC = $2q^2 - 28q + 222$, min. when $q = 7$, MC = 124

(b) $\text{AVC} = \frac{2}{3}q^2 - 14q + 222$, min. when $q = 10.5$, AVC = 148.5

(c) AFC = $50q^{-1}$, min. when $q \to \infty$, AFC $\to 0$ (d) TR = $200q - 2q^2$, max. when $q = 50$, TR = 5,000 (e) MR = $200 - 4q$, no turning point, end-point max. when $q = 0$

(f) $\pi = -\frac{2}{3}q^3 + 12q^2 - 22q - 50$, max. when $q = 11, \pi = 272.67$

$\left(\pi \text{ min. when } q = 1, \pi = -60\frac{2}{3}\right)$

2. (a) 16 (b) 8 (c) 12 3. (a) No turning point but end-point min. when $q = 0$ (b) No turning point but end-point min. when $q = 0$ (c) No turning point but end-point max. q when $p = 0$

10.5 1. π max. when $q = 4$ (theoretical min. when $q = -1.67$ not realistic) 2. Max. when $q = 10$ 3. π max. when $q = 12.67$, gives $\pi = -48.8$ 4. 5,075 when $q = 10$ 5. 27 6. when $q = 37$

10.6 1. 15 orders of 400 2. 560 3. 480 4. 140

10.7 1. (i) (a) $q = 90 - 0.2t$, $p = 270 + 0.4t$ (b) and (c) $q = 90$, $p = 270$ (ii) (a) $q = 250 - 1.25t$, $p = 125 + 0.375t$ (b) and (c) $q = 250$, $p = 125$ (iii) (a) $q = 25 - 0.9615t$, $p = 160 + 0.385t$ (b) and (c) $q = 25$, $p = 160$ (Note: there is no tax impact for (b) and (c) in all cases) 2. 2.$q = 100$, $p = 380$ (no tax impact)

CHAPTER 11

11.1 1. (a) $3 + 8x$, $16 + 4z$ (b) $42x^2z^2$, $28x^3z$ (c) $4z + 6x - 3z^3$, $4x - 9x^{-2}z^2$ 2. $MP_L = 4.8K^{0.4}L^{-0.6}$, falls as L increases 3. $MP_K = 12K^{-0.7}L^{0.3}R^{0.4}$, $MP_L = 12K^{0.3}L^{-0.7}R^{0.4}$, $MP_R = 16K^{0.3}L^{0.3}R^{-0.6}$ 4. MPL = 0.7, does not decline as L increases 5. No 6. $1.2x_j^{-0.7}$

11.2 1. (a) 0.228 (b) falls to 0.224 (c) inferior as $\partial q/\partial m < 0$ (d) elasticity with respect to $p_s = 0.379$ and so a 1% increase in both prices would cause a percentage rise in q of $0.379 - 0.228 = 0.151\%$ 2. (a) yes, MU_A and MU_B will rise at first but then fall (b) no, MU_A falls but MU_B continually rises, therefore law not obeyed (c) yes, both MU_A and MU_B continually fall 3. No, MU will never reach zero for finite values of A or B 4. 3,738.46; balance of payments changes from 4.23 deficit to 68.85 surplus 5. $25 + 0.6q_1^2 + 2.4q_1q_2$ 6. 0.45; 1.81818; 55

11.3 1. $-2K^{0.6}L^{-1.5}$, $2.4K^{-0.4}L^{-0.5}$ 2. $Q_{LL} = 6.4$, MPL has constant slope; $Q_{LK} = 35 + 2.8K$, position of MP_L will rise as K rises; $Q_{KK} = 2.8L$, MP_K has constant slope, actual value varies with L; $Q_{KL} = 35 + 2.8K$, rise in L will increase MP_K, effect depends on level of K
3. $TC_{11} = 0.008q_3^2, TC_{22} = 0, TC_{33} = 0.008q_1^2$
$TC_{12} = 1.2q_3, TC_{21}, TC_{23} = 9 + 1.2q_1 = TC_{32}$
$TC_{31} = 0.016q_1q_3 + 1.2q_2 = TC_{13}$

11.4 1. $q_1 = 12.46$; $q_2 = 36.55$ 2. $p_1 = 97.60$, $p_2 = 101.81$ 3. $q_1 = 0$, $q_2 = 501.55$ (mathematical answer gives $q_1 = -1{,}292.24$, $q_2 = 1{,}701.77$ so rework without market 1) 4. £575.81 when $q_1 = 47.86$ and $q_2 = 39.01$ 5. $q_1 = 266.67$, $q_2 = 333.33$ 6. $q_1 = 1{,}580.2$, $q_2 = 1{,}791.8$ 7. $K = 2{,}644.2$, $L = 3{,}718.5$ 8. £29,869.47 when K = 1,493.47 and L = 2,489.12 9. Because max. π = £18,137.95 when $K = 2{,}176.5$ and $L = 2{,}015.22$ 10. $K = 10{,}149.1$, $L = 9{,}743.1$

11.5 1. (a) $12K^{-0.4}L^{0.4}\mathrm{d}K + 8K^{0.6} L^{-0.6}\mathrm{d}L$ (b) $14.4K^{-0.7}L^{0.2}R^{0.4}\mathrm{d}K + 9.6^{0.3} L^{-0.8}R^{0.4}\mathrm{d}L + 19.2K^{0.3}L^{0.2}R^{-0.6}\mathrm{d}R$ (c) $(4.8K^{-0.2} + 1.6KL^2)\mathrm{d}K + (3.5L^{-0.3} + 1.6K^2 L)\mathrm{d}L$ 2. (a) yes (b) no, surplus (c) no, surplus 3. $40x^{-0.6}z^{-0.45} + 12x^{0.4} z^{-0.7}$ 4. $\dfrac{\partial Q_A}{\partial P_A}+\dfrac{\partial Q_A}{\partial M}\dfrac{\mathrm{d}M}{\mathrm{d}P_A}$

CHAPTER 12

12.1 1. $K = 12.6$, $L = 21$ 2. $K = 500$, $L = 2{,}500$ 3. $A = 6$, $B = 4$ 4. 141.42 when $K = 25$, $L = 50$ 5. (a) $K = 1{,}000$, $L = 50$ (b) $K = 400$, $L = 20$ 6. 1,950 when $K = 60$, $L = 120$ 7. $L = 241$, $K = 201$, TC = £3,617

12.2 See answers to 12.1

12.3 1. See answers to 12.1 2. $L = 38.8$, $K = 20.7$, TC = £3,104.50 3. C_1 = £480,621, C_2 = £213,609 4. $L = 19.04$, $K = 8.18$, TC = £1,145.30

12.4 1. $x = 30$, $y = 30$, $z = 90$ 2. 877.8 when $K = 15$, $L = 45$, $R = 13$ 3. $x = 50$, $y = 100$, $z = 150$ 4. 79,602.1 when $x = 300$, $y = 300$, $z = 1{,}875$ 5. $K = 26.7$, $L = 33.3$, $R = 8.9$, $M = 55.6$ 6. $L = 60$, $K = 45$, $R = 40$

CHAPTER 13

13.1 1. 9 2. Proof 3. $M(1 + i)^2$ 4. $0.6x(3 + 0.6x^2)^{-0.5}$ 5. $0.5(6 + x)^{-0.5}$ 6. $\mathrm{MRP_L} = 60L^{-0.5} - 8$, $L = 16$ 7. 169 units 8. £8 9. 0.000868

13.2 1. $(6x + 7)^{-0.5}(39x^2 + 36.4x - 5.7)$ 2. 12 3. $76.5L^{-0.5}(0.5K^{0.8} + 3L^{0.5})^{-0.4}$ 4. 312.5 5. £190 6. (a) $-0.05(60 - 0.1q)^{-0.5}$ (b) slope rate of change $= -0.0025(60 - 0.1q)^{-0.5} < 0$ when $q < 600$ (c) 400

13.3 1. $(24 + 6.4x - 4.5x^{1.5} - 3x^{2.5})(8 - 6x^{1.5})^{-1.5}$ 2. $(18{,}000 + 360q)(25 + q)^{-1.5}$ 3. -0.113 4. $q = 1{,}333\frac{1}{3}$, $\mathrm{d}^2\mathrm{TR}/\mathrm{d}q^2 = -0.00367$ 5. $L = 4.8$, $H = 7.2$ 6. Adapt proof in text for MC and AC to AVC = $\mathrm{TVC}(q)^{-1}$

13.4 1. (a) $12.5x^2 + C$ (b) $5x + 0.6x^2 + 0.05x^3 + C$ (c) $24x^5 - 15x^4 + C$ (d) $42x + 18x^{-1} + C$ (e) $60x^{1.5} + 220x^{-0.2} + C$ 2. (a) $4q + 0.05q^2$ (b) $42q - 9q^2 + 2q^3$ (c) $35q + 0.3q^3$ (d) $62q - 8q^2 + 0.5q^3$ (e) $185q - 12q^2 + 0.3q^4$ 3. (a) $40q - 2q^2$ (b) $600q - 12.5q^2$

13.5 1. (a) £750,000 (b) £81,750 (c) £250,000 (d) £67,750 2. £49,600 3. (a) £195,419.40 (b) $T = 4.3267$ (approx. 4 yrs + 4 months) 4. (a) £4,601.20 (b) $T = 14.1311$ (approx. 14 yrs + 2 months)

13.6 1. (a) $\frac{1}{24}(21+6x)^4+C$ (b) $\frac{1}{4-3m}+C$ (c) $0.25(8x+6x^5)^4+C$
(d) $\frac{-1}{3\left(x^3-6\right)^3}+C$ (e) $\frac{-1}{4\left(5y^8-2y\right)^2}+C$ 2. (a) $\frac{1}{10}x^{10}+\frac{2}{7}x^7+C$
(b) $-x^{-1}-x^{-3}+C$ (c) $x^2(0.5x^2+3)+C$
3. (a) $\frac{-1}{2\left(3+4x\right)}+C$, solve by substitution
(b) $\frac{2}{3}y\left(y+3\right)^{\frac{3}{2}}-\frac{4}{15}\left(y+3\right)^{\frac{5}{2}}+C$, by parts and substitution
(c) $\frac{1}{12}\left(7+2x\right)^6+C$, by substitution

CHAPTER 14

14.1 1. 20 2. No production in period 4 3. (a) Unstable (b) stable
14.2 1. $P_t = 4 + 0.25(-2)^t$ 2. Stable, 118.54 3. 404.64
14.3 1. 2,790.625; yes 2. 39,946.789 3. 492.57 4. 1,848.259
14.4 1. 2,460.79 2. No, 1,976.67 < 1,980 3.
$P_t^X = 562 - 63(0.83)^t$, 555.25.

CHAPTER 15

15.1 1. 64.44 million 2. 61,062 units 3. 16.8 million tonnes
4. Usage in million units: (a) 94.6, yes (b) 137.6, yes (c) 200.2, no
(d) 291.31, no 5. 56,609 units 6. €31,308.07
15.2 1. 2%; 9.84 million; no 2. 9%, 401,767,300 barrels
3. £122,197.54 4. 587 5. £165,201
15.3 1. 0.48% 2. 2.05%; 3.49% 3. 0.83%, 621.43 million tones
4. €6,446.39 million 5. 8.8% 6. 5.83% 7. 6.18%
8. 9% discrete (equivalent to 8.62% continuous)
15.4 1. (a) $200e^{0.2t}$, 1477.81 (b) $45e^{1.2t}$, 7323965.61 (c) $14e^{-0.4t}$, 0.26
(d) $40e^{1.32t}$, 21614597.49 (e) $128e^{-0.03t}$, 99.69 2. 10%, 6.77
15.5 1. (a) $-20e^{0.4t}+200$ (b) 52.22 (c) diverge 2. (a) $-19.2e^{-1.5t}+32$
(b) 31.99 (c) converge 3. (a) $-20e^{-0.75t}+120$ (b) 119.53 (c) converge
4. (a) $75e^{0.08t}-300$ (b) −188.11 (c) diverge
15.6 1. $7e^{-0.325t}+30$, stable 2. $-7.25e^{-0.96t}+26.25$, difference 0.01
3. Yes, as predicted spot price is \$27.56 4. $32.54e^{-0.347t}+17.46$, £23.20,
3 periods 5. \$44.01

15.7 1. $25e^{-0.2t} + 180$, 183.38 2. $-63.33e^{-0.195t} + 1583.33$, 1574.32
3. $12.22e^{-0.036t} + 2027.78$, 2036.3 4. $-9.49e^{-0.226t} + 141.49$, 140.495
5. $-18.154e^{-0.176t} + 346.15$, 343.015

CHAPTER 16

16.1 1. $\begin{bmatrix} 2 & 8 & 23 \\ 3 & 5 & 26 \end{bmatrix}$ 2. Yes 3. (a) yes, $\begin{bmatrix} 6 & 70 \\ 27 & 23 \end{bmatrix}$

(b) no (c) yes, $\begin{bmatrix} 14 \\ 3 \\ 14 \\ -3 \\ 2 \end{bmatrix}$ 4. $\begin{bmatrix} 1.4 & 0.6 & 0.2 & 0.8 \\ 1.2 & 0.6 & 1.6 & 0.5 \\ 0.8 & 0.24 & 0. & 0 \end{bmatrix}$ in £ m

16.2 1. $\begin{bmatrix} 27 & 29 \end{bmatrix}$ 2. (a) $\begin{bmatrix} 98 & 84 \\ 163 & 134 \end{bmatrix}$ (b) $\begin{bmatrix} 71 & 24 & 13 \\ 68 & 8 & 17 \end{bmatrix}$

(c) not possible 3. $\boldsymbol{PR} = \begin{bmatrix} 52 & 30 & 56 & 43 \\ 62 & 32 & 71 & 55 \end{bmatrix}$

16.3 1. (a) $\begin{bmatrix} 17.5 & 45 & 19 \\ 61 & 130 & 84 \end{bmatrix}$ (b) $\begin{bmatrix} 136 & 67.5 & 90 & 40 \\ 130 & 47.5 & 51 & 44 \\ 59 & 10 & 9 & 36 \end{bmatrix}$

(c) $\begin{bmatrix} 460 & 579 & 299 & 400 & 2110 & 3181 \\ 291 & 1077 & 240 & 418 & 2155 & 1166 \\ 907 & 4505 & 400 & 1030 & 9932 & 7386 \\ 114 & 20 & 466 & 560 & 151 & 318 \\ 93 & 133 & 165 & 213 & 135 & -215 \end{bmatrix}$ 2. 130.52

16.4 1. (a) $\begin{bmatrix} 5 & 4 & 9 \\ 2 & 1 & 4 \\ 2 & 5 & 4 \end{bmatrix}\begin{bmatrix} x \\ y \\ z \end{bmatrix} = \begin{bmatrix} 95 \\ 32 \\ 61 \end{bmatrix}$

(b) $\begin{bmatrix} 6 & 4 & 8 \\ 3 & 2 & 4 \\ 1 & -8 & 2 \end{bmatrix}\begin{bmatrix} x \\ y \\ z \end{bmatrix} = \begin{bmatrix} 56 \\ 28 \\ 34 \end{bmatrix}$

(c) $\begin{bmatrix} 5 & 4 & 2 \\ 9 & 4 & 0 \\ 2 & 4 & 4 \end{bmatrix}\begin{bmatrix} x \\ y \\ z \end{bmatrix} = \begin{bmatrix} 95 \\ 32 \\ 61 \end{bmatrix}$ (d) not possible

2. (b) and (c) 3. (b) not square, (c) rows linearly dependent

16.5 1. (a) 2 (b) 0 (c) 56 (d) 137 (e) 119

16.6 1. (a) −39 (b) 15 (c) −4 (d) 28 (e) 50 (f) 4 2. |**A**| = 636, |**B**| = −101, |**C**| = −4,462

16.7 1. $\begin{bmatrix} 0.16 & -0.3 \\ -0.2 & 0.5 \end{bmatrix}$ 2. $C = \begin{bmatrix} 3 & 4 & -5 \\ 2 & 6 & -5 \\ -8 & -19 & 20 \end{bmatrix}$,

$AdjA = \begin{bmatrix} 3 & 2 & -8 \\ 4 & 6 & -19 \\ -5 & -5 & 20 \end{bmatrix}$, $A^{-1} = \begin{bmatrix} 0.6 & 0.4 & -1.6 \\ 0.8 & 1.2 & -3.8 \\ -1 & -1 & 4 \end{bmatrix}$

3. $\begin{bmatrix} -0.5 & 0.5 & -0.5 & -0.5 \\ 0.1075 & -0.0215 & 0.1505 & -0.0538 \\ 1.7742 & -1.3548 & 0.4839 & 1.6129 \\ -0.3978 & 0.2796 & 0.043 & -0.3011 \end{bmatrix}$

16.8 1. $x = 5, y = 8$ 2. $x_1 = 10, x_2 = 6, x_3 = 2$ 3. $\beta_1 = -0.5, \beta_2 = 1, \beta_3 = 0.4, Q = 9.5$ 4. $\beta_1 = -300, \beta_2 = 75, \beta_3 = 400, \beta_4 = -100, \beta_5 = 0.2, \beta_6 = 10, Q = 8370$

16.9 1. $x = 3, y = 7$ 2. 6 3. See 16.8 answers.

16.10 1. $q_1 = 389.6$, $q_2 = 62.3$, max SOC met as $|H_1| = -0.6$, $|H_2| = 0.448$
2. $q_1 = 216.8$, $q_2 = 435.8$, max SOC met as $|H_1| = -0.5$, $|H_2| = 0.19$
3. $q_1 = 6.485$, $q_2 = 2.376$, $q_3 = 5.4$, max SOC met as $|H_1| = -16.8$, $|H_2| = 21.2$, $|H_3| = -16.8$
4. $q_1 = 5.2$, $q_2 = 35.4$, $q_3 = 20.8$, max SOC met as $|H_1| = -4.08$, $|H_2| = 0.4832$, $|H_3| = -0.0225$

16.11 1. $K = 25$, $L = 100$, max SOC met as $|H_B| = 0.72$ 2. $K = 16$, $L = 32$, min SOC met as $|H_B| = -14.618$ 3. $X = 25$, $Y = 100$, max SOC met as $|H_B| = 4.543$ 4. $K = 121.93$, $L = 243.86$, min SOC met as $|H_B| = -0.945$

16.12 1. (c) $Y = 2415$, $Z = 1725$ 2. (a) $Y = 550$, $Z = 350$, $R = 270$ (b) $Y = 570$, $Z = 390$, $R = 288$ 3. (a) $A = 6812.5$, $B = 2781$, $R = 1300$
(b) $Y = 7487.5$, $Z = 2969$, $R = 1420$

16.13 1. (a) $F = 1617$, $G = 1382$, $H = 1223$ (b) $L = 1290$ (c) $F = 1737$, $G = 1510$, $H = 1345$ 2. $Q - 1134$, $R = 722$, $T = 1118$, $L = 935$ 3. $Q = 330$, $R = 204$, $T = 263$, $L = 185$ 4. Cannot invert the singular I–A matrix 5. $V = 2823.7$, $W = 1422$, $X = 1121.7$, $Y = 1545.3$, $Z = 1444.2$, $R = 1495$ 6. $V = 3056.7$, $W = 3863.3$, $X = 2226$, $Y = 1906.2$, $Z = 4844.7$, $R = 2071.3$ 7.

AGR	MIN	I&S	ENG	VEH	ARM	CON	TRAN	SER	Lab
202.4	80.3	170.3	203.5	209.9	60.0	164.5	198.*s*1	189.7	586

Index

absolute value 60
accumulated value of an investment 189–95, 474–6
addition 8–10; of algebraic terms 38; of fractions 13; of matrices 509; of negative numbers 22
adjoint matrix 532
agriculture and time lags in supply 434–6
algebra 4, 33–61; matrix 507–55
annual equivalent rate (AER) 196–201
annual percentage rate (APR) 197–8, 242–8
annuity 221–2, 224–7; income from 231–3; perpetual 227–31; present value of 224–7
antilogs: of natural logarithm 482
arc elasticity 15–17
area 42; derived by integration 420–4; *see also* feasible area
arithmetic 8–15
asset valuation 263–74
average cost 88–9; minimum point 181, 287, 411–12
average fixed cost 88–9
average propensity to consume 81
average variable cost 88–9; minimum point 287
axes 69, 72

balance of payments 329–31
base for logarithms 29–30, 481–2
Bertrand duopoly model 465–9
bond valuation 263–9
bordered Hessian 551–4
border preserving principal minor 554
brackets 11, 40–2
break-even chart 111, 305
budget constraint 81–4, 119–20, 163; and maximum output 364–8, 371–4

calculators 3–4, 27–8
calculus 275–80
cancellation 14
capital formation 426–7
cardinal utility 326
cartels 102, 140
Cartesian axes 69
cause and effect 67–8
ceteris paribus 323
chain rule 395–8; applications of 398–402
clean price of bonds 266
Cobb-Douglas production function 99–101, 320–2; convexity to origin 378–9; derivation of marginal product 318, 320–2; optimization of 364–7, 371–4; product exhaustion 356–60; returns to scale 100, 347, 358–60
cobweb model 434–43; difference equation solution 444–52; iterative solution 434–43; spreadsheet solution 439–40, 459–60; stability 440, 442–3, 446–52
coefficient vector 517–19
cofactor matrix 532
cofactors 529
collusion 140
column vector 508

common ratio of geometric series 222
comparative statics 123–31; effect of taxes 128–30, 312–17
complementary solution: to difference equations 444–6; to differential equations 493–5
complements 332
composite functions 88–91; differentiating *see* chain rule
compound interest 191–5
concavity to origin 378
constant in functions 86, 168; when differentiated 279
constant of integration 415
constant returns to scale 100, 357–60
constrained maximization: in linear programming 147–56; by substitution 364–70; using bordered Hessian matrix 551–4; using Lagrange multiplier 371–7
constrained minimization: in linear programming 159–64; by substitution 369–70; using Lagrange multiplier 380–4
constrained optimization *see* constrained maximization; constrained minimization
constraint *see* budget constraint; in linear programming; Lagrangian
consumer surplus 423–5
consumer utility *see* utility
consumption function 113, 130–2
continuous functions 190
continuous growth 190, 474–85; compared to discreet growth 476, 485–6; derivation of rate 477–8; discounting 479–80
convergence 437, 496
convergent series 228–9
convexity to origin 378
coordinates 69
corner solutions 305–6, 378
cost minimization: subject to output constraint 369–70, 380–2, 390–3; subject to quality constraints 159–64
costs *see* average, marginal and total
Cournot model 465
Cramer's rule 542–3
cross partial derivatives 334–9
cubic function 86, 183–6, 286–7
cubic roots 27
cubic values 23–4
current yield on a bond 266
curved functions 86–7
decaying rates of growth 484
decimals 19–22, 191–2
decreasing returns to scale 100, 347, 359–60
definite integrals 420–3; applications of 422–7
definite solutions 447, 490–1, 493–5
degree of a polynomial 89
demand for labour 396, 398–400
demand function 50, 64, 68, 72; and elasticity 15–17; and marginal revenue 128, 282, 418–19; and supply equilibrium 108–9, 124–5, 434–42; and total revenue 89–90, 282
denominator 12
dependent variables 64–5, 97, 131
deposit accounts 190, 199
derivatives: first-order 276–81; partial 319–20; second-order 301–2; second-order partial 323, 334–9; total 361–2
determinants 525–7, 530–31; using spreadsheet 531
difference equations 435–6; and Bertrand duopoly 465–9; and cobweb model 438–9, 444–52; and distributed lag 463–4; and Keynesian model 453–4, 456–64
differential calculus 275–80
differential equations 488–95; homogeneous 489–92; non-homogeneous 489, 493–5; using a spreadsheet 496–7
differentials 355–6
differentiation: basic rules 277–80; to derive MC from TC 285–6; to derive MR from TR 281–4; from first principles 280–1; of logarithmic and exponential functions 477–8, 487–8; partial 319–22; using chain rule 395–8; using product rule 403–8; using quotient rule 408–10
diminishing marginal productivity 319–22
diminishing marginal utility 326–8
dirty price of bonds 266, 268
discounting 207; of continuous return flows 479–80; factors 209, 220
discount rate 200–1
discrete functions 190
distributed lags 463–4
disturbances to equilibrium *see* cobweb model; dynamics
divergent series 228–9
dividend discount model (DDM) 271–3
dividend pricing approach 270–1

division 8; of algebraic expressions 48–9; of decimals 20–1; of negative numbers 23; of powers 24
domain 64, 67, 305–6
drawdown pensions 233–7
duopoly price adjustment 465–9
dynamics: continuous adjustments 474–86; in cobweb model 433–43; in Keynesian model 453–6

e: as base for natural logarithms 481–2; derivation of 472–3
econometrics 3, 519, 539
economic model, estimating parameters of 539–41
elasticity: arc 15–17; income 325; point 78–79, 291–3, 324–5, 401
elements of a matrix 508, 515
elimination of unknowns 116
end point solutions 305–6
equality constraints 166–7
equating to solve simultaneous equations 110–11
equations: Lagrange 371–2; polynomial 89, 182–8; quadratic 168–77; simple linear solution 50–4; *see also* differential equations; simultaneous linear equations
equilibrium: adjustment mechanism between 434–8, 447–8; supply and demand equilibrium 108–9, 124–5
Euler's theorem 356–60
evaluation of expressions 36–7
Excel spreadsheets: cobweb model 439–40, 443; inverse matrix 535–6; investment appraisal 210–14, 217–18, 220, 245–6, 268–9; matrix determinants 531; matrix multiplication 516–18; plotting functions 92–6; solving simultaneous equations 538–9
exhaustion of total product 356–60
exogenous variables 64, 125; changes in *see* reduced form
exponential functions 471–3; differentiation of 477–8; and growth rates 474–8; natural 472–3
exponents 24–5
export multiplier 329
extraction rates 257–60

face value of a bond 263–4
factorization 44–7, 175–6
feasible area 149, 156
final value of investment: with continuous growth 472–6; with discrete growth 192–5; *see also* capital formation
first degree price discrimination 135–6, 138
first-order conditions for maximization/minimization 299–300, 340
first-order difference equations 435–6, 444
first-order differential equations 488–95
first-order partial derivatives 319–20
fitting linear functions 73–5
fixed costs *see* total and average fixed cost
fractions 12–15; decimal 19–22, 191–2
functions 50, 63–105; cubic 86, 183–6, 286–7; exponential 471–3; inverse 66–8, 72, 290–1; linear 69–72; polynomial 89, 182–8; quadratic 169–82; two independent variables 97–9; *see also* Cobb-Douglas production function

general form of a function 64
general solution: difference equations 445–6; differential equations 490–1, 493–5
geometric series 222–4; sum of 223–4; infinite 228–9
Gordon growth model 271–3
government expenditure multiplier 329–31
gradient *see* slope
graphs 69–75, 85–90, 92–6, 171–5
growth: continuous 190, 474–6, 477–9, 483–5; discrete 190, 257–62; discrete and continuous compared 190, 476, 485–6

Hawkins-Simon conditions 565–6
Hessian matrix 545–50; bordered 551–5; higher order 550; third order 548–9
Hicks's income and substitution effects 374–7
homogeneity of functions 100–1
horizontal lines and zero slope 78, 91
horizontal summation: of MC functions 104–5; of MR functions 102–4
hours of work/leisure 412–14

identities 50
identity matrix 523
income effect 374–7
income elasticity 325
increasing returns to scale 100, 347, 358–60

indefinite integrals 414–17; applications of 417–20
independent variables 64–5, 97, 131
indifference curves 375
indirect functional relationships 361–2
individual labour supply 412–14
inequalities 59–61, 148–9
infinite geometric series 228–9
infinitesimally small changes 283, 292, 325, 353, 355–6
inflation 56–8
inflexion points 301, 304
initial sum: calculation of 202, 478–9
input-output analysis 555–66; coefficients 556; multiple industry using spreadsheet 561–5; using matrix algebra 558–9, 562–5
inputs: optimal combination 119–20, 369–70, 380–3; optimal order size 309–11; requirements in linear programming 160
integers 13
integrals *see* integration
integration 414–20; definite 420–3; indefinite 415–20; of 487; by parts 430–2; by substitution 428–30
intercepts 72, 83–4, 364
interest 191–2; annual equivalent rate (AER) 196–201; changes in 194–5; compound 191–5; continuous and discrete compared 476; daily 199; determination of 204–6; monthly rates 197–8, 239–40; part year 196–7; rate on loans 244–47; simple 191
internal rate of return (IRR) 216–21; deficiencies of 218–21; Excel command 217–18, 220
inventory control 309–11
inverse function rule 324
inverse functions 66–8, 72, 290–1
inverse matrix 532–7; using Excel 535–6
investment 193–4; appraisal 207–21; as a function of time 426–7; final value 193–4, 472–6; IRR 216–21; IRR and NPV methods compared 218–19
isoquants 97–8, 320–2, 355–6; and output maximization 378–9

Keynesian macroeconomic model 113, 120, 130–1, 296–7, 328–31; continuous growth 502–6; difference equations 453–4, 456–64; distributed lag 463–4; lagged 453–6; multiplier 296–7, 328–31; spreadsheet solution 457–8

labour: individual supply 412–14; profit maximizing level to employ 398–400
Lagrange multiplier: maximization with two variables 371–7; minimization with two variables 380–5; second order conditions 377–9, 551–2; using matrix algebra 551–5; with more than two variables 385–93
lags: in consumption function 453–4; distributed 463–4; in supply/production 435
Laplace expansion 530–1
Laspeyre price index 57
leisure hours 412–14
Leontief input-output analysis *see* input-out analysis
linear dependence 524
linear equations 50; in matrix format 524–4, 536–9; solution of single 50–4; *see also* simultaneous linear equations
linear functions 69–72; fitting equations to 73–5
linear programming 147–68; constrained maximization 147–56; constrained minimization 159–64; equality constraints 166–7; mixed constraints 166–7; more than two variables 168; multiple solutions 153–5; solutions on axes 155–6
loan repayments 241–7; spreadsheet calculation of interest rate 245–6
local maxima/minima 304
logarithms 29–31; in investment time period calculation 202–4; for linear regression 101; natural 481–6
long run 181; equilibrium 445

macroeconomic model *see* Keynesian macroeconomic model
managerial economics 147
marginal cost: cutting AVC 287; cutting minimum AC 287, 411–12; by differentiation 285–6, 331–2; horizontal summation 104–5, 140–5; integral of 417–18; intersection with MR 307; of labour 398–9
marginal product 119; by partial differentiation 327–30, 344–5
marginal propensity to consume 296–7, 328
marginal propensity to import 329

marginal rate of technical substitution 355–6
marginal revenue: by differentiation 281–4, 332–3; horizontal summation 102–4; integral of 418–20; interrelated demand functions 332–3; for non-linear demand functions 283–4
marginal revenue product 396, 398–400
marginal utility 326, 368–9
market clearing price 124–7, 435–8, 497–501
matrices 507–8; addition and subtraction 509; adjoint 532; of cofactors 532; division 521; identity 523; inverse 532–7; multiplication 512–17; transposition 531
maximization: by differentiation 299–301; linear programming 147–56; of profit 287–9, 307–8, 313–17, 341–2, 344–51, 419–20; with two independent variables 340–51; *see also* constrained maximisation
minimization: by differentiation 303; linear programming 159–64; *see also* constrained minimization
minimum point: of AC 181, 287, 411–12; of AVC 287
minors 528–9
mixed constraints 166–7
mixed derivatives *see* cross partial derivatives
monopoly 68, 127–30, 133–4; multiplant 104–5, 140–5, 344–5, 348–9
monotonic functions 66–7
monthly interest rates 197–8, 239–40
mortgage repayments 242–4, 247–8
multiple operations 9–10
multiplication 8–9; of algebraic terms 40–3; of decimal fractions 20; of matrices 512–17; of negative numbers 23; of pairs of brackets 41–3; of powers 24–6; using natural logs 482
multiplier: export 329; government expenditure 329–31; investment 329, 453–4

Nash equilibrium 465
national income determination 113, 120–1, 296–7, 328–31, 453–64, 502–5
natural exponential function 472–3
natural logarithms 481–2; applications of 483–6
naturally ordered principal minors 552
necessary conditions for maximization 299–300, 304
negative area below axis 423
negative definite Hessian 545
negative numbers 22–3, 178; and inequality signs 60
negative slope 77
net investment function 426–7
net present value (NPV) 207–14, 218–19
nominal interest rate 196–7, 200
non-linear functions 86–7, 90–1
non-negativity restrictions 67, 149
non-singular matrix 526
normal goods 377
null matrix 524
numerator 12

objective functions 148, 159, 313, 363, 365–6, 377–80, 386
oligopoly 465–9
opportunity cost 84
optimal input determination 119–20, 369–70, 380–3
optimization 2; constrained 363–93; unconstrained 299–317, 340–51; using matrix algebra 545–50
order of a matrix 508
ordinal utility 326
oscillating prices 446
output *see* total product
output maximization subject to budget constraint: by substitution 364–8; using Lagrange multiplier 371–4, 386–90; using spreadsheet 183–6

Paasche price index 58
parallel functions 83, 108–9, 150
parameters 3, 74, 101, 125–6
parameters estimation 539–41
partial derivatives 319–20; cross 334–9; second-order 323, 334–9
partial differentiation 319–39
particular solution: to difference equation 444–5; to differential equation 493–6
pensions 231–7; annuity income from 232–3; drawdown 233–7; funds/pots 232–6
percentages 21–2, 191–2
perfect competition 178–82, 346–7; and cobweb model 434–5
perfect price discrimination 135–6, 138
perpetual annuities 227–31
per unit taxes 121–2, 126–8, 294–6, 313–15

plotting: linear functions 69–72; quadratic functions 171–5; using spreadsheets 92–6, 173–5
point elasticity 78–79, 291–3, 324–5, 401; linear demand schedule 79; non-linear demand schedule 261–2, 324–5, 401
pollution control 383–5
polynomial functions 89, 182–8; solution by spreadsheet 183–8
population growth rate 258, 475, 479, 484
positive definite Hessian 545
positive value restrictions 70–1, 86–7, 149, 272
powers 23–6, 61, 85–7; fractional 26–8, 86–7; negative 25–6, 86–7
present value 202, 207–8, 397; of annuities 224–7; of continuous stream of returns 479–80; of perpetual annuities 229–31; *see also* net present value
price determination: in competitive markets 107–8, 124–7; in dynamic models 434–43, 497–501; under Bertrand model 465–9
price discrimination 102–3, 133–8, 143–5; first degree (perfect) 135–6, 138; with multiplant monopoly 143–5; second degree 135, 137–8; third degree 133–5
price elasticity of demand 15–17, 78–79, 291–3, 324–5, 401
price indices/price index 57–8
price ratio and slope of budget line 83–4
principal minors 545; border preserving 554; naturally ordered 552
producer surplus 423–5
product exhaustion 356–60
product matrix 513
product rule 403–8
production functions 64–5, 97–101, 318–22; Cobb-Douglas 99–101, 320–2
profit function 148–50, 313, 315–16, 337
profit maximization 287–9, 307–8, 313–17, 341–2, 344–51, 419–20; and demand for labour 398–400; by differentiation 287–9, 341–2, 344–51, 419–20; linear programming 147–8, 150–6; second-order conditions 308, 342

quadratic equations 168–77; factorization 175–6; graphical solution 171–5; no solution to 172, 177–8; simultaneous 178–82
quadratic formula 177–8
quadratic functions 169–82; slope of 275–6
qualitative predictions 2
quality constraints 159–64
quotient rule 408–10; applications of 409–14

range 64
rate of change of a function 275–6, 319; or marginal function 334–5
reaction functions 465–9
rectangular hyperbola 98
reduced form 125–32
reorder costs 309–11
repayments on loans 241–8
representation 33–4
resource allocation 2, 363
resource depletion rates 257–62, 476
respecifying functions 66–8, 290–1
returns to scale 100, 347, 358–60; and product exhaustion 358–9
revenue *see* total revenue
roots 26–8
rounding errors 4, 14
row operations 114–15, 116–18
row vectors 508

sales revenue *see* total revenue
sales tax *see* tax
savings schemes 249–52; sinking fund 253–6
scalar: division and multiplication of matrices 509–11
second degree price discrimination 135, 137–8
second-order conditions 304–5; Lagrangian 377–9, 551–2; for a maximum 300–2, 340–1; for a minimum 303, 340–1; using bordered Hessian matrix 551–5; using Hessian matrix 545–50; with two independent variables 340–1, 544–5
second-order derivatives 301–2
second-order partial derivatives 323, 334–9; applications of 341–51
securities 263
sequences *see* geometric series
share valuation 269–74
short run 179, 435–7
short run supply schedule 436–7
simple interest 191
simplex method 168
simplification 2, 38–49
simultaneous linear equations 107–22; equating to same variable solution 110–11; graphical

solution 108–9; more than two unknowns 116–18; row operations 114–15; solution using Cramer's rule 542–3; solution using matrix algebra 521–4, 536–9; substitution method 112–13; using matrix algebra 536–41
simultaneous quadratic equations 178–82
singular matrix 526
sinking fund 253–6
slack 151, 153, 156, 162
slope 76–9; budget constraint 81–4; demand and MR functions 281–3; by differentiation 276–83; non-linear functions 85–7; and point elasticity 291–3; rate of change of 275–6, 319; zero and maximization 299–300
solving simple equations 50–4
specific form of a function 64
specific solution *see* definite solution
spreadsheets *see* Excel
square roots 26–8, 177–8
stability 440, 442, 446–52, 496, 503–5
stationary points 300–2, 304
statistics 3
stock control *see* inventory control
straight lines *see* linear functions
substitutes 343
substitution 112–13, 364; integration by 428–30
substitution effect 374–7
subtraction 8; of algebraic terms 39; with inequalities 60–1; of matrices 509
sufficient conditions for maximization 300
summation: finite geometric series 223–4; infinite geometric series 229
summation sign ∑ 55–6, 210–11
summing functions horizontally 102–5
supply and demand equilibrium *see* equilibrium
surplus: consumer 423–5; producer 423–5

tangents 90–1
tax: lump sum 315–16; per unit sales 121–2, 126–8, 294–6, 313–15; profits 316–17; proportional 128–30
tax yield maximization 294–6, 406–8
third degree price discrimination 133–5
time periods: adjustment between *see* dynamics; for investment 202–4
total cost 50, 111; as cubic function 183, 186–7, 286–7; by integration 417–18; in linear programming 159–60; relationship with MC 285–6
total derivatives 361–2
total differentials 355–6
total fixed cost 50
total product 318–19; maximisation of *see* output maximization
total revenue 89–90, 111, 170, 172, 281–5; by integrating marginal revenue 418–20; maximization 299–300, 302, 343
total variable cost 50; by integrating marginal cost 417–18
transpose matrix 531
treasury bills 200–1
turning points 300–2, 304
two independent variable functions 97–100

unconstrained optimization 299–317, 340–51; using Hessian matrix 545–50; with two independent variables 340–51, 544–5
uneconomic region 321
unstable equilibrium 442, 449–51
utility 326; marginal 326, 368–9; maximisation 368–9; maximising combination of work and leisure 412–14

valuation *see* asset valuation
variable cost *see* total and average variable cost
variables: dependent 64–5, 97, 131; independent 64–5, 97, 131; unknown 112–18
vectors 508; multiplication of 512
vertical axis 69–72

wage rates 412–14
weak inequalities 59–60, 148–9
welfare measure 423
working hours 412–14

yield 266–9; current 266; to maturity (YTM) 267–9; tax yield 294–6, 406–8

zero matrix 524
zero slope 78, 91, 299–300

For Product Safety Concerns and Information, please contact our EU representative GPSR@taylorandfrancis.com Taylor & Francis Verlag GmbH, Kaufingerstraße 24, 80331 München, Germany

Printed by Integrated Books International, United States of America